Gert R. Strobl

The Physics of Polymers

Concepts for Understanding Their Structures and Behavior

With 218 Figures

Springer

Professor Dr. Gert R. Strobl

Fakultät für Physik
der Albert-Ludwigs-Universität
D-79104 Freiburg im Breisgau
Germany

ISBN 3-540-60768-4 Springer-Verlag Berlin Heidelberg NewYork

Strobl, Gert R.: The physics of polymers: concepts for understanding their structures and behavior/
Gert R. Strobl. - Berlin; Heidelberg; New York; Barcelona; Budapest; Hong Kong; London; Milan;
Paris; Santa Clara; Singapore; Tokyo: Springer, 1996
ISBN 3-540-60768-4

Typesetting: Camera ready by author
SPIN: 10516794 02/3020 - 5 4 3 2 1 0 - Printed on acid-free paper

Preface

In our faculty, we offer to the graduate students in physics a course on 'Condensed Matter Physics' which goes beyond the usual lectures on solid state physics, by also including the physics of simple liquids, liquid crystals and polymers. While there is a large selection of textbooks on solid state physics and also a choice of excellent treatises on the physics of liquids and liquid crystals, a book on a comparable level covering the major parts of the physics of polymers apparently does not exist. The desire is to have a textbook on polymer physics which, ideally, would stand in line with the 'Kittel', the 'Egelstaff' and de Gennes' books on the physics of liquid crystals, to cite only some of the best known volumes. This book is a first attempt to comply with these needs and to fill the gap. Certainly the aim is high, too high to be reached with this first approach, but hopefully other polymer physicist will also take on the task in future and then do better, once a frame is set and a first trial exists.

For me personally, writing such a textbook was indeed highly valuable and a worthwhile experience. In a time when science has such a strong tendency for diversification, there is a great danger of losing contact even with the neighboring branches and simultaneously the ability to see and assess the relevance of one's own activities. Students have this sensitivity and often have a better feeling about the importance of a topic. When teaching students as a lecturer, it is of primary importance always to provide the motivation and to make clear the role and relevance of a certain problem. Indeed, for me this amounts to a true check which helps me to discriminate between the major phenomena and secondary effects. Senior scientists with time tend to become acquainted with complicated, sometimes even artificial concepts; the young student, however, being confronted for the first time with an explanation, reacts naturally and distinguishes intuitively between reasonable, illuminating concepts and less attractive complicated ways of thinking. Hence, writing a textbook also means to put the state of the art of polymer physics to the test. If it is possible to present this field coherently and to explain convincingly the main properties with the aid of clear and appealing concepts, then it is in good shape. It is my impression, already gained in the lectures and now further corroborated during writing, that this is the case. The level of understanding is quite satisfactory and compares well with the understanding of simple liquids or liquid crystals. Therefore, the goal to write a coherent textbook on polymer physics can be reached, I am only rather uncertain if I have succeeded in demonstrating it.

As I am not sufficiently familiar from own experience with all the topics treated in the various chapters I am certainly not in a position to eliminate all errors. Hopefully, the ones I have made, are only minor ones. In any case, I would be grateful for reactions and comments from readers and any indication of faults in the treatment. Some incorrect conclusions have already been eliminated, after comments by Professor M.H. Wagner (Stuttgart) and Dr. L. Könczöl (Freiburg), who were kind enough to go through chapters 7 and 8 and I wish to thank them here once again for their advice.

Even if all of us in the scientific community use the English language, for a non-native speaker, the writing of a book is a different matter. As I do not like to read something in bad German I guess that Anglo-American scientists must feel the same. I received help at the beginning of my writing from Dr. Sandra Gilmour, who was working at this time as a postdoctoral student in Freiburg, and would like to express my gratitude again. Then, after completion, the manuscript was thoroughly revised by the copy editor, but he remarked that 'the sentence structure is very German which often makes it sound strange to a native speaker'. So I can only hope that this does not amount to a problem in understanding and offer my apologies.

In the first version produced two years ago the manuscript was dictated immediately after given lectures. This is the reason for the 'pedagogical style' of the writing. The emphasis is on the various concepts which have successfully established the present-day understanding of polymer physics. The focus is on the major phenomena, both in the formation of structures and the behavior under forces applied externally, mainly mechanical ones. This implies that many further effects, although important in certain cases, remain untouched. Hence, this textbook does not represent a comprehensive treatise and, therefore, should be better considered as an 'interface', providing help to enter into the various fields of polymer science, emanating from a basis which shows the interrelations. The recommendations given under the 'Further Readings' at the end of each chapter, the selected works included as figures, and the bibliography supplied at the end are meant to open the way for more detailed studies.

One active area of research is completely missing. These are the optical and electrical properties, with effects such as the high conductivity of doped conjugated polymers, electro-luminescence in polymeric light emitting diodes, or the ferro- and piezoelectricity of poly(vinylidene fluoride), to cite only a few examples. There is no good reason for this omission, only that I did not want to overload the book with another topic of different character which, besides, mostly employs concepts which are known from the physics of semi-conductors and low molar mass molecules.

As already mentioned, this book is primarily written for students of physics and physicists wishing to enter into polymer science for the first time. Interested macromolecular chemists and chemical engineers may also find it useful. The prerequisite for an understanding is not a special one, all that is needed is a background in phenomenological and statistical thermodynamics on the level of the respective courses in physical chemistry, together with the related mathematical knowledge.

Of course, I will be happy if the book finds many readers. It is a matter of fact that polymer physics is largely unknown to the majority of physicists. As a consequence, it is only rarely included in university courses on condensed matter behavior. This is difficult to comprehend considering the widespread uses of polymeric materials and in view of the appealing physical concepts developed for the description of their properties. It is therefore my wish that this book will contribute a little to change the present situation by helping to widen the physicists' general knowledge with a better understanding of the physics of polymers.

Freiburg, November 1995 *Gert Strobl*

Contents

Chapter 1

The Constitution and Architecture of Chains

Polymers, also known as 'macromolecules', are built up of a large number of molecular units which are linked together by covalent bonds. Usually they represent organic compounds, containing carbon atoms together with hydrogen, oxygen, nitrogen, and halogens etc. In this first chapter we briefly survey the main characteristics of their chemical constitution and molecular architecture and introduce the notions employed for their description, using examples for the explanation.

Let us begin with a look at polyethylene, which has a particularly simple structure. It is depicted in Fig. 1.1. The *structure unit* or *monomeric unit* building up the chain is the CH_2-(methylene-)group and their number determines the *degree of polymerization*, denoted by the symbol N. Macromolecules are generally obtained by a polymerization process starting from reactive low molar mass compounds. The name 'polyethylene' indicates that here the process is usually based on ethylene.

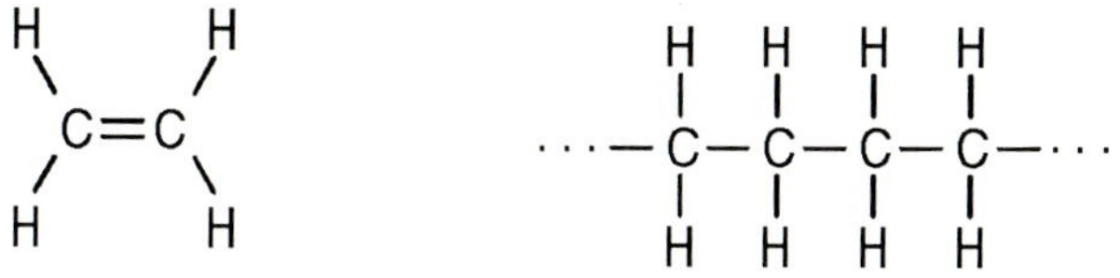

```
H     H               H   H   H   H
 \   /                |   |   |   |
  C=C           ···—C—C—C—C—···
 /   \                |   |   |   |
H     H               H   H   H   H
```

Fig. 1.1. Ethylene and polyethylene

Figure 1.2 presents, as a second example, the chemical composition of another common polymer, that of polystyrene. Here phenyl-groups are attached as *side-groups* to the C-C-*backbone chain*.

Further common polymers are compiled in Table 1.1. The list gives their chemical constitutions and includes also the commonly used short forms.

Rather than leading to polymers with a unique degree of polymerization, reactions usually result in a mixture of macromolecules with various molecular weights. Therefore, for a full characterization, the *molecular weight distribution function* has to be determined, and this is usually accomplished by gel permeation chromatography. We choose the symbol M for the molecular

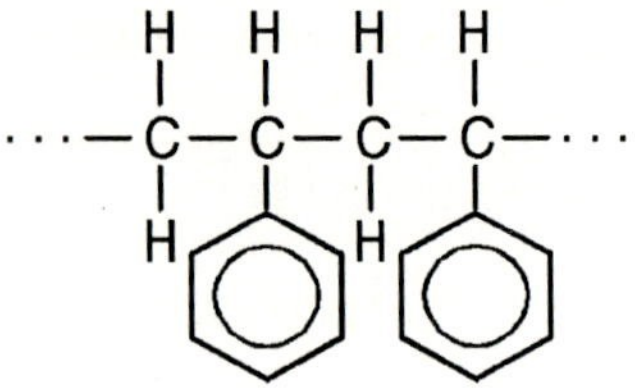

Fig. 1.2. Polystyrene

weight and introduce the distribution function $p(M)$ as a number density, adopting the definition that the product

$$p(M)\mathrm{d}M$$

gives the fraction of polymers with molecular weights in the range from M to $M + \mathrm{d}M$. As a distribution function $p(M)$ must be normalized:

$$\int p(M)\mathrm{d}M = 1 \tag{1.1}$$

The average molecular weight follows by

$$\overline{M}_{\mathrm{n}} = \int p(M)M\mathrm{d}M \tag{1.2}$$

In this description we treat M as a continuous variable although, strictly speaking, M changes in discrete steps, corresponding to the molecular weight of the monomer. For the normally given high degrees of polymerization, this discrete character does not become apparent and can be ignored.

Instead of using the *number average* $\overline{M}_{\mathrm{n}}$, the *weight average* of the molecular weight, $\overline{M}_{\mathrm{w}}$, may also be employed. $\overline{M}_{\mathrm{w}}$ is given by

$$\overline{M}_{\mathrm{w}} = \frac{\int p(M)M \cdot M\mathrm{d}M}{\int p(M)M\mathrm{d}M} \tag{1.3}$$

The origin of Eq. (1.3) is obvious. Just recognize that the function

$$p'(M) = \frac{pM}{\int p(M)M\mathrm{d}M} \tag{1.4}$$

describes the molecular weight distribution in terms of weight fractions.

For molecular weight distributions with a finite width $\overline{M}_{\mathrm{w}}$ is always larger than $\overline{M}_{\mathrm{n}}$. The ratio of the two mean values may be used to specifiy the width of the distribution. One introduces the *polydispersity coefficient* U defined as

$$U := \frac{\overline{M}_{\mathrm{w}}}{\overline{M}_{\mathrm{n}}} - 1 \tag{1.5}$$

Table 1.1. Chemical structure of some common polymers

Structure	Polymer
$-[CH_2-CH(CH_3)]_n-$	polypropylene 'PP'
$-[CH_2-C(CH_3)_2]_n-$	polyisobutylene 'PIB'
$-[CH_2-CH(C(=O)-OH)]_n-$	poly(acrylicacid)
$-[CH_2-C(CH_3)(C(=O)-O-CH_3)]_n-$	poly(methylmethacrylate) 'PMMA'
$-[CH_2-CH(O-C(=O)-CH_3)]_n-$	poly(vinylacetate) 'PVAc'
$-[CH_2-CH(O-CH_3)]_n-$	poly(vinylmethylether) 'PVME'
$-[CH=CH-CH_2-CH_2]_n-$	polybutadiene 'PB'
$-[C(CH_3)=CH-CH_2-CH_2]_n-$	polyisoprene 'PI'

Table 1.1. Chemical structure of some common polymers (continued)

Structure	Name
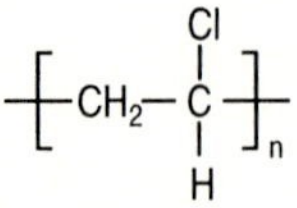 $\left[-CH_2-C(Cl)(H)- \right]_n$	poly(vinyl chloride) 'PVC'
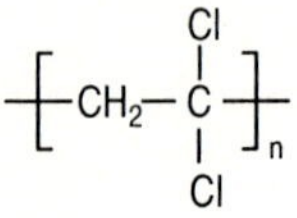 $\left[-CH_2-C(Cl)(Cl)- \right]_n$	poly(vinylidene chloride)
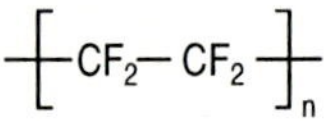 $\left[-CF_2-CF_2- \right]_n$	poly(tetrafluoroethylene) 'PTFE'
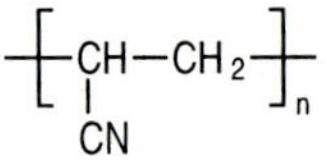 $\left[-CH(CN)-CH_2- \right]_n$	poly(acrylonitrile) 'PAN'
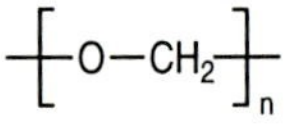 $\left[-O-CH_2- \right]_n$	poly(oxymethylene) 'POM'
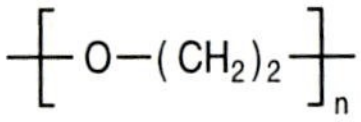 $\left[-O-(CH_2)_2- \right]_n$	poly(ethyleneoxide) 'PEO'
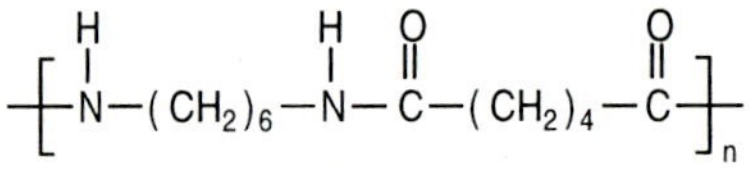 $\left[-N(H)-(CH_2)_6-N(H)-C(=O)-(CH_2)_4-C(=O)- \right]_n$	poly(hexamethylene adipamide) nylon 6,6
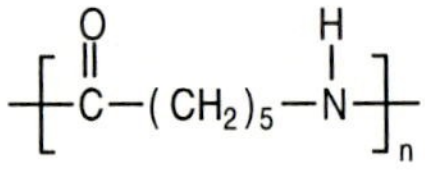 $\left[-C(=O)-(CH_2)_5-N(H)- \right]_n$	poly(ε-caprolactam) nylon 6

Table 1.1. Chemical structure of some common polymers (continued)

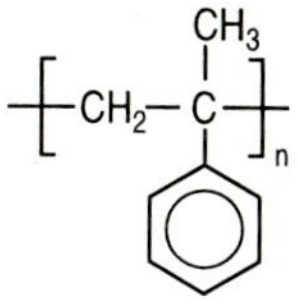	poly(α-methylstyrene)
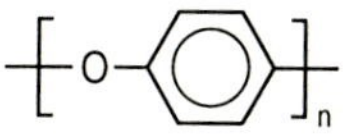	poly(phenylene oxide) 'PPO'
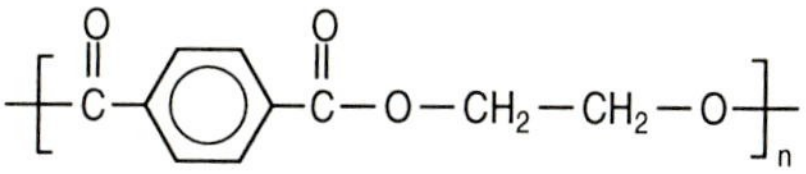	poly(ethylene terephthalate) 'PET'
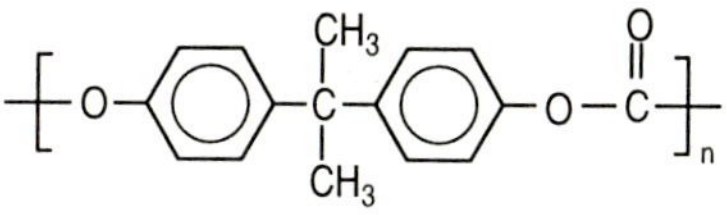	'polycarbonate' 'PC'
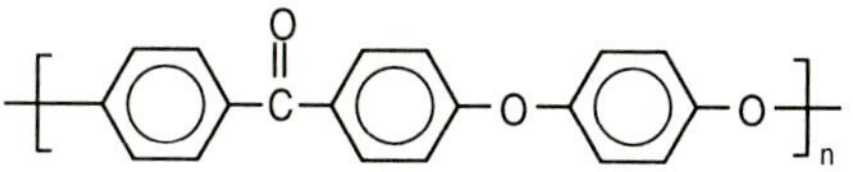	'poly(ether ether ketone)' 'PEEK'
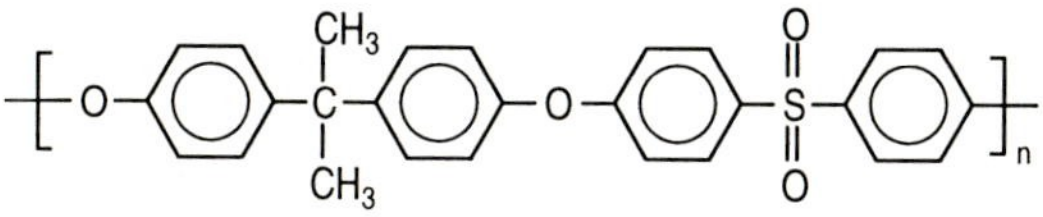	'polysulfone'
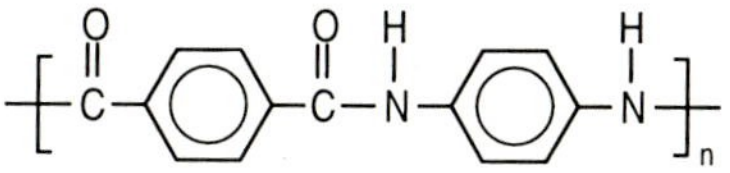	poly(p-phenylene- terephthalamide) 'Kevlar'

Table 1.1. Chemical structure of some common polymers (continued)

Structure	Name
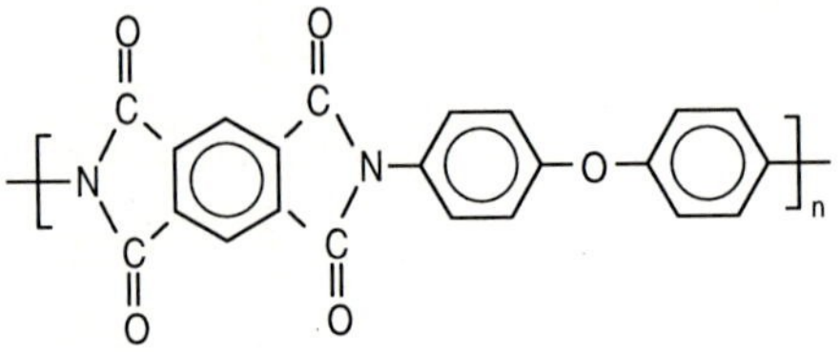 $-[N(CO)_2C_6H_2(CO)_2N-C_6H_4-O-C_6H_4]_n-$	'polyimide'
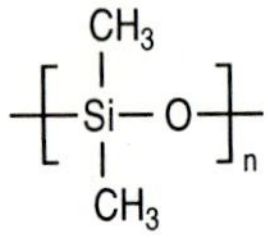 $-[Si(CH_3)_2-O]_n-$	poly(dimethylsiloxan) 'silicon rubber'
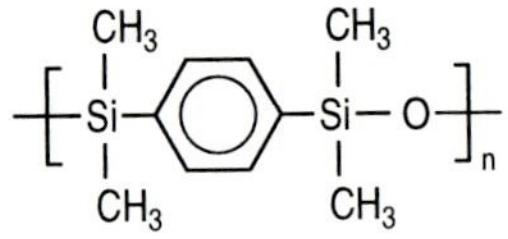 $-[Si(CH_3)_2-C_6H_4-Si(CH_3)_2-O]_n-$	poly(tetramethyl-p-silphenylene-siloxane) 'TMPS'

U does indeed measure the polydispersity, as it can be directly related to the variance of $p(M)$. We have

$$\begin{aligned} \langle \Delta M^2 \rangle &= \int p(M)(M - \overline{M}_{\rm n})^2 {\rm d}M \\ &= \int p(M) M^2 {\rm d}M - \overline{M}_{\rm n}^2 \end{aligned} \tag{1.6}$$

or, according to Eq. (1.3)

$$\langle \Delta M^2 \rangle = \overline{M}_{\rm w} \cdot \overline{M}_{\rm n} - \overline{M}_{\rm n}^2 \tag{1.7}$$

and therefore

$$\frac{\langle \Delta M^2 \rangle}{\overline{M}_{\rm n}^2} = U \tag{1.8}$$

U becomes zero only for a perfectly '*monodisperse*' sample, i.e. a sample with a uniform molecular weight.

Molecular weight distribution functions may vary greatly between different polymeric compounds. Distributions depend on the method of synthesis used in the polymerization process, and most methods belong to either of two general classes. In the first class of processes, known as *step polymerizations*,

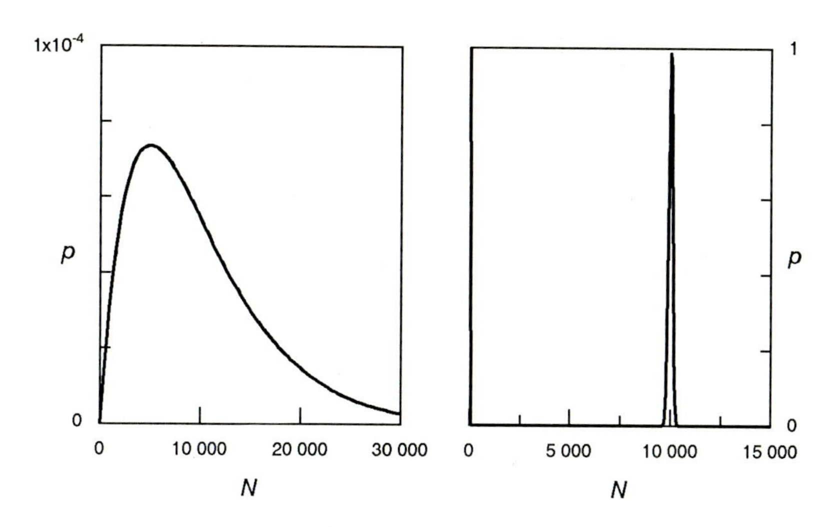

Fig. 1.3. Molecular weight distributions of the Schulz-Zimm type for $\beta = 2$ (*left*) and of the Poisson-type (*right*). Both correspond to the same number average degree of polymerization, $\overline{N}_{\mathrm{n}} = 10^4$

monomers react in such a way that groups of them which are already linked together can be coupled with other groups. In the second class, called *chain polymerizations*, reactive centers which react only with monomers are created at the beginning and become shifted after the reaction to the new end of the chain thus growing. Figure 1.3 shows, in idealized form, distribution functions resulting from the two different synthetic routes. For step polymerizations, distributions are broad, and often a good representation is achieved by the 'Schulz-Zimm'-distribution. The latter is usually formulated in terms of the degree of polymerization rather than the molecular weight and is given by the equation

$$p(N) = \frac{1}{\Gamma(\beta)} \left(\frac{\beta}{\overline{N}_{\mathrm{n}}} \right)^{\beta} \cdot N^{\beta - 1} \exp - \frac{\beta N}{\overline{N}_{\mathrm{n}}} \tag{1.9}$$

The function includes two parameters: β determines the shape, and $\overline{N}_{\mathrm{n}}$ denotes the number average of the degree of polymerization; Γ is the gamma function. A straightforward evaluation shows that the polydispersity index U is given by

$$U = \frac{1}{\beta}$$

For values $\beta \approx 2$, Eq. (1.9) provides a good data fit in many practical cases.

Much narrower distributions can be obtained for chain polymerizations. Typical is here a Poisson-distribution, given by

$$p(N) = \exp - \overline{N}_{\mathrm{n}} \cdot \frac{(\overline{N}_{\mathrm{n}})^N}{\Gamma(N+1)} \approx \exp - \overline{N}_{\mathrm{n}} \left(\frac{\overline{N}_{\mathrm{n}} \mathrm{e}}{N} \right)^N \tag{1.10}$$

As only one parameter, $\overline{N}_{\rm n}$, is included, U is no more independent. A straightforward calculation yields

$$U = \frac{1}{(\overline{N}_{\rm n})^{1/2}} \tag{1.11}$$

The two distribution functions are presented in Fig. 1.3, choosing $\beta = 2$ for the Schulz-Zimm distribution, and equal values of $\overline{N}_{\rm n}(= 10^4)$ in both cases.

Properties of polymer systems are generally affected by the shape of $p(M)$. This dependence is of considerable technical importance, and variations of $p(M)$ may often be used to improve and optimize the performance of materials. These are specific questions, and in what follows we shall mostly omit discussion of them. For the sake of simplicity, sharp molecular weight distributions will always be assumed, i.e. distributions like that shown on the right-hand side of Fig. 1.3. The degree of polymerization, N, then becomes a well-defined quantity.

Coupling of the units in polyethylene is unambiguous. For polystyrene, however, this is no longer the case, as styrene is composed of two different groups, CH_2 and C_7H_6. In principle, either group can be attached to the growing chain and, in addition, the phenyl-group can be placed on either side of the C-C-backbone. Variations may result in steric disorder along the chain. The notion used to describe steric order is *tacticity*. Polymers with a unique way of coupling of the monomeric units are called *isotactic*, and contrasted to those with an irregular steric structure which are addressed as *atactic*. If the coupling varies, but in a regular way, polymer chains are called *syndiotactic*. All three types are found for polystyrene, depending on the process chosen for the synthesis.

Polyethylene and polystyrene are built up of one type of monomeric unit only. This is not a necessity, and large variations in the chemical structure may be achieved by combination of different monomers. The procedure is known as *copolymerization*. To give an example: Ethylene and propylene monomers can be copolymerized, which leads to 'ethylene-propylene copolymers'. For the coupling of the two monomeric units in the chain two limiting cases exist. In the first, the coupling is statistical and determined by the probabilities of attachment of the two monomers on a growing chain. Chains of this type are called *statistical copolymers*. They can exhibit short-range order with preferred sequences, thus being different from a random mixing, but possess no order in the chemical composition over the long range. The second limiting case is realized by *block copolymers*. These are obtained by coupling long macromolecular sequences of uniform composition, and depending on the number of sequences, di-, tri- or multiblock copolymers may be prepared. The structures of the two types are sketched in Fig. 1.4, using the ethylene-propylene system as an example.

The chains discussed so far have all a linear topology. There exists a large group of polymers with a different architecture, and some typical forms are sketched in Fig. 1.5. For example, a polymer may include *short- and long-chain branches* in a statistical distribution. A well-known representative of

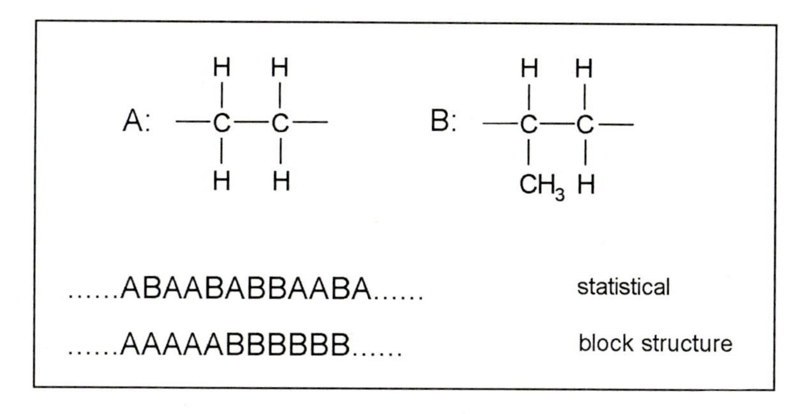

Fig. 1.4. Ethylene-propylene copolymers

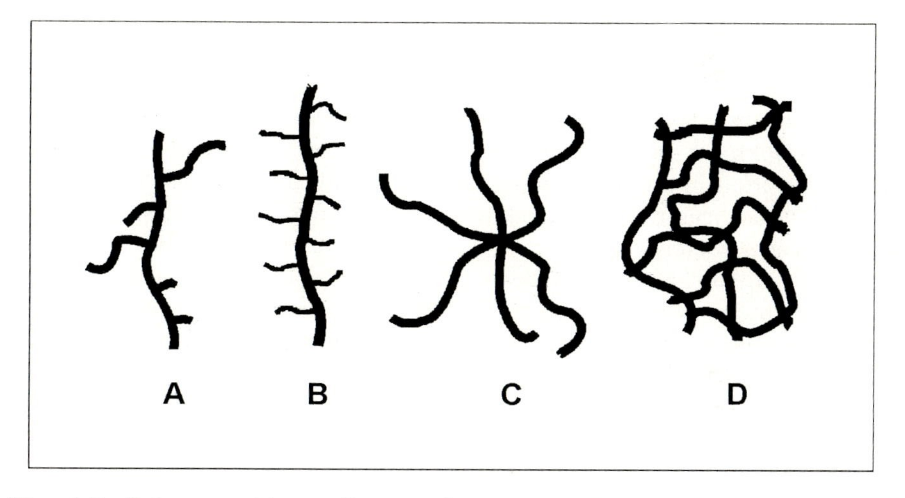

Fig. 1.5. Polymers with non-linear architectures: polymer with short-chain and long-chain branches (A), polymer with grafted oligomer side chains (B), star polymer (C), network of cross-linked chains (D)

this architecture is 'low density polyethylene', which incorporates, as a result of the polymerization process, alkyl-branches of different length in random fashion. Typical values for the branching ratio, i.e. the fraction of branched units, are in the order of several percent. If more extended 'oligomeric' chains are attached to a backbone chain with different composition, *grafted-chain* polymers are obtained. A quite exotic species are *star polymers*, where several polymer chains emanate from one common multifunctional center.

A qualitative change in properties is achieved by coupling all the polymer chains of a sample together, thereby building up a three-dimensional *network*.

This is the basic structure of *rubbers*. A rubbers in fact represents one huge macromolecule of macroscopic dimensions, with properties depending on the cross-link density and the functionality of the junction units.

Finally, at the end of this short first chapter, let us briefly recall the manifold of uses of polymers in daily life. First there are solid polymeric materials in various forms, 'commodity' polymers for widespread applications, as well as speciality polymers for specific utilizations. In industry these are called *thermoplastics*, expressing that they can be shaped and brought into forms of choice by thermo-mechanical treatments at elevated temperatures. The price to be paid for the advantage of using comparatively simple processing techniques is the limitation in the temperature range of use, which contrasts with the much larger application range of ceramics and metals. *Polymer fibers* comprise a second large class of materials, and they are mostly used for the production of textiles and woven products. Fibers are generally obtained by spinning processes carried out on the melt or concentrated solutions at elevated temperatures, which is followed by a fixing accomplished by rapid cooling. Again, the temperature range for uses is limited. If a fiber is heated to too high temperatures it shrinks. The rubbers, technically addressed as *elastomers*, constitute the third class of polymeric materials. Both synthetic and natural products are utilized. The essential step in rubber production is the cross-linking process. 'Natural rubber', for example, is obtained by heating *cis*-polyisoprene in the presence of sulfur. This 'vulcanization process' creates cross-links between the polyisoprene chains, composed of short sequences of sulfur atoms. For high cross-link densities the large deformability characteristic of a rubber gets lost, and one obtains stiff solids. This is the class of *duromers* or *thermosets*, also known as *resins*. The various adhesives based on the mixing of two reacting components belong to this class of polymers. Shapes are rather stable for these compounds and remain unaffected on heating, up to the point of chemical decomposition.

With the exception of natural rubber, all the above polymers are synthetic products. Although this book will deal with the properties of synthetic materials only, we have to be aware of the decisive role played by polymers in nature. Control of life processes is based on two polymer species, nucleic acids and proteins. The specific property of these polymers is that they form stable microscopic objects, mainly as the result of the action of intramolecular hydrogen bonds. The stable, specifically ordered surface of the proteins provides the high selectivity and catalytic potential used in biochemical reactions; selectivity and catalytic activity disappear when the globular molecular shape is destroyed at elevated temperatures or upon the addition of an active chemical agent. The synthetic polymers discussed in this book do not possess the potential to form a unique molecular conformation as single chains and, therefore, do not show any biochemical activity.

The large variability in the chemical constitution and architecture of macromolecules opens a broad route to the preparation of materials with a wide spectrum of different properties. Chemistry, however, is not the only factor responsible for the actual behavior. It is a specific feature of polymers that one finds a particularly strong impact of the thermo-mechanical processes experienced during manufacture since these control the final formation of structures. An understanding of these processes is a necessary prerequisite for a successfull utilization of polymers, and the promotion of knowledge to levels as high as possible therefore constitutes one of the main aims of polymer physics.

1.1 Further Reading

F.W. Billmeyer: *Textbook on Polymer Science*, John Wiley & Sons, 1984

J.M.G. Cowie: *Polymers: Chemistry and Physics of Modern Materials*, International Textbook Co, 1973

P.J. Flory: *Principles of Polymer Chemistry*, Cornell University Press, 1953

P. Munk: *Introduction to Macromolecular Science*, John Wiley & Sons, 1989

Chapter 2

Single Chain Conformations

Condensed matter is composed of strongly interacting molecules, and discussions of the bulk properties of simple low molar mass compounds therefore focus from the beginning on the role of the interaction forces between different molecules in establishing thermal equilibrium. In dealing with polymeric systems, one encounters a different situation. As each macromolecule possesses a huge number of internal degrees of freedom, the analysis of the properties of the individual polymer becomes an important first point of concern. It is obvious that understanding of single chain behavior is a necessary prerequisite for treatments of aggregate properties, but in fact, it implies even more and leads in many cases to a major step forward. There are, of course, effects which are dominated by the intermolecular forces, like the phase behavior of binary polymer mixtures, or the flow properties of polymer melts, but other important phenomena, in particular essential parts of the viscoelasticity, are much under the control of the dynamic properties of the individual molecules. It is therefore quite natural and also necessary, to start a lecture series on polymer physics with a discussion of the conformational states of single chains.

2.1 Rotational Isomeric States

Let us choose polyethylene (PE) as an example and consider its full steric structure. The latter is shown in Fig. 2.1. A polymer chain like polyethylene possesses a great internal flexibility and is able to change its conformation totally. Basically, the number of degrees of freedom of the chain is given by three times the number of atoms, and it is convenient, to split them up into two different classes. The first group concerns changes in valence angles and bond lengths, as they occur during molecular vibrations, with frequencies in the infrared range. These movements are limited and do not affect the overall form of the chain. The second group of motions is of different character, in that they have the potential to alter the form. These are the rotations about the C-C-bonds, which can convert the stretched chain of Fig. 2.1 into a coil and can accomplish the transitions between all the different conformational states. Clearly, in dealing with the conformational properties of a given polymer, only the latter group of degrees of freedom has to be considered. A discussion of

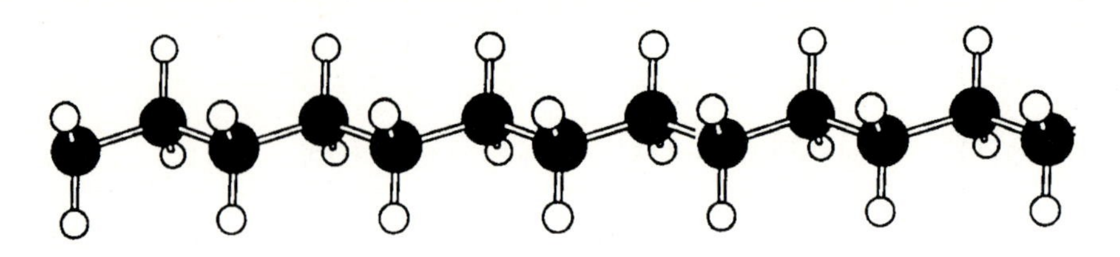

Fig. 2.1. Steric structure of polyethylene. Rotations about the C-C-bonds result in a change in the conformation

the conformational states of a given macromolecule therefore requires first of all an analysis of the bond rotation potentials.

To begin with, we first go back to a related low molar mass molecule and consider the rotational potential of ethane. Here a rotation about the central C-C-bond is possible, and one can anticipate the general form of the rotational potential. Interaction energies may be described as a superposition of a part which directly relates to the rotational state of the C-C-bond and 'non-bonded' interactions between the hydrogen atoms. The latter are for the given distances of repulsive nature. Figure 2.2 will help us to describe the situation. The staggered conformation of ethane, shown at the bottom on the right, corresponds to the minimum in the potential energy since it is associated with the largest distances between the hydrogens. Owing to the three-fold symmetry of the two methyl groups the rotational potential $\tilde{u}(\varphi)$ (φ denotes the rotation angle) exhibits a 120°-periodicity. Therefore, in first approximation, employing only the lowest order Fourier contribution, it can be described by

$$\tilde{u} = \tilde{u}_0(1 - \cos 3\varphi) \tag{2.1}$$

This rotational potential is indicated in the figure by the broken line, and shows three energy minima with equal potential energies.

Next, we consider the rotational potential of butane. The replacement of one hydrogen atom by a methyl-group for both carbon atoms removes the three-fold symmetry. As a consequence the potential energy function $\tilde{u}(\varphi)$ gets a shape like that indicated by the continuous curve in Fig. 2.2. The minimum occurs for the staggered conformation depicted in the upper part on the right, where the distance between the two methyl-groups is at the maximum. There still exist local minima in the potential energy at 120° and 240°, but now at an elevated level. The maximum of $\tilde{u}(\varphi)$ is expected for 180°, when the two methyl groups closely approach each other.

Particular terms are used to address the three energy minima. The conformational state with the lowest energy at $\varphi = 0°$ is called the '*trans*'-conformation. The other two minima at 120° and 240° are called '*gauche*' and distinguished by adding a plus or minus sign. Note that in the *trans* state the three C-C-bonds of butane lie in one plane; the *gauche*-states are non-planar.

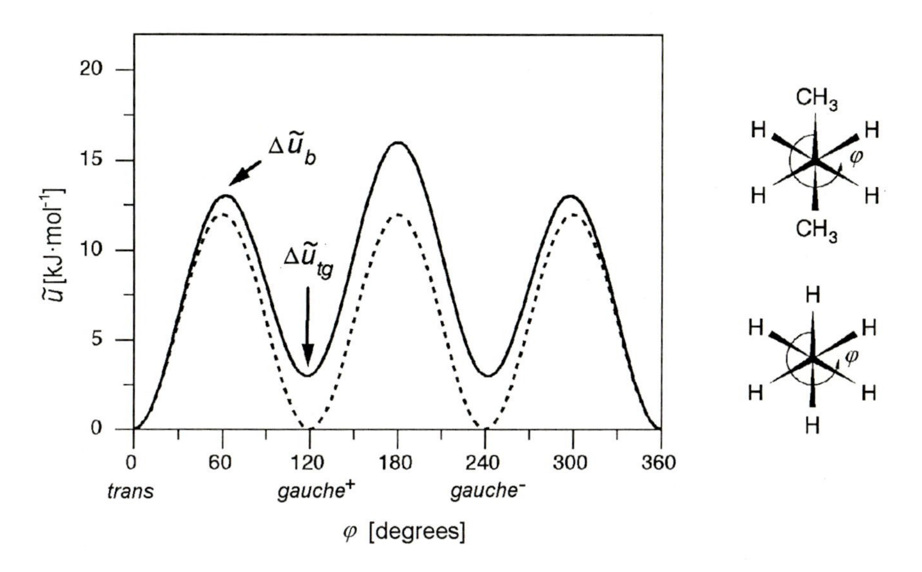

Fig. 2.2. Potential energies associated with the rotation of the central C-C-bond for ethane (*broken line*) and butane (*continuous line*). The sketches show the two molecules in views along the C-C-bond

The internal dynamics of butane depends on the energy difference $\Delta\tilde{u}_{\mathrm{tg}}$ between the *trans*- and the *gauche*-states and the height of the barriers, $\Delta\tilde{u}_{\mathrm{b}}$, between the local minima. One can envisage two limiting cases. For $\tilde{R}T \gg \Delta\tilde{u}_{\mathrm{b}}$ (the symbol $\tilde{R}$ stands for the perfect gas constant) rotations about the C-C-bonds are quasi-free, and the details of $\tilde{u}(\varphi)$ with its minima and maxima become irrelevant. In the opposite case, $\tilde{R}T \ll \Delta\tilde{u}_{\mathrm{tg}}$, the molecules settle down in the lowest energy state, i.e. in the *trans*-conformation, and only librate about the equilibrium position.

The prerequisite for a judgement of the actual situation, say at ambient temperature, is therefore a knowledge of the two energy differences $\Delta\tilde{u}_{\mathrm{tg}}$ and $\Delta\tilde{u}_{\mathrm{b}}$. These can be determined by spectroscopic and calorimetric experiments. $\Delta\tilde{u}_{\mathrm{tg}}$ can be derived from temperature dependent Raman-scattering experiments, due to the vibrational spectra for the *trans*- and the *gauche*-states being different. One selects two bands associated with the *gauche*- or the *trans*-conformation respectively and measures their intensities, I_{g} and I_{t}, as a function of temperature. The ratio $I_{\mathrm{g}}/I_{\mathrm{t}}$ changes with the populations of the two states, according to

$$\frac{I_{\mathrm{g}}}{I_{\mathrm{t}}} \sim \exp -\frac{\Delta\tilde{u}_{\mathrm{tg}}}{\tilde{R}T} \tag{2.2}$$

An Arrhenius-plot of $\ln(I_g/I_t)$ versus $1/T$ thus yields $\Delta\tilde{u}_{tg}$. Experiments were performed on different n-alkanes and gave values in the range

$$\Delta\tilde{u}_{tg} \simeq 2 - 3 \text{ kJ} \cdot \text{mol}^{-1}$$

The fraction ϕ_g of molecules in the two *gauche*-states follows from

$$\phi_g = \frac{2\exp(-\Delta\tilde{u}_{tg}/\tilde{R}T)}{1 + 2\exp(-\Delta\tilde{u}_{tg}/\tilde{R}T)} \tag{2.3}$$

which leads to

$$\phi_g \simeq 0.5$$

Hence, *trans*- and *gauche*- states are populated with similar probabilities.

The barrier height $\Delta\tilde{u}_b$ can be deduced, for example, from measurements of the heat capacity of ethane. It turns out that data can only be described if the internal degree of freedom associated with the C-C-bond rotation is accounted for in addition to the translational and rotational degrees of freedom of the whole molecule. A fit of the data yields $\Delta\tilde{u}_b$, with the result

$$\Delta\tilde{u}_b \simeq 12 \text{ kJ} \cdot \text{mol}^{-1}$$

This is the barrier height for ethane. For butane, one expects somewhat larger values.

Looking at these results we may resume that

$$\Delta\tilde{u}_b \gg \tilde{R}T \simeq \Delta\tilde{u}_{tg}$$

Under these conditions, internal dynamics of the butane molecule may be envisaged as follows. Most of the time, the molecule is in either of the three conformational states and just vibrates about the respective energy minimum. From time to time, the molecule collects sufficient thermal energy so that the barrier can be passed over and the conformation changes. As the transitions take place rapidly compared to the times of stay near to a minimum, a sample of butane resembles a mixture of different 'rotational isomers'. For each molecule there exist three 'rotational isomeric states'. They are all accessible and populated according to the available thermal energy.

We now turn to polyethylene and discuss its case by starting from butane and considering the effect of a replacement of the two methyl-endgroups by longer chain sequences. The result is qualitatively clear. One expects modifications in the details of $\tilde{u}(\varphi)$ which depend also on the conformations of the two sequences, however, the overall form of the rotational potential energy of a given C-C-bond will remain unchanged. There still exists an energy minimum for the *trans*-conformation and local minima for two *gauche*-states. Also the values of $\Delta\tilde{u}_{tg}$ and $\Delta\tilde{u}_b$ should not alter significantly. We may thus conclude that the conformation of polyethylene can again be described in terms of rotational isomeric states and that there are still three rotational isomeric states per bond, corresponding to *trans*, *gauche*$^+$ and *gauche*$^-$. A polyethylene chain with a degree of polymerization N therefore possesses 3^N different conformational states. In order to address one specific conformation, the rotational isomeric states for all bonds have to be given. This may be done, for example, in the form

$$(\varphi_1, \varphi_2, \dots, \varphi_N) \tag{2.4}$$

whereby $\varphi_i \,\widehat{=}\, trans, gauche^+$ or $gauche^-$.

Two different situations are found in polymer systems and we shall deal with them separately: In the crystalline state, chains adopt unique conformations which represent helices with straight axes. In the fluid state, on the other hand, all rotational isomeric states are populated, with probabilities determined by the temperature and the respective energies.

2.2 Helices

Polymers can form crystals, not like low molar mass compounds under all circumstances, but for many species under the prerequisite that the cooling from the molten state occurs slowly enough to enable the necessary rearrangements of the chains. The building principle is obvious. As a basic requirement, chains must adopt a straight, perfectly ordered form. Then a lattice can be constructed by orienting the chains uniformly in one direction and packing them laterally in regular manner. The thus-obtained lattice with three-dimensional order has the monomeric units as structure units. The specific property is the strong anisotropy in the binding forces, with valence forces in one and weak van der Waals forces in the two other directions.

Crystal lattices at low temperatures generally represent the structure with the minimum internal energy. For polymer crystals one expects that the main contribution should be furnished by the intramolecular energies as determined by the bond rotations. Structure determinations in combination with energetic calculations, which were carried out for several polymers, do indeed support this view. They indicate that the conformation adopted by a polymer in the crystalline phase equals, or comes very close to, the lowest energy rotational isomeric state of the single chain. Conditions are simple if the coupling between successive bonds is only weak. Then rotations take place independently and each bond settles down in the energy minimum.

To have equal conformations in all monomeric units is of course not a peculiarity of chains with independent bonds. In the general case, one also finds, in the lowest energy state of a chain, a uniform conformation of the monomers. What is the general structure which then emerges? It is important to recognize that this is of the helical type. To see and to illustrate this, we should consider some examples.

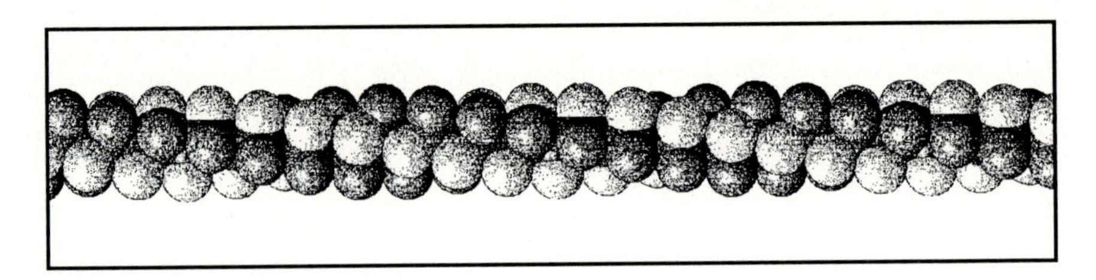

Fig. 2.3. PTFE in the crystalline state. The conformation corresponds to a 13/6-helix

We begin with polyethylene. Here, in order to have the energy minimum, all C-C-bonds must be in the *trans*-state. Crystalline polyethylene thus adopts the '*all-trans*'-conformation, which is the structure shown in Fig. 2.1.

While polyethylene with its planar zig-zag structure is not a helix in the usual sense, poly(tetrafluoroethylene) (PTFE) in the crystalline state possesses the typical wound appearance. This chain is shown in Fig. 2.3. Poly(tetrafluoroethylene) is obtained from polyethylene by a replacement of all hydrogen atoms by fluorines. The reason for the resulting structure change is easy to see. The replacement of hydrogen atoms by the much larger fluorines increases the interaction energy between the CF_2-groups of second nearest neighbors. When we start off from an *all-trans*-form, a uniform twist of the chain diminishes the repulsive F-F interaction energies but, at the same time, the bond rotational energy increases. There exists an energy minimum at a finite torsion angle, $\varphi_{\min} = 16.5°$.

Figure 2.4a presents as a further example a helix of poly(oxymethylene) (POM). It corresponds to an *all-gauche*-conformation. Different from poly(tetrafluoroethylene), here a *trans*-rotational isomeric state also exists, but it does not represent the energy minimum.

For some polymers, one does not observe a unique helical form but two or three different ones which possess very similar energies. Poly(oxymethylene) actually gives an example for such 'polymorphism'. Here one can also find the helix shown in Fig. 2.4b. It also represents an '*all-gauche*'-conformation, but the torsion angle, which was near to 60° for the first modification, now has increased to 77°. Which of the two helices is formed depends on the crystallization conditions. In principle, at a given temperature, only one of the modifications can be stable, the other one being metastable. Annealing can induce a transformation to the stable state, but often this transformation is kinetically hindered, and then it may become difficult to identify the stable modification.

For the description of a given helix, we have a natural basis. One refers to the screw symmetry and just specifies the screw operation which maps the molecule onto itself. Screw operations comprise a turn about a certain angle, say $\Delta\zeta$, together with a simultaneous longitudinal shift, say Δz. These two values which describe the move from one monomer unit to the adjacent one

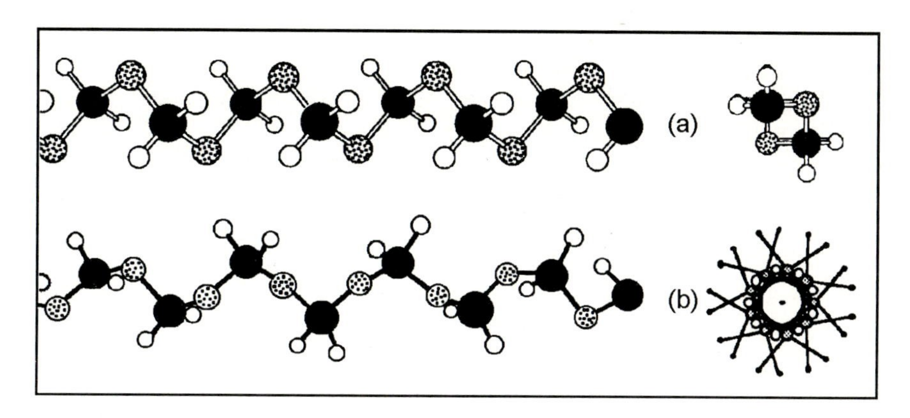

Fig. 2.4. Two different helices formed by POM: 2/1-helix (*a*) and 9/5-helix (*b*). Side views (*left*) and views along the helix axis (*right*)

constitute the 'external helix parameters'. For polyethylene we have $\Delta\zeta = 180°$ and $\Delta z = 1.27$ Å, for poly(tetrafluoroethylene) $\Delta\zeta = 166°$ and $\Delta z = 1.31$ Å, for the two polymorphic forms of poly(oxymethylene) $\Delta\zeta = 180°$, $\Delta z = 1.78$ Å and $\Delta\zeta = 200°$, $\Delta z = 1.93$ Å respectively. The external helix parameters are functions of the 'internal helix parameters', the latter being given by the bond rotational angles. The dependencies may be formulated on the basis of geometrical considerations, assuming constant values for both bond lengths and valence angles.

Several polymers form helices which are simple in the sense that $\Delta\zeta$ is a fraction of 360°, like 180°, 120° or 90°. These are called '$m/1$'-helices, m giving the number of monomeric units arranged along one 360°-turn. Examples are polypropylene, polystyrene or poly-1-butene, which all form 3/1-helices. The *all-trans*-conformation of polyethylene and the *all-gauche*-conformation of poly(oxymethylene) correspond to 2/1-helices.

The next general helical form is given by the 'm/n'-helices. The name is meant to indicate that m monomeric units are equally distributed over n turns.

Particularly interesting is the observation that often polymers with $n > 1$ cannot be described as a m/n-helix with m and n being small numbers. Poly(tetrafluoroethylene) and the second form of poly(oxymethylene) represent such examples. The helix of poly(tetrafluoroethylene) is usually called 13/6-helix, thereby indicating that 13 CF_2-units are distributed over 6 turns. This results in the measured value of $\Delta\zeta$, as

$$360° \cdot 6 : 13 = 166°$$

Likewise, the second modification of poly(oxymethylene) also requires a non-simple notation for the description of the conformation, namely that of a 9/5-helix. If we recall the reason for the winding of chains, these results are not surprising. The helix conformation represents the minimum of the intramolecular energy and this may arise as a result of competing forces. The internal helix parameters thus determined do not always have to be associated with a simple external helical structure.

Interestingly enough, measurements with higher accuracy showed that even the descriptions of the poly(tetrafluoroethylene)-chain as a 13/6-helix and of the poly(oxymethylene)-chain as a 9/5-helix are only approximations. Refinements of data necessitated a change in the characteristic ratio m/n towards higher numbers. In the case of poly(tetrafluoroethylene) the evaluation of electron diffraction data indicated a 473:219-helix. What does this mean? Indeed, these observations can be understood as indicating a qualitative structural change. Rather than accepting helix descriptions with large numbers n and m as representing exact structures, it might be more appropriate to assume that these helices possess no strict periodicity at all. This would mean that poly(tetrafluoroethylene) and poly(oxymethylene), in the crystalline state, both form 'irrational' helices. As a consequence, the lattice lacks periodicity in one direction. We rather find, in chain direction, two independent length scales, as given by the height Δz per monomeric unit and the height of one 360°-turn respectively. Structures like this are generally addressed as 'incommensurate'. They are found also in non-polymeric materials as, for example, certain anorganic ferromagnetic compounds, where the positional and the magnetic order may show different periodicities.

2.3 Coils

The huge number of rotational isomeric states which a polymer chain may adopt becomes effective in fluid phases. Polymers in solution or the melt change between the different states, and these are populated according to the laws of Boltzmann statistics. As the large majority of conformations are coil-like, it is said that polymers in the fluid state represent 'random coils'.

At first, one might think that any treatment of the properties of a polymer has to emanate from its microscopic chemical structure since this determines the rotational isomeric states. We then would have to consider in detail the effects of bond lengths, bond angles, rotational potentials, the presence and length of sidegroups, etc. Treatments of this kind are necessarily specific and vary between different compounds. It is now a most important fact that one can omit the consideration of all these structural details in many discussions. Indeed, the dependence on the chemical constitution vanishes if structural properties are discussed for a lowered resolution, corresponding to length scales in the order of some nanometers. In such a 'coarse-grained' picture, polymer chains become equivalent to each other and then also exhibit a common behavior.

Figure 2.5 shows a polymer coil as it might look like at limited resolution. We would observe a worm-like chain with a continuous appearance. For its

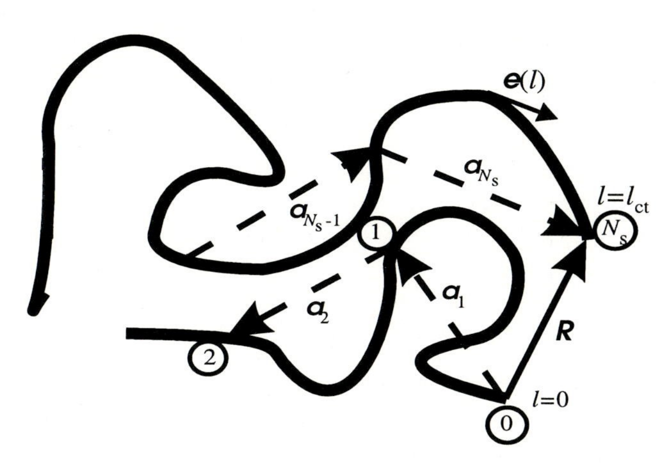

Fig. 2.5. Polymer chain in low resolution (contour length l_{ct}, local chain direction $\boldsymbol{e}(l)$) together with an associated chain of N_{s} freely jointed segments, connecting the junction points 0 to N_{s}

representation, we choose a curvilinear coordinate l, running from $l = 0$ at one end to $l = l_{\mathrm{ct}}$ at the other end, and describe the varying local chain direction by unit vectors $\boldsymbol{e}(l)$.

What can be said about such a chain? A first point of interest is the internal chain flexibility. Chains may be stiff, i.e. oppose a strong bending, or also highly flexible, thus facilitating a coiling. Searching for a parameter which provides a measure for the chain flexibility we can start from the 'orientational correlation function'. This function, denoted K_{or}, describes the correlation between the chain directions at two points with a curvilinear distance Δl. It is defined as

$$K_{\mathrm{or}}(\Delta l) := \langle \boldsymbol{e}(l)\boldsymbol{e}(l+\Delta l)\rangle \tag{2.5}$$

Here and in the following, the brackets indicate an ensemble average which includes all chain conformations with their statistical weights. For homopolymers, K_{or} is independent of the position l, and we restrict our attention to this case. Figure 2.6 shows the general shape expected for K_{or}. Owing to the finite flexibility of the chain, orientational correlations must vanish for sufficiently large distances Δl. Therefore, K_{or} tends asymptotically to zero

$$K_{\mathrm{or}}(\Delta l \to \infty) \to \langle \boldsymbol{e}(l)\rangle \langle \boldsymbol{e}(l+\Delta l)\rangle = 0 \tag{2.6}$$

We are looking for a parameter which measures the chain stiffness. As a suitable choice, one can take the integral width of K_{or}. This is known in the literature as the 'persistence length', and we denote it with l_{ps}.

The second, main point of concern are the global chain properties, as described by statistical means. More specifically, we inquire about the distribution of chain conformations. As already indicated, for the coarse-grained chain this problem can be solved immediately. Distribution functions may be

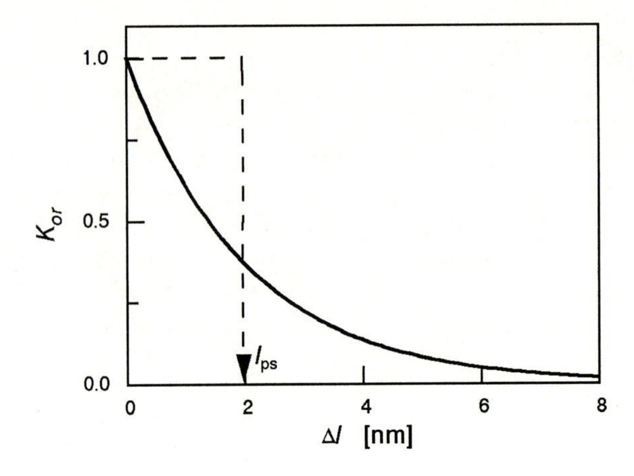

Fig. 2.6. Schematic representation of the orientational correlation function of a chain. The integral width determines the persistence length l_{ps}

directly deduced, with the aid of the following special procedure. We split the chain into subchains of uniform length, this length being much larger than l_{ps}. Then, as indicated in Fig. 2.5, we associate with the chain a sequence of vectors

$$(\boldsymbol{a}_1, \boldsymbol{a}_2, \ldots, \boldsymbol{a}_{N_s})$$

which connect the junction points of the subchains. We thus have created a segment chain composed of N_s straight units. Now, rather than discussing the statistical properties of the worm-like chain, we consider the distribution functions of the segment chain. Obviously, with regard to the global properties both agree with each other. Quantities of interest are the distribution functions for the vectors connecting any two junction points. One pair is of special interest, namely the two end points of the chain, being connected by the end-to-end distance vector $\boldsymbol{R}$. Regarding the given situation, the distribution function for $\boldsymbol{R}$ can be directly formulated. As we have chosen the subchains large compared to the persistence length, successive steps $\boldsymbol{a}_i$ of the segment chain show no orientational correlations. For this 'freely jointed segments chain' we have therefore a situation which is formally identical with the case of the movement of a Brownian particle suspended in a liquid. The latter performs a diffusive motion set up by perfectly uncorrelated steps. So do the segments of the chain! The distribution function for the displacement of a Brownian particle is well-known. It equals a Gaussian function and this, therefore, represents also the solution of our polymer problem.

Hence, it appears at first that the system can be treated quite easily. However, this is not the full truth. On reconsidering the situation more carefully, we must recognize that an important point was disregarded: The comparison with the Brownian motion is only allowed, if the volume of the monomer units is neglected since, in contrast to the diffusing Brownian particle, a chain of monomers with non-vanishing sizes cannot occupy a given location twice. What are the consequences? They are drastic indeed since the existence of an 'excluded volume' alters the situation not just a little, but completely. For monomers of finite size, the stated equivalence between the polymer problem and the diffusion case is no more valid. Since the excluded volume forces are effective between any two monomers at arbitrary distances along the chain and thus, are of long-range nature, there results a qualitative change in behavior. It is intuitively clear that the excluded volume interaction must result in a chain expansion but how can this effect be quantified? Up to 1972, this amounted to a major problem but then it was solved in one step by de Gennes, who noted that the problem is formally equivalent to a solvable problem in the physics of critical phenomena. With the basic solution on hand, excluded volume effects were analysed further in experiments, theories and computer simulations. Studies led to a second most important conclusion: It became clear that all 'expanded' chains have properties in common, to be described by simple power laws.

Since all monomer units have of course a finite size, at this point one might wonder if chains with Gaussian properties exist at all. In fact, they do exist in two cases. First, one finds solvents known as 'theta-solvents', which produce conditions for the interaction between the monomer units resulting in an effective vanishing of the excluded volume interactions. The second case is even more astonishing and may come as a real surprise. Flory predicted in the early years of polymer science on theoretical grounds that chains in the melt should behave as if they would have a zero volume. In fact, much later this was proven as being valid, utilizing neutron scattering.

Hence in summary, it can be stated that polymer chains possess, on length scales in the order of some nanometers, properties which are independent of the chemical structure. They may be grouped in two 'universality classes' with common characteristic behaviors

- Gaussian or 'ideal' chains, for vanishing excluded volume interactions
- 'expanded chains', otherwise.

Strictly speaking, we have to restrict this statement somewhat. In fact, the prediction that all coils may be assigned to either of two universality classes holds only for very large molecular weights, strictly in the limit $M \to \infty$. Real chains, in particular those with moderate molecular weights, often exhibit intermediate structures, representing 'mixed states ', or belonging to 'cross-over regions'. Examples which follow will also illustrate this behavior.

2.3.1 Ideal Chains

Let us now at first treat in more detail the ideal chains, on the basis of the introduced freely jointed segments model. We choose numbers, from 0 to N_s, for the junction points, as indicated in Fig. 2.5. The distance between the junction points i and j is given by

$$\boldsymbol{r}_{ij} = \sum_{l=i+1}^{j} \boldsymbol{a}_l , \tag{2.7}$$

and the end-to-end vector $\boldsymbol{R}$ by

$$\boldsymbol{R} = \sum_{l=1}^{N_s} \boldsymbol{a}_l \tag{2.8}$$

Of interest are the properties of the distribution function

$$p(\boldsymbol{R})\mathrm{d}^3\boldsymbol{R}$$

which expresses the probability that the end-to-end vector points into the volume element $\mathrm{d}^3\boldsymbol{R}$ at a distance $\boldsymbol{R}$. The distribution function is isotropic, depending only on

$$R := |\boldsymbol{R}|$$

and is normalized

$$\int p(\boldsymbol{R})\mathrm{d}^3\boldsymbol{R} = 1 \tag{2.9}$$

As explained above, $p(\boldsymbol{R})$ is identical with the distribution function for the displacement of a Brownian particle, after N_s uncorrelated steps. The latter is derived in many of the textbooks of statistical physics: It equals a Gaussian function with the form

$$p(\boldsymbol{R}) = \left(\frac{3}{2\pi\langle R^2\rangle}\right)^{3/2} \exp -\frac{3R^2}{2\langle R^2\rangle} \tag{2.10}$$

Figure 2.7 depicts this functional dependence.

As is important to notice, Eq. (2.10) includes one parameter only, namely the mean squared end-to-end distance $\langle R^2\rangle$, related to $p(\boldsymbol{R})$ by

$$\langle R^2\rangle = \int_{R=0}^{\infty} p(\boldsymbol{R})R^2 4\pi R^2 \mathrm{d}R \tag{2.11}$$

$\langle R^2\rangle$ can be directly calculated for our chain of freely jointed segments, by

$$\langle R^2\rangle = \left\langle \left(\sum_{l=1}^{N_s} \boldsymbol{a}_l\right)^2 \right\rangle = \left\langle \sum_{l,l'=1}^{N_s} \boldsymbol{a}_l \cdot \boldsymbol{a}_{l'} \right\rangle \tag{2.12}$$

Since

$$\langle \boldsymbol{a}_l \cdot \boldsymbol{a}_{l'}\rangle = \langle |\boldsymbol{a}_l|^2\rangle \cdot \delta_{ll'} \tag{2.13}$$

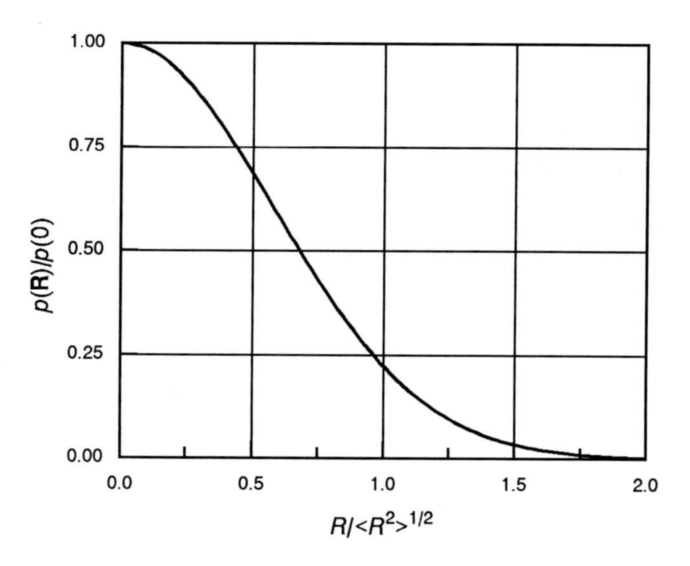

Fig. 2.7. Gaussian distribution function of the end-to-end distance vector of an ideal chain

we obtain

$$\langle R^2 \rangle = N_s \langle | \boldsymbol{a}_l |^2 \rangle \tag{2.14}$$

When dealing with polymer chains, one requires a parameter for estimating the size of the volume occupied by a polymer chain in the fluid phase. A suitable measure is provided by the quantity R_0, defined as

$$R_0 := \langle R^2 \rangle^{1/2} \tag{2.15}$$

Furthermore, if we introduce a mean segment length a_s

$$a_s := \langle | \boldsymbol{a}_l |^2 \rangle^{1/2} \tag{2.16}$$

which agrees with the mean diameter of the subchains, we obtain as an important first result the relation between the size of a polymer and the segment number N_s

$$R_0 = a_s N_s^{1/2} \tag{2.17}$$

Equivalent equations hold for the distribution functions of the internal distance vectors $\boldsymbol{r}_{ij}$ between two junction points in the chain, which follow as

$$p(\boldsymbol{r}_{ij}) = \left(\frac{3}{2\pi \langle r_{ij}^2 \rangle} \right)^{3/2} \exp - \frac{3r_{ij}}{2 \langle r_{ij}^2 \rangle} \tag{2.18}$$

with

$$\langle r_{ij}^2 \rangle = | i - j | \langle | \boldsymbol{a}_l |^2 \rangle = | i - j | a_s^2 \tag{2.19}$$

The coarse-grained representation of a given polymer by the freely jointed segments model is not unambiguous. There is only the requirement to select the subchain so that its contour length is large compared to the persistence length l_{ps}. Different choices of subchains imply different values of N_{s} and a_{s}. All pairs of values, however, have to lead to the same value of R_0^2. For two different choices, with parameters $N_{\mathrm{s}}, a_{\mathrm{s}}$ and $N_{\mathrm{s}}', a_{\mathrm{s}}'$, we have

$$N_{\mathrm{s}} a_{\mathrm{s}}^2 = R_0^2 = N_{\mathrm{s}}' a_{\mathrm{s}}'^2 \tag{2.20}$$

and therefore

$$\frac{N_{\mathrm{s}}}{N_{\mathrm{s}}'} = \frac{a_{\mathrm{s}}'^2}{a_{\mathrm{s}}^2} \tag{2.21}$$

An identical equation holds for the number of segments, $N_{\mathrm{s}}(i,j)$, between the points i,j in the chain. A change of the subchain size from a_{s} to a_{s}' has to be accompanied by changes of the segment numbers according to

$$\frac{N_{\mathrm{s}}(i,j)}{N_{\mathrm{s}}'(i,j)} = \frac{a_{\mathrm{s}}'^2}{a_{\mathrm{s}}^2} \tag{2.22}$$

The possibility of a rescaling of the representative freely jointed chain, as formulated by these equations, expresses an important basic property of ideal polymer chains, namely their 'self-similarity'. Self-similarity here means that independent of the chosen length scale, i.e. the resolution, an ideal chain always exhibits the same internal structure, one for which all internal distance vectors are distributed like Gaussian variables. A change of the length scale leaves this structure's characteristics invariant.

Self-similarity is the basic property of 'fractal objects' and ideal chains do indeed represent a nice example. The fractal dimension can be directly derived. If one proceeds n_{s} segmental steps, starting from a point in the interior of the chain, there results on average a displacement in the order of

$$r(n_{\mathrm{s}}) \simeq a_{\mathrm{s}} n_{\mathrm{s}}^{1/2} \tag{2.23}$$

In reverse, the number of monomers n contained in a sphere with radius r may be estimated as

$$n \sim n_{\mathrm{s}} \sim r^2 \tag{2.24}$$

Hence, the fractal dimension d, being defined as the exponent in the general relation

$$n \sim r^{\nu} \tag{2.25}$$

between the mass of an object and its diameter, follows as

$$d = 2 \tag{2.26}$$

We see that the polymer chain which occupies a region in three dimensions, fills this volume only partially, corresponding to a fractal dimension $d = 2$. As one consequence, the monomer density c_{m} in the volume of size R_0 occupied by a chain decreases with an increasing degree of polymerization, N, like

$$c_{\mathrm{m}}(N) \simeq \frac{N}{R_0^3} \sim \frac{1}{N^{1/2}} \tag{2.27}$$

It is important to recognize that self-similarity holds only for a finite range. Clearly, the upper bound is set by the size of the molecule, R_0. On the other hand, there is a lower limit, being given by the persistence length.

It is possible to remove the arbitrariness in the choice of the freely jointed segments-model by imposing a second condition. For this purpose, the length of the real chain in the fully extended straight form, denoted R_{max}, may be employed, and a second condition formulated as

$$R_{\mathrm{max}} = N_{\mathrm{s}} \cdot a_{\mathrm{s}} \tag{2.28}$$

It implies that the model chain and the real chain agree not only in size, as assured by the equality

$$\langle R^2 \rangle = N_{\mathrm{s}} a_{\mathrm{s}}^2 \tag{2.29}$$

but also in the contour length. Both equations together yield a unique value for the segment length a_{s} which is known as the 'Kuhn-length' a_{K}, and given by

$$a_{\mathrm{s}} = a_{\mathrm{K}} := \frac{\langle R^2 \rangle}{R_{\mathrm{max}}} \tag{2.30}$$

Evidently, the Kuhn-length characterizes the stiffness of a given polymer chain. Stiffer chains have larger values of a_{K}, and for a perfectly stiff chain one obtains $a_{\mathrm{K}} = R_{\mathrm{max}}$. One may anticipate and does find indeed a close correspondence between the persistence length and the Kuhn-length, with similar values for both quantities, i.e.

$$a_{\mathrm{K}} \simeq l_{\mathrm{ps}} \tag{2.31}$$

A related parameter, called the 'characteristic ratio', was introduced by Flory and is defined as

$$C_\infty := \frac{\langle R^2 \rangle}{N a_{\mathrm{b}}^2} \tag{2.32}$$

Here a_{b}^2 stands for the sum of the squares of the lengths of the backbone bonds of one monomer unit

$$a_{\mathrm{b}}^2 = \sum_i a_i^2 \tag{2.33}$$

For low degrees of polymerization, this ratio is not a constant but varies with N. For large values of N, an asymptotic value is reached and the latter is referred to as C_∞. In the hypothetical case of freely jointed bonds forming the chain backbone, C_∞ would equal unity whereas for real polymers, having fixed valence angles and restrictions in the rotations about the C-C-bonds, the values are in the range of 4 to 12. As will be explained in a later section (Eq. (6.165)), viscosity measurements on polymers dissolved in theta-solvents may be used for a determination of C_∞. Table 2.1 contains values of C_∞ for some selected polymers obtained by this method. Solvent and temperature are always indicated, since C_∞ may vary between different theta systems.

Table 2.1. Characteristic ratios C_∞ of some selected polymers, derived from viscosity measurements under theta-conditions at the indicated temperatures (data from Flory [1])

Polymer	Solvent	T[°C]	C_∞
polyethylene	dodecanol-1	138	6.7
polystyrene (atactic)	cyclohexane	35	10.2
polypropylene (atactic)	cyclohexane	92	6.8
polyisobutylene	benzene	24	6.6
poly(vinylacetate) (atactic)	*i*-pentanone-hexane	25	8.9
poly(methylmethacrylate) (atactic)	various solvents	4–70	6.9
poly(oxyethylene)	aqueous K_2SO_4	35	4.0
poly(dimethylsiloxane)	butanone	20	6.2

With the aid of C_∞ a 'scaling law' can be formulated which relates the size R_0 of a polymer to the degree of polymerization N. It reads

$$R_0^2 = \langle R^2 \rangle = C_\infty a_\mathrm{b}^2 N \tag{2.34}$$

or

$$R_0 = a_0 N^{1/2} \tag{2.35}$$

with

$$a_0 := a_\mathrm{b} C_\infty^{1/2} \tag{2.36}$$

Equation (2.35) tells us how R_0 'scales' with N: If we double N then R_0 increases by a factor $2^{1/2}$. As we see, 'scaling' is a property of power law dependencies, and the scaling factor follows from the exponent.

Exact Distribution Function for Finite Chain Lengths

When we expressed Eq. (2.18) for the distribution function $p(\boldsymbol{R})$, we did this based on the assumption that the random walk carried out by a chain of freely jointed segments should be equivalent to the motion of a Brownian particle. We pointed out that the equivalence is lost in the presence of excluded volume forces, however, this is not the only possible deficiency in the treatment. Checking the properties for large values of R we find that the Gaussian function never vanishes and actually extends to infinity. For the model chain, on the other hand, an upper limit exists, and it is reached for

$$R = R_\mathrm{max} = N_\mathrm{s} a_\mathrm{s} \tag{2.37}$$

At this point, the chain is completely stretched and cannot be extended further. It is possible to analyse the problem and to derive the exact distribution function for the end-to-end distance vector of a finite segment chain. The result may then be compared to the Gaussian function.

The distribution function $p(\boldsymbol{R})$ of a chain of N_s freely jointed segments with length a_s may be expressed as

$$p(\boldsymbol{R}) = \frac{1}{Z} \int\limits_{\sum_l \boldsymbol{a}_l = \boldsymbol{R}} \mathrm{d}^3\boldsymbol{a}_1 \cdot \mathrm{d}^3\boldsymbol{a}_2 \ldots \mathrm{d}^3\boldsymbol{a}_{N_s} \tag{2.38}$$

where Z denotes the partition function

$$Z = \int\limits_{\boldsymbol{a}_1} \mathrm{d}^3\boldsymbol{a}_1 \cdot \int\limits_{\boldsymbol{a}_2} \mathrm{d}^3\boldsymbol{a}_2 \ldots \int\limits_{\boldsymbol{a}_{N_s}} \mathrm{d}^3\boldsymbol{a}_{N_s} = \left(4\pi a_s^2\right)^{N_s} \tag{2.39}$$

The integral in Eq. (2.38) includes all chain conformations $\{\boldsymbol{a}_l\}$ which lead to an end-to-end distance vector $\boldsymbol{R}$.

One can derive an exact expression for the Fourier transform of p

$$p(\boldsymbol{q}) = \int \exp \mathrm{i}\boldsymbol{q}\boldsymbol{R} \cdot p(\boldsymbol{R}) \mathrm{d}^3\boldsymbol{R} \tag{2.40}$$

As the integration over all values of $\boldsymbol{R}$ removes the bounds in the integral Eq. (2.38), all orientations of all segments become included, and $p(\boldsymbol{q})$ may be written as

$$\begin{aligned} p(\boldsymbol{q}) &= \frac{1}{(4\pi a_s^2)^{N_s}} \cdot \int\limits_{\{\boldsymbol{a}_l\}} \exp \mathrm{i}(\boldsymbol{q} \cdot \sum_l \boldsymbol{a}_l) \mathrm{d}^3\boldsymbol{a}_1 \mathrm{d}^3\boldsymbol{a}_2 \ldots \mathrm{d}^3\boldsymbol{a}_{N_s} \\ &= \frac{1}{(4\pi a_s^2)^{N_s}} \left(\int \exp \mathrm{i}(\boldsymbol{q} \cdot \boldsymbol{a}) \mathrm{d}^3\boldsymbol{a} \right)^{N_s} \end{aligned} \tag{2.41}$$

The integral over all orientations of one segment can be evaluated. Adopting spherical coordinates for $\boldsymbol{a}$ ($|\boldsymbol{a}| = a_s, \vartheta, \varphi$), with the axis ($\vartheta = 0$) oriented parallel to $\boldsymbol{q}$, we have

$$\begin{aligned} \int \exp \mathrm{i}(\boldsymbol{q} \cdot \boldsymbol{a}) \mathrm{d}^3\boldsymbol{a} &= a_s^2 \int\limits_{\vartheta,\varphi} \exp(\mathrm{i}q a_s \cos\vartheta) \cdot \sin\vartheta \mathrm{d}\vartheta \mathrm{d}\varphi \\ &= a_s^2 2\pi \int\limits_{x:=\cos\vartheta=-1}^{1} \exp \mathrm{i}x q a_s \cdot \mathrm{d}x \\ &= \frac{a_s 4\pi}{q} \sin q a_s \end{aligned} \tag{2.42}$$

This leads us to an analytical expression for $p(\boldsymbol{q})$:

$$p(\boldsymbol{q}) = \left(\frac{\sin q a_s}{q a_s} \right)^{N_s} \tag{2.43}$$

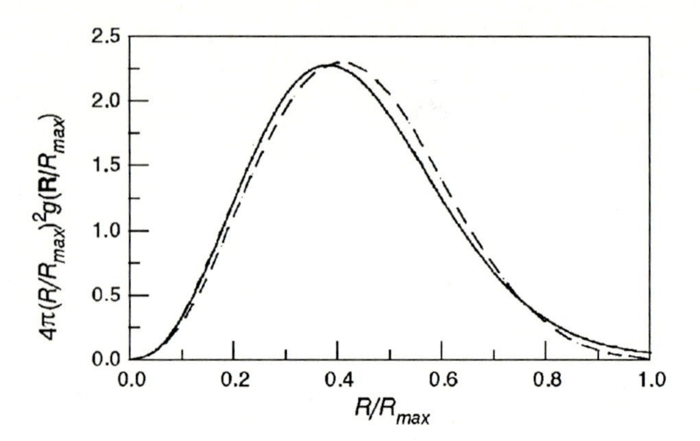

Fig. 2.8. Exact distribution function for the end-to-end distance vector of a chain of 5 freely jointed segments (*broken line*) compared to the Gaussian distribution function with the same value of $\langle R^2 \rangle$ (*continuous line*)

The distribution function $p(\boldsymbol{R})$ follows as the reverse Fourier transform

$$p(\boldsymbol{R}) = \frac{1}{(2\pi)^3} \int \exp -\mathrm{i}\boldsymbol{q}\boldsymbol{R} \left(\frac{\sin q a_\mathrm{s}}{q a_\mathrm{s}} \right)^{N_\mathrm{s}} \mathrm{d}^3\boldsymbol{q} \tag{2.44}$$

Equation (2.44) can be evaluated numerically, and Fig. 2.8 shows, as an example, the result of a calculation for $N_\mathrm{s} = 5$ which corresponds to a rather short chain. It is compared to the Gaussian function for the same value of $\langle R^2 \rangle (= N_\mathrm{s} a_\mathrm{s}^2 = R_\mathrm{max}^2 / N_\mathrm{s})$. As we see, both equations produce almost identical results for $R < R_\mathrm{max}$. In addition, the contributions of the Gaussian distribution function for $R > R_\mathrm{max}$ appear to be negligible.

We may conclude from this result that in treatments of polymer chains in solution or the melt, use of the Gaussian distribution function is always permitted. However, situations exist which require the application of the exact equation. Rubber elasticity, or treatments of yielding properties of polymers are affected by the limits in extensibility given for finite chains. Here, the Gaussian approximation can be used for small deformations only, and dealing with large deformations necessitates an introduction of the exact expression.

Brownian Chain

Here the question may arise as to which system does the Gaussian distribution function Eq. (2.10) really belong. There is a formal answer: It is associated with a limiting configuration of a segment chain, having an infinite number of segments

$$N_\mathrm{s} \longrightarrow \infty$$

and vanishing segment lengths a_s

$$a_\mathrm{s} \longrightarrow 0$$

both parameters being coupled, so that the mean-squared end-to-end distance remains constant

$$N_{\mathrm{s}}a_{\mathrm{s}}^2 = R_0^2 = \mathrm{const}$$

This constitutes a purely mathematical procedure, resulting in a mathematical object rather than a real polymer chain. Note that the contour length of this object tends to infinity,

$$l_{\mathrm{ct}} = N_{\mathrm{s}}a_{\mathrm{s}} = \frac{R_0^2}{a_{\mathrm{s}}} \longrightarrow \infty \quad \text{for} \quad a_{\mathrm{s}} \longrightarrow 0 \tag{2.45}$$

and this property also emerges in the Gaussian distribution function.

The object we have thus created is usually addressed as the 'Brownian chain', and it is characterized by just one parameter, namely the mean-squared end-to-end distance $\langle R^2 \rangle$. Although being defined as a mathematical limiting structure, the Brownian chain may indeed serve as a representative of real polymer chains in a certain region which is well-defined, namely for length scales which are larger than the persistence length and smaller than the size R_0 of the chain. The Brownian chain represents correctly the fractal properties and differs from the real chain in that it extends these fractal properties down to zero distances.

In polymer theories, one proceeds even one step further and introduces the 'infinite Brownian chain', which is associated with the passage to the limit $\langle R^2 \rangle \to \infty$. By this procedure, the upper bound for the self-similarity is also removed, and we have now an object which is self-similar on all length scales. This is exactly the situation of physical systems at critical points. Hence, the 'infinite Brownian chain' represents a perfect critical object and the consequence are far-reaching. Application of all the effective theoretical tools developed for the study of critical phenomena now becomes possible also for polymer systems. In particular, scaling laws may be derived which tell us how certain structure properties scale with the degree of polymerization. As mentioned above, scaling laws always have the mathematical form of a power law, and we have already met one example in Eq. (2.35)

$$R_0 = a_0 N^{1/2}$$

We derived this relation directly, by simple arguments, but it could have been assessed as well on general grounds, referring to fundamental properties of critical systems. On these grounds, scaling laws may be deduced for other properties and conditions, thereby proceeding much in the same manner as, for example, in discussions of the critical behavior of a ferromagnet near the Curie temperature.

We will leave this discussion now as any further extension would be definitely outside our scope. Nevertheless, it might have become clear that the introduction of mathematical objects such as the infinite Brownian chain may be very helpful. Although they are not real, they may be employed with success, as starting points for series expansions which lead us right back into the world of real polymer systems.

Debye-Structure Function

Much insight into the structure of polymer systems is provided by scattering experiments, and so we have to be concerned about the scattering properties of single ideal polymer chains. As we shall see, the associated scattering function can be formulated explicitly and then applied in the evaluation of experimental data. Our treatment is based on general relations of scattering theory, in particular on Eq. (A.21) given in the Appendix and readers who lack this knowledge should first study Sect. A.1.1 for a brief introduction.

Let us first consider the pair distribution function for the segments of a freely jointed chain. For N_s segments in a chain, this pair distribution function, $g_s(\boldsymbol{r})$, is given by

$$g_s(\boldsymbol{r}) = \frac{1}{N_s} \sum_{m=-(N_s-1)}^{N_s-1} (N_s - |m|) \left(\frac{3}{2\pi |m| a_s^2} \right)^{3/2} \exp\left(-\frac{3r^2}{2|m|a_s^2} \right) \quad (2.46)$$

Here, $g_s(\boldsymbol{r})$ is obtained by a summation over all pairs of segments, whereby we identify their distances with those of the (lower) adjacent junction points (the contribution for $m = 0$ equals a δ-function). There exist $N_s - |m|$ pairs with a distance m, and all these pairs have a common distribution function, as given by Eqs. (2.18) and (2.19). Approximating the sum by an integral

$$\sum_{m=-(N_s-1)}^{N_s-1} \longrightarrow \int_{m=-(N_s-1)}^{N_s-1} dm \approx 2 \int_{m=0}^{N_s} dm$$

and introducing the following substitutions

$$u' = \frac{ma_s^2}{r^2} \qquad u = \frac{N_s a_s^2}{r^2} = \frac{R_0^2}{r^2} \quad (2.47)$$

one obtains

$$g_s(\boldsymbol{r}) = 2\frac{N_s}{R_0^3} \cdot \frac{R_0}{r} \int_{u'=0}^{u} \left(1 - \frac{u'}{u}\right) \left(\frac{3}{2\pi u'}\right)^{3/2} \exp -\frac{3}{2u'} \cdot du' \quad (2.48)$$

$$:= \frac{N_s}{R_0^3} \cdot \tilde{g}_0 \left(\frac{r}{R_0}\right) \quad (2.49)$$

The pair distribution function for the monomers follows as

$$g(\boldsymbol{r}) = \frac{N}{N_s} g_s(\boldsymbol{r}) \quad (2.50)$$

giving

$$g(\boldsymbol{r}) = \frac{N}{R_0^3} \tilde{g}_0 \left(\frac{r}{R_0}\right) \quad (2.51)$$

According to this result, the pair distribution function of an ideal chain is given by a general function $\tilde{g}_0$ which depends on the dimensionless quantity

r/R_0. In this reduced representation all ideal chains become equivalent. Note that the integral value of $\tilde{g}_0$ is unity. Since

$$\int g \mathrm{d}^3\boldsymbol{r} = N \tag{2.52}$$

we have

$$\int \tilde{g}_0 \left(\frac{r}{R_0}\right) \mathrm{d}^3 \left(\frac{\boldsymbol{r}}{R_0}\right) = 1 \tag{2.53}$$

Quite characteristic is the behavior for $r/R_0 \ll 1$, i.e. for distances within the macromolecule. The integral in Eq. (2.48) may be replaced in this limit by its value for $u \to \infty$, and this leads to

$$g(\boldsymbol{r}) \sim \tilde{g}_0 \left(\frac{r}{R_0} \ll 1\right) \sim \frac{1}{r} \tag{2.54}$$

Equation (2.54) expresses a power law behavior. In the last section we addressed the self-similar nature of ideal polymer chains. The power law reflects exactly this property, since the function $g(r) \sim 1/r$ maintains its shape if we alter the unit length employed in the description of $\boldsymbol{r}$.

Next we derive the structure function of an ideal chain. It can be measured in diluted states, i.e. for low average monomer densities

$$\langle c_{\mathrm{m}} \rangle \approx 0 \tag{2.55}$$

Using Eq. (2.50), the general Eq. (A.21) becomes

$$S(\boldsymbol{q}) = \frac{N}{N_{\mathrm{s}}} \int \exp \mathrm{i}\boldsymbol{q}\boldsymbol{r} \cdot g_{\mathrm{s}}(\boldsymbol{r}) \mathrm{d}^3\boldsymbol{r} \tag{2.56}$$

If we take Eq. (2.46) and carry out the Fourier transformation for the Gaussian functions we obtain

$$S(\boldsymbol{q}) = \frac{2N}{N_{\mathrm{s}}^2} \int_{m=0}^{N_{\mathrm{s}}} (N_{\mathrm{s}} - m) \exp\left(-\frac{m a_{\mathrm{s}}^2 q^2}{6}\right) \mathrm{d}m \tag{2.57}$$

Hereby, we again replaced the summation by an integration. The substitutions

$$v' = \frac{m a_{\mathrm{s}}^2 q^2}{6} \qquad v = \frac{N_{\mathrm{s}} a_{\mathrm{s}}^2 q^2}{6} = \frac{R_0^2 q^2}{6} \tag{2.58}$$

lead to

$$S(\boldsymbol{q}) = N \frac{2}{v} \cdot \int_{v'=0}^{v} \left(1 - \frac{v'}{v}\right) \exp -v' \cdot \mathrm{d}v' \tag{2.59}$$

The integral can be evaluated, and the result then is usually presented as

$$S(\boldsymbol{q}) = N \cdot S_{\mathrm{D}}(q) \tag{2.60}$$

with

$$S_{\mathrm{D}}\left(v = \frac{R_0^2 q^2}{6}\right) = \frac{2}{v^2}(\exp -v + v - 1) \tag{2.61}$$

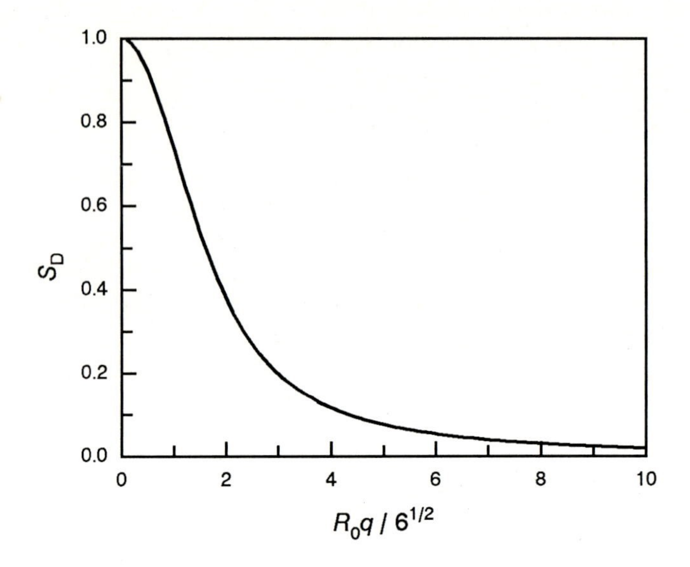

Fig. 2.9. Debye-structure function of an ideal chain with size R_0

S_{D} is known as the 'Debye-structure function' of an ideal chain, and a plot is shown in Fig. 2.9. In correspondence to the pair distribution function, the Debye-structure function can also be expressed in a reduced form, with v as general variable. Both the equations for the pair distribution function and for the scattering law indicate that all ideal chains are similar to each other, differing only in the length scale as expressed by R_0.

Some properties of S_{D} are noteworthy and furnish the basis for a straightforward analysis of experiments. First consider the limiting behavior for $v \to 0$, focussing on data near to the origin of reciprocal space, i.e. in the range of small scattering angles. Series expansion of Eq. (2.61) gives

$$S_{\mathrm{D}}(v \to 0) = 1 - \frac{v}{3} \cdots \tag{2.62}$$

Experimental results are usually presented in plots of S^{-1} versus q^2, corresponding to the equivalent formulation

$$S^{-1}(q^2) = N^{-1}\left(1 + q^2\frac{R_0^2}{18} + \cdots\right) \tag{2.63}$$

Equation (2.63) exemplifies 'Guinier's law' which generally holds for scattering experiments on dilute colloidal systems. As explained in the Appendix, Sect. A.3.1, analysis of small angle scattering data enables the determination of the colloidal mass, here given by the molecular weight or the degree of polymerization, and of the size of the colloid, here represented by R_0. The general relation formulated by Eq. (A.71) in the Appendix

$$S(q^2) = N\left(1 - q^2\frac{R_g^2}{3} + \cdots\right)$$

or

$$S^{-1}(q^2) = N^{-1}\left(1 + q^2\frac{R_g^2}{3} + \cdots\right)$$

includes the 'radius of gyration', denoted R_g. For a chain of N monomers R_g is defined as

$$R_g^2 := \frac{1}{N}\sum_{i=1}^{N}\langle|\boldsymbol{r}_i - \boldsymbol{r}_c|^2\rangle$$

whereby $\boldsymbol{r}_c$ denotes the location of the center of gravity

$$\boldsymbol{r}_c = \frac{1}{N}\sum_{i=1}^{N}\boldsymbol{r}_i$$

As is also shown in the Appendix, this is equivalent to

$$R_g^2 = \frac{1}{2N^2}\sum_{i,j=1}^{N}\langle|\boldsymbol{r}_i - \boldsymbol{r}_j|^2\rangle$$

Comparison of Eqs. (2.63), (A.71) tells us that for an ideal chain we have

$$R_{g,0}^2 = \frac{R_0^2}{6} \tag{2.64}$$

According to Eq. (2.61), the Debye-structure function exhibits a characteristic asymptotic behavior

$$S_D(v \to \infty) = \frac{2}{v} = \frac{12}{q^2 R_0^2} \tag{2.65}$$

We find here again a power law, $S_D \sim 1/q^2$ and indeed, it just represents the Fourier-transform of the previously discussed power law $g(r) \sim 1/r$.

Quite instructive is a plot of measured data in the form Sq^2 versus q, known as the 'Kratky-plot'. This is depicted schematically in Fig. 2.10. The plateau region, representing the part of the curve where $S_D \sim 1/q^2$, contains information about the internal structure of the chain. As the plateau value is given by

$$Sq^2(q^2R_0^2 \gg 1) = \frac{12N}{R_0^2} \tag{2.66}$$

$$= \frac{12}{C_\infty \cdot a_b^2} \tag{2.67}$$

when applying Eq. (2.32), it can be used for a determination of the characteristic ratio C_∞, which expresses an intrinsic, molecular weight independent property.

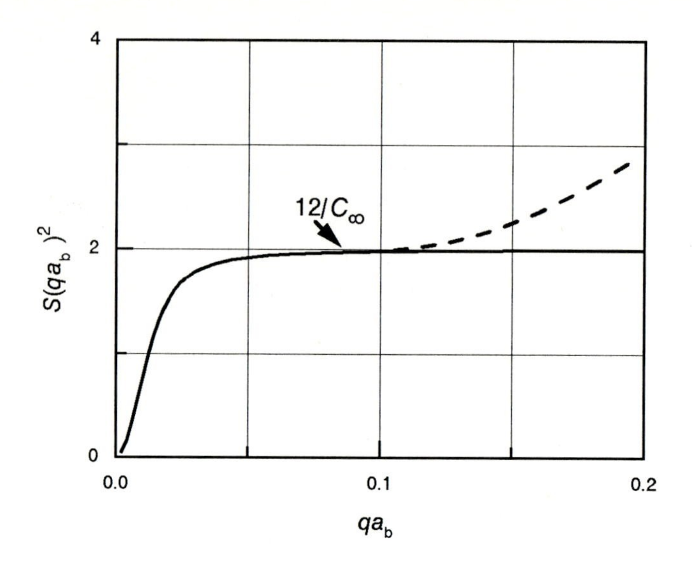

Fig. 2.10. Kratky-plot of the scattering-function of an ideal chain. The internal rigidity of the chain results in a change to a rod-like scattering at higher q's

The range of the self-similar internal structure ends when r approaches R_0. Around $q \simeq 1/R_0$ the curve deviates from the plateau and decays to zero. As mentioned earlier, universal behavior is also limited towards small distances . Specific polymer properties emerge, if one approaches distances in the order of the persistence length. For this reason, for polymers at higher q's deviations from the Debye-structure function show up, beginning around $q \simeq l_{\mathrm{ps}}^{-1}$. As indicated in Fig. 2.10, one observes an increase in the slope, and the structure function changes into that of a rod. It can be shown that the latter is generally given by

$$S_{\mathrm{rod}} \sim \frac{1}{q} \tag{2.68}$$

and this implies

$$Sq^2 \sim q \tag{2.69}$$

The point of cross-over from coil-like to rod-like scattering may be used for an estimation of the persistence length.

Theta-Solutions and Polymer Melts

As proved by experiments, ideal chains are found in 'theta-solutions' and in the melt. It is possible to provide qualitative explanations for this peculiar behavior, and we begin with the solutions.

For an explanation of what 'theta-conditions' in polymer solutions mean, it may help to recall the properties of real gases. For these systems the relation between pressure p, (molar) volume $\tilde{v}$ and temperature can be represented by the van-der-Waals equation

$$\left(p + \frac{\tilde{a}}{\tilde{v}^2}\right)(\tilde{v} - \tilde{b}) = \tilde{R}T \tag{2.70}$$

It includes two parameters, $\tilde{a}$ and $\tilde{b}$, which account for attractive van-der-Waals forces and repulsive hard core interactions respectively. There exists one temperature, called 'Boyle'-temperature, where the attractive and repulsive forces compensate each other, so that an apparent ideal gas behavior results. To locate this point a virial expansion in terms of the particle density $c_{\mathrm{m}} = \mathrm{N_L}/\tilde{v}$ ($\mathrm{N_L}$: Avogadro-Loschmidt number) can be used, in the form

$$p = kT(c_{\mathrm{m}} + A_2 c_{\mathrm{m}}^2 + A_3 c_{\mathrm{m}}^3 \dots) \tag{2.71}$$

The Boyle-temperature is that point at which the second virial coefficient A_2 vanishes, so that second order corrections to the ideal equation of state don not exist. The virial expansion of the van-der-Waals equation is

$$\begin{aligned} p\tilde{v} &= \tilde{R}T - \frac{\tilde{a}}{\tilde{v}} + \tilde{b}p + \mathcal{O}\left(\frac{1}{\tilde{v}^2}\right) \\ &= \tilde{R}T - \frac{\tilde{a}}{\tilde{v}} + \frac{\tilde{b}\tilde{R}T}{\tilde{v}} + \mathcal{O}\left(\frac{1}{\tilde{v}^2}\right) \\ &= \tilde{R}T\left(1 + \frac{1}{\tilde{v}}\left(\tilde{b} - \frac{\tilde{a}}{\tilde{R}T}\right)\right) + \mathcal{O}\left(\frac{1}{\tilde{v}^2}\right) \end{aligned} \tag{2.72}$$

or, in terms of c_{m}

$$p = kT(c_{\mathrm{m}} + c_{\mathrm{m}}^2 \cdot (\frac{\tilde{b}}{\mathrm{N_L}} - \frac{\tilde{a}}{\mathrm{N_L}^2 kT}) + \dots) \tag{2.73}$$

Hence, the second virial coefficient is given by

$$A_2 = \frac{\tilde{b}}{\mathrm{N_L}} - \frac{\tilde{a}}{\mathrm{N_L}^2 kT} \tag{2.74}$$

and we obtain for the Boyle-temperature, T_{B}, the result

$$T_{\mathrm{B}} = \frac{\tilde{a}}{\mathrm{N_L} k \tilde{b}} = \frac{\tilde{a}}{\tilde{R}\tilde{b}} \tag{2.75}$$

Another formulation for A_2 and derivation of T_{B} follows from the general theory of real gases. A_2 may be directly deduced from the pair interaction potential between the gas molecules, $u(r)$, by

$$A_2 = \frac{1}{2}\int_{r=0}^{\infty}\left(1 - \exp -\frac{u(r)}{kT}\right) 4\pi r^2 \mathrm{d}r \tag{2.76}$$

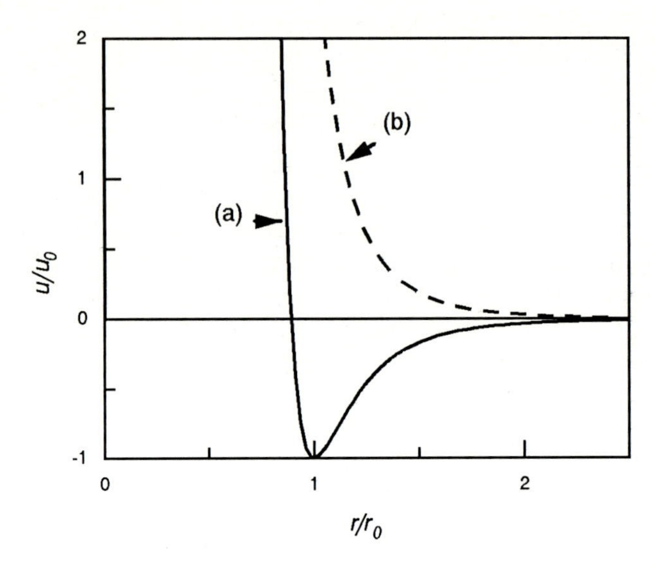

Fig. 2.11. Pair interaction potential for two monomers in a poor solvent (a) and in a good solvent (b). The potential (a) is also representative for a van-der-Waals gas

Curve (a) in Fig. 2.11 gives an example for the form of $u(r)$ which results from the superposition of the repulsive hard core interaction, originating from Pauli's exclusion principle opposing an overlap of electron wave functions, and the attractive dispersive forces of the van-der-Waals type. At the Boyle-temperature this integral vanishes.

An analogous situation is encountered in a theta-solution, with the quantity of interest now being the effective interaction potential between two solute molecules, or in the polymer case, between two monomers. The curve (b) in Fig. 2.11 represents the situation in a good solvent, where the potential is repulsive at all distances. Each solute molecule is surrounded by a hydrate-shell of solvent molecules and this shell has to be destroyed when two solvent molecules are to approach each other. The situation in a poor solvent is different, due to there being a preference for solute-solute contacts. Here, the solute molecules effectively attract each other and repulsion occurs only at short distances, then for the same reason as for the real gases, namely the presence of hard core interactions. For poor solvents, therefore, $u(r)$ has an appearance similar to the pair interaction potential in a van-der-Waals gas and a shape like curve (a) in Fig. 2.11.

On dealing with solutions, the osmotic pressure Π rather than the pressure p becomes the quantity of interest. Its virial expansion may be written as

$$\Pi = kT(A_1 c_{\mathrm{m}} + A_2 c_{\mathrm{m}}^2 + A_3 c_{\mathrm{m}}^3 + \ldots) \tag{2.77}$$

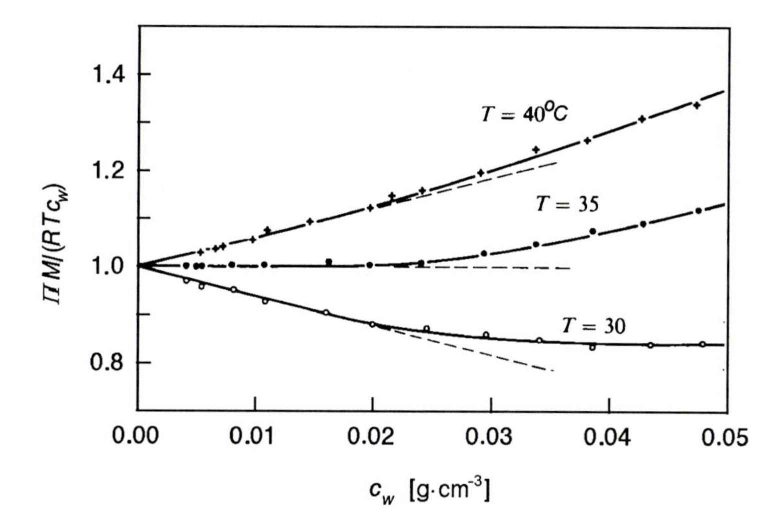

Fig. 2.12. Concentration and temperature dependence of the osmotic pressure measured for solutions of PS ($\overline{M}_n = 1.3 \cdot 10^5$) in cyclohexane. Data from Strazielle in Ref. [2]

where c_m now denotes the number density of monomers in the solution. If we identify $u(r)$ in Eq. (2.76) with the effective monomer-monomer interaction potential we obtain the second virial coefficient for the osmotic pressure. We can now define what 'theta-conditions' mean: For a solution at the theta-point, A_2 vanishes. A solution then becomes quasi-ideal; neither excluded volume interactions nor attractive forces emerge, since both compensate each other. Under this condition, monomers appear to have zero volumes, and polymers exhibit ideal chain behavior.

Figure 2.12 presents measurements of osmotic pressures for solutions of polystyrene in cyclohexane, carried out under variation of the weight fraction of the dissolved polymer, $c_w \sim c_m$, and of the temperature. As can be seen, A_2 vanishes at 35 °C, which therefore represents the theta-point of this system. As indicated by the positive values of A_2, for temperatures above 35 °C the repulsive forces become dominant. On the other hand, we observe negative values of A_2 for $T < 35\,°C$, indicative for the actual presence of effective attractive forces between the monomers. On further cooling, the attractive forces then lead to a separation of solvent and solute, and the sample becomes turbid.

Figure 2.13 shows the result of a light scattering experiment on the same system, a dilute solution of polystyrene ($\overline{M}_n = 8.79 \cdot 10^6$) in cyclohexane. The measurement was conducted exactly at the theta-point. As we have learned, ideal chains scatter according to the Debye-structure function, with the asymptotic limit $S_D \sim 1/q^2$. The data display the product

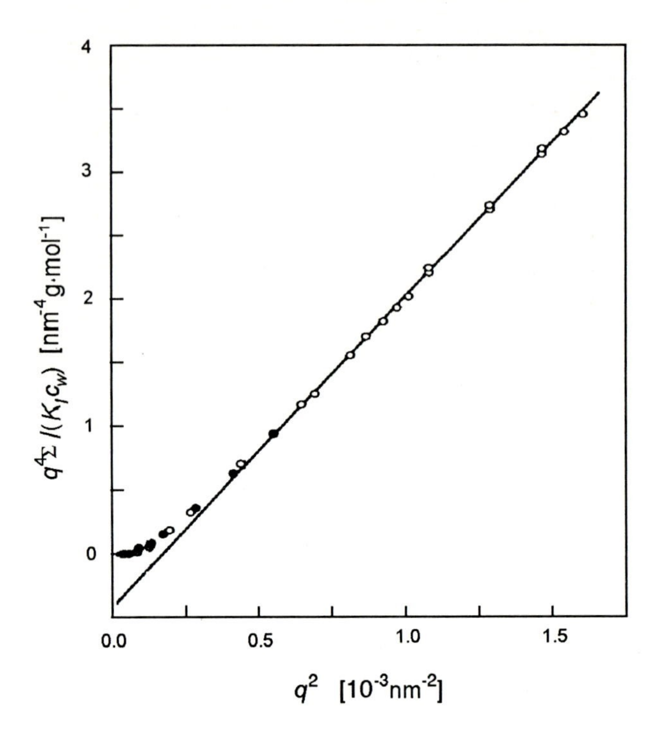

Fig. 2.13. Result of a light scattering experiment on a dilute solution of PS ($\overline{M}_{\rm n} = 8.78 \cdot 10^6$) in cyclohexane, carried out under theta-conditions ($T = 34.7\,°$C). Data from Miyaki et al. [3]

$$\Sigma(q)q^4 \sim S_{\rm D}q^4 \sim \exp - \frac{R_0^2 q^2}{6} + \frac{R_0^2 q^2}{6} - 1$$

plotted versus q^2. $\Sigma(q)$ is the 'Rayleigh ratio', and $K_{\rm l}$ represents the 'contrast factor for light'; definitions and equations are given in the Appendix, in Eqs. (A.4), (A.50), (A.51). The observed linear dependence at high q's demonstrates that the chains in this system do indeed show ideal behavior.

Next, let us turn to the situation in a polymer melt. The arguments to be presented are even more qualitative than those given for the theta-point, but they nevertheless address the basic features correctly. We begin with considering the conditions experienced by one isolated polymer chain. The density distribution, averaged over all conformations of the macromolecule, has an appearance similar to the bell-shaped curve in Fig. 2.14. There is a central maximum followed by continuous decays. Let us begin with an ideal chain and then consider the changes introduced by the excluded volume forces. These create a potential energy which is sensed by each monomer. In a simplified approximate treatment, this potential, denoted $\psi_{\rm m}^{\rm e}$, may be represented by

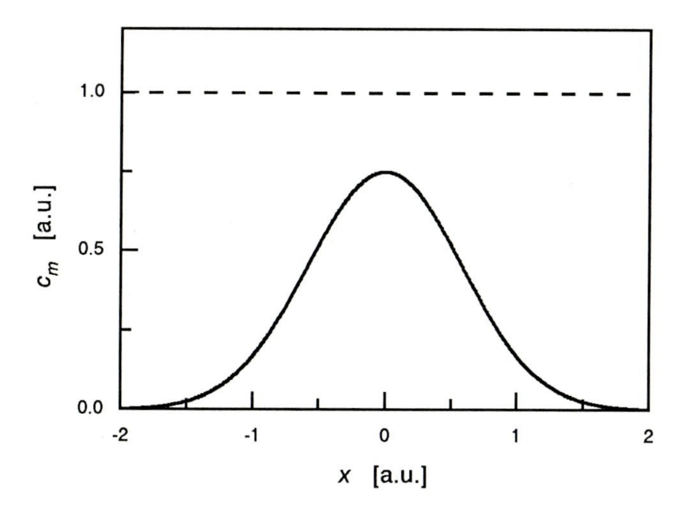

Fig. 2.14. Monomer density distribution for an individual chain along a line through the center and the constant overall density in a melt (*broken line*)

the expression

$$\psi_{\mathrm{m}}^{\mathrm{e}} = kTv_{\mathrm{e}}c_{\mathrm{m}} \tag{2.78}$$

This is an empirical equation of the 'mean-field' type, based on the assumption that $\psi_{\mathrm{m}}^{\mathrm{e}}$ should be proportional to the local density of monomers. The magnitude of the excluded volume interactions is described by the volume-like parameter v_{e}, with typical values in the order of 0.01–1 nm^3. The factor kT is explicitly included, not only for dimensional reasons, but also in order to stress that excluded volume energies, like hard core interactions in general, are of entropic nature (entropic forces are always proportional to T, as is exemplified by the pressure exerted by an ideal gas, or the restoring force in an ideal rubber, to be discussed in a later chapter). If the local potential experienced by a monomer is given by Eq. (2.78), then forces arise for all non-uniform density distributions. For the coil under discussion, forces in radial direction result since everywhere, with the exception of the center at $x = 0$, we have $\mathrm{d}c_{\mathrm{m}}/\mathrm{d}|x| < 0$. The obvious consequence is an expansion of the chain.

Envisage now the situation given in the melt. In contrast to an isolated polymer molecule, the monomer concentration c_{m} here is constant. For the excluded volume force onto a monomer, it is irrelevant whether the contacting other monomers are parts of the same chain or of other chains. The determining quantity is the total concentration and the latter does not vary. Hence, no forces arise and the polymer chain does not expand. In the literature one often finds a particular formulation for addressing this effect. As the concentration gradient given for an isolated chain is compensated for by the presence of monomers from the other chains, one says that the latter ones 'screen' the

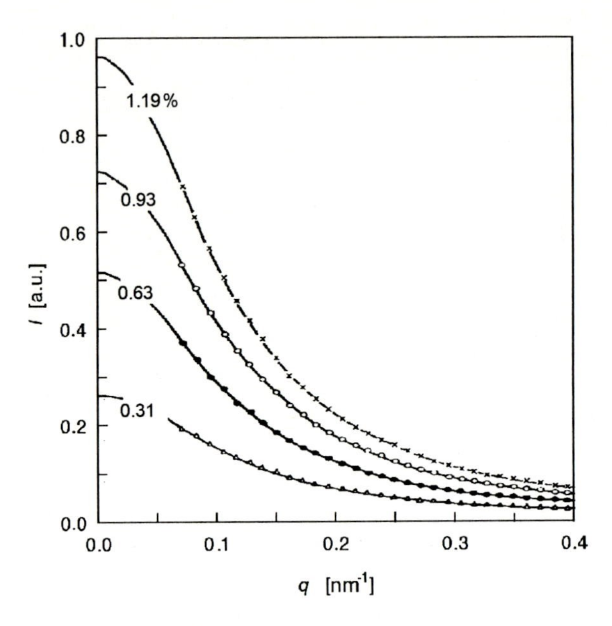

Fig. 2.15. Scattering curves obtained in a neutron scattering experiment on solutions of PMMA (mol fractions as indicated) in d-PMMA. Data from Kirste et al. [4]

intramolecular excluded volume interactions. Here we leave it with this short remark. Further comments on this picture and the origin of the saying will follow at a later stage when we discuss the properties of semi-dilute solutions.

Neutron scattering experiments allowed this prediction for melts to be verified. There is no way to investigate the conformation of single chains in a melt by conventional scattering experiments since a melt just represents a densely packed ensemble of monomers which shows some short-range order. In order to make single chains visible, one has to label them, i.e. supply them with contrast, so that they stand out from the background produced by the majority chains. Choosing neutrons for the scattering experiment, there is a preparatory technique to achieve this aim. Experiments can make use of the fact that neutrons possess different cross-sections for protons and deuterons. On the other hand, substitution of hydrogens by deuterium for part of the chains in a sample leaves the interaction forces and thus the chain conformations essentially unchanged (minor effects exist, but in melts they appear to be negligible). Thus, for a determination of the single chain conformation in a melt, a dilute solution of deuterated chains in a protonated matrix (or reversely, of protonated chains in a deuterated matrix) can be prepared and investigated by neutron scattering. Figure 2.15 presents the results of one of the first experiments, carried out on a solution of poly(methyl methacrylate) (PMMA) in deuterated PMMA. The continuous lines, which give perfect data-fits, represent Debye-structure functions (with minor corrections to account for polydispersity effects), thus proving the ideal behavior of chains. In addition, intensities were found to be proportional to the weight-fraction of deuterated chains

$$I(q) \sim c_{\mathrm{w}} S_{\mathrm{D}}(q) \tag{2.79}$$

As will be discussed in the next chapter, exactly this proportionality is indicative for a vanishing second virial coefficient. Hence, deuterated chains dissolved in a normal melt of the same polymer represent a theta-system, fully equivalent to a theta-solution with a low molar mass liquid. Therefore, as this second criterion is also fulfilled, we have unambiguous evidence for indeed having ideal chains in melts.

2.3.2 Expanded Chains

Expanded chains are found in dilute solution in good solvents. The effective interaction energy between two monomers is always repulsive here, and, as a consequence, chains become expanded. Expansion will come to an end at some finite value since it is associated with a decreasing conformational entropy. The reason for this decrease is easily seen by noting that the number of accessable rotational isomeric states decreases with increasing chain extension. The decrease produces a retracting force which balances, at equilibrium, the repulsive excluded volume forces.

The formal problem treated by polymer theory in the analysis of excluded volume effects is the analysis of the properties of 'self-avoiding random walks'. As already mentioned, it is not possible to come to a satisfactory solution if one takes an ideal random walk as a starting point and then introduces the excluded volume forces as a perturbation. Since excluded volume forces are effective between any pair of monomers, for arbitrary distances Δl along the chain, they are of long-range nature and as a consequence ideal chains and expanded chains become qualitatively different. The distribution function for the end-to-end vector and, in particular, the scaling law, which relates the size of the volume occupied by a polymer with its degree of polymerization, differ qualitatively from Eqs. (2.10) and (2.35) valid for ideal chains. Derivations are due to des Cloiseaux and de Gennes. Solutions were obtained using field theoretical techniques and renormalization group methods, and we just cite here the results.

The distribution function $p(\boldsymbol{R})$ has a general shape as indicated in Fig. 2.16. When compared to the properties of ideal chains there is first a change in the asymptotic behavior at large R. It is now given by

$$p(|\boldsymbol{R}| \to \infty) \sim \exp - \left(\frac{R}{R_{\mathrm{F}}} \right)^{5/2} \tag{2.80}$$

rather than by $p \sim \exp -(R/R_0)^2$. An even more drastic modification occurs around $R = 0$. Whereas ideal chains there reach the maximum, expanded

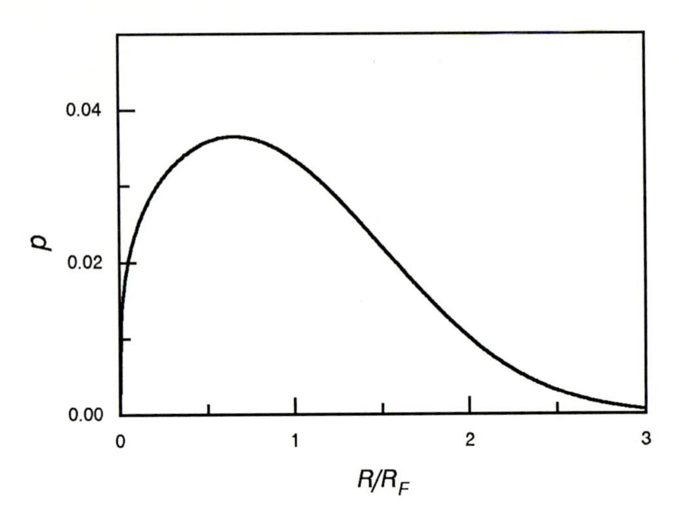

Fig. 2.16. Distribution function for the end-to-end vector of an expanded chain in the asymptotic limit of large degrees of polymerization (R_{F} : size of the chain)

chains show a steep decrease down to zero, indicating that a return of a self-avoiding random walk to its starting point is highly improbable. The shape near to the origin equals the power law

$$p(|\boldsymbol{R}| \to 0) \sim \left(\frac{R}{R_{\mathrm{F}}}\right)^{0.275} \tag{2.81}$$

The described distribution function refers to the asymptotic limit of large degrees of polymerization. It is important to note that, as for ideal chains, $p(\boldsymbol{R})$ includes one parameter only, now the quantity R_{F}, called in the literature 'Flory-radius'. R_{F} is a measure for the diameter of the volume occupied by the expanded polymer chain, with the identical definition as for ideal chains

$$R_{\mathrm{F}} := \langle R^2 \rangle^{1/2} \tag{2.82}$$

Of central importance for the discussion of the properties of expanded chains is the relation between R_{F} and the degree of polymerization N. It is given by the scaling law

$$R_{\mathrm{F}} = a_{\mathrm{F}} N^{3/5} \tag{2.83}$$

The value of the exponent, $\nu = 3/5$, expresses the difference to the ideal chains, where $\nu = 1/2$. Strictly speaking, $\nu = 0.6$ is not the exact value, as the calculation by renormalization group methods gives $\nu = 0.588$, but for the description of experimental results the rounded value is accurate enough. The second parameter included in the equation, a_{F}, denotes the effective length per monomer. It depends on the microstructure of the chain, i.e. on bond lengths, volume angles and the rotational angles of the isomeric states, and, in particular, on the strength of the excluded volume force.

Although the rigorous solution was not presented until 1972, Flory offered arguments in support of the scaling law Eq. (2.83) much earlier. The point of concern is the equilibrium conformation of a chain, being the result of a balance between repulsive excluded volume forces and retracting forces arising from the decreasing conformational entropy. Above, with Eq. (2.78), we have already introduced Flory's expression for the potential produced by the excluded volume forces

$$\psi_{\mathrm{m}}^{\mathrm{e}} = kTv_{\mathrm{e}}c_{\mathrm{m}}$$

$\psi_{\mathrm{m}}^{\mathrm{e}}$ describes, in a mean-field approximation, the potential experienced by a monomer as a result of the interaction with the other monomers in its neighborhood. Consequently, we may represent the contribution of the excluded volume interactions to the free energy density by

$$\frac{\Delta\mathcal{F}^{\mathrm{e}}}{\Delta V} = \frac{c_{\mathrm{m}}\psi_{\mathrm{m}}^{\mathrm{e}}}{2} = \frac{kTv_{\mathrm{e}}c_{\mathrm{m}}^2}{2} \tag{2.84}$$

whereby the division by 2 eliminates the twofold counting of each pair of monomers. The excluded volume contribution to the free energy of one chain, denoted $f_{\mathrm{p}}^{\mathrm{e}}$, is obtained by an integration over the occupied volume, hereby calculating the average over all conformations

$$f_{\mathrm{p}}^{\mathrm{e}} = \frac{1}{2}\int kTv_{\mathrm{e}}\,\langle c_{\mathrm{m}}^2(\boldsymbol{r})\rangle \mathrm{d}^3\boldsymbol{r} \tag{2.85}$$

In the mean-field treatment one approximates the square average $\langle c_{\mathrm{m}}^2\rangle$ by $\langle c_{\mathrm{m}}\rangle^2$. If we choose for the description of the mean local density $\langle c_{\mathrm{m}}(\boldsymbol{r})\rangle$ a Gaussian function, with a radius of gyration R_{g} and the maximum at the center of gravity $\boldsymbol{r}_{\mathrm{c}}$

$$\langle c_{\mathrm{m}}(\boldsymbol{r})\rangle = N\left(\frac{3}{2\pi R_{\mathrm{g}}}\right)^{3/2} \exp -\frac{3\,|\boldsymbol{r}-\boldsymbol{r}_{\mathrm{c}}|^2}{2R_{\mathrm{g}}^2} \tag{2.86}$$

we obtain

$$\begin{aligned} f_{\mathrm{p}}^{\mathrm{e}} &= \frac{kT}{2}v_{\mathrm{e}}N^2\left(\frac{3}{4\pi R_{\mathrm{g}}^2}\right)^{3/2}\left(\frac{6}{2\pi R_{\mathrm{g}}^2}\right)^{3/2}\int \exp -\frac{6\,|\boldsymbol{r}-\boldsymbol{r}_{\mathrm{c}}|^2}{2R_{\mathrm{g}}^2}\mathrm{d}^3\boldsymbol{r} \\ &= \frac{kT}{2}v_{\mathrm{e}}N^2\left(\frac{3}{4\pi R_{\mathrm{g}}^2}\right)^{3/2} \end{aligned} \tag{2.87}$$

As we are interested in the change of $f_{\mathrm{p}}^{\mathrm{e}}$ following from an expansion, we choose the ideal state with vanishing excluded volume forces, where $R_{\mathrm{g}} = R_{\mathrm{g},0}$ and $R_0^2 = a_0^2 N$, as our reference and write

$$f_{\mathrm{p}}^{\mathrm{e}} = \frac{kT}{2}v_{\mathrm{e}}\left(\frac{3}{4\pi}\right)^{3/2}\frac{R_0^4}{a_0^4}\frac{1}{R_{\mathrm{g},0}^3}\cdot\left(\frac{R_{\mathrm{g},0}}{R_{\mathrm{g}}}\right)^3 \tag{2.88}$$

or, by replacement of R_g against R assuming $R \sim R_g$ and applying Eq. (2.64)

$$f_p^e = \frac{kT}{2} v_e \left(\frac{18}{4\pi}\right)^{3/2} \frac{R_0}{a_0^4} \left(\frac{R_0}{R}\right)^3 \tag{2.89}$$

Now we introduce a parameter 'z' , defined as

$$z := \left(\frac{3}{2\pi}\right)^{3/2} \frac{v_e}{a_0^4} R_0 \tag{2.90}$$

and express f_p^e as

$$f_p^e = \frac{kT}{2} 3^{3/2} \cdot z \left(\frac{R_0}{R}\right)^3 \tag{2.91}$$

The parameter 'z' is dimensionless and determines according to Eq. (2.91) the excluded volume energy associated with a single chain.

Next, we require a formula for the second part of the free energy, f_p^s, the one originating from the conformational entropy. The following expression appears suitable

$$f_p^s = \beta kT \left(\left(\frac{R_0}{R}\right)^2 + \left(\frac{R}{R_0}\right)^2 \right) \tag{2.92}$$

This is an empirical equation which accounts for the fact that for ideal chains, i.e. vanishing excluded volume interactions, the coil size in thermal equilibrium equals R_0; β is a dimensionless coefficient of order unity. The first term gives the repulsion experienced on squeezing a polymer chain, the second term represents the retracting force built up on a coil expansion. As only the second term appears relevant for the case under discussion we ignore the first term and write

$$f_p^s \approx \beta kT \left(\frac{R}{R_0}\right)^2 \tag{2.93}$$

Combination of the expressions for the two contributions yields the free energy of a chain as a function of R

$$f_p\left(\frac{R}{R_0}\right) = f_p^e + f_p^s = \frac{kT}{2} 3^{3/2} z \left(\frac{R_0}{R}\right)^3 + \beta kT \left(\frac{R}{R_0}\right)^2 \tag{2.94}$$

We now calculate the equilibrium value of R at the minimum of the free energy, where

$$\frac{\mathrm{d} f_p}{\mathrm{d}(R/R_0)} = 0 \tag{2.95}$$

This leads us to

$$z \left(\frac{R_0}{R}\right)^4 \simeq \left(\frac{R}{R_0}\right) \tag{2.96}$$

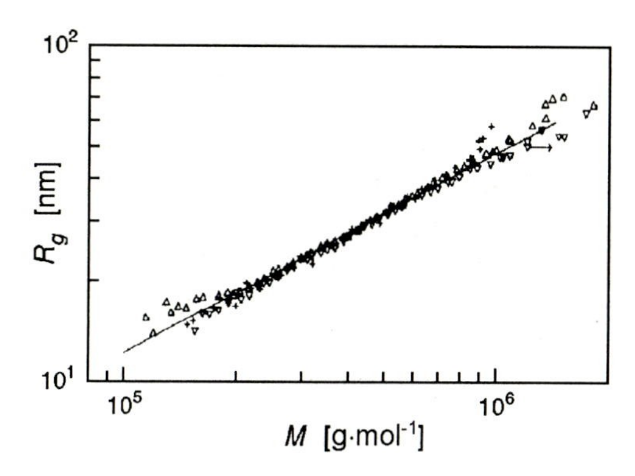

Fig. 2.17. Relation between the radius of gyration R_g and the molecular weight M, observed in light scattering experiments on dilute solutions of PS in toluene. The continuous straight line corresponds to the scaling law Eq. (2.83). Data from Wintermantel et al. [5]

and to the equation

$$R \simeq R_0 z^{1/5} \simeq \left(\frac{v_e}{a_0^4}\right)^{1/5} R_0^{6/5} = (v_e a_0^2)^{1/5} N^{3/5} \tag{2.97}$$

This is Flory's result. We identify R with the Flory radius R_F and write

$$R_F = a_F N^{3/5} \tag{2.98}$$

with

$$a_F \simeq (v_e a_0^2)^{1/5} \tag{2.99}$$

Here and generally we use the symbol '$\simeq$' in all 'order of magnitude equations' which correctly include all variables but omit the exact numerical front factor.

As we see, Flory's simple mean-field treatment leads to the same result as the exact analysis by renormalization group methods. In fact, this comes as a real surprise because there is no good reason to expect that a mean-field treatment which in principle is not allowed for a single chain in view of the pronounced concentration variations would give the correct result. However, it does, and so it appears that different faults in the treatment mutually cancel out.

The scaling law Eq. (2.83) is indeed in full accord with experiments, and Fig. 2.17 presents one example. It shows the results of light scattering experiments on dilute solutions of polystyrene in toluene. These can be used for determining the radius of gyration $R_g \sim R_F$ and the molecular weight $M \sim N$. Both parameters follow from a measurement of the scattering intensity in the low angle range, by applying Guinier's law Eq. (A.71). The straight line in Fig. 2.17 agrees exactly with the scaling law Eq. (2.83). The set of data was obtained in only two runs on two polydisperse samples. In the measurements, a fractionation of the samples by gel permeation chromatography was combined with a simultaneous registration of low angle light scattering curves.

It is possible to deduce from the scaling law Eq. (2.83) the fractal dimension of expanded chains, as well as characteristic properties of the pair distribution function and the structure function. The fractal dimension d follows from the same argument as applied above for the ideal chains, by estimating the average number of monomers included in a sphere of radius r which is now given by

$$n(r) \sim r^{5/3} \tag{2.100}$$

This power law implies that we have

$$d = 5/3 \tag{2.101}$$

The pair distribution function in the chain interior may be derived similarly. For a given pair distribution function $g(\boldsymbol{r})$ the average number of monomers in a sphere of radius r can be calculated in general by

$$n(r) = \int_{r'=0}^{r} g(\boldsymbol{r}') 4\pi r'^2 \mathrm{d}r' \tag{2.102}$$

On the other hand, we know that

$$n(r) \simeq \left(\frac{r}{a_{\mathrm{F}}}\right)^{5/3} \tag{2.103}$$

Equating the two expressions and taking the first derivatives on both sides gives

$$g(\boldsymbol{r}) r^2 \sim r^{2/3} \tag{2.104}$$

hence

$$g(\boldsymbol{r}) \sim r^{-4/3} \tag{2.105}$$

Equation (2.105) describes the pair distribution function for distances which lie within the chain interior, i.e. for $r < R_{\mathrm{F}}$. Note the difference to the respective relation for ideal chains, $g(\boldsymbol{r}) \sim r^{-1}$ (Eq. (2.54)).

More detailed considerations show that, as in the case of the ideal chains, the pair distribution function of expanded chains may also be presented in a reduced form

$$g(\boldsymbol{r}) = \frac{N}{R_{\mathrm{F}}^3} \tilde{g}_{\mathrm{F}} \left(\frac{r}{R_{\mathrm{F}}}\right) \tag{2.106}$$

Again one parameter only, R_{F}, enters into this expression, and $\tilde{g}_{\mathrm{F}}$ is a general function. Fourier transformation of the pair distribution function gives the structure function $S_{\mathrm{F}}(\boldsymbol{q})$

$$S_{\mathrm{F}}(\boldsymbol{q}) = \frac{N}{R_{\mathrm{F}}^3} \int \exp \mathrm{i}\boldsymbol{q}\boldsymbol{r} \cdot \tilde{g}_{\mathrm{F}} \mathrm{d}^3\boldsymbol{r} \tag{2.107}$$

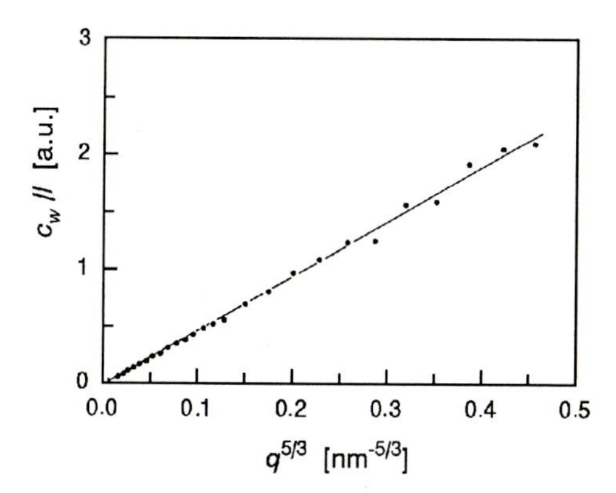

Fig. 2.18. Intensities obtained in a neutron scattering experiment on dilute solutions of (deuterated) PS ($\overline{M}_{\rm w} = 1.1 \cdot 10^6$) in CS_2 ($c_{\rm w} = 10^{-3}{\rm g}\cdot{\rm cm}^{-3}$). The straight line corresponds to the scattering law characteristic for expanded chains. Data from Farnoux [6].

A straightforward evaluation of the integral yields, for the asymptotic behavior at large q's in an inverse correspondence to Eq. (2.105), the power law

$$S_{\rm F}(\boldsymbol{q}) \sim q^{-5/3} \qquad (2.108)$$

Note that the fractal dimension $d = 5/3$ shows up directly in the asymptotically valid power law. Recall that this was also found for the ideal chains, where we had $S_{\rm D} \sim q^{-2}$. Indeed, we meet here a general relationship: Scattering laws measured for fractal objects exhibit the fractal dimension directly in the exponent

$$S(q) \sim q^{-d} \qquad (2.109)$$

Figure 2.18 exemplifies the predicted behavior and depicts the neutron scattering curve measured for a dilute solution of polystyrene in CS_2 which is a good solvent. The plot of I^{-1} versus $q^{5/3}$ gives a straight line, in agreement with Eq. (2.108).

In fact, such a result, with the power law characteristic for expanded chains extending over the full q-range of the measurement, is not always found, but only for very high molecular weights and a really good solvent. In general the range, where Eqs. (2.105), (2.108) apply, is limited. There is first an upper limit for r and correspondingly a lower limit for q, being given by the size of the chain; the validity of Eq. (2.108) ends on approaching $q \simeq 1/R_{\rm F}$. Near to $q = 0$, the structure function depends only on the degree of polymerization and the radius of gyration of the expanded chain, as described by Guinier's law. The relation between $R_{\rm g}$ and $R_{\rm F}$ differs slightly from the respective relation valid for ideal chains, Eq. (2.64). Computer simulations suggest a relation of the form

$$R_{\rm g,F}^2 = \frac{R_{\rm F}^2}{6.66} \qquad (2.110)$$

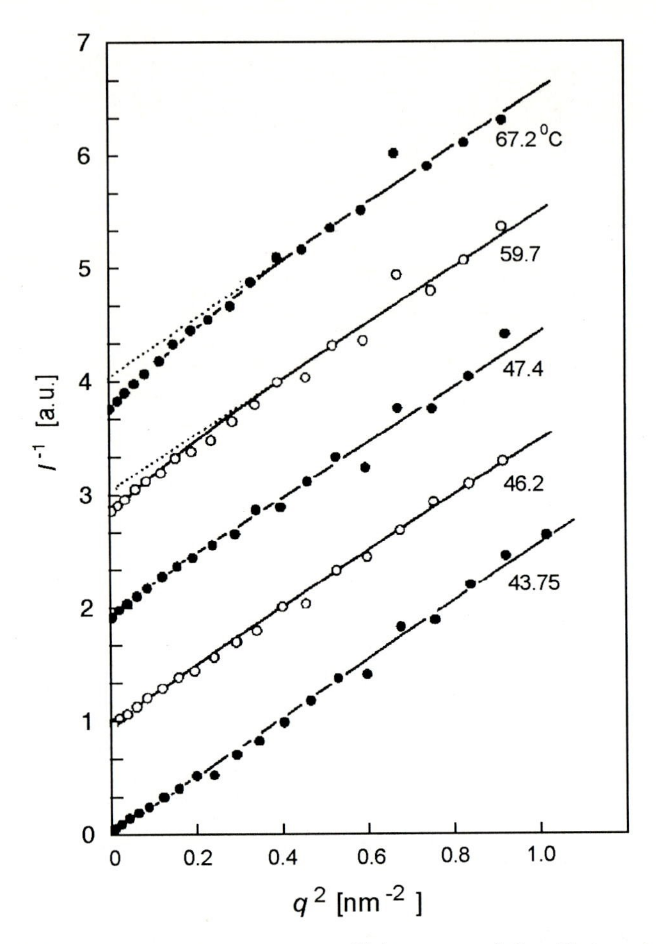

Fig. 2.19. Neutron scattering intensities $I(q)$ measured for dilute solutions of PS in cyclohexane ($\overline{M}_{\mathrm{w}} = 3.8 \cdot 10^6; qR \gg 1$) at the indicated temperatures above the theta-point ($T = 35\,°\mathrm{C}$). The straight lines correspond to the scattering law of ideal chains (subsequent curves are shifted upwards by constant amounts). Data from Farnoux et al. [7]

Of particular interest are the observations in neutron scattering experiments on solutions of polystyrene in cyclohexane presented in Fig. 2.19. These findings point to a second limitation for the power laws Eqs. (2.105), (2.108), now towards small distances r, corresponding to high values of q. Curves enable us to follow the changes in the internal chain structure which are introduced, if starting from an ideal conformation at the theta-point ($T = 35\,°\mathrm{C}$) by increasing the temperature the excluded volume interaction is 'switched on', and then becomes further intensified. We see that, in the q-range of the experiment up to temperatures of about $50\,°\mathrm{C}$, chains remain ideal, as is

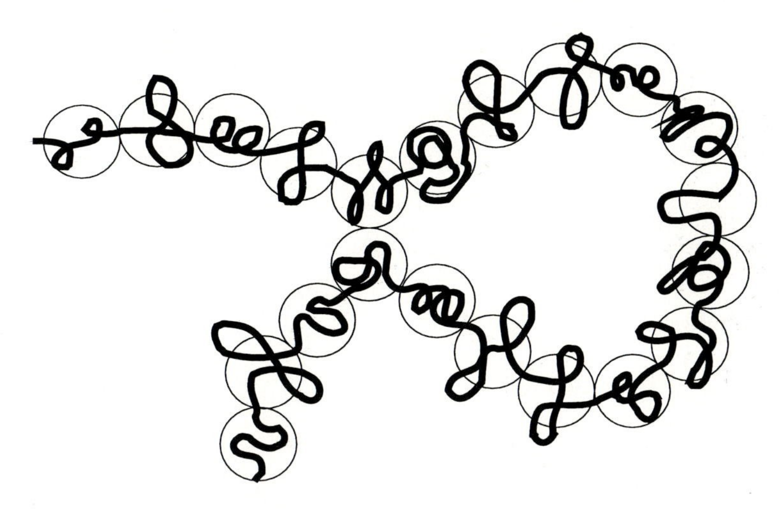

Fig. 2.20. Model of a composite chain with different fractal dimensions for $r < \xi_t(d = 2)$ and $r > \xi_t(d = 5/3)$. The cross-over distance corresponds to the size of the pearls.

demonstrated by the scattering law $I^{-1} \sim q^2$. Then, at higher temperatures deviations from ideal behavior become apparent on the low-q side and indicate a cross-over to the scattering behavior of an expanded chain, $I^{-1} \sim q^{5/3} < q^2$. This is a most interesting observation, as it tells us that the chain structure actually becomes dependent on the length scale: While for short distances the structure is still ideal, we find for larger distances the properties of expanded chains. In other words, the fractal dimension of the chain here depends on the resolution; for low resolutions there is $d = 5/3$, for high resolutions $d = 2$. The cross-over occurs at a certain distance, called in the literature 'thermic correlation length'. Its value, ξ_t, can be derived from the value of q where the change takes place in the scattering curve, by $\xi_t \simeq 1/q$. Obviously ξ_t must be determined by the excluded volume forces, i.e. by the parameter v_e, with ξ_t shifting to lower values on increasing v_e.

The functional dependence $\xi_t(v_e)$ may be derived using simple 'scaling arguments'. The experimental observation suggests modeling a chain with non-vanishing excluded volume interactions as indicated in Fig. 2.20. Here, the polymer is represented by a chain of N_{su} subunits with size ξ_t, each subunit being composed of n_{su} monomers. We have

$$N = N_{su} n_{su} \tag{2.111}$$

Since within each subunit, chain sequences exhibit ideal behavior, we can write

$$\xi_t^2 = a_0^2 n_{su} \tag{2.112}$$

Excluded volume interactions become effective for distances larger than ξ_t. We may account for this behavior in the model by assuming that the subunits cannot interpenetrate each other. The chain of subunits then displays the properties of an expanded chain, and we express its size with the aid of the scaling law Eq. (2.83), identifying a_{F} with ξ_{t} and N with N_{su}

$$R_{\mathrm{F}} \simeq \xi_{\mathrm{t}} N_{\mathrm{su}}^{3/5} \tag{2.113}$$

This leads us to

$$R_{\mathrm{F}}^{5/3} \simeq \xi_{\mathrm{t}}^{5/3} \frac{N}{n_{\mathrm{su}}} \tag{2.114}$$

and, in combination with Eq. (2.112), to

$$R_{\mathrm{F}}^{5/3} = \xi_{\mathrm{t}}^{5/3} \frac{a_0^2}{\xi_{\mathrm{t}}^2} N = \frac{a_0^2}{\xi_{\mathrm{t}}^{1/3}} N \tag{2.115}$$

On the other hand, we may apply the scaling law Eq. (2.83) for expanded chains also directly, together with Eq. (2.99)

$$R_{\mathrm{F}}^{5/3} = a_{\mathrm{F}}^{5/3} N = v_{\mathrm{e}}^{1/3} a_0^{2/3} N \tag{2.116}$$

Comparison shows us

$$\xi_{\mathrm{t}} \simeq \frac{a_0^4}{v_{\mathrm{e}}} \tag{2.117}$$

Thus, we have a reciprocal relation between ξ_{t} and v_{e}.

We have previously introduced the dimensionless parameter 'z' as (Eq. (2.90))

$$\mathrm{z} = \left(\frac{3}{2\pi}\right)^{3/2} \frac{v_{\mathrm{e}}}{a_0^4} R_0$$

Making use of Eq. (2.117), we realize what 'z' actually means. We obtain

$$\mathrm{z} \simeq \frac{R_0}{\xi_{\mathrm{t}}} = \frac{a_0 N^{1/2}}{a_0 n_{\mathrm{su}}^{1/2}} = (N_{\mathrm{su}})^{1/2} \tag{2.118}$$

and thus may conclude that z^2 gives the number of subunits of the model chain.

Above, it was pointed out that samples with ideal behavior are well represented by the asymptotic limit given by the Brownian chain, provided that the number of segments is large enough. An analogous property is found for expanded chains. Here, a mathematical object also exists which furnishes a good representation of real chains. It is constructed in an analogous manner by carrying out a passage to the limit $N \to \infty, a_{\mathrm{F}} \to 0, \xi_{\mathrm{t}} \to 0$, thereby keeping R_{F} constant. This mathematical object, which is a continuous curve with infinite contour length as the Brownian chain, has been named 'Kuhnian chain' by Janninck and des Cloiseaux. It constitutes an object with a fractal dimension $d = 5/3$ for all distances $r \ll R_{\mathrm{F}}$, down to $r \to 0$. Polymers in good solvents are well represented by the Kuhnian chain provided that the number of subunits is large, i.e. z $\gg$ 1.

It is always useful to check for the number of independent parameters. For the Brownian chain there is one parameter only, namely R_0. For the expanded chain in general, we find two parameters, R_F and ξ_t, but in the Kuhnian limit $\xi_t \to 0$ we return again to the simple one-parameter case.

2.4 The Ising-Chain

A discussion of the properties of a polymer chain on the basis of the global scaling laws alone would be incomplete. Description of specific properties of a given polymer molecule, as for example its internal energy or entropy, requires a different approach. For this purpose one needs a treatment which takes account of the energetics of the chain. As was explained at the begin of this chapter, chain conformations may be described microscopically in terms of the accessible rotational isomeric states. Now we shall see that this representation of a polymer corresponds exactly to the one-dimensional Ising-model, the 'Ising-chain', an important concept in general statistical mechanics. As the tools for the treatment of Ising-chains are well-known, Birshtein and Ptitsyn, and Flory adapted the Ising-model to the polymer problem. This adaption, addressed in the literature as 'rotational isomeric state (RIS)-model', opens a straightforward way to calculate the thermodynamic functions and the specific structural properties of a given polymer chain.

The general Ising-chain is set up by an array of interacting particles, with each particle being able to change between a certain number of different states. In the simplest case, interactions are restricted to adjacent pairs. Then the total energy of the chain equals the sum of the interaction energies between neighbors, and is for n particles given by

$$u = \sum_{i=2}^{n} u(\varphi_{i-1}, \varphi_i) \tag{2.119}$$

Here φ_i denotes the state of particle i and $u(\varphi_{i-1}, \varphi_i)$ is the pair interaction energy.

The relation to a polymer chain becomes clear when one considers that the energy of one conformational state is a function of the rotational isomeric states of all N_b backbone bonds. The latter correspond to the 'particles' of the Ising-chain. Conditions would be trivial if all bonds were energetically independent since then the chain energy would be equal to N_b times the mean energy of a single bond. In reality, however, adjacent bonds may well affect each other. This is nicely exemplified by polyethylene, where the 'pentane effect', indicated in Fig. 2.21, becomes effective. The depicted conformation represents the sequence *trans-gauche*$^+$*-gauche*$^-$*-trans*, and pentane is the shortest n-alkane, for which this sequence may be built up. As we see, a

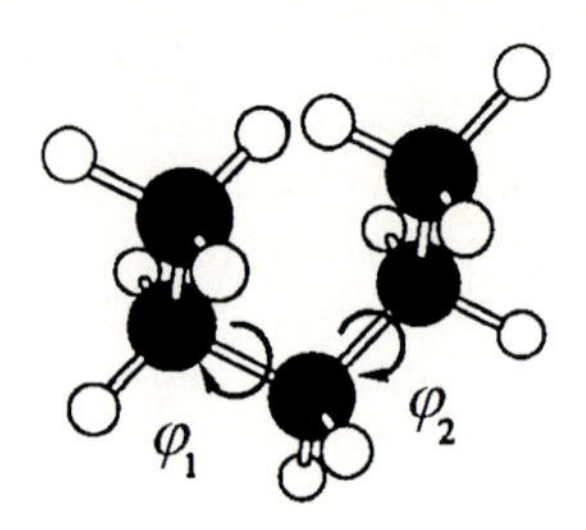

Fig. 2.21. The conformation of pentane associated with a sequence *trans-gauche*$^+$-*gauche*$^-$-*trans*. A sharp fold with elevated energy is formed

sharp fold is formed, and it is clear that this requires more energy than is necessary to form two independent *gauche*-states. Since the Ising-model deals with energies depending on the states of both partners in a pair, it can take account of this situation.

The main task in the computation of thermodynamic functions is the calculation of the partition function, denoted Z. In our case, its basic form can be formulated directly, as

$$Z = \sum_{\{\varphi_i\}} \exp - \frac{u\{\varphi_i\}}{kT} \tag{2.120}$$

The summation includes all conformational states, here shortly designated by $\{\varphi_i\}$, each state being determined by specifying the conformations of all bonds

$$\{\varphi_i\} \hat{=} (\varphi_1, \varphi_2, \ldots, \varphi_{N_\mathrm{b}})$$

The energy for each conformational state of the chain follows from Eq. (2.119).

Knowing the partition function, we can employ general laws of thermodynamics in order to deduce the free energy per polymer chain, using

$$f_\mathrm{p} = -kT \ln Z \tag{2.121}$$

the entropy per chain, by

$$s_\mathrm{p} = -\frac{\partial f_\mathrm{p}}{\partial T} \tag{2.122}$$

and the internal energy per polymer, by

$$e_\mathrm{p} = f_\mathrm{p} + T s_\mathrm{p} \tag{2.123}$$

The partition function Z can be evaluated in straightforward manner. We write

$$\begin{aligned} Z &= \sum_{\varphi_2} \cdots \sum_{\varphi_{N_\mathrm{b}-1}} \exp - \frac{1}{kT} \sum_{i=2}^{N_\mathrm{b}} u(\varphi_{i-1}, \varphi_i) \\ &= \sum_{\varphi_2} \cdots \sum_{\varphi_{N_\mathrm{b}-1}} \prod_{i=2}^{N_\mathrm{b}} \exp - \frac{1}{kT} u(\varphi_{i-1}, \varphi_i) \end{aligned} \tag{2.124}$$

or, introducing the 'statistical weights'

$$t(\varphi_{i-1},\varphi_i) := \exp -\frac{1}{kT}u(\varphi_{i-1},\varphi_i) \tag{2.125}$$

briefly

$$Z = \sum_{\varphi_2}\cdots\sum_{\varphi_{N_b-1}}\prod_{i=2}^{N_b} t(\varphi_{i-1},\varphi_i) \tag{2.126}$$

For explaining further, let us select polyethylene as an example. Here, for the three states per bond, nine different values t_{ij} exist. We collect them in a matrix $\mathbf{T}$

$$\mathbf{T} = \begin{pmatrix} 1 & w_0 & w_0 \\ 1 & w_0 & w_1 w_0 \\ 1 & w_1 w_0 & w_0 \end{pmatrix} \tag{2.127}$$

Thereby we attribute to the different rotational isomeric states the following indices

$$\textit{trans} \hat{=} 1, \qquad \textit{gauche}^+ \hat{=} 2, \qquad \textit{gauche}^- \hat{=} 3$$

The matrix includes two parameters, denoted w_0 and w_1. To understand the matrix structure, imagine that a specific conformation is formed by subsequently arranging all bonds, emanating from the lowest energy state *all-trans*. We start at $i = 2$ and then proceed up to the end, $i = N_b - 1$ (the step to the last bond, $i = N_b$, can be omitted, since for this bond without a further neighbor, no energy contribution arises). The coefficients of the matrix give the statistical weights associated with each step:

1. Since no energy is required if the *trans*-state is maintained we have

$$u(i,1) = 0 \rightarrow t_{i1} = 1$$

2. Formation of a *gauche*$^+$-state after a *trans*- or *gauche*$^+$-state requires an energy Δu_{tg} and thus carries a statistical weight

$$w_0 := \exp -\Delta u_{tg}/kT < 1 \tag{2.128}$$

3. Increased energies of formation are associated with the 'hairpin-bend'-conformations *gauche*$^+$-*gauche*$^-$ and *gauche*$^-$-*gauche*$^+$, resulting in lowered statistical weights, as expressed by the product $w_1 w_0$ with

$$w_1 < 1$$

The evaluation of the partition function

$$\begin{aligned} Z \;=\; & \sum_{\varphi_2} t(\varphi_t,\varphi_2)\cdot\sum_{\varphi_3} t(\varphi_2,\varphi_3)\cdots\sum_{\varphi_{N_b-2}} t(\varphi_{N_b-3},\varphi_{N_b-2}) \\ & \cdot\sum_{\varphi_{N_b-1}} t(\varphi_{N_b-2},\varphi_{N_b-1})\cdot t(\varphi_{N_b-1},\varphi_t) \end{aligned} \tag{2.129}$$

can be rationalized using matrix multiplication rules. We can repeatedly apply the general summation rule

$$\sum_l t_{il} t_{lj} = (\mathbf{T}^2)_{ij} \tag{2.130}$$

for successive reductions:

$$\begin{aligned} Z &= \sum_{\varphi_2} t(\varphi_t, \varphi_2) \cdot \sum_{\varphi_3} t(\varphi_2, \varphi_3) \cdots \sum_{\varphi_{N_\mathrm{b}-2}} t(\varphi_{N_\mathrm{b}-3}, \varphi_{N_\mathrm{b}-2}) \cdot (\mathbf{T}^2)_{\varphi_{N_\mathrm{b}-2}, \varphi_t} \\ &= \sum_{\varphi_2} t(\varphi_t, \varphi_2) \cdot \sum_{\varphi_3} t(\varphi_2, \varphi_3) \cdots \sum_{\varphi_{N_\mathrm{b}-3}} t(\varphi_{N_\mathrm{b}-4}, \varphi_{N_\mathrm{b}-3}) \cdot (\mathbf{T}^3)_{\varphi_{N_\mathrm{b}-3}, \varphi_t} \\ &\ \ \cdot \\ &\ \ \cdot \\ &= (\mathbf{T}^{N_\mathrm{b}-1})_{11} \end{aligned} \tag{2.131}$$

Hence, Z can be obtained by calculating the power $(N_\mathrm{b}-1)$ of $\mathbf{T}$ and extracting the 11-coefficient. The task of calculating the power $(N_\mathrm{b}-1)$ of $\mathbf{T}$ can be much simplified if $\mathbf{T}$ is first transformed into a diagonal form. This can be achieved as usually, by solving the set of homogeneous linear equations

$$\sum_j T_{ij} A_j = \lambda A_i \tag{2.132}$$

i.e. evaluating the determinant

$$|\,\mathbf{T} - \lambda \mathbf{1}\,| = 0 \tag{2.133}$$

There are three eigenvalues, λ_1, λ_2 and λ_3, and they set up a diagonal matrix $\mathbf{\Lambda}$. The matrix $\mathbf{A}$ which transforms $\mathbf{T}$ into $\mathbf{\Lambda}$

$$\mathbf{\Lambda} = \mathbf{A}^{-1}\mathbf{T}\mathbf{A} \tag{2.134}$$

is composed of the three eigenvectors, $(A_{1,j}), (A_{2,j}), (A_{3,j})$. With the aid of $\mathbf{\Lambda}$, the matrix multiplication becomes very simple:

$$\begin{aligned} \mathbf{T}^{N_\mathrm{b}-1} &= (\mathbf{A}\mathbf{\Lambda}\mathbf{A}^{-1})^{N_\mathrm{b}-1} \\ &= \mathbf{A}\mathbf{\Lambda}\mathbf{A}^{-1}\mathbf{A}\mathbf{\Lambda}\mathbf{A}^{-1} \cdots \mathbf{A}\mathbf{\Lambda}\mathbf{A}^{-1} \\ &= \mathbf{A}\mathbf{\Lambda}^{N_\mathrm{b}-1}\mathbf{A}^{-1} \end{aligned} \tag{2.135}$$

We employ this equation and obtain an explicit expression for the partition function Z

$$Z = A_{11} \cdot (A^{-1})_{11} \cdot \lambda_1^{N_\mathrm{b}-1} + A_{12} \cdot (A^{-1})_{21} \cdot \lambda_2^{N_\mathrm{b}-1} + A_{13}(A^{-1})_{31} \cdot \lambda_3^{N_\mathrm{b}-1} \tag{2.136}$$

Usually all three eigenvalues are different and one, say λ_1, is the largest:

$$\lambda_1 > \lambda_2, \lambda_3$$

Since N_b is huge , the partition function is well approximated by

$$Z \approx A_{11} \cdot (A^{-1})_{11} \lambda_1^{N_b - 1} \tag{2.137}$$

The free energy then follows as

$$f_p = -kT \left((N_b - 1) \cdot \ln \lambda_1 + \ln \left(A_{11} A_{11}^{-1} \right) \right) \tag{2.138}$$

For a polymer, where $N_b \gg 1$, we can ignore the constant second term. This leads us to a simple expression for the free energy per bond:

$$\frac{f_p}{N_b} = -kT \ln \lambda_1 \tag{2.139}$$

The entropy and the internal energy per bond follow as

$$\frac{s_p}{N_b} = k \ln \lambda_1 + \frac{kT}{\lambda_1} \frac{\partial \lambda_1}{\partial T} \tag{2.140}$$

and

$$\frac{e_p}{N_b} = \frac{f_p}{N_b} + T \frac{s_p}{N_b} \tag{2.141}$$

As we see, in the framework of the RIS-model it is a simple matter to derive the thermodynamic functions for a given polymer chain, the only requirement being a knowledge of the matrix $\mathbf{T}$ of the statistical weights.

Let us carry out the calculation for polyethylene. The determinant equation to be solved is

$$\begin{vmatrix} 1 - \lambda & w_0 & w_0 \\ 1 & w_0 - \lambda & w_0 w_1 \\ 1 & w_0 w_1 & w_0 - \lambda \end{vmatrix} = 1 \tag{2.142}$$

This is a third order equation, but an evaluation shows that it factorizes, having the form

$$(w_0 - \lambda - w_0 w_1)[\lambda^2 - \lambda(w_0 + w_0 w_1 + 1) + w_0(w_1 - 1)] = 1 \tag{2.143}$$

Therefore, the solutions can be given analytically. The three eigenvalues are

$$\begin{aligned} \lambda_{1/2} &= \frac{1}{2} \left[(w_0 + w_0 w_1 + 1) \pm \sqrt{(w_0 + w_0 w_1 + 1)^2 + 4 w_0 (1 - w_1)} \right] \\ \lambda_3 &= w_0 (1 - w_1) \end{aligned} \tag{2.144}$$

As is obvious, the largest eigenvalue is λ_1.

It is instructive to consider the numerical results for polyethylene in a computation for its melting point, $T_f = 415$ K. We choose for the energy required to form a *gauche*-state after a *trans*-state the value $\Delta \tilde{u}_{tg} = 2\ \mathrm{kJ \cdot mol^{-1}}$ (compare Sect. 2.1) and obtain $w_0 = 0.56$. The second statistical weight which is needed, the product $w_0 w_1$, has so far not been experimentally determined. Estimates for the related energy of the hairpin-bend states have been obtained by potential energy calculations, using empirical expressions for the non-bonded interaction energies. Values in the range of 7 kJ · mol^{-1} are thus indicated, corresponding to a statistical weight $w_0 w_1 = 0.13$. With these values the following results are obtained for the thermodynamic functions, expressed per mol of CH_2-units:

$$\begin{aligned}
\tilde{f}_{\mathrm{m}} &= \mathrm{N_L} f_{\mathrm{p}}/N_{\mathrm{b}} = 2.28\ \mathrm{kJ \cdot mol^{-1}} \\
\tilde{s}_{\mathrm{m}} &= \mathrm{N_L} s_{\mathrm{p}}/N_{\mathrm{b}} = 8.25\ \mathrm{J \cdot K^{-1} mol^{-1}} \\
\tilde{e}_{\mathrm{m}} &= \mathrm{N_L} e_{\mathrm{p}}/N_{\mathrm{b}} = 1.14\ \mathrm{kJ \cdot mol^{-1}}
\end{aligned}$$

It is interesting to compare these results with the measured heat of fusion and the entropy of fusion

$$\begin{aligned}
\Delta \tilde{h}^{\mathrm{f}}_{\mathrm{m}} &= 4.10\ \mathrm{kJ \cdot mol^{-1}} \\
\Delta \tilde{s}^{\mathrm{f}}_{\mathrm{m}} &= \Delta \tilde{h}^{\mathrm{f}}_{\mathrm{m}}/T_{\mathrm{f}} = 9.9\ \mathrm{J \cdot K^{-1} mol^{-1}}
\end{aligned}$$

We notice that the experimental heat of fusion, $\Delta \tilde{h}^{\mathrm{f}}_{\mathrm{m}}$, is much larger than can be accounted for by the change in the intramolecular conformational energy, as given by $\tilde{e}_{\mathrm{m}}$. Hence, the major part of the heat of fusion seems to be related to a change in the intermolecular energy, i.e. to the increase in the specific volume (which amounts to 15%). With regard to the change in entropy, conclusions are different. Here the major part is indeed contributed by the change of the conformation from the *all-trans* state into the coiled form, with only the smaller rest being due to the increase in free volume.

In Sect. 2.1 we carried out a first estimate of the fraction of *trans*- and *gauche*-states in polyethylene. In this estimate, independence of the rotational isomeric states of different bonds was implicitly assumed. We now may check for the modification introduced by the pentane effect, as the Ising-model also provides us with equations for the fractions of the different pairs of conformational states. We rewrite the partition function

$$Z = \sum_{\{\varphi_i\}} t(\varphi_1, \varphi_2) \ldots t(\varphi_{N_{\mathrm{b}}-1}, \varphi_{N_{\mathrm{b}}})$$

and choose a special form which collects all conformations with l pairs of type (i,j). These pairs produce a factor t^l_{ij}. We extract this factor and denote the remainder $\Omega(i,j;l)$

$$Z = \sum_{s=0}^{N_{\mathrm{b}}-1} (t_{ij})^l \cdots \Omega(i,j;l) \tag{2.145}$$

The probability for one specific conformation, $p\{\varphi_i\}$, is given by

$$p\{\varphi_i\} = \frac{\exp -u\{\varphi_i\}/kT}{Z} \tag{2.146}$$

We first derive the probability $p(i,j;l)$ that l pairs of type (i,j) occur in the chain. The necessary summation over the corresponding states of the chain is already implied in our formulation, and we can write

$$p(i,j;l) = \frac{t_{ij}^{l} \cdot \Omega(i,j,l)}{Z} \tag{2.147}$$

From this result there follows that the average number of pairs (i,j), denoted $\langle n_{ij} \rangle$ is given by

$$\langle n_{ij} \rangle = \sum_{l=0}^{N_\mathrm{b}-1} l \cdot \frac{t_{ij}^{l} \cdot \Omega(i,j,l)}{Z} \tag{2.148}$$

or, using the above relations, by

$$\langle n_{ij} \rangle = \frac{t_{ij}}{Z} \cdot \frac{\partial Z}{\partial t_{ij}} = \frac{\partial \ln Z}{\partial \ln t_{ij}} = (N_\mathrm{b} - 1)\frac{\partial \ln \lambda_1}{\partial \ln t_{ij}} \tag{2.149}$$

The probability for a sequence (i,j) in adjacent bonds, denoted ϕ_{ij}, is

$$\phi_{ij} = \frac{\langle n_{ij} \rangle}{N_\mathrm{b} - 1} = \frac{\partial \ln \lambda_1}{\partial \ln t_{ij}} \tag{2.150}$$

Insertion of the statistical weights w_0 and $w_0 w_1$ yields

$$\begin{aligned} \phi_{\mathrm{tt}} &= 0.29 \\ \phi_{\mathrm{g^+t}} &= \phi_{\mathrm{tg^+}} = \phi_{\mathrm{g^-t}} = \phi_{\mathrm{tg^-}} = 0.14 \\ \phi_{\mathrm{g^+g^+}} &= \phi_{\mathrm{g^-g^-}} = 0.06 \\ \phi_{\mathrm{g^+g^-}} &= \phi_{\mathrm{g^-g^+}} = 0.015 \end{aligned}$$

The pentane effect shows up quite clearly, as the fraction of pairs with sequences *gauche*$^+$-*gauche*$^-$ and *gauche*$^-$-*gauche*$^+$ is rather low.

Finally, the fractions of *trans*- and *gauche*-conformational states in the polyethylene chains are obtained by

$$\phi_i = \sum_j \phi_{ij} \tag{2.151}$$

resulting in

$$\begin{aligned} \phi_t &= 0.60 \\ \phi_{\mathrm{g^+}} &= \phi_{\mathrm{g^-}} = 0.20 \end{aligned}$$

As expected, compared to the estimate based on the assumption of independent rotational isomeric states, the fraction of *trans*-states is increased.

The RIS-model also enables a computation of the characteristic ratio C_∞ to be made, if the stereochemical properties of the chain are included into the considerations. The calculations are more tedious but, using the algebraic

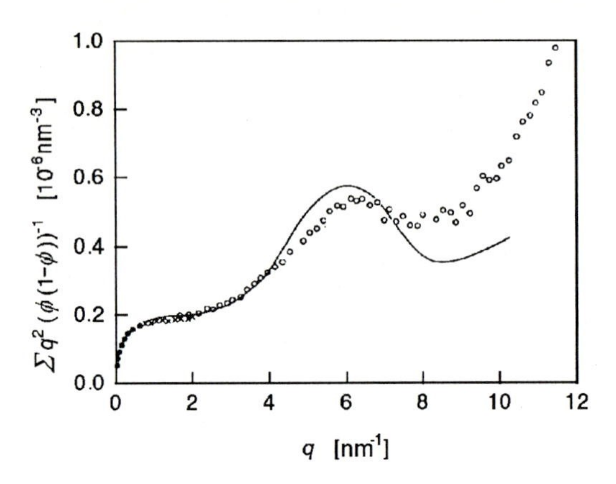

Fig. 2.22. Neutron scattering experiment on mixtures of PC and d-PC. The continuous curve has been calculated on the basis of the RIS-model. Data from Gawrisch et al. [8]

properties of matrices, they can still be carried out in straightforward manner. As it turns out, the experimental value for polyethylene, $C_\infty = 6.7$, is reproduced for reasonable assumptions about the molecular parameters, namely a C-C-C-valence angle of $112\,°$ and *gauche*-rotational angles $\varphi_{g+} = 127.5\,°$ and $\varphi_{g-} = 232.5\,°$.

One may even advance one step further and calculate structure factors of specific chains numerically, for a comparison with the results of scattering experiments. Figure 2.22 presents, as an example, neutron scattering data of polycarbonate obtained for mixtures of deuterated and protonated species. The experiment covers a large range of q's, and results are represented in the form of a Kratky-plot. We observe a plateau, characteristic for ideal chains, and then a rise at higher q's, for distances shorter than the persistence length where the microscopic chain structure takes over control. The peculiar shape of the curve in this range reflects specific properties of polycarbonate and indeed, these can be reproduced by calculations on the basis of the RIS-model. The continuous curve represents the theoretical results, and even if the agreement is not perfect, it describes the main characteristics qualitatively correctly.

Calculations based on the RIS-model now exist for the majority of common polymers, thus providing a quantitative representation of the energetic and structural properties of single macromolecules. The prerequisite is a knowledge about the energies $u(\varphi_{i-1}, \varphi_i)$ associated with the different pairs of conformational states. Information about these values has improved steadily with the number of carefully analyzed experiments. Clearly, the model does not account for the excluded volume interaction, but it provides a microscopic understanding for all situations with ideal chain behavior.

2.5 Further Reading

R.H. Boyd, P.J. Phillips: *The Science of Polymer Molecules*, Cambridge University Press, 1993

P.J. Flory: *Statistical Mechanics of Chain Molecules*, John Wiley & Sons, 1969.

P.-G. de Gennes: *Scaling Concepts in Polymer Physics*, Cornell University Press, 1979

A.Y. Grosberg, A.R. Khokhlov: *Statistical Physics of Macromolecules*, AIP Press, 1994

W.L. Mattice, U.W. Suter *Conformational Theory of Large Molecules - The Rotational Isomeric State Model in Macromolecular Systems*, Wiley & Sons, 1994

H. Tadokoro: *Structure of Crystalline Polymers*, John Wiley & Sons, 1979

Chapter 3

Liquid Equilibrium States

After having considered the structural behavior of single chains we turn now to the collective properties of polymers in bulk phases and discuss in this chapter liquid states of order. Liquid polymers are in thermal equilibrium, so that statistical thermodynamics can be applied. At first view one might think that theoretical analysis presents a formidable problem since each polymer may interact with many other chains. This multitude of interactions of course can create a complex situation, however, cases also exist, where conditions allow for a facilitated treatment. Important representatives for simpler behavior are melts and liquid polymer mixtures, and the basic reason is easy to see: As here each monomer encounters, on average, the same surroundings, the chain as a whole experiences in summary a 'mean field', thus fulfilling the requirements for an application of a well established theoretical scheme, the 'mean-field treatment'. We shall deal with this approach in the second part of this chapter, when discussing the properties of polymer mixtures.

Polymer solutions, on the other hand, cannot be treated under the mean-field assumptions. Take expanded chains in dilute solution, for example. Here, we find considerable variations in the monomer density, in the solution altogether and also within each chain where we have a maximum at the center, followed by decays to the edges. Replacement of the spatially varying interaction energy density by a mean value, corresponding to an equal interaction of each monomer with a constant mean field, here is not allowed. In the previous chapter, when discussing expanded chains, we indicated how theory can comply with such a situation, scaling arguments and renormalization group treatments providing solutions. It was another great achievement of the French school around des Cloiseaux and de Gennes to show that this treatment can be further extended to deal also with situations where chains begin to overlap. This became known as the regime of 'semi-dilute'-solutions, and we shall present in the next section some major results meant to provide a first basic understanding.

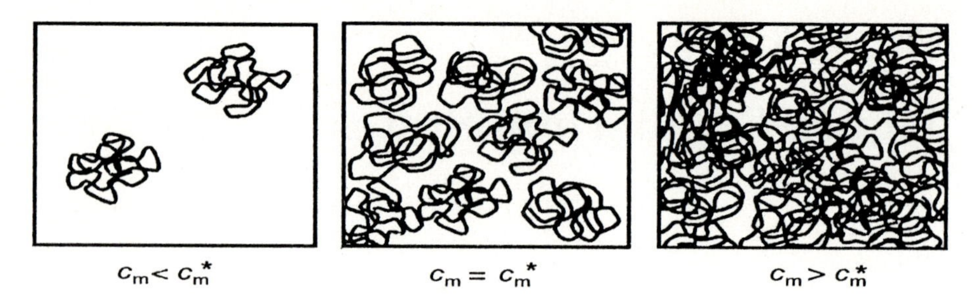

Fig. 3.1. Polymer solution: dilute regime ($c_{\rm m} < c_{\rm m}^*$) and semi-dilute regime with overlapping chains ($c_{\rm m} > c_{\rm m}^*$). The cross-over occurs for $c_{\rm m} = c_{\rm m}^*$, when the volumes occupied by the individual chains just cover the sample volume

3.1 Dilute and Semi-Dilute Polymer Solutions

Polymer chains in dilute solutions are isolated and interact with each other only during brief times of encounter. Increasing the polymer concentration in a solvent leads to a change at a certain stage. As is schematically indicated in Fig. 3.1, a limit is reached when the polymer molecules become closely packed because then they begin to interpenetrate.

The monomer concentration at this limit can be estimated, by regarding that for close-packed polymers this concentration must agree with the mean concentration in a single chain. For expanded chains we find for the 'critical concentration at the overlap limit', $c_{\rm m}^*$, the expression

$$c_{\rm m}^* \simeq \frac{N}{R_{\rm F}^3} \tag{3.1}$$

Since we need a unique expression for the further treatments, we replace the above estimate by an exact equation and write, per definition

$$c_{\rm m}^* := \frac{N}{R_{\rm F}^3} \tag{3.2}$$

It is interesting to check how the location of $c_{\rm m}^*$ changes with the degree of polymerization. For a good solvent, the scaling law Eq. (2.83) applies and we have

$$c_{\rm m}^* = \frac{N}{a_{\rm F}^3 N^{9/5}} = \frac{1}{a_{\rm F}^3 N^{4/5}} \tag{3.3}$$

The volume fraction ϕ of polymers in a solution is generally

$$\phi = v_{\rm m} c_{\rm m} \tag{3.4}$$

where v_{m} designates the monomer volume. The volume fraction ϕ^* associated with the critical concentration is therefore

$$\phi^* = v_{\mathrm{m}} c^*_{\mathrm{m}} = \frac{v_{\mathrm{m}}}{a_{\mathrm{F}}^3 N^{4/5}} \tag{3.5}$$

Since a_{F} is always much larger than the actual length of a monomer, we have

$$\frac{v_{\mathrm{m}}}{a_{\mathrm{F}}^3} < 1$$

and therefore

$$\phi^* < N^{-4/5} \tag{3.6}$$

We learn from this estimate that for a typical polymer, say with $N \simeq 10^4$, interpenetration of chains already begins at volume fractions below 0.001. In order to have dilute conditions, polymer volume fractions have to be really low, below 10^{-4}. On the other hand, solutions with polymer volume fractions in the order of $10^{-3} - 10^{-1}$, which can still be considered as low, are already clearly affected by the chain interpenetration. To set this special class of solutions apart from both the dilute and the concentrated solutions a new name was introduced: They are called 'semi-dilute'.

3.1.1 Osmotic Pressure

In discussions of solution behavior, the osmotic pressure Π becomes a property of primary interest. Π depends on the temperature and the concentration of the solute. In this section we will discuss the form of this dependence and begin with considering dilute polymer solutions.

As for low molar mass solutes, for polymers a virial expansion can also be used to give Π in the limiting range of low concentrations

$$\Pi = kT(A_1 c_{\mathrm{m}} + A_2 c_{\mathrm{m}}^2 + \ldots) \tag{3.7}$$

This is a series expansion in powers of the solute concentration, and the A_i's are the 'i-th virial coefficients'. For an ideal solution of low molar mass molecules all higher order virial coefficients beginning with the second virial coefficient A_2 vanish, and we have furthermore

$$A_1 = 1 \tag{3.8}$$

The dependence $\Pi(c_{\mathrm{m}})$ then agrees exactly with the pressure-concentration dependence of an ideal gas. Dissolved polymers do not constitute ideal solutions in this strict sense, even if all higher order virial coefficients vanish, because the first virial coefficient is not unity but given by

$$A_1 = 1/N \tag{3.9}$$

The reason for the change is easily revealed. Regard that the osmotic pressure is exerted by the translational motion of the centers of mass of the polymers only and remains unaffected by the other internal degrees of freedom of the chains. This implies that, for polymer solutions, the polymer density

$$c_{\mathrm{p}} = \frac{c_{\mathrm{m}}}{N} \tag{3.10}$$

rather than the monomer density, is the quantity which controls the osmotic pressure. The virial expansion therefore has to be expressed as

$$\begin{aligned} \frac{\Pi}{kT} &= \frac{c_{\mathrm{m}}}{N} + A_2 c_{\mathrm{m}}^2 + \ldots \\ &= \frac{c_{\mathrm{m}}}{N}(1 + N A_2 c_{\mathrm{m}} + \ldots) \end{aligned} \tag{3.11}$$

Equation (3.11) is the virial expansion valid for a dilute polymer solution. One may also interpret the second order term. It describes an increase in osmotic pressure due to the contacts between the dissolved polymer molecules, which occur with a probability proportional to $c_{\mathrm{p}}^2 \sim c_{\mathrm{m}}^2$.

For experimental studies, another form is more convenient, where the number density of monomers, c_{m}, is replaced by the fraction by weight, c_{w}, using

$$c_{\mathrm{w}} = \frac{c_{\mathrm{m}}}{\mathrm{N_L}} M_{\mathrm{m}} \tag{3.12}$$

(M_{m} describes the molar mass of a monomer). In addition, one introduces a modified second virial coefficient, $\tilde{A}_2$, defined as

$$\tilde{A}_2 := \frac{\mathrm{N_L} A_2}{M_{\mathrm{m}}^2} \tag{3.13}$$

Equation (3.11) can then be rewritten as

$$\frac{\Pi}{\tilde{R}T} = c_{\mathrm{w}} \left(\frac{1}{M} + \tilde{A}_2 c_{\mathrm{w}} + \ldots \right) \tag{3.14}$$

Figure 3.2 presents a typical experimental result. It shows the dependence of the osmotic pressure on c_{w} for a series of poly(α-methylstyrenes), with different molecular weights, dissolved in toluene. First, it may be noticed that the limiting values $c_{\mathrm{w}} \to 0$ change with the molecular weight. Actually, they agree with the relation

$$\lim_{c_{\mathrm{w}} \to 0} \frac{\Pi}{\tilde{R}T c_{\mathrm{w}}} = \frac{1}{M} \tag{3.15}$$

Secondly, since $\Pi/\tilde{R}Tc_{\mathrm{w}}$ increases with c_{w}, we have evidence for a non-vanishing positive second virial coefficient as it is indicative for a good solvent. The third observation is that, in the limit of high concentrations, the molecular weight dependence of Π vanishes. This occurs when the polymer chains

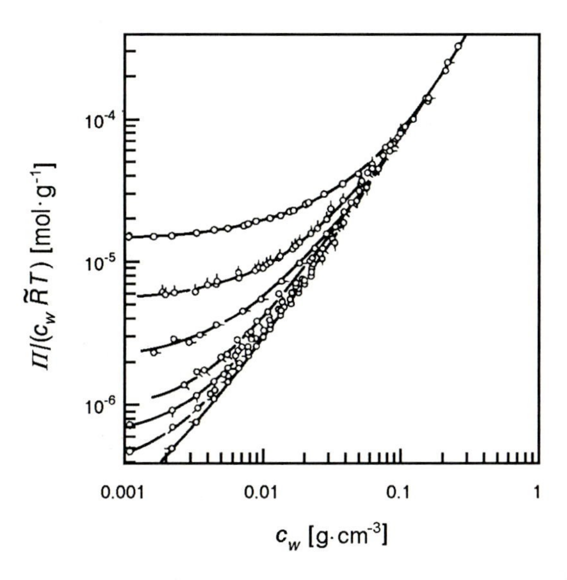

Fig. 3.2. Osmotic pressures measured for samples of poly(α-methylstyrene) dissolved in toluene (25 °C). Molecular weights vary between $M = 7 \cdot 10^4$ (*uppermost curve*) and $M = 7.47 \cdot 10^6$ (*lowest curve*). Data taken from Noda et al.[9]

interpenetrate each other and an entanglement network builds up. The observation tells us that, once the entanglement network has formed, the single chain properties become irrelevant.

An especially interesting result is presented in Fig. 3.3. It deals with the same set of data plotted here in a special form. We employ c_{w}^*, the polymer weight fraction at the overlap-limit, and replace c_{w} by a dimensionless reduced variable x , called the 'overlap ratio'

$$\mathrm{x} := \frac{c_{\mathrm{w}}}{c_{\mathrm{w}}^*} = \frac{c_{\mathrm{m}}}{c_{\mathrm{m}}^*} \tag{3.16}$$

c_{m}^* or c_{w}^* follow from Eq. (3.2) when R_{F} is determined by light scattering or, as will be explained below, also directly from A_2. Along the ordinate, the expression $\Pi M/(\tilde{R}Tc_{\mathrm{w}})$ is plotted. For this 'reduced osmotic pressure' we have a common limit for $c_{\mathrm{w}} \to 0$, independent of M:

$$\lim_{c_{\mathrm{w}} \to 0} \frac{\Pi M}{\tilde{R}Tc_{\mathrm{w}}} = 1 \tag{3.17}$$

The result of this redrawing procedure is quite remarkable: The curves for all samples coincide.

The conclusions which can be drawn from these observations are far-reaching and important. Polymers in solution interact with each other and details of this interaction become apparent in the concentration dependence

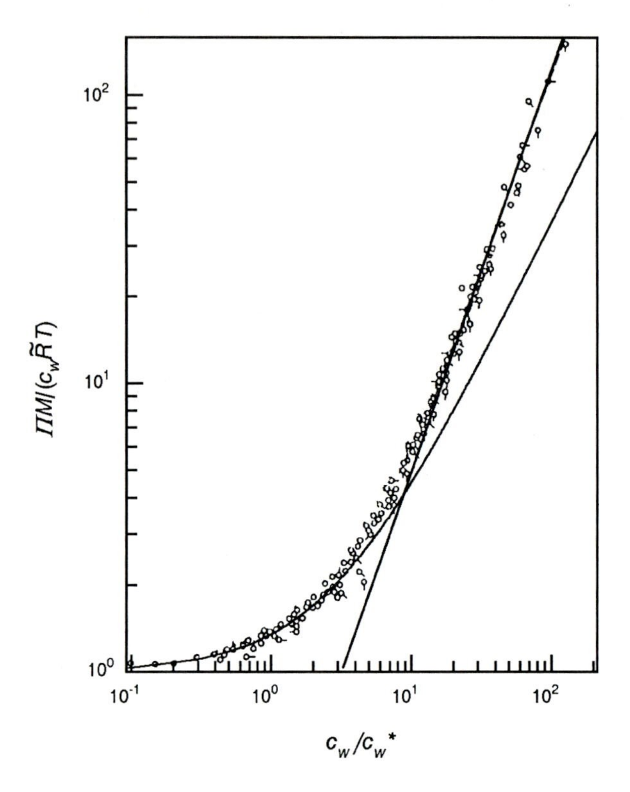

Fig. 3.3. Data of Fig. 3.2, presented in a plot of the reduced osmotic pressure versus the overlap ratio. The *continuous lines* correspond to the theoretical results Eqs. (3.26), (3.41)

of the osmotic pressure. The results presented in Figs. 3.2 and 3.3 strongly suggest that the interaction obeys general laws which are valid for all polymers and solvents.

Let us consider the given situation and search for an equation for the osmotic pressure. First we have to inquire about the independent variables in the system. In the case of a dilute solution, we have three of them apart from the trivial T. To specify the single chain properties, we must know the Flory radius R_F and the thermic correlation length ξ_t or the parameter z. The third variable is the number density of polymers c_p. The question arises if there are any more parameters which have to be accounted for when leaving the dilute range and coming into the regime of the semi-dilute solutions. The answer is no for physical reasons. We encounter just one class of interactions throughout, the excluded volume forces, which do not differentiate between monomers within a chain and on different chains. Their effect is implied in the values of R_F and z. As the cross-over to the semi-dilute regime does not bring in any new forces, there is also no further parameter. This remains true as long as the monomer concentration in the solution is sufficiently low so that one has to account for binary interactions only. At higher concetrations, where ternary interactions become progressively important, the situation changes.

Having identified the independent variables, we can now formulate for Π the following functional dependence

$$\frac{\Pi}{kT} = c_{\mathrm{p}} \mathrm{F}(R_{\mathrm{F}}, \xi_{\mathrm{t}}, c_{\mathrm{p}}) \tag{3.18}$$

F is a universal function with properties to be discussed. With the extraction of the factor c_{p}, we fix the limit $\mathrm{F}(c_{\mathrm{p}} \to 0)$. As we expect ideal properties for $c_{\mathrm{p}} \to 0$, we have

$$\mathrm{F}(c_{\mathrm{p}} \to 0) = 1 \tag{3.19}$$

The experimental result depicted in Fig. 3.3 provides us with a hint with regard to the form of F. It suggests that R_{F} and c_{p} are included in a coupled manner, namely as the product $c_{\mathrm{p}} R_{\mathrm{F}}^3$ which is identical with the overlap ratio

$$\mathrm{x} := \frac{c_{\mathrm{m}}}{c_{\mathrm{m}}^*} = c_{\mathrm{p}} R_{\mathrm{F}}^3 \tag{3.20}$$

We therefore write, and this was first proposed by des Cloiseaux

$$\frac{\Pi}{kT} = c_{\mathrm{p}} \mathrm{F}_{\Pi}(\mathrm{x}, \mathrm{z}) \tag{3.21}$$

Here ξ_{t} is substituted by z.

For low concentrations, we can use a series expansion in powers of x for the function F_{Π}

$$\mathrm{F}_{\Pi} = 1 + h(\mathrm{z})\mathrm{x} + \ldots \tag{3.22}$$

Since chains become ideal for $\mathrm{z} \to 0$, i.e. $\xi_{\mathrm{t}} \to \infty$, we must have

$$h(\mathrm{z} = 0) = 0 \tag{3.23}$$

On the other hand, one can carry out the passage to the Kuhnian limit $\mathrm{z} \to \infty$ as realized in good solvents. Theory shows that there is a well-defined limiting value, $h(\mathrm{z} \to \infty)$, and a corresponding limiting function, which now depends on x only

$$\mathrm{F}_{\Pi}(\mathrm{x}, \mathrm{z} \to \infty) := \mathrm{F}_{\Pi}(\mathrm{x})$$

If we employ this limiting function, we obtain a general equation for the osmotic pressure exerted by polymers in good solvents

$$\frac{\Pi}{kTc_{\mathrm{p}}} = \frac{\Pi M}{\tilde{R}Tc_{\mathrm{w}}} = \mathrm{F}_{\Pi}\left(\mathrm{x} = \frac{c_{\mathrm{w}}}{c_{\mathrm{w}}^*}\right) \tag{3.24}$$

According to the derivation, Eq. (3.24) is valid for both dilute and semi-dilute solutions. We now understand the experimental curve in Fig. 3.3: It represents exactly the universal function $\mathrm{F}_{\Pi}(\mathrm{x})$.

The value of the expansion coefficient in Eq. (3.22) in the Kuhnian limit, $h(\mathrm{z} \to \infty)$, can be calculated using renormalization group methods, with the result

$$h(\mathrm{z} \to \infty) = 0.353 \qquad (3.25)$$

A check is displayed in Fig. 3.3, by inclusion of the curve corresponding to

$$\mathrm{F(x)} = 1 + 0.353\mathrm{x} \qquad (3.26)$$

Comparison shows an excellent agreement with the data for $\mathrm{x} < 2$.

Discussion of Eq. (3.21) enables some direct conclusions. First, consider the dilute case, $\mathrm{x} \ll 1$, where the virial expansion is valid. We write for $\mathrm{F}_{\Pi}(\mathrm{x},\mathrm{z})$

$$\mathrm{F}_{\Pi}(\mathrm{x},\mathrm{z}) = 1 + h(\mathrm{z})\mathrm{x} + \ldots \qquad (3.27)$$

Comparison of Eqs. (3.21),(3.27) and Eq. (3.11) gives a relation between the concentration at chain overlap c^*_{m} and the second virial coefficient A_2:

$$c^*_{\mathrm{m}} = \frac{h(\mathrm{z})}{NA_2} \qquad (3.28)$$

Use of Eq. (3.2) yields

$$A_2 = \frac{h(\mathrm{z})R^3_{\mathrm{F}}}{N^2} \qquad (3.29)$$

The last equation relates the second virial coefficient to the Flory radius and the degree of polymerization of the chains.

It is instructive to introduce this relation in Eq. (3.11), also replacing c_{m} by the polymer density c_{p}. We obtain

$$\frac{\Pi}{kT} = c_{\mathrm{p}} + h(\mathrm{z})R^3_{\mathrm{F}} \cdot c^2_{\mathrm{p}} + \ldots \qquad (3.30)$$

Equation (3.30) formulates an interesting result. It reveals that the increase in the osmotic pressure over the ideal behavior, as described in lowest order by the second term on the right-hand side, may be understood as being caused by repulsive hard core interactions between the polymer chains which occupy volumes in the order of $h(\mathrm{z})R^3_{\mathrm{F}}$. To see it, just compare Eq. (3.30) with Eq. (2.73) valid for a van der Waal's gas. For a gas with hardcore interactions only, i.e. $\tilde{a} = 0$, the second virial coefficient equals the excluded volume per molecule $\tilde{b}/\mathrm{N_L}$. Therefore, we may attribute the same meaning to the equivalent coefficient in Eq. (3.30). Our result thus indicates that polymer chains in solution behave like hard spheres, with the radius of the sphere depending on R_{F}, and additionally on z, i.e. on the solvent quality. For good solvents, in the Kuhnian limit $h(\mathrm{z} \to \infty) = 0.353$, the radius is similar to R_{F}.

This strong repulsion is understandable since an overlap between two coils produces many contacts between the monomers. We can estimate the related energy, utilizing Eq. (2.78). For monomer density distributions $c_{\mathrm{m}}(\boldsymbol{r}-\boldsymbol{r}_{\mathrm{c,i}})$ about the centers of gravity $\boldsymbol{r}_{\mathrm{c},1}$ and $\boldsymbol{r}_{\mathrm{c},2}$ of two polymers, it is given by

$$f^{\mathrm{e}}(\boldsymbol{r}_{\mathrm{c},1},\boldsymbol{r}_{\mathrm{c},2}) = kTv_{\mathrm{e}}\int\langle c_{\mathrm{m}}(\boldsymbol{r}-\boldsymbol{r}_{\mathrm{c},1})c_{\mathrm{m}}(\boldsymbol{r}-\boldsymbol{r}_{\mathrm{c},2})\rangle \mathrm{d}^3\boldsymbol{r} \tag{3.31}$$

Assuming Gaussian density distributions, we obtain in the limit of a complete overlap, $\boldsymbol{r}_{\mathrm{c},1}=\boldsymbol{r}_{\mathrm{c},2}$, an expression identical to the internal excluded volume interaction energy as given by Eq. (2.91), apart from a factor 1/2. Omitting the numerical prefactor of order unity we can write

$$f^{\mathrm{e}}(\boldsymbol{r}_{\mathrm{c},1}=\boldsymbol{r}_{\mathrm{c},2}) \simeq kT\mathrm{z}\left(\frac{R_0}{R_{\mathrm{F}}}\right)^3 \tag{3.32}$$

This has to be compared with the mean kinetic energies associated with the translational motion of the centers of mass

$$\langle u_{\mathrm{kin}}\rangle = \frac{3}{2}kT \tag{3.33}$$

Even for an only partial overlap, to a degree β, we have

$$\frac{\langle u_{\mathrm{kin}}\rangle}{f^{\mathrm{e}}} \simeq \frac{1}{\beta\mathrm{z}} \ll 1 \tag{3.34}$$

if, as is the case for good solvents and standard degrees of polymerization,

$$\mathrm{z} = N_{\mathrm{su}}^{1/2} \gg 1 \tag{3.35}$$

As a consequence, in a dilute solution of expanded chains, interpenetration of two polymer molecules is largely suppressed so that they do indeed resemble hard spheres.

Equation (3.29) enables us to deduce the molecular weight dependence of A_2. For good solvents we again may set $h(\mathrm{z}) = h(\mathrm{z}\to\infty) = 0.353$, then obtaining

$$\begin{aligned} A_2 &= 0.353\frac{R_{\mathrm{F}}^3}{N^2} \\ &= 0.353\frac{a_{\mathrm{F}}^3 N^{9/5}}{N^2} = 0.353 a_{\mathrm{F}}^3 N^{-1/5} \end{aligned} \tag{3.36}$$

Figure 3.4 presents experimental results, obtained for solutions of fractions of PS in benzene. They do in fact agree with the power law $\tilde{A}_2 \sim A_2 \sim N^{-1/5} \sim M^{-1/5}$.

Next we consider the other limit, $\mathrm{x} \gg 1$, associated with an entangled semi-dilute solution far above the overlap threshold. Here, the degree of polymerization N must become irrelevant, as is also demonstrated by the data in

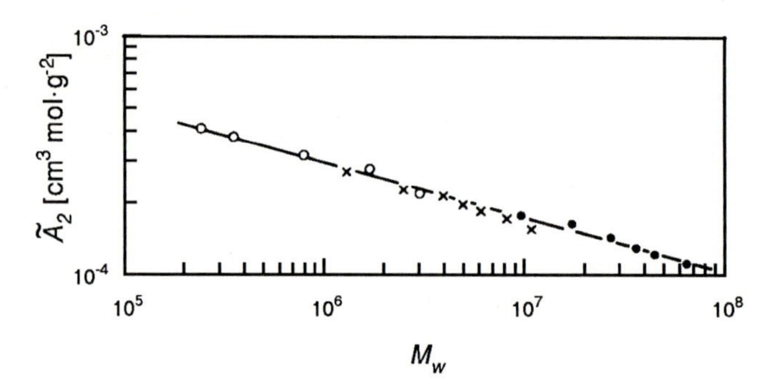

Fig. 3.4. Molecular weight dependence of the second virial coefficient, obtained for solutions of fractions of polystyrene in benzene. Data from Cotton [10]

Fig. 3.2. For des Cloiseaux's expression, Eq. (3.21), this condition implies a stringent requirement, since N is included in both the factor c_{p} and F_{Π}

$$\frac{\Pi}{kT} = \frac{c_{\mathrm{m}}}{N}\mathrm{F}_{\Pi}(\mathrm{x},\mathrm{z}) \tag{3.37}$$

The dependence on N contributed by c_{p} can only be eliminated if F_{Π} shows asymptotically an appropriate power law behavior. Assuming

$$\mathrm{F}_{\Pi}(\mathrm{x},\mathrm{z}) \sim \mathrm{x}^k \tag{3.38}$$

and therefore

$$\mathrm{F}_{\Pi}(\mathrm{x},\mathrm{z}) \sim c_{\mathrm{m}}^k \frac{R_{\mathrm{F}}^{3k}}{N^k} \sim c_{\mathrm{m}}^k N^{4k/5} \tag{3.39}$$

we obtain at first

$$\frac{\Pi}{kT} \sim c_{\mathrm{m}}^{k+1} N^{(4k/5)-1} \tag{3.40}$$

The exponent of N has to vanish, and this occurs for $k = 5/4$. By introduction of this value, we obtain the concentration dependence of the osmotic pressure in the semi-dilute regime:

$$\frac{\Pi}{kT} \sim c_{\mathrm{m}}^{9/4} \tag{3.41}$$

Figures 3.2 and 3.3 confirm this prediction. The slope of the continuous straight lines drawn through the data in the limit of high concentrations exactly corresponds to Eq. (3.41). Note the qualitative difference to the dilute regime, where $\Pi \sim c_{\mathrm{m}}^2$. We may conclude that the entanglements further enhance the osmotic pressure.

The Zimm-Diagram

Light scattering experiments permit further checks of the theoretical predictions since they can be used for a determination of both the osmotic compressibility and the second virial coefficient. As explained in Sect. A.3.2 in the

Appendix, a general equation in scattering theory relates the osmotic compressibility $(\partial c_{\mathrm{m}}/\partial \Pi)_T$ to the scattering in the forward direction (Eq. (A.81)):

$$S(q=0) = kT \left(\frac{\partial c_{\mathrm{m}}}{\partial \Pi} \right)_T$$

Taking the reciprocal expressions on both sides gives

$$S^{-1}(q=0) = \frac{1}{kT} \left(\frac{\partial \Pi}{\partial c_{\mathrm{m}}} \right)_T \tag{3.42}$$

$(\partial \Pi/\partial c_{\mathrm{m}})_T$ represents the 'osmotic modulus'.

This relation is very useful. First, we can employ the virial expansion Eq. (3.11) valid in the dilute range for a calculation of the osmotic modulus. The result is

$$S^{-1}(q=0, c_{\mathrm{m}} \to 0) = \frac{1}{N} + 2A_2 c_{\mathrm{m}} \tag{3.43}$$

In a second step, Eq. (3.43) can be combined with Guinier's law, Eq. (A.71), valid but for dilute solutions. The latter relates the curvature at $q=0$ to the radius of gyration of the polymer, by

$$S(q \to 0, c_{\mathrm{m}} \to 0) = N \left(1 - \frac{R_{\mathrm{g}}^2 q^2}{3} + \ldots \right)$$

Taking again the reciprocals, we have

$$S^{-1}(q \to 0, c_{\mathrm{m}} \to 0) = N^{-1} \left(1 + \frac{R_{\mathrm{g}}^2 q^2}{3} + \ldots \right) \tag{3.44}$$

The combination is achieved by writing the reciprocal scattering function as a product of both expressions

$$S^{-1}(q \to 0, c_{\mathrm{m}} \to 0) = \frac{1}{N} \left(1 + \frac{R_{\mathrm{g}}^2 q^2}{3} + \ldots \right) (1 + 2A_2 N c_{\mathrm{m}} + \ldots) \tag{3.45}$$

Equation (3.45) describes correctly the dependence of S^{-1} on q and c_{m} within the limit of small values of both parameters, and it is well-known as the basis of 'Zimm-plots'. To make use of the equation, scattering experiments on polymer solutions, and these are mostly carried out by light, have to be conducted under variation of both the scattering angle and the concentration. Extrapolation of q and c_{m} to zero then permits a determination of three parameters of the polymer to be made: One can deduce the radius of gyration R_{g}, the degree of polymerization N and the second virial coefficient A_2.

The extrapolations are carried out in a peculiar manner. Figure 3.5 shows, as an example, data obtained for a dilute solution of polystyrene in toluene. The reciprocal of the scattering function is plotted as a function of the sum $q^2 + \beta c_{\mathrm{w}}$, where β is a conveniently chosen constant. If we would utilize

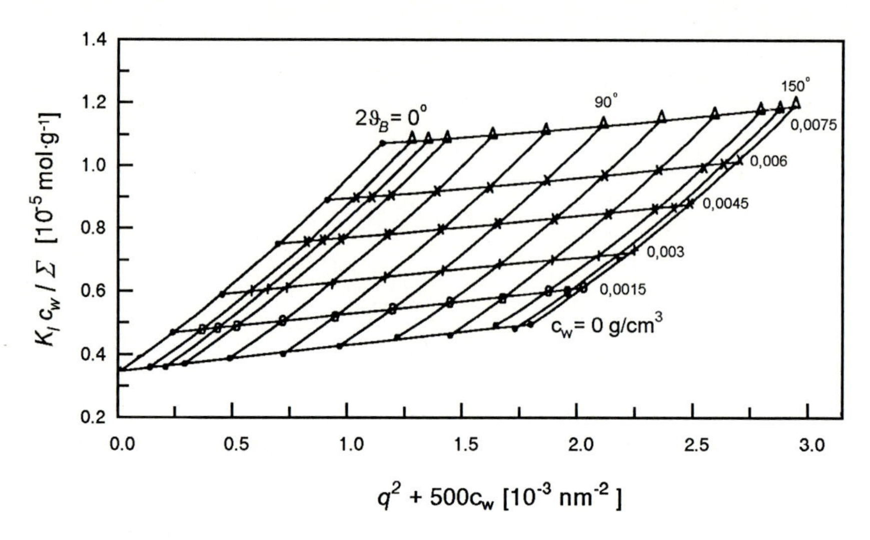

Fig. 3.5. Light scattering experiments on solutions of PS ($M = 2.8 \cdot 10^5$) in toluene at 25 °C. Results are presented in a Zimm-plot, enabling an extrapolation to $c_{\rm w} = 0$ and $q = 0$. Data from Lechner et al.[11]

Eq. (3.45), then the slopes $\mathrm{d}S^{-1}/\mathrm{d}q$ and $\mathrm{d}S^{-1}/\mathrm{d}c_{\rm m}$ of the two lines at the origin would give $R_{\rm g}^2/3$ and $2A_2$; the limiting value $S^{-1}(q=0, c_{\rm m}=0)$ would furnish N. Actually, in experiments directly measurable quantities such as the Raleigh ratio Σ and the weight concentration $c_{\rm w}$ are usually employed, rather than S and $c_{\rm m}$. Corresponding substitutions can be carried out in Eq. (3.45) which is then converted to

$$\frac{c_{\rm w}K_{\rm l}}{\Sigma} = \left(1 + \frac{R_{\rm g}^2 q^2}{3} + \ldots\right)\left(\frac{1}{M} + 2\tilde{A}_2 c_{\rm w} + \ldots\right) \tag{3.46}$$

with

$$K_{\rm l} = 4\pi^2 {\rm n}^2 \left(\frac{\rm dn}{{\rm d}c_{\rm w}}\right)^2 / \left({\rm N_L}\lambda_0^4\right)$$

Here n denotes the index of refraction and λ_0 is the wavelength in a vacuum (compare Appendix, Eqs. (A.50) and (A.51)). One obtains the modified second virial coefficient $\tilde{A}_2$ defined by Eq. (3.13).

Equation (A.81) also permits us to make a check of the osmotic pressure equation deeper in the semi-dilute regime. Introduction of Eq. (3.21) gives

$$\begin{aligned} S^{-1}(q=0, c_{\rm m}) &= \frac{\rm d}{{\rm d}c_{\rm m}} \cdot \frac{c_{\rm m}}{N} {\rm F}_\Pi({\rm x}, {\rm z}) \\ &= \frac{1}{N}{\rm F}_\Pi({\rm x},{\rm z}) + \frac{c_{\rm m}}{N} \cdot \frac{1}{c_{\rm m}^*}\frac{\partial {\rm F}_\Pi}{\partial {\rm x}}({\rm x},{\rm z}) \end{aligned} \tag{3.47}$$

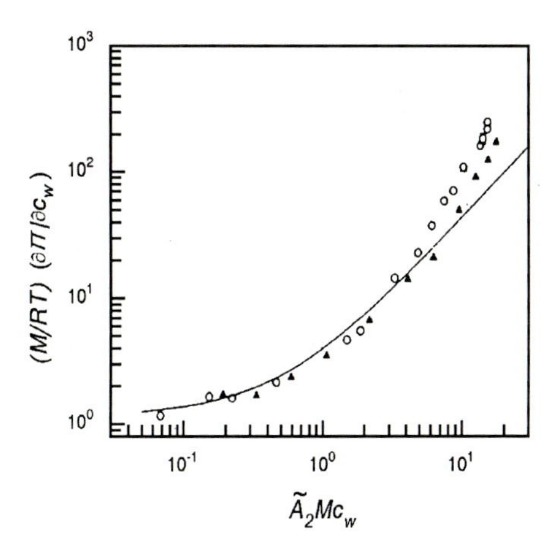

Fig. 3.6. Reduced osmotic moduli of two fractions of PS ($M = 2.3 \cdot 10^4$ and $4.7 \cdot 10^4$) dissolved in toluene, as derived from light scattering experiments. In the dilute and semi-dilute range results agree with the theoretical prediction as given by the *continuous curve* [12]

For good solvents as represented by the Kuhnian limit $\mathrm{z} \to \infty$ we obtain

$$S^{-1}(q = 0, c_{\mathrm{m}}) = \frac{1}{N}(\mathrm{F}_{\Pi}(\mathrm{x}) + \mathrm{x}\frac{\partial}{\partial \mathrm{x}}\mathrm{F}_{\Pi}(\mathrm{x})) := \frac{1}{N}\mathrm{F}_{\Pi}{}'(\mathrm{x}) \tag{3.48}$$

Here, $\mathrm{F}_{\Pi}{}'(\mathrm{x})$ denotes another general function. Hence, one expects to find a unique curve when plotting $NS^{-1}(q = 0)$ versus x. Light scattering experiments were carried out for various polymer solutions and they indeed confirm this prediction. Figure 3.6 gives an experimental result in a slightly modified representation. As an alternative to the overlap ratio x one may also use the quantity x′ defined as

$$\mathrm{x}' := A_2 N c_{\mathrm{m}} \tag{3.49}$$

According to Eq. (3.28), x′ is proportional to x

$$\mathrm{x}' = h(\mathrm{z} = 0) \cdot \mathrm{x} = 0.353\mathrm{x} \tag{3.50}$$

Using the concentration by weight c_{w}, the molecular weight M and the modified second virial coefficient $\tilde{A}_2$ rather than c_{m},N and A_2 leaves the plotted variables unchanged, since

$$NS^{-1}(q = 0) = N\frac{1}{kT}\frac{\partial \Pi}{\partial c_{\mathrm{m}}} = M\frac{1}{\tilde{R}T}\frac{\partial \Pi}{\partial c_{\mathrm{w}}} \tag{3.51}$$

and

$$\mathrm{x}' = A_2 N c_{\mathrm{m}} = \tilde{A}_2 M c_{\mathrm{w}} \tag{3.52}$$

The arguments presented provide information on the limiting behavior of the osmotic modulus only, for $\mathrm{x} \ll 1$ and $\mathrm{x} \gg 1$. It is possible to derive the full shape of the curve encompassing the dilute regime, the cross-over region, and the semi-dilute range, using renormalization group methods. Figure 3.6 includes a theoretically deduced curve and demonstrates good agreement with the experimental results. Deviations show up at higher concentrations, thereby pointing at the limitations of the treatment. As already mentioned, at higher concentrations it is no longer sufficient to consider only binary interactions, and we must also include higher order contributions. The situatation then becomes much more involved and it appears that universality is lost.

3.1.2 Screening Effect

In the previous chapter, we considered the structures of single chains in the dilute regime. Now we may inquire, how these become altered in semi-dilute solutions. Discussions can be based on the pair distribution function of the individual chains, thereby focusing on the structure of single chains in states where chains overlap and interpenetrate. We choose for this intramolecular pair correlation function a symbol with a hat, $\hat{g}(\boldsymbol{r})$, to distinguish it from the general pair distribution function $g(\boldsymbol{r})$ which includes monomers from all chains.

For $\hat{g}(\boldsymbol{r})$, we can assess the behavior for both limits, dilute solutions and the melt. As explained earlier, in Sect.2.3.2, we find for isolated expanded chains $\hat{g} \sim r^{-4/3}$, for distances in the range $\xi_{\mathrm{t}} < r < R_{\mathrm{F}}$. On the other hand, one observes in the melt ideal chain behavior, i.e. $\hat{g} \sim 1/r$, for $r < R_0$. Therefore, a change has to occur and indeed, it is possible to describe it in qualitative terms. Explanations were first provided by Edwards, in a theory which envisages a 'screening effect'. The view is that, similar to the screening of the long-range Coulomb forces in electrolytes caused by the presence of mobile ions as described by the classical Debye-Hückel theory, the long-range excluded volume forces acting within an isolated chain are screened if monomers of other polymer molecules interfere. We have already addressed this effect from a different point of view when discussing the chain structure in melts referring to Fig. 2.14. There we argued that the presence of foreign monomers reduces and finally completely removes the concentration gradient of the monomers belonging to one chain, thus blocking the chain expansion. In Edwards' treatment this process corresponds formally to a screening, put into effect by the contacts with foreign monomers.

In entangled solutions screening becomes effective at a characteristic distance, called the 'screening-length', and denoted ξ_{s}. Figure 3.7 provides an experimental example of the evidence. We see the scattering intensity measured for a semi-dilute solution of polystyrene in CS_2, in a plot of I^{-1} versus $q^{5/3}$. We are interested in the single chain structure factor in the semi-dilute solution, as given by the Fourier-transform of $\hat{g}$

$$\hat{S}(\boldsymbol{q}) = \int \exp \mathrm{i}\boldsymbol{q}\boldsymbol{r} \cdot \hat{g}(\boldsymbol{r}) \mathrm{d}^3\boldsymbol{r} \tag{3.53}$$

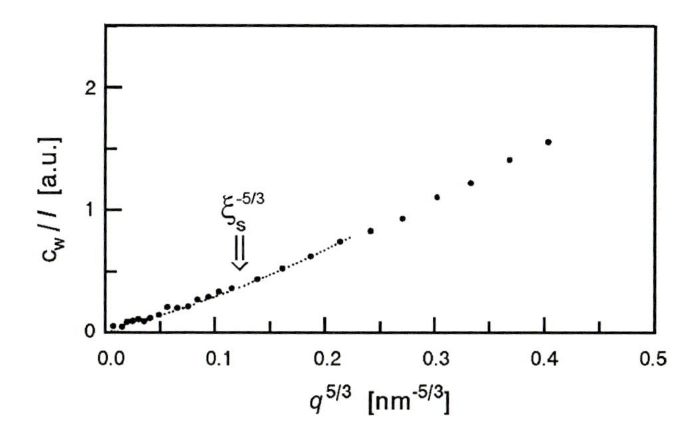

Fig. 3.7. Result of a neutron scattering experiment on a semi-dilute solution of a mixture of deuterated and protonated PS ($\overline{M}_w = 1.1 \cdot 10^6$) in CS_2 ($c_w = 0.15\ \mathrm{g \cdot cm^{-3}}$). Intensities reflect the structure factor of individual chains. The cross-over from the scattering of an expanded chain to that of an ideal chain at $q^{5/3} \simeq \xi_s^{-5/3}$ is indicated. Data from Farnoux [6]

$\hat{S}(\boldsymbol{q})$ can be measured, if the dissolved polystyrene includes a small fraction of deuterated molecules. Due to the large difference in the scattering length of protons and deuterium, the deuterated chains dominate the scattering pattern which then indeed may be described as

$$I(\boldsymbol{q}) \sim \hat{S}(\boldsymbol{q}) \tag{3.54}$$

For a dilute solution, one observes the scattering function of expanded chains, $I \sim q^{-5/3}$, which corresponds to the straight line shown previously in Fig. 2.18. Now we notice a change at low q's, indicative of a cross-over from the scattering behavior of an expanded chain to that of an ideal one, with $I^{-1} \sim q^2 = (q^{5/3})^{6/5}$. The cross-over occurs around a certain q, related to ξ_s by $q \simeq \xi_s^{-1}$.

Combining all the information collected so far, we can predict, for the single chain pair distribution function in a semi-dilute solution, an overall shape as indicated in Fig. 3.8. For the presentation we choose a plot of $4\pi r^2 \hat{g}(\boldsymbol{r})$ versus r. The curve is a composite of different functions in four ranges, with cross-overs at the persistence length l_{ps}, the thermic correlation length ξ_t and the screening length ξ_s. Up to $r \simeq \xi_s$ we find the properties of the expanded chain which

- resembles a rigid rod for $r < l_{ps}$, with $4\pi r^2 \hat{g}$ =const
- exhibits at first ideal behavior, in the range $l_{ps} < r < \xi_t$ ($4\pi r^2 \hat{g} \sim r$)
- and then shows excluded volume effects as indicated by the scaling-law

$$\hat{g} \sim r^{-4/3} \longrightarrow 4\pi r^2 \hat{g} \sim r^{2/3}$$

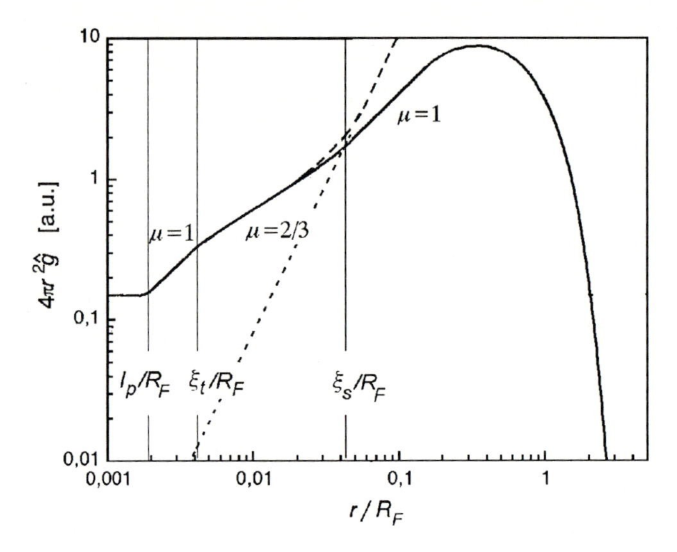

Fig. 3.8. Pair distribution function of an individual chain in a semi-dilute solution, exhibiting different regions with specific power laws $4\pi r^2 \hat{g} \sim (r/R_F)^\mu$. R_F denotes the Flory-radius in the dilute state. The *dotted line* gives the function $4\pi r^2 \langle c_m \rangle$. The *dashed line* indicates the pair distribution function for all monomers, $4\pi r^2 g$, which deviates from $4\pi r^2 \hat{g}$ for $r > \xi_s$.

This pertains up to the cross-over at $r \simeq \xi_s$, and then we enter again into an ideal regime

$$4\pi r^2 \hat{g} \sim r$$

The final range is determined by the size of the chain.

With the knowledge about the single chain pair distribution function, one can also predict the general shape of the pair distribution function for all monomers, $g(\boldsymbol{r})$. The behavior of $g(\boldsymbol{r})$ in the limits of small and large distances is obvious. Since for small r's correlations are mostly of intramolecular nature we have $g \approx \hat{g}$. On the other hand, we have a non-vanishing asymptotic value for large r's, given by $g = \langle c_m \rangle$. Indeed, with the aid of the screening length, we may express the behavior more accurately, as

$$g(\boldsymbol{r}) \approx \hat{g}(\boldsymbol{r}) \quad \text{for} \quad r \ll \xi_s \tag{3.55}$$

and

$$g(\boldsymbol{r}) \approx \langle c_m \rangle \quad \text{for} \quad r \gg \xi_s \quad , \tag{3.56}$$

Why the change from the one to the other limit must take place around ξ_s, is intuitively clear. ξ_s corresponds essentially to the distance between entanglement points, as these are the points where monomers interact with other chains which is the cause for the alteration of the chain structure. In the literature, ξ_s is therefore often addressed as 'mesh-size' to emphasize that it can also be interpreted as the diameter of the meshes of the entanglement network built up by the chains in a semi-dilute solution. It is furthermore plausible to assume that ξ_s approximately equals the distance r, where $\hat{g}(r)$ comes down to values in the order of the mean monomer density in the sample

$$\hat{g}(\xi_s) \simeq \langle c_m \rangle \tag{3.57}$$

since this condition implies similar weights of intra- und intermolecular contributions to $g(\boldsymbol{r})$. Figure 3.8 shows also these properties. $4\pi r^2 g$ is indicated by a dashed line which deviates from $4\pi r^2 \hat{g}$ at $r \simeq \xi_s$. The plot includes also the function $4\pi r^2 \langle c_m \rangle$, given by the straight dotted line. It crosses $4\pi r^2 \hat{g}$ at $r \simeq \xi_s$.

While studies of $\hat{g}$ require neutron scattering experiments on partially deuterated samples, information on $g(\boldsymbol{r})$ can be obtained by standard X-ray (or neutron) scattering experiments on normal solutions. Measurements yield the screening length ξ_s, using a simple straightforward procedure which may be explained as follows. Generally the structure function for a polymer solution is given by Eq. (A.25) which is valid for all isotropic systems (see Appendix, Sect. A.1.1)

$$S(q) = \int_{r=0}^{\infty} \frac{\sin qr}{qr} 4\pi r^2 (g(r) - \langle c_m \rangle) \mathrm{d}r$$

For the small angle range we may use a series expansion in powers of q

$$S(q) = \int \left(1 - \frac{1}{3} q^2 r^2 + \ldots \right) 4\pi r^2 (g(r) - \langle c_m \rangle) \mathrm{d}r \tag{3.58}$$

which yields for the curvature at $q = 0$

$$\frac{\mathrm{d}^2 S}{\mathrm{d}q^2} = -\frac{2}{3} \int r^2 \cdot 4\pi r^2 (g(r) - \langle c_m \rangle) \mathrm{d}r \tag{3.59}$$

We can represent this result writing

$$\frac{\mathrm{d}^2 S}{\mathrm{d}q^2} := -2\xi^2 \cdot S(0) \tag{3.60}$$

We have introduced here a parameter ξ. According to the definition, ξ^2 is one third of the second moment of the function $4\pi r^2 (g(r) - \langle c_m \rangle)$, and ξ is therefore a measure for the width of this function. On the other hand, as follows from a look at Fig. 3.8, this width essentially agrees with the screening length. Hence, we may identify ξ with ξ_s. So far, there has been no precise definition of ξ_s; Eq. (3.60) provides us with one.

Based on Eq. (3.60), $S(q)$ may now be represented in the small angle range by

$$S(q) = S(0)(1 - \xi_s^2 q^2 + \ldots) \tag{3.61}$$

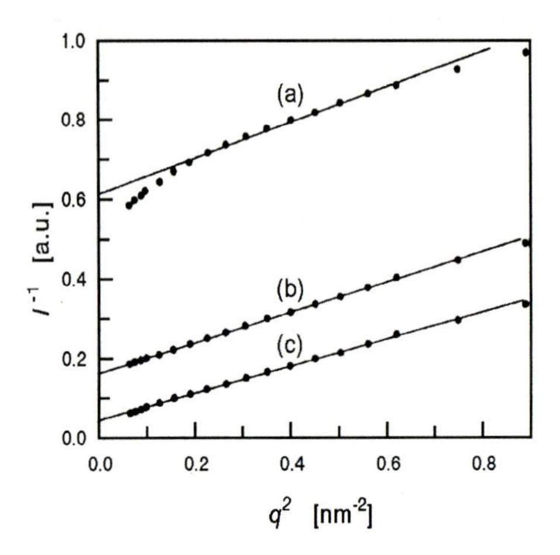

Fig. 3.9. Solutions of PS ($M = 5 \cdot 10^4$) in toluene, with concentrations $c_{\mathrm{w}} =$ 0.0105 (a), 0.0953 (b) and 0.229 $\mathrm{g \cdot cm^{-3}}$ (c). Intensities measured by SAXS. Data from Hamada et al.[13]

or, using the reciprocal function, by

$$S^{-1}(q) = S^{-1}(0)(1 + \xi_{\mathrm{s}}^2 q^2 + \ldots). \tag{3.62}$$

Figure 3.9 presents experimental results obtained from small angle X-ray scattering ('SAXS'-) experiments on semi-dilute solutions of polystyrene in toluene, choosing three different concentrations. They agree with Eq. (3.62) and enable a determination of the concentration dependence of ξ_{s} to be made.

One can predict this dependence for the semi-dilute range, using again scaling arguments. We anticipate that in a good solvent we have only one relevant parameter, namely the overlap ratio x, and write

$$\xi_{\mathrm{s}} = R_{\mathrm{F}} \mathrm{F}_{\xi_{\mathrm{s}}}(\mathrm{x}) \tag{3.63}$$

The parameter R_{F} is included as a prefactor, in order to set the length scale. The limiting value of $\mathrm{F}_{\xi_{\mathrm{s}}}$ for $\mathrm{x} \to 0$ is necessarily unity

$$\mathrm{F}_{\xi_{\mathrm{s}}}(\mathrm{x} \to 0) = 1 \tag{3.64}$$

Within the semi-dilute range we expect power law behavior again

$$\mathrm{F}_{\xi_{\mathrm{s}}}(\mathrm{x} \to \infty) \sim \mathrm{x}^k \tag{3.65}$$

In this range, where the chains overlap strongly, ξ_{s} has to be independent of N, and this can only be accomplished by a power law in connection with an

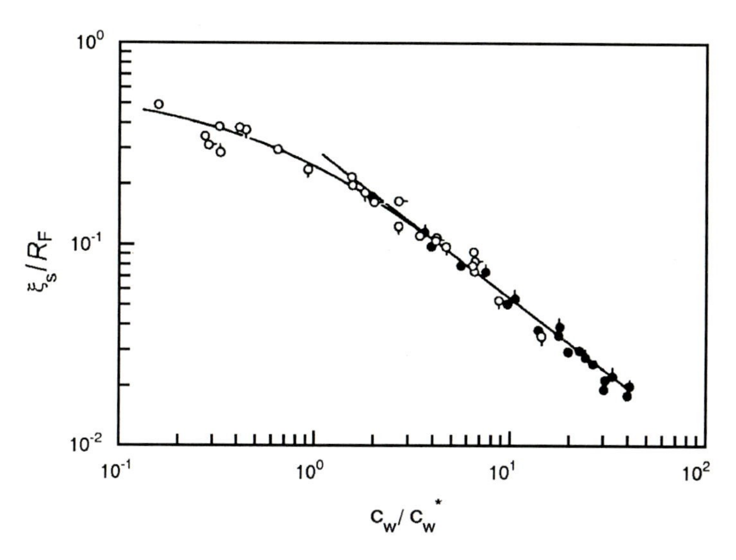

Fig. 3.10. Same system as in Fig. 3.9. Values derived for the concentration dependence of the screening length ξ_s. Data from Hamada et al.[13]

appropriate exponent. Application of Eqs. (2.83), (3.3) leads to

$$\xi_s \sim N^{3/5} \cdot N^{4k/5} \cdot c_w^k \tag{3.66}$$

Independence of ξ_s with regard to N requires that

$$k = -3/4 \tag{3.67}$$

Hence, we obtain

$$\xi_s \simeq R_F \left(\frac{c_w}{c_w^*} \right)^{-3/4} \tag{3.68}$$

Figure 3.10 depicts the concentration dependence of the screening length for the same system as presented in Fig. 3.9. The decrease, as observed for higher concentrations, agrees exactly with the scaling law Eq. (3.68).

Similar arguments may be used to derive the concentration dependence of the mean squared end-to-end distance of a chain. If we start with an expanded chain in dilute solution, we expect a shrinkage back to the size of an ideal chain when screening becomes effective on increasing the concentration. We assume a dependence

$$\langle R^2 \rangle^{1/2} = R_F \mathrm{F}_R(\mathrm{x}) \tag{3.69}$$

with

$$\mathrm{F}_R(\mathrm{x} \to 0) = 1 \tag{3.70}$$

For higher concentrations in the semi-dilute regime we again expect a power law

$$\mathrm{F}_R(\mathrm{x} \to \infty) \sim \mathrm{x}^k \tag{3.71}$$

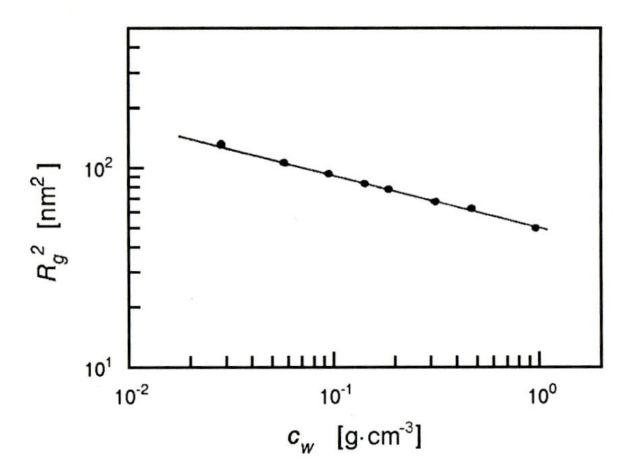

Fig. 3.11. PS ($M = 1.14 \cdot 10^5$) dissolved in CS_2. Shrinkage of the radius of gyration with increasing polymer concentration. Data from Daoud et al.[14]

thereby obtaining

$$\langle R^2 \rangle^{1/2} \sim N^{3/5} \cdot c_{\mathrm{w}}^k \cdot N^{4k/5} \tag{3.72}$$

On the other hand, since chains behave ideally for distances which are large compared to the screening length, we have

$$\langle R^2 \rangle^{1/2} \sim N^{1/2} \tag{3.73}$$

Comparison yields the exponent, with the value

$$k = -1/8 \tag{3.74}$$

Hence, we can formulate the equation for the dependence of the coil diameter on the degree of polymerization and the concentration

$$\langle R^2 \rangle^{1/2} \sim N^{1/2} c_{\mathrm{w}}^{-1/8} \tag{3.75}$$

This is proved to be valid by the results shown in Fig. 3.11. They were obtained in neutron scattering experiments on solutions of a mixture of protonated and deuterated polystyrenes in CS_2.

Considering these results, we can now see how the change from the expanded chains in dilute solutions to the ideal chains in a melt is accomplished: With increasing concentration the screening length decreases continuously, and if it comes down to the thermic correlation length ξ_{t} all excluded volume effects disappear. Simultaneously the chain size shrinks, from the Flory radius R_{F} in the dilute solution down to the radius R_0 of the Gaussian chains in the melt.

3.2 Polymer Mixtures

A large part of applications oriented research is devoted to the study of polymer blends, since mixing opens a route for a combination of different properties. Take, for example, the mechanical performance of polymeric products. In many cases one is searching for materials which combine high stiffness with resistance to fracture. For the majority of common polymers these two requirements cannot be realized simultaneously, because an increase in stiffness, i.e. the elastic moduli, is usually associated with samples becoming more brittle and decreasing in strength. Using mixtures offers a chance to achieve good results for both properties. 'High-impact polystyrene', a mixture of polystyrene and polybutadiene, represents a prominent example. Whereas polystyrene is stiff but brittle, a blending with rubbers furnishes a tough material which still retains a satisfactory stiffness. Mixing here results in a two-phase structure with rubber particles of spherical shape being incorporated in the matrix of polystyrene. Materials are 'tough', if fracture energies are high due to yield processes preceding the ultimate failure and these become initiated at the surfaces of the rubber spheres, where stresses are intensified. On the other hand, inclusion of rubber particles in the polystyrene matrix results only in a moderate reduction in stiffness. Hence, the blending yields a material with properties which are in many situations superior to pure polystyrene. There are other cases, where an improvement of the mechanical properties is achieved by a homogeneous mixture of two polymers, rather than a two-phase structure. A well-known example is again given by polystyrene, when blended with poly(phenyleneoxide). In this case a homogeneous phase is formed, and as it turns out in mechanical tests, it also exhibits a satisfactory toughness together with a high elastic modulus.

It is generally very difficult or even impossible to predict the mechanical properties of a mixture, however, this is only the second step. The first problem is an understanding of the mixing properties, i.e. a knowledge, under which conditions two polymeric compounds will form either a homogeneous phase or a two-phase structure. In the latter case it is important to see how structures develop and how this can be controlled. This section deals with these topics. We shall first discuss the thermodynamics of mixing of two polymers and derive formulae which can be used for the setting-up of phase diagrams. Subsequently we shall be concerned with the kinetics of unmixing and here in particular with a special mode known as 'spinodal decomposition'.

3.2.1 Flory-Huggins Treatment of Compatibility

Flory and Huggins devised a general scheme which enables one to deal with the mixing properties of a pair of polymers. It provides a basic understanding of the occurrence of different types of phase diagrams, in dependence on temperature and the molecular weights.

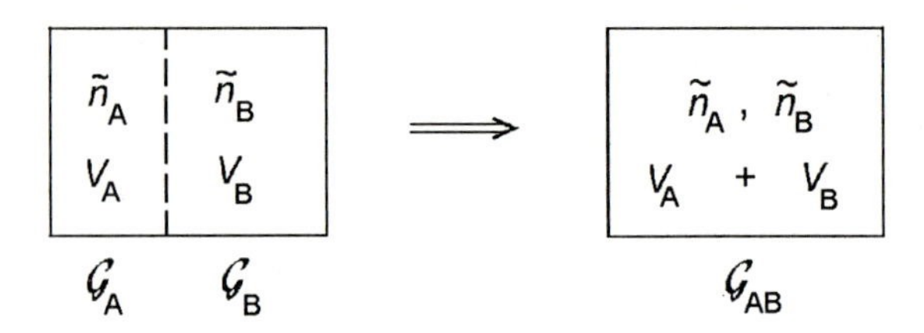

Fig. 3.12. Variables used in the description of the process of mixing of two polymers, denoted A and B

The mixing properties of two components may generally be discussed by considering the change in the Gibbs free energy. Figure 3.12 addresses the situation and introduces the relevant thermodynamic variables. Let us assume that we have $\tilde{n}_{\mathrm{A}}$ moles of polymer A, contained in a volume V_{A}, and $\tilde{n}_{\mathrm{B}}$ moles of polymer B, contained in a volume V_{B}. Mixing may be initiated by removing the boundary between the two compartments, so that both components can expand into the full volume, of size $V = V_{\mathrm{A}} + V_{\mathrm{B}}$. In order to find out whether a mixing would indeed occur, the change in the Gibbs free energy has to be considered. This change, called the 'Gibbs free energy of mixing' and denoted with $\Delta\mathcal{G}_{\mathrm{mix}}$, is given by

$$\Delta\mathcal{G}_{\mathrm{mix}} = \mathcal{G}_{\mathrm{AB}} - (\mathcal{G}_{\mathrm{A}} + \mathcal{G}_{\mathrm{B}}) \tag{3.76}$$

where $\mathcal{G}_{\mathrm{A}}, \mathcal{G}_{\mathrm{B}}$ and $\mathcal{G}_{\mathrm{AB}}$ denote the Gibbs free energies of the compounds A and B in separate states and the mixed state respectively.

The Flory-Huggins treatment represents $\Delta\mathcal{G}_{\mathrm{mix}}$ as a sum of two contributions

$$\Delta\mathcal{G}_{\mathrm{mix}} = -T\Delta\mathcal{S}_{\mathrm{t}} + \Delta\mathcal{G}_{\mathrm{loc}} \tag{3.77}$$

which describe the two main aspects of the mixing process. Firstly, mixing leads to an increase of the entropy associated with the motion of the centers of mass of all polymer molecules, and secondly, it may change the local interactions and motions of the monomers. We call the latter part $\Delta\mathcal{G}_{\mathrm{loc}}$, and the increase in the 'translational entropy' $\Delta\mathcal{S}_{\mathrm{t}}$. $\Delta\mathcal{S}_{\mathrm{t}}$ and the related decrease $-T\Delta\mathcal{S}_{\mathrm{t}}$ in the Gibbs free energy always favor a mixing. $\Delta\mathcal{G}_{\mathrm{loc}}$, on the other hand, may act favorably or unfavorably, depending on the character of the monomer-monomer pair interactions. In most cases, and as can be verified, for van-der-Waals interactions generally, attractive energies between equal monomers are stronger than those between unlike pairs. This behavior implies $\Delta\mathcal{G}_{\mathrm{loc}} > 0$ and therefore opposes a mixing. As a free energy, $\Delta\mathcal{G}_{\mathrm{loc}}$ also accounts for changes in the entropy due to local effects. For example, mixing can lead to an altered total volume , i.e. to $V \neq V_{\mathrm{A}} + V_{\mathrm{B}}$. A shrinkage or an expansion of the total volume results correspondingly in a change in the local mobility of the monomers, hence in a change of entropy, to be included in $\Delta\mathcal{G}_{\mathrm{loc}}$.

The decomposition of $\Delta\mathcal{G}_{\rm mix}$ in these two contributions points to the two main aspects of the mixing process but this alone would not be of much value. What is needed, for practical uses, are explicit expressions for $\Delta\mathcal{S}_{\rm t}$ and $\Delta\mathcal{G}_{\rm loc}$, so that the sum of the two contributions can be calculated. The Flory-Huggins treatment is based on approximate equations for both parts. We formulate them first and then discuss their origins and the implications. The equations have the following forms:

1. The increase in the translational entropy is described by

$$\frac{\Delta\mathcal{S}_{\rm t}}{\tilde{R}} = \tilde{n}_{\rm A}\ln\frac{V}{V_{\rm A}} + \tilde{n}_{\rm B}\ln\frac{V}{V_{\rm B}} \tag{3.78}$$

Introducing the volume fractions $\phi_{\rm A}$ and $\phi_{\rm B}$ of the two components in the mixture, given by

$$\phi_A = \frac{V_A}{V} \quad \text{and} \quad \phi_B = \frac{V_B}{V} \tag{3.79}$$

$\Delta\mathcal{S}_{\rm t}$ can be written as

$$\frac{\Delta\mathcal{S}_{\rm t}}{\tilde{R}} = -\tilde{n}_{\rm A}\ln\phi_{\rm A} - \tilde{n}_{\rm B}\ln\phi_{\rm B} \tag{3.80}$$

2. The change in the local interactions is expressed by the equation

$$\Delta\mathcal{G}_{\rm loc} = \tilde{R}T\frac{V}{\tilde{v}_{\rm c}}\chi\phi_{\rm A}\phi_{\rm B} \tag{3.81}$$

It includes two parameters. The less important one is $\tilde{v}_{\rm c}$, denoting the (molar) volume of a reference unit common to both polymers. Principally it can be chosen arbitrarily, but usually it is identified with the volume occupied by one of the monomeric units. The decisive factor is the 'Flory-Huggins parameter' χ. It is dimensionless and determines in an empirical manner the change in the local free energy per reference unit.

What is the physical background of these expressions? There are numerous discussions in the literature, mainly based on Flory's and Huggins' original derivations. As the full treatment lies outside our scope, we present here only a simplified view which nevertheless may aid in providing a basic understanding. The view emanates from a mean-field description. We consider the actual system of interpenetrating interacting chains which comprise the fluid mixture as being equivalent to a system of independent chains which interact with a common uniform 'mean-field' set up by the many-chain system as a whole. The interaction of a given chain with all other chains, as represented in an integral form by the mean-field, has two effects. The first one has been discussed earlier: The contacts with other chains 'screen' the intramolecular excluded volume interactions, thus leading to ideal chain behavior. The Flory-Huggins treatment assumes that this effect is maintained in a mixture, with unchanged conformational distributions. The second effect has already been mentioned in the introduction to this chapter. Being in contact with a large number of other chains, a given chain in a binary mixture effectively integrates over the varying monomer-monomer interactions and thus probes their average value. The change in the monomer-monomer interactions following from a mixing may therefore be expressed as change of the mean-field, with uniform values for all units of the A- and B-chains respectively.

Equations (3.80) and (3.81) are in agreement with this picture, as can be easily verified. In order to formulate the increase in the translational entropy for $\tilde{n}_{\mathrm{A}}$ moles of independent A-chains, expanding from an initial volume V_{A} to a final volume V, and $\tilde{n}_{\mathrm{B}}$ moles of B-chains, expanding from V_{B} to V, we may just apply the standard equations used for perfect gases and these lead exactly to Eq. (3.80). As the single chain conformational distributions should not change on mixing, we have no further contribution to the entropy (Flory addressed in his original treatment Eq. (3.80) correspondingly as the change in 'configurational entropy', rather than associating it with the centers of mass motions only).

Regarding the expression for $\Delta\mathcal{G}_{\mathrm{loc}}$, we may first note that Eq. (3.81) represents the simplest formula which fulfills the requirement, that $\Delta\mathcal{G}_{\mathrm{loc}}$ must vanish for $\phi_{\mathrm{A}} \to 0$ and $\phi_{\mathrm{B}} \to 0$. More about the background may be learned if we consider the change in the interaction energy following from a transfer of an A-chain from the separated state into the mixture. Each chain probes the average value of the varying contact energies with the adjacent foreign monomers and the increase in the potential energy per reference unit may be written as

$$\frac{z_{\mathrm{eff}}}{2}\phi_{\mathrm{B}}kT\chi'$$

Here, the 'effective coordination number' z_{eff} gives the number of nearest neighbors (in reference units) on other chains, and the division by 2 is necessary to avoid a double counting of the pair contacts. An increase in the local Gibbs free energy results only if an AB-pair is formed, and this occurs with a probability equal to the volume fraction of the B's, ϕ_{B}. The product $kT\chi'$ is meant to specify this energy increase employing a dimensionless parameter χ'. For the potential experienced by the units of the B-chains in the mixture we write correspondingly

$$\frac{z_{\mathrm{eff}}}{2}\phi_{\mathrm{A}}kT\chi'$$

Fig. 3.13. Lattice model of a polymer mixture. Structure units of equal size setting up the two species of polymers occupy a regular lattice

with the identical parameter χ'. To obtain $\Delta\mathcal{G}_{\rm loc}$, which refers to the total system, we have to add the contributions of all A- and B-chains, weighted according to the respective fraction. This leads us to

$$\begin{aligned}\Delta\mathcal{G}_{\rm loc} &= \frac{V}{\tilde{v}_{\rm c}}\mathrm{N_L}\cdot\frac{z_{\rm eff}}{2}(\phi_{\rm A}\phi_{\rm B}+\phi_{\rm B}\phi_{\rm A})kT\chi' \\ &= \tilde{R}T\frac{V}{\tilde{v}_{\rm c}}\phi_{\rm A}\phi_{\rm B}z_{\rm eff}\chi' \end{aligned} \tag{3.82}$$

The prefactor $V\mathrm{N_L}/\tilde{v}_{\rm c}$ gives the number of reference units in the system. As we can see, Eq. (3.82) is equivalent to Eq. (3.81) if we set

$$\chi = z_{\rm eff}\chi' \tag{3.83}$$

Originally the χ-parameter was introduced to account for the contact energies only. However, its meaning can be generalized and this is in fact necessary. Experiments indicate that $\Delta\mathcal{G}_{\rm loc}$ often includes an entropic part, so that we have in general

$$\Delta\mathcal{G}_{\rm loc} = \Delta\mathcal{H}_{\rm mix} - T\Delta\mathcal{S}_{\rm loc} \tag{3.84}$$

The enthalpic part $\Delta\mathcal{H}_{\rm mix}$ shows up in the heat of mixing which is positive for endothermal and negative for exothermal systems. The entropic part $\Delta\mathcal{S}_{\rm loc}$ is usually due to changes in the mobility as has already been mentioned.

One particular concept employed in the original works must be mentioned, since it is still important. In the theoretical developments, Flory used a 'lattice model', constructed as shown schematically in Fig. 3.13. The A- and B-units of the two polymer species both have the same volume $v_{\rm c}$ and occupy the 'cells' of a regular lattice with coordination number z. It is assumed that the interaction energies are purely enthalpic and effective between nearest neighbors only. Excess contributions $kT\chi'$ which add to the interaction energies in the separated state arise for all pairs of unlike monomers. The parameter $\chi = (z-2)\chi'$ was devised to deal with this model and depends therefore on the size of the cell. Flory evaluated this model with the tools of statistical thermodynamics. Using approximations, he arrived at Eqs. (3.80), (3.81).

Although a modelling of a liquid polymer mixture on a lattice may first look rather artificial, it makes sense because it retains the important aspects of both the entropic and enthalpic part of $\Delta\mathcal{G}_{\mathrm{mix}}$. In recent years, lattice models have gained a renewed importance as a concept which is suitable for computer simulations. Numerical investigations make it possible to check and assess the validity range of the Flory-Huggins treatment. In fact, limitations exist and, as analytical calculations are difficult, simulations are very helpful and important. We shall present one example in a later section.

Application of the two expressions for $\Delta\mathcal{S}_{\mathrm{t}}$ and $\Delta\mathcal{G}_{\mathrm{loc}}$, Eqs. (3.80) and (3.81), results in the Flory-Huggins formulation for the Gibbs free energy of mixing of polymer blends

$$\Delta\mathcal{G}_{\mathrm{mix}} = \tilde{R}TV\left(\frac{\phi_{\mathrm{A}}}{\tilde{v}_{\mathrm{A}}}\ln\phi_{\mathrm{A}} + \frac{\phi_{\mathrm{B}}}{\tilde{v}_{\mathrm{B}}}\ln\phi_{\mathrm{B}} + \frac{\chi}{\tilde{v}_{\mathrm{c}}}\phi_{\mathrm{A}}\phi_{\mathrm{B}}\right) \tag{3.85}$$

$$= \tilde{R}T\tilde{n}_{\mathrm{c}}\left(\frac{\phi_{\mathrm{A}}}{N_{\mathrm{A}}}\ln\phi_{\mathrm{A}} + \frac{\phi_{\mathrm{B}}}{N_{\mathrm{B}}}\ln\phi_{\mathrm{B}} + \chi\phi_{\mathrm{A}}\phi_{\mathrm{B}}\right) \tag{3.86}$$

Here, we have introduced the molar volumes of the polymers, $\tilde{v}_{\mathrm{A}}$ and $\tilde{v}_{\mathrm{B}}$, using

$$\tilde{n}_{\mathrm{A}} = V\frac{\phi_{\mathrm{A}}}{\tilde{v}_{\mathrm{A}}} \quad \text{and} \quad \tilde{n}_{\mathrm{B}} = V\frac{\phi_{\mathrm{B}}}{\tilde{v}_{\mathrm{B}}} \tag{3.87}$$

and the molar volume of the reference unit, given by

$$\tilde{n}_{\mathrm{c}} = \frac{V}{\tilde{v}_{\mathrm{c}}} \tag{3.88}$$

The second equation follows when we replace the molar volumes by the degrees of polymerization, expressed in terms of the numbers of structure units. If we choose the same volume, equal to the reference volume $\tilde{v}_{\mathrm{c}}$, for both the A- and B-structure units we have

$$N_{\mathrm{A}} = \frac{\tilde{v}_{\mathrm{A}}}{\tilde{v}_{\mathrm{c}}} \quad \text{and} \quad N_{\mathrm{B}} = \frac{\tilde{v}_{\mathrm{B}}}{\tilde{v}_{\mathrm{c}}} \tag{3.89}$$

ϕ_{A} and ϕ_{B} add up to unity

$$\phi_{\mathrm{A}} + \phi_{\mathrm{B}} = 1 \tag{3.90}$$

Equations (3.85) and (3.86) are famous and widely used. They are the basis from which the majority of discussions of the properties of polymer mixtures emanates.

Having established the Flory-Huggins equation, we now consider the consequences for polymer mixtures. Starting from $\Delta\mathcal{G}_{\mathrm{mix}}$, the entropy of mixing,

$\Delta\mathcal{S}_{\text{mix}}$, follows as

$$\begin{aligned}\Delta\mathcal{S}_{\text{mix}} &= -\frac{\partial\Delta\mathcal{G}_{\text{mix}}}{\partial T} \\ &= -\tilde{R}V\left(\frac{\phi_{\text{A}}}{\tilde{v}_{\text{A}}}\ln\phi_{\text{A}} + \frac{\phi_{\text{B}}}{\tilde{v}_{\text{B}}}\ln\phi_{\text{B}} + \frac{\phi_{\text{A}}\phi_{\text{B}}}{\tilde{v}_{\text{c}}}\frac{\partial(\chi T)}{\partial T}\right) \end{aligned} \quad (3.91)$$

and the enthalpy of mixing, $\Delta\mathcal{H}_{\text{mix}}$, as

$$\Delta\mathcal{H}_{\text{mix}} = \Delta\mathcal{G}_{\text{mix}} + T\Delta\mathcal{S}_{\text{mix}} = \tilde{R}T\frac{V}{\tilde{v}_{\text{c}}}\phi_{\text{A}}\phi_{\text{B}}\left(\chi - \frac{\partial(\chi T)}{\partial T}\right) \quad (3.92)$$

These expressions show that the χ-parameter includes an entropic contribution given by

$$\chi_{\mathcal{S}} = \frac{\partial}{\partial T}(\chi T) \quad (3.93)$$

and an enthalpic part

$$\chi_{\mathcal{H}} = \chi - \frac{\partial(\chi T)}{\partial T} = -T\frac{\partial\chi}{\partial T} \quad (3.94)$$

both setting up χ as

$$\chi = \chi_{\mathcal{H}} + \chi_{\mathcal{S}} \quad (3.95)$$

Equation (3.93) indicates that for purely enthalpic local interactions, χ must have a temperature dependence

$$\chi \sim \frac{1}{T} \quad (3.96)$$

In this case, the increase in entropy is associated with the translational entropy only

$$\Delta\mathcal{S}_{\text{mix}} = \Delta\mathcal{S}_{\text{t}} \quad (3.97)$$

and the heat of mixing is given by

$$\Delta\mathcal{H}_{\text{mix}} = \tilde{R}T\frac{V}{\tilde{v}_{\text{c}}}\chi\phi_{\text{A}}\phi_{\text{B}} = \tilde{R}T\tilde{n}_{\text{c}}\chi\phi_{\text{A}}\phi_{\text{B}} \quad (3.98)$$

The Flory-Huggins equation provides the basis for a general discussion of the miscibility properties of a pair of polymers. As we shall see, this can be achieved in a transparent manner and leads to clear conclusions. To start with, we recall that, as a necessary requirement, mixing must be accompanied by a decrease of the Gibbs free energy. For liquid mixtures of low molar mass molecules this is mainly achieved by the large increase in the translational entropy. For these systems the increase in $\Delta\mathcal{S}_{\text{t}}$ can accomplish miscibility even in the case of unfavorable AB-interaction energies, i.e. for mixtures with an endothermal heat of mixing. In polymers we find a qualitatively different situation. The Flory-Huggins equation teaches us that, for polymer mixtures, the increase in the translational entropy $\Delta\mathcal{S}_{\text{t}}$ is extremely small and vanishes in the limit of infinite molecular weights, i.e. $\tilde{v}_{\text{A}}, \tilde{v}_{\text{B}} \to \infty$. The consequences are obvious:

- Positive values of χ necessarily lead to incompatibility. Since the entropic part, χ_S, appears to be mostly positive, one may also state that no polymer mixtures exist with a positive heat of mixing.
- If the χ-parameter is negative, then mixing takes place.

The reason for this behavior becomes clear if we regard miscibility as the result of a competition between the basic part of the osmotic pressure forces emerging from the translational motion of the polymers and the forces acting between the monomers. The osmotic pressure, which always favors miscibility, depends on the polymer density $c_{\rm p}$, whereas the part of the pressure produced by the monomer-monomer interactions may be attractive or repulsive and is a function of the monomer density $c_{\rm m}$. Since $c_{\rm p}/c_{\rm m} = 1/N$, the osmotic pressure part is extremely small compared to the effect of the monomer-monomer forces. Hence, mutual compatibility of two polymers, i.e. their potential to form a homogeneous mixture, is almost exclusively determined by the local interactions. Endothermal conditions are the rule between two different polymers, exothermal conditions the exception. Hence, the majority of pairs of polymers cannot form homogeneous mixtures. Compatibility is only found if there are special interactions between the A- and B-monomers, as they may arise in the form of dipole-dipole forces, hydrogen bonds or special donor-acceptor interactions.

All these conclusions refer to the limit of large degrees of polymerization. It is important to see that the Flory-Huggins equation permits one to consider how the compatibility changes if the degrees of polymerization are reduced and become moderate or small. For the sake of simplicity, we choose for a discussion the case of a 'symmetric' mixture with equal degrees of polymerization for both components, i.e.

$$N_{\rm A} = N_{\rm B} = N \tag{3.99}$$

Using

$$\frac{\tilde{n}_{\rm c}}{N} = \tilde{n}_{\rm A} + \tilde{n}_{\rm B} \tag{3.100}$$

we obtain

$$\Delta\mathcal{G}_{\rm mix} = \tilde{R}T(\tilde{n}_{\rm A} + \tilde{n}_{\rm B})(\phi_{\rm A}\ln\phi_{\rm A} + \phi_{\rm B}\ln\phi_{\rm B} + \chi N\phi_{\rm A}\phi_{\rm B}) \tag{3.101}$$

Note that there is only one relevant parameter, namely the product $N\chi$. The dependence of $\Delta\mathcal{G}_{\rm mix}$ on $\phi_{\rm A}$ is shown in Fig. 3.14, as computed for different values of χN.

A discussion of these curves enables us to reach some important conclusions. For a vanishing χ, one has negative values of $\Delta\mathcal{G}_{\rm mix}$ for all $\phi_{\rm A}$, with a minimum at $\phi_{\rm A} = 0.5$. In this case, we have perfect miscibility caused by the small entropic forces related with $\Delta\mathcal{S}_{\rm t}$. For negative values of χN, we have a further decrease of $\Delta\mathcal{G}_{\rm mix}$ and therefore also perfect miscibility.

A change in behavior is observed for positive values of χN. The curves alter their shape, and for parameters χN above a critical value

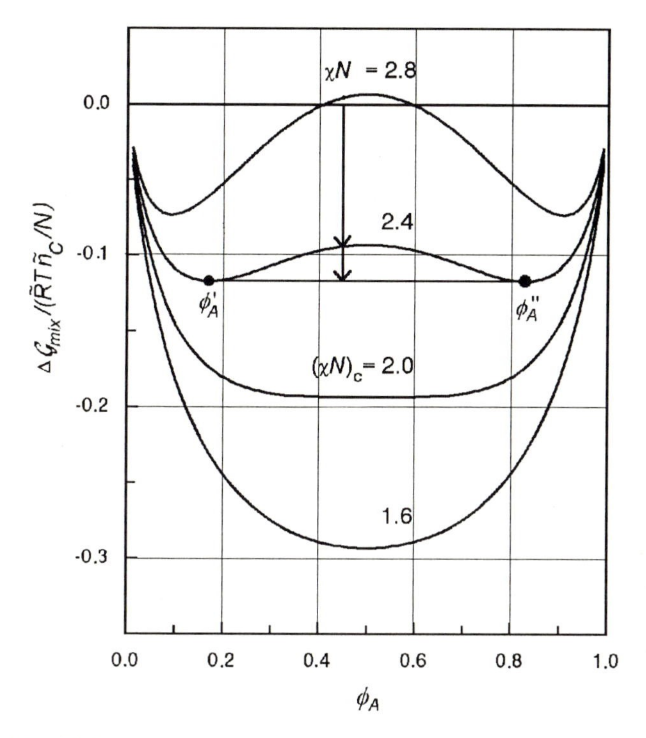

Fig. 3.14. Gibbs free energy of mixing of a symmetric binary polymer mixture ($N_A = N_B = N$), as described by the Flory-Huggins equation

$$(\chi N) > (\chi N)_c$$

a maximum rather than a minimum emerges at $\phi_A = 0.5$. This change leads us into a different situation. Even if $\Delta\mathcal{G}_{mix}$ is always negative, there does not form always a homogeneous mixture. To understand the new conditions, let us look for example at the curve for $\chi N = 2.4$ and consider a blend with $\phi_A = 0.45$. There the two arrows are drawn. The first arrow indicates that a homogeneous mixing of A and B would lead to a decrease in the Gibbs free energy, when compared to two separate one-component phases. However, as shown by the second arrow, the Gibbs free energy can be further reduced, if again a two-phase structure is formed, now being composed of two mixed phases, with compositions ϕ_A' and ϕ_A''. The specific feature in the selected curve responsible for this peculiar behavior is the occurrence of the two minima at ϕ_A' and ϕ_A'', as these enable the further decrease of the Gibbs free energy. For which values of ϕ_A can this decrease be achieved? Not for all values, because there is an obvious restriction: The overall volume fraction of the A-chains has to be in the range

$$\phi_A' \leq \phi_A \leq \phi_A''$$

Outside this central range, for $\phi_A < \phi_A'$ and $\phi_A > \phi_A''$, a separation into the two phases with the minimum Gibbs free energies is impossible and one homogeneous phase is formed.

For a given ϕ_A we can calculate the fractions ϕ_1, ϕ_2 of the two coexisting mixed phases. As we have

$$\phi_A = \phi_1 \cdot \phi'_A + (1 - \phi_1)\phi''_A \tag{3.102}$$

we find

$$\phi_1 = \frac{\phi''_A - \phi_A}{\phi''_A - \phi'_A} \tag{3.103}$$

and

$$\phi_2 = 1 - \phi_1 = \frac{\phi_A - \phi'_A}{\phi''_A - \phi'_A} \tag{3.104}$$

Hence in conclusion, for curves $\Delta\mathcal{G}_{mix}(\phi_A)$ which exhibit two minima and a maximum in-between, mixing properties depend on the value of ϕ_A. Miscibility is found for low and high values of ϕ_A only, and in the central region there is a 'miscibility gap'.

One can determine the critical value of χN which separates the range of perfect mixing, i.e. compatibility through all compositions, from the range with a miscibility gap. Clearly, for the critical value of χN, the curvature at $\phi_A = 0.5$ must vanish

$$\frac{\partial^2 \Delta\mathcal{G}_{mix}(\phi_A = 0.5)}{\partial \phi_A^2} = 0 \tag{3.105}$$

The first derivative of $\Delta\mathcal{G}_{mix}$ is given by

$$\frac{1}{(\tilde{n}_A + \tilde{n}_B)\tilde{R}T}\frac{\partial \Delta\mathcal{G}_{mix}}{\partial \phi_A} = \ln\phi_A + 1 - \ln(1 - \phi_A) - 1 + \chi N(1 - 2\phi_A) \tag{3.106}$$

and the second derivative by

$$\frac{1}{(\tilde{n}_A + \tilde{n}_B)\tilde{R}T}\frac{\partial^2 \Delta\mathcal{G}_{mix}}{\partial \phi_A^2} = \frac{1}{\phi_A} + \frac{1}{1 - \phi_A} - 2\chi N \tag{3.107}$$

Our condition is fulfilled for

$$\chi N = 2 \tag{3.108}$$

Hence, we expect full compatibility for

$$\chi < \chi_c = \frac{2}{N} \tag{3.109}$$

and a miscibility gap for

$$\chi > \chi_c \tag{3.110}$$

Equations (3.109), (3.110) describe the effect of the molecular weight on the compatibility of a pair of polymers. In the limit $N \to \infty$ we have

$$\chi_c \to 0$$

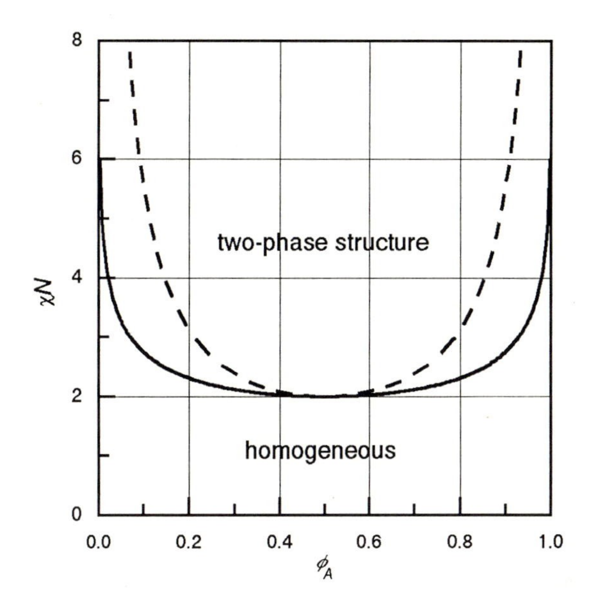

Fig. 3.15. Phase diagram of a symmetric polymer mixture ($N_{\mathrm{A}} = N_{\mathrm{B}} = N$). In addition to the binodal (*continuous line*) the spinodal is shown (*broken line*)

This agrees with our previous conclusion that for positive values of χ polymers of average and high molecular weight do not mix at all.

The properties of symmetric polymer mixtures are summarized in the phase diagram shown in Fig. 3.15. It depicts the two regions associated with homogeneous and two-phase structures in a plot which uses the sample composition, as expressed by the volume fraction ϕ_{A}, and the parameter χN as variables. The boundary between the one-phase and the two-phase region is called 'binodal'. It is determined by the compositions ϕ'_{A} and ϕ''_{A} of the equilibrium phases with minimum Gibbs free energies in the miscibility gap. ϕ'_{A} and ϕ''_{A} follow for a given value of χN from

$$\frac{\partial \Delta \mathcal{G}_{\mathrm{mix}}}{\partial \phi_{\mathrm{A}}} = 0 \tag{3.111}$$

Using Eq. (3.106) we obtain an analytical expression for the binodal

$$\chi N = \frac{1}{1 - 2\phi_{\mathrm{A}}} \cdot \ln \frac{1 - \phi_{\mathrm{A}}}{\phi_{\mathrm{A}}} \tag{3.112}$$

The derived phase diagram is universal in the sense that it is valid for all symmetric polymer mixtures. It indicates a miscibility gap for $\chi N > 2$ and enables us to make in this range a determination of χN if the compositions of the two coexisting phases are known.

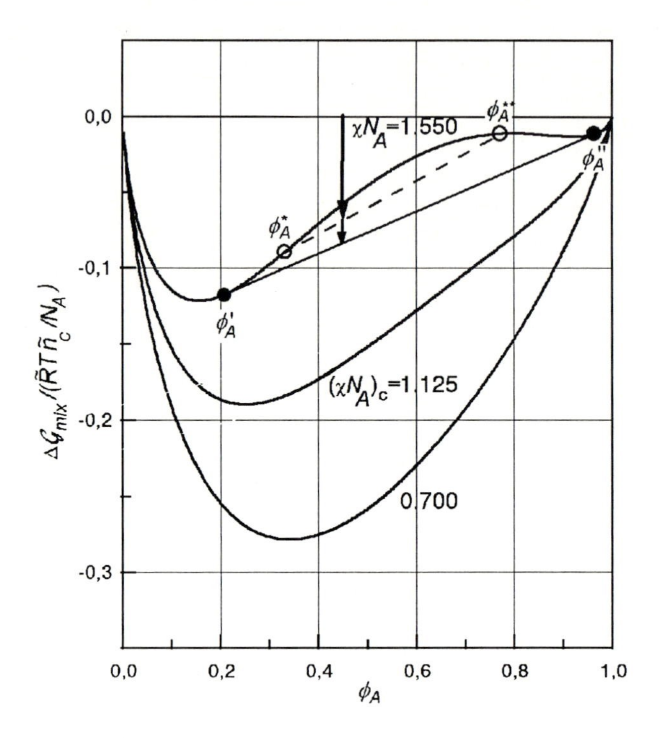

Fig. 3.16. Gibbs free energy of mixing of an asymmetric polymer mixture with $N_B = 4N_A$, calculated for the indicated values of χN_A. The points of contact with the common tangent, located at ϕ'_A and ϕ''_A, determine the compositions of the equilibrium phases on the binodal. The critical values are $(\chi N_A)_c = 9/8$ and $\phi_c = 2/3$

For mixtures of polymers with different degrees of polymerization, i.e. $N_A \neq N_B$, the phase diagram loses its symmetrical shape. Figure 3.16 depicts $\Delta\mathcal{G}_{mix}(\phi_A)$ for a mixture with $N_B = 4N_A$, as computed on the basis of the Flory-Huggins equation. Straightforward analysis shows that, in this general case, the critical value of χ is given by

$$\chi_c = \frac{1}{2}\left(\frac{1}{\sqrt{N_A}} + \frac{1}{\sqrt{N_B}}\right)^2 \tag{3.113}$$

The critical point, where the miscibility gap begins, is located at

$$\phi_{A,c} = \frac{\sqrt{N_B}}{\sqrt{N_A} + \sqrt{N_B}} \tag{3.114}$$

The points along the binodal can be determined by the construction of the common tangent as indicated in the figure. The explanation for this procedure is simple and we refer here to the two arrows drawn at $\phi_{\mathrm{A}} = 0.45$, and the curve calculated for $\chi N_{\mathrm{A}} = 1.550$. First, consider the change in $\Delta\mathcal{G}_{\mathrm{mix}}$ if starting-off from separate states, two arbitrary mixed phases with composition ϕ_{A}^{*} and ϕ_{A}^{**} are formed. $\Delta\mathcal{G}_{\mathrm{mix}}$ is given by the point at $\phi_{\mathrm{A}} = 0.45$ on the straight line which connects $\Delta\mathcal{G}_{\mathrm{mix}}(\phi_{\mathrm{A}}^{*})$ and $\Delta\mathcal{G}_{\mathrm{mix}}(\phi_{\mathrm{A}}^{**})$. This is seen when we first write down the obvious linear relation

$$\Delta\mathcal{G}_{\mathrm{mix}}(\phi_{\mathrm{A}}) = \phi_1 \Delta\mathcal{G}_{\mathrm{mix}}(\phi_{\mathrm{A}}^{*}) + \phi_2 \Delta\mathcal{G}_{\mathrm{mix}}(\phi_{\mathrm{A}}^{**}) \tag{3.115}$$

where ϕ_1 and ϕ_2 denote the volume fractions of the two mixed phases. Recalling that ϕ_1 and ϕ_2 are given by Eqs. (3.103) and (3.104), we obtain the expression

$$\Delta\mathcal{G}_{\mathrm{mix}}(\phi_{\mathrm{A}}) = \frac{\phi_{\mathrm{A}}^{**} - \phi_{\mathrm{A}}}{\phi_{\mathrm{A}}^{**} - \phi_{\mathrm{A}}^{*}} \Delta\mathcal{G}_{\mathrm{mix}}(\phi_{\mathrm{A}}^{*}) + \frac{\phi_{\mathrm{A}} - \phi_{\mathrm{A}}^{*}}{\phi_{\mathrm{A}}^{**} - \phi_{\mathrm{A}}^{*}} \Delta\mathcal{G}_{\mathrm{mix}}(\phi_{\mathrm{A}}^{**}) \tag{3.116}$$

which indeed describes a straight line connecting $\Delta\mathcal{G}_{\mathrm{mix}}(\phi_{\mathrm{A}}^{*})$ and $\Delta\mathcal{G}_{\mathrm{mix}}(\phi_{\mathrm{A}}^{**})$. So far, the choice of ϕ_{A}^{*} and ϕ_{A}^{**} has been arbitrary, but we know that on separating into two mixed phases, the system seeks to maximize the gain in Gibbs free energy. The common tangent now represents that connecting line between any pair of points on the curve which is at the lowest possible level. A transition to this line therefore gives the largest possible change $\Delta\mathcal{G}_{\mathrm{mix}}$. It is associated with the formation of two phases with compositions ϕ_{A}' and ϕ_{A}'', as given by the points of contact with the common tangent. The binodal is set up by these points, and a determination may be based on the described geometrical procedure.

Phase Diagrams: Upper and Lower Miscibility Gap

Phase diagrams of polymer blends under atmospheric pressure are usually presented in terms of the variables ϕ_{A} and T. Emanating from the discussed universal phase diagram in terms of χ and ϕ_{A}, these can be obtained by introducing into the consideration the temperature dependence of the Flory-Huggins parameter. This function, $\chi(T)$, then solely determines the appearance. For different types of temperature dependencies $\chi(T)$, different classes of phase diagrams emerge and we discuss them in this section.

Let us first consider an endothermal polymer mixture with negligible entropic contributions to the local Gibbs free energy, i.e. a system with $\chi = \chi_H > 0$. Here the temperature dependence of χ is given by Eq. (3.96)

$$\chi \sim \frac{1}{T}$$

The consequences for the phase behavior are evident. Perfect miscibility can principally exist at high temperatures, provided that the molecular weights of the components are low enough. The increase of χ with decreasing temperature necessarily results in a termination of this region and the formation of

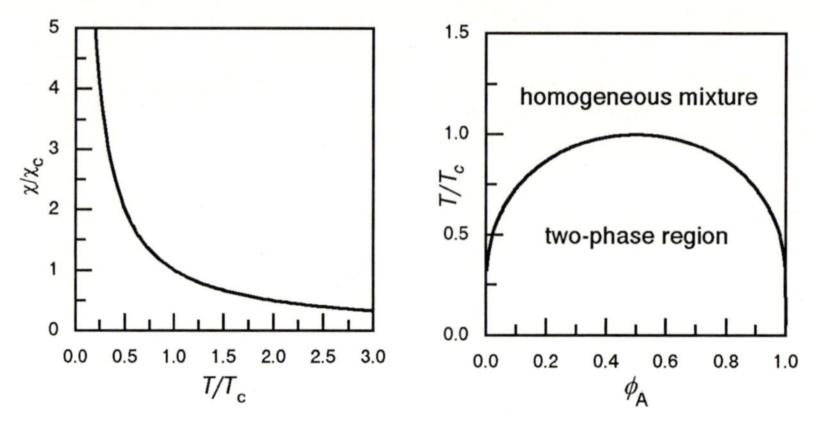

Fig. 3.17. Endothermal symmetrical mixture with a constant heat of mixing. Temperature dependence of the Flory-Huggins parameter (*left*) and phase diagram showing a lower miscibility gap (*right*)

a miscibility gap, found when $\chi > \chi_c$. For a symmetric mixture we obtained $\chi_c = 2/N$ (Eq. (3.110)). If χ_c is reached at a temperature T_c, we can write

$$\chi = \frac{2}{N} \cdot \frac{T_c}{T} \tag{3.117}$$

The resulting phase diagram is shown in Fig. 3.17, together with the temperature dependence of χ. The binodal follows from Eq. (3.112), as

$$\frac{T}{T_c} = \frac{2(1-2\phi_A)}{\ln((1-\phi_A)/\phi_A)} \tag{3.118}$$

It marks the boundary between the homogeneous state at high temperatures and the two-phase region at low temperatures. Upon cooling a homogeneous mixture, phase separation at first sets in for samples with the 'critical composition', $\phi_A = 0.5$, at the temperature T_c. For the other samples demixing occurs at lower temperatures, as described by the binodal. We observe here a 'lower miscibility gap'. A second name is also used in the literature: T_c is called the 'upper critical dissolution temperature', shortly 'UCDT'. The latter name refers to the structural changes induced when coming from the two-phase region, where one observes a dissolution and merging of the two phases.

Experiments show that exothermal polymer blends sometimes have an 'upper miscibility gap', i.e. one which is open towards high temperatures. One may wonder why a mixture which is homogeneous at ambient temperature separates in two phases upon heating, and we shall have to think about possible physical mechanisms. At first, however, we should discuss the formal prerequisites. On the right of Fig. 3.18 there are phase diagrams of symmetric

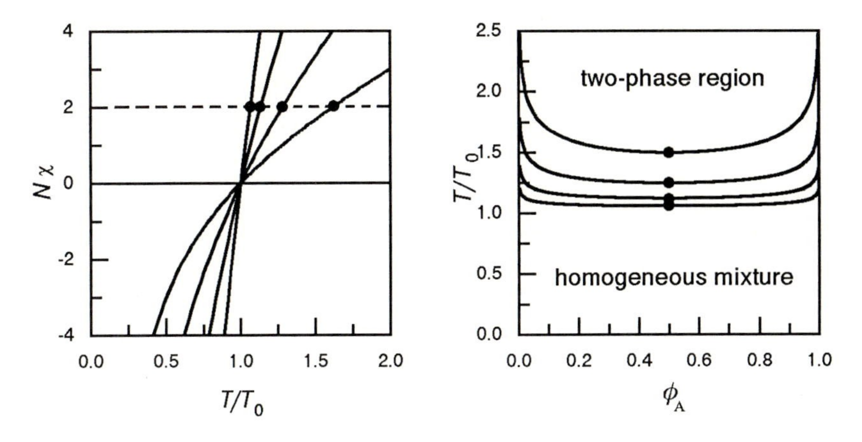

Fig. 3.18. Phase diagram of an exothermal symmetric polymer mixture with an upper miscibility gap. The binodals correspond to the different functions $N\chi(T)$ shown on the left, associated with an increase in the molecular weight by factors 2,4 and 8. Critical points are determined by $N\chi(T_c/T_0) = 2$, as indicated by the filled points in the drawings

polymer mixtures which display an upper miscibility gap. The various depicted binodals are associated with different molecular weights. The curved binodals relate to polymers with low or moderate molar masses. For medium and high molecular weights, the phase boundary becomes a horizontal line, and phase separation then occurs for $\chi \geq 0$ independent of ϕ_A. The latter result agrees with the general criterion for phase separations in polymer systemes with high molar masses. It is therefore not particular to the symmetric system, but would be obtained in the general case, $N_A \neq N_B$, as well.

The temperature dependencies $\chi(T)$ which lead to these diagrams are shown on the left-hand side of Fig. 3.18. Their main common property is a change of the Flory-Huggins parameter from negative to positive values. The crossing of the zero line takes place at a certain temperature, denoted T_0. Coming from low temperatures, unmixing sets in for $T = T_c$ with

$$N\chi(T_c) = 2$$

In the limit of high degrees of polymerization we have $\chi(T_c) \to 0$ and therefore $T_c \to T_0$. We see that the prerequisite for an upper miscibility gap, or a 'lower critical solution temperature', abbreviated 'LCST', as it is alternatively called, is a negative value of χ at low temperatures, followed by an increase to values above zero.

One can envisage two different mechanisms as possible explanations for such behavior. First, there can be a competition between attractive forces between specific groups incorporated in the two polymers on one side and repulsive interactions between the remaining units on the other side. In copolymer systems with pairs of specific comonomers which are capable of forming stable bonds, these conditions may arise. With increasing temperature the fraction of closed bonds decreases and the repulsive forces finally dominate. For such a system, χ may indeed be negative for low temperatures and positive for high ones.

The second conceivable mechanism has already been mentioned. Sometimes it is observed that a homogeneous mixing of two polymers results in a volume shrinkage. The related decrease in the 'free volume' available for local motions of the monomers may lead to a reduced mobility and hence a lowering of the entropy. The effect usually increases with temperature and finally overcompensate the initially dominating attractive interactions.

For mixtures of polymers with low molecular weights there is also the possibility that both a lower and an upper miscibility gap appear. In this case χ crosses the critical value χ_{c} twice, first during a decrease in the low temperature range and then, after passing through a minimum, during the subsequent increase at higher temperatures. Such a temperature dependence reflects the presence of both a decreasing endothermal contribution and an increasing entropic part.

As we can see, the Flory-Huggins treatment is able to account for the various general shapes of existing phase diagrams. This does not mean, however, that one can reproduce measured phase diagrams in a quantitative manner. To comply strictly with the Flory-Huggins theory, the representation of measured binodals has to be accomplished with one temperature dependent function $\chi(T)$ only. As a matter of fact, this is rarely the case. Nevertheless, data can be formally described if one allows for a ϕ_{A}-dependence of χ. As long as the variations remain small, one can consider the deviations as perturbations and still feel safe on the grounds of the Flory-Higgins treatment. For some systems, however, the variations with ϕ_{A} are large. Then the basis gets lost and the meaning of χ becomes rather unclear. Even then the Flory-Huggins equation is sometimes employed but only as a means to carry out interpolations and extrapolations and to relate different sets of data. That deviations arise is not unexpected. The mean-field treatment, on which the Flory-Huggins theory is founded, is only an approximation with varying quality.

Let us look at two examples. Figure 3.19 presents phase diagrams of mixtures of different polystyrenes with polybutadiene, all of them having moderate to low molecular weights (M = 2000-4000). The temperature points on the curves are measured 'cloud points'. As samples are transparent in the homogeneous phase and become turbid when demixing starts, the cloudiness can be used for a determination of the binodal. For an accurate detection one can use measurements of the intensity of scattered or transmitted light. We are dealing here with an endothermal system which exhibits a lower miscibility gap. Note that T_{c}, as given by the highest point of each curve, decreases with decreasing molecular weight in accordance with the theoretical prediction. The curves, which provide a satisfactory data fit, were obtained on the basis of the Flory-Huggins theory assuming a weakly ϕ_{A}-dependent χ.

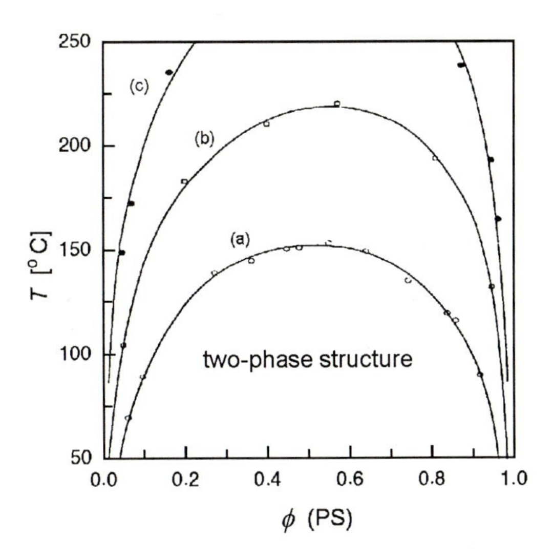

Fig. 3.19. Phase diagrams for different PS/PB-mixtures, exhibiting lower miscibility gaps. (a) $M(\text{PS}) = 2250$, $M(\text{PB}) = 2350$; (b) $M(\text{PS}) = 3500$, $M(\text{PB}) = 2350$; (c) $M(\text{PS}) = 5200$, $M(\text{PB}) = 2350$. Data from Roe and Zin [15]

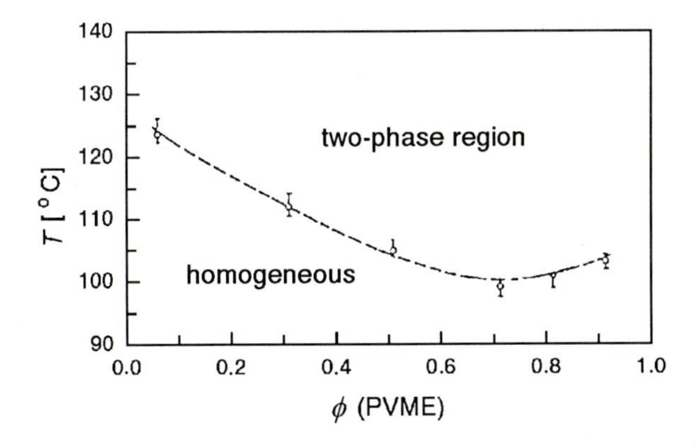

Fig. 3.20. Phase diagram of mixtures of PS ($M = 2\cdot 10^5$) and PVME ($M = 4.7\cdot 10^4$), showing an upper miscibility gap. Data from Hashimoto et al.[16]

Figure 3.20 shows, as a second example, a phase diagram obtained for mixtures of polystyrene and poly(vinylmethylether) (PVME). One observes here that homogeneous mixtures are obtained in the temperature range below 100 °C, and that there is an upper miscibility gap. The phase diagram depicted in the figure was obtained for polymers with molecular weights

$M(\mathrm{PS}) = 2 \cdot 10^5$, $M(\mathrm{PVME}) = 4.7 \cdot 10^4$. For molecular weights in this range the contribution of the translational entropy becomes very small indeed and mixing properties are mostly controlled by χ. The curved appearance of the binodal, which contrasts with the result of the model calculation in Fig. 3.18 where we obtained for polymers with medium or high molecular weights a nearly horizontal line, is indicative of a pronounced compositional dependence of χ. This represents a case where the Flory-Huggins treatment does not provide a comprehensive description. Interactions in this mixture are of a complex nature and apparently change with the sample composition, so that it becomes impossible to represent them by just one constant.

3.2.2 Phase Separation Mechanisms

As we have seen, binary polymer mixtures can vary in structure with temperature, forming either a homogeneous phase or in a miscibility gap a two-phase structure. We now have to discuss the processes which are effective during a change, i.e. the mechanisms of phase transition.

Phase separation is induced, when a sample is transferred from the one-phase region into a miscibility gap. Usually, this is accomplished by a change in temperature, upward or downward depending on the system under study. The evolution of the two-phase structure subsequent to a temperature jump can often be continuously monitored and resolved in real-time, owing to the high viscosity of polymers which slows down the rate of unmixing. If necessary for detailed studies, the process may also be stopped at any stage by quenching samples to temperatures below the glass transition. Suitable methods for observations are light microscopy or scattering experiments.

Figure 3.21 presents as an example two micrographs obtained with a light microscope using an interference technique, showing two-phase structures observed for mixtures of polystyrene and partially brominated polystyrene, with both species having equal degrees of polymerization ($N = 200$). The two components show perfect miscibility at temperatures above 220 °C and below this temperature a miscibility gap. Phase separation here was induced by a temperature jump from 230 °C to 200 °C, for two mixtures of different composition, $\phi(\mathrm{PS}) = 0.8$ and $\phi(\mathrm{PS}) = 0.5$. We observe two structure patterns which do not just vary in length scale, but differ in the general characteristics: The picture on the left shows spherical precipitates in a matrix, whereas the pattern on the right exhibits interpenetrating continuously extending domains. The diverse evidence suggests that different mechanisms were effective during phase separation. The structures with spherical precipitates are indicative of 'nucleation and growth', and the pattern with two structurally equivalent interpenetrating phases reflects a 'spinodal decomposition'. In fact, this example is quite typical and is representative of the results of investigations on various polymer mixtures. The finding is that structure evolution in the early stages of unmixing is generally controlled by either of these two mechanisms.

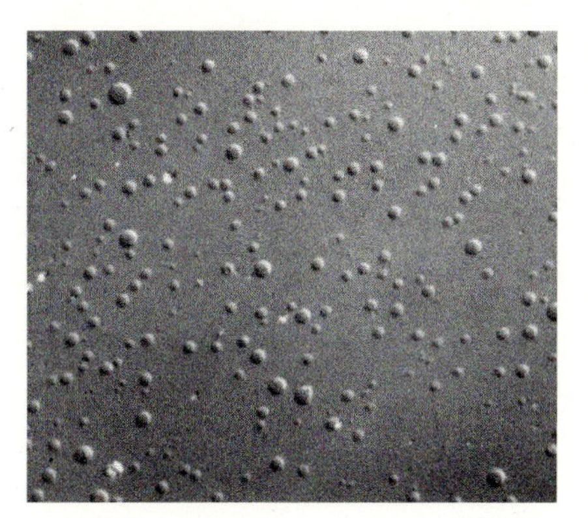
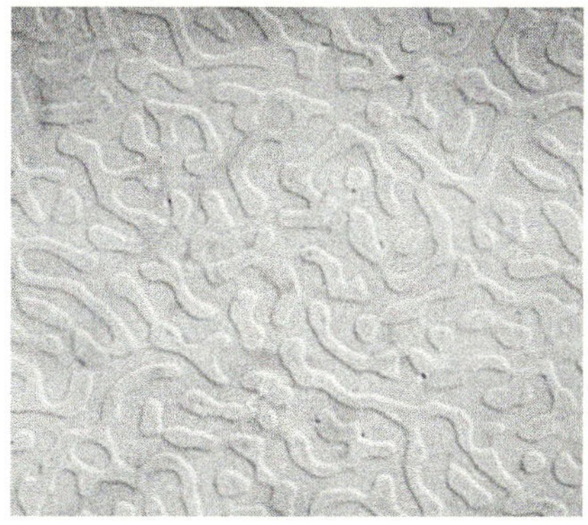

Fig. 3.21. Structure patterns emerging during phase separation in PS/PBr$_x$S-mixtures. *left*: Pattern indicating phase separation by nucleation and growth (ϕ(PS) = 0.8); *right* : Pattern suggesting phase separation by spinodal decomposition (ϕ(PS) = 0.5) [17]

The cause for the occurrence of two different modes of phase separation becomes revealed when we consider the shape of the curve $\Delta\mathcal{G}_{\text{mix}}(\phi_{\text{A}})$. As ϕ_{A} is the only independent variable, in the following we will omit the subscript A, i.e. replace ϕ_{A} by the shorter symbol ϕ. Figure 3.22 depicts in the upper part functions $\Delta\mathcal{G}_{\text{mix}}(\phi)$ computed for three different values of χ which belong to the one-phase region (χ_{i}), the two-phase region (χ_{f}) and the critical point (χ_{c}). The lower part of the figure gives the phase diagram, with the positions of $\chi_{\text{i}}, \chi_{\text{f}}$ and χ_{c} being indicated. The arrows '1' and '2' indicate two jumps which transfer a polymer mixture from the homogeneous phase into the two-phase region. Immediately after the jump, the structure is still homogeneous but of course no longer stable. What is different in the two cases, is the character of the instability. The difference shows up when we consider the consequences of a sponteneous local concentration fluctuation, as it could be thermally induced directly after the jump. Figure 3.23 represents such a fluctuation schematically, being set up by an increase $\delta\phi$ in the concentration of A-chains in one half of a small volume $\text{d}^3\boldsymbol{r}$ and a corresponding decrease in the other half. The fluctuation leads to a change in the Gibbs free energy, described as

$$\delta\mathcal{G} = \frac{1}{2}(g(\phi_0 + \delta\phi) + g(\phi_0 - \delta\phi))\text{d}^3\boldsymbol{r} - g(\phi_0)\text{d}^3\boldsymbol{r} \qquad (3.119)$$

Here, we have introduced the free energy density, i.e. the Gibbs free energy per unit volume, denoted $g(\phi)$. Series expansion of $g(\phi)$ up to the second order in ϕ yields for $\delta\mathcal{G}$ the expression

$$\delta\mathcal{G} = \frac{1}{2}\frac{\partial^2 g}{\partial\phi^2}(\phi_0)\delta\phi^2\text{d}^3\boldsymbol{r} \qquad (3.120)$$

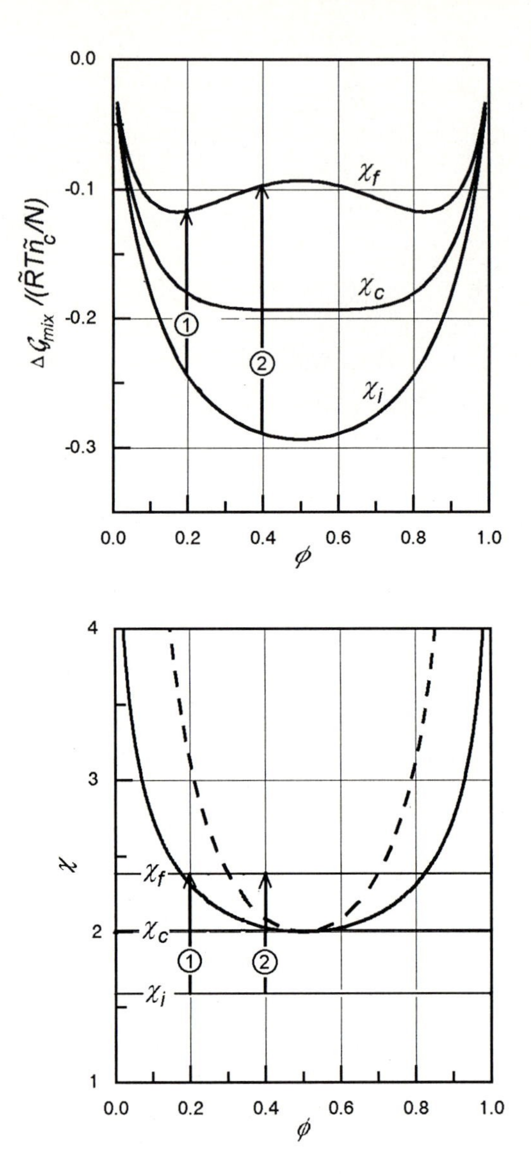

Fig. 3.22. Temperature jumps that transfer a symmetric binary polymer mixture from the homogeneous state into the two-phase region. Depending upon the location in the two-phase region, phase separation occurs either by nucleation and growth ('1') or by spinodal decomposition ('2')

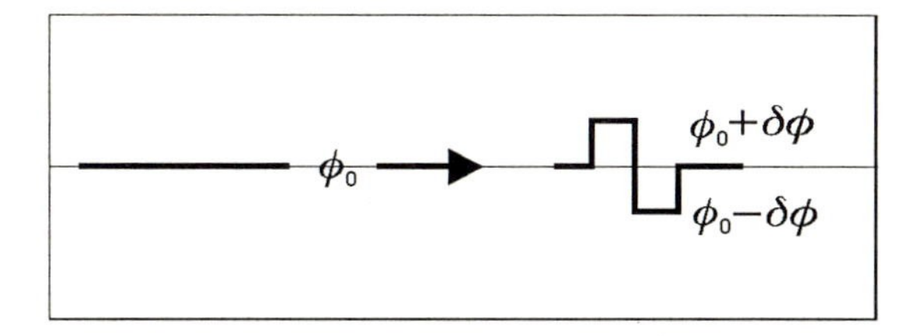

Fig. 3.23. Local concentration fluctuation

We calculate $\partial^2 g/\partial\phi^2$ with the aid of the Flory-Huggins equation, i.e. write

$$\frac{\partial^2 g}{\partial\phi^2} = \frac{1}{V}\frac{\partial^2 \Delta\mathcal{G}_{\text{mix}}}{\partial\phi^2} \tag{3.121}$$

with $\Delta\mathcal{G}_{\text{mix}}$ being given by Eq. (3.85). Then the change $\delta\mathcal{G}$ associated with the local fluctuation is

$$\delta\mathcal{G} = \frac{1}{2}\frac{1}{V}\frac{\partial^2 \Delta\mathcal{G}_{\text{mix}}}{\partial\phi^2}(\phi_0)\cdot\delta\phi^2 \mathrm{d}^3\boldsymbol{r} \tag{3.122}$$

This is a most interesting result. It tells us that, depending on the sign of the curvature $\partial^2\Delta\mathcal{G}_{\text{mix}}/\partial\phi^2$, the fluctuation may either lead to an increase, or a decrease in the Gibbs free energy. In stable states, there always has to be an increase to ensure that a spontaneous local association of monomers A disintegrates again. This situation is found for jump '1'. It leads to a situation where the structure is still stable with regard to spontaneous concentration fluctuations provided that they remain sufficiently small. Jump '2' represents a qualitatively different case. Since the curvature here is negative, the Gibbs free energy decreases immediately, even for an infinitesimally small fluctuation, and no restoring force arises. On the contrary, there is a tendency for further growth of the fluctuation amplitude. Hence, by the temperature jump '2', an initial structure is prepared which is perfectly unstable.

It is exactly the latter situation which results in a spinodal decomposition. The process is sketched at the bottom of Fig. 3.24. The drawing indicates that a spinodal decomposition implies a continuous growth of the amplitude of a concentration fluctuation, starting from infinitesimal values and ensuing up to the final state of two equilibrium phases with compositions ϕ' and ϕ''. The principles which govern this process have been studied in numerous investigations and clarified to a large extent. We shall discuss its properties in detail in the next section. At this point, we leave it with one short remark with reference to the figure. There the arrows indicate the directions of flow of the A-chains. The normal situation is found for nucleation and growth, where the flow is directed as usual, towards decreasing concentrations of the A's. In spinodal decompositions the flow direction is reversed. The A-chains diffuse towards higher concentrations, which corresponds formally to a negative diffusion coefficient.

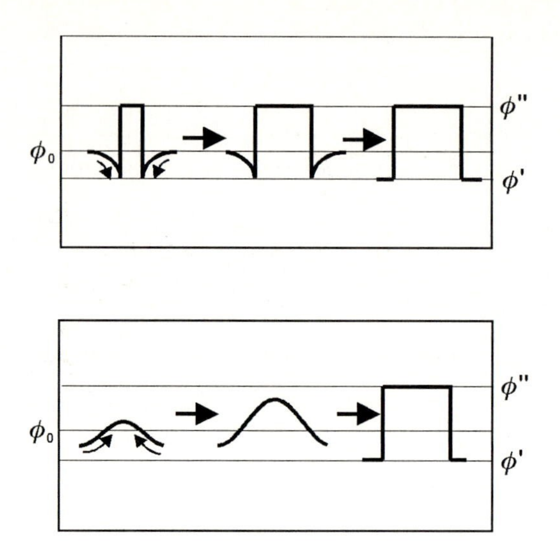

Fig. 3.24. Mechanisms of phase separation: Nucleation and growth (*top*) and spinodal decomposition (*bottom*). The curved small arrows indicate the direction of the diffusive motion of the A-chains

The upper half of the figure shows the process which starts subsequent to the temperature jump '1'. As small fluctuations decay again, the only way to achieve a gain in the Gibbs free energy is a large fluctuation which leads directly to the formation of a nucleus of the new equilibrium phase with composition ϕ''. After it has formed it can increase in size. Growth is accomplished by regular diffusion of the chains since there exists, as indicated in the drawing, a zone with a reduced ϕ at the surface of the particle which attracts a stream of A-chains.

The process of nucleation and growth is not peculiar to polymers, but observed in many materials, and we consider it only briefly. The specific point making up the difference to the case of a spinodal decomposition is the existence of an activation barrier. The reason for its occurrence is easily recognized. Figure 3.25 shows the change of the Gibbs free energy, $\Delta\mathcal{G}$, following from the formation of a spherical precipitate of the new equilibrium phase. $\Delta\mathcal{G}$ depends on the radius r of the precipitate, as described by the equation

$$\Delta\mathcal{G}(r) = -\frac{4\pi}{3}r^3\Delta g + 4\pi r^2\sigma \tag{3.123}$$

with

$$\Delta g = g(\phi_0) - g(\phi'') \tag{3.124}$$

Equation (3.123) emanates from the view that $\Delta\mathcal{G}$ is set up by two contributions, one being related to the gain in the bulk Gibbs free energy of the precipitate, the other to the effect of the interface between particle and ma-

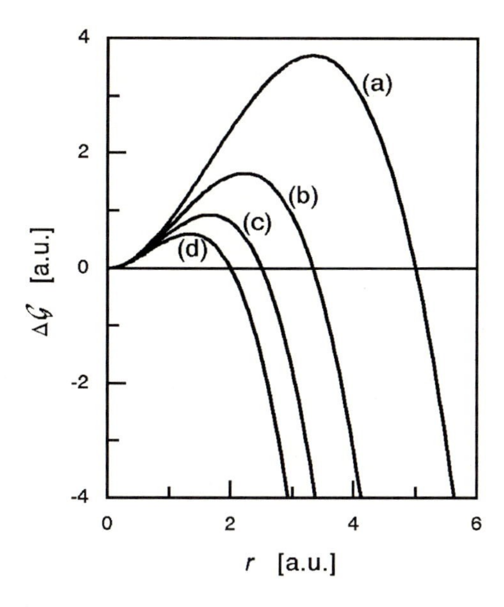

Fig. 3.25. Activation barrier encountered during formation of a spherical nucleus. Curves (*a*)-(*d*) correspond to a sequence 2:3:4:5 of values for $\Delta g/\sigma$.

trix. This interface is associated with an excess free energy, and the symbol σ stands for the excess free energy per unit area.

Since the building up of the interface causes an increase in the free energy, a barrier $\Delta\mathcal{G}_{\mathrm{b}}$ develops, which first has to be overcome before growth can set in. The passage over this barrier constitutes the nucleation step. Representing an activated process, it occurs with a rate given by the Arrhenius equation

$$\nu_{\mathrm{nuc}} \sim \exp -\frac{\Delta\mathcal{G}_{\mathrm{b}}}{kT} \tag{3.125}$$

whereby $\Delta\mathcal{G}_{\mathrm{b}}$ is the barrier height

$$\Delta\mathcal{G}_{\mathrm{b}} = \frac{16\pi}{3}\frac{\sigma^3}{(\Delta g)^2} \tag{3.126}$$

Equation (3.126) follows from Eq. (3.123) when searching for the maximum. $\Delta\mathcal{G}_{\mathrm{b}}$ increases with decreasing distance from the binodal, where we have $\Delta g = 0$. The change is illustrated by the curves in Fig. 3.25 which were calculated for different values of the ratio $\Delta g/\sigma$. We learn from this behavior that, in order to achieve reasonable rates, nucleation requires a certain degree of supercooling (or overheating, if there is an upper miscibility gap).

Nucleation and growth occurs if the unmixing is induced at a temperature near to the binodal, where the system is still stable with regard to small concentration fluctuations. Further away from the binodal this restricted 'metastability' gets lost and spinodal decomposition sets in. Transition from one to the other growth regime occurs in the range of the 'spinodal', which is defined as the locus of those points in the phase diagram where the stabilizing restoring forces vanish. According to the previous arguments this occurs for

$$\frac{\partial^2 \Delta \mathcal{G}_{\text{mix}}}{\partial \phi^2} = 0 \tag{3.127}$$

Equation (3.127) determines for each ϕ a certain value χ, and we choose for the resulting spinodal curve the designation $\chi_{\text{sp}}(\phi)$. In the case of a symmetric mixture with a degree of polymerization N for both species, we can use Eq. (3.107) for a determination. The spinodal follows as

$$\chi_{\text{sp}} = \frac{1}{2N\phi_{\text{A}}(1-\phi_{\text{A}})} \tag{3.128}$$

It is this line which is included in Fig. 3.22. For $N_{\text{A}} \neq N_{\text{B}}$ we start from Eq. (3.85) and obtain

$$\frac{\partial^2 \Delta \mathcal{G}_{\text{mix}}}{\partial \phi^2} \sim \frac{1}{N_{\text{A}}\phi} + \frac{1}{N_{\text{B}}(1-\phi)} + \frac{\partial^2}{\partial \phi^2}\chi\phi(1-\phi) \tag{3.129}$$

In this case the spinodal is given by the function

$$2\chi_{\text{sp}} = \frac{1}{N_{\text{A}}\phi} + \frac{1}{N_{\text{B}}(1-\phi)} \tag{3.130}$$

As mentioned earlier, reality in polymer mixtures often differs from the Flory-Higgins model, in that a ϕ-dependent χ is required. Then we have to write for the equation of the spinodal

$$\frac{1}{N_{\text{A}}\phi} + \frac{1}{N_{\text{B}}(1-\phi)} = -\frac{\partial^2}{\partial \phi^2}(\chi(\phi)\phi(1-\phi)) := 2\Lambda \tag{3.131}$$

Here we have introduced another function, Λ, which is related to χ by

$$\Lambda = \chi - (1-2\phi)\frac{\partial \chi}{\partial \phi} - \frac{1}{2}\phi(1-\phi)\frac{\partial^2 \chi}{\partial \phi^2} \tag{3.132}$$

We see that the situation now has become more involved. As we shall learn in the next section, from an experimental determination of the spinodal, Λ follows, rather than χ.

It might appear at first that the spinodal marks a sharp transition between two growth regimes but this is not true. Activation barriers for the nucleation are continuously lowered when approaching the spinodal and thus may loose their effectiveness already prior to the final arrival. As a consequence, the transition from the nucleation and growth regime to the region of spinodal decompositions is actually diffuse and there is no way to employ it for an accurate determination of the spinodal. There is, however, another effect for which the spinodal is significant and well-defined: The distance from the spinodal controls the concentration fluctuations in the homogeneous phase. The next section deals in detail with this interesting relationship.

3.2.3 Critical Fluctuations and Spinodal Decomposition

The critical point of a polymer mixture, as given by the critical temperature T_c jointly with the critical composition ϕ_c, is the locus of a second-order phase transition. Second order phase transitions have general properties which are found independent of the particular system, it may be a ferromagnetic or ferroelectric solid near its Curie temperature, a gas near to the critical point, or, as in our case, a mixture. As one general law, the approach of a critical point is always accompanied by a strong increase of the local fluctuations of the order parameter associated with the transition. For our mixture, the order parameter is given by the composition, as specified for example by the volume fraction of A-chains. So far, we have been concerned with the overall concentrations of the A- and B-chains in the sample only. On microscopic scales, concentrations are not uniform but show fluctuations about the mean value, owing to the action of random thermal forces. According to the general scenario of critical phase transitions, one expects a steep growth of these fluctuations on approaching T_c.

The most convenient technique for a verification are scattering experiments, as these probe the fluctuations directly. Figure 3.26 presents, as an example, results obtained by neutron scattering for a mixture of (deuterated) polystyrene and poly(vinylmethylether). As mentioned earlier, this system shows an upper miscibility gap (Fig. 3.20). Measurements were carried out for a mixture with the critical composition at a series of temperatures in the one-phase region. The figure depicts the reciprocals of the scattering intensities in plots versus q^2. We notice that approaching the critical point indeed leads to an overall increase of the intensities, with the strongest growth being

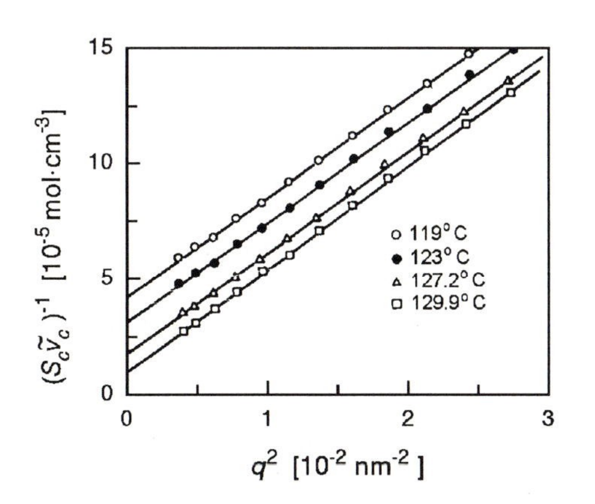

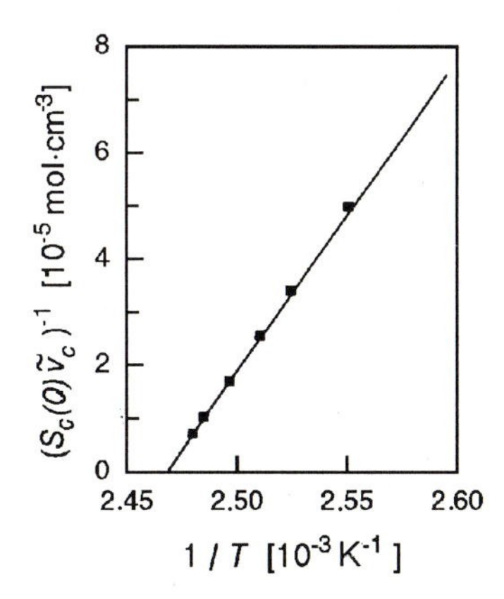

Fig. 3.26. Results of neutron scattering experiments on a (0.13 : 0.87) - mixture of d-PS ($M = 3.8 \cdot 10^5$) and PVME ($M = 6.4 \cdot 10^4$). S_c denotes the scattering function Eq. (3.153) referring to structure units with a molar volume $\tilde{v}_c$. Intensities increase on approaching the critical point (*left*). Extrapolation of $S(q \to 0)$ to the point of divergence yields the critical temperature (*right*). Data from Schwahn et al. [18]

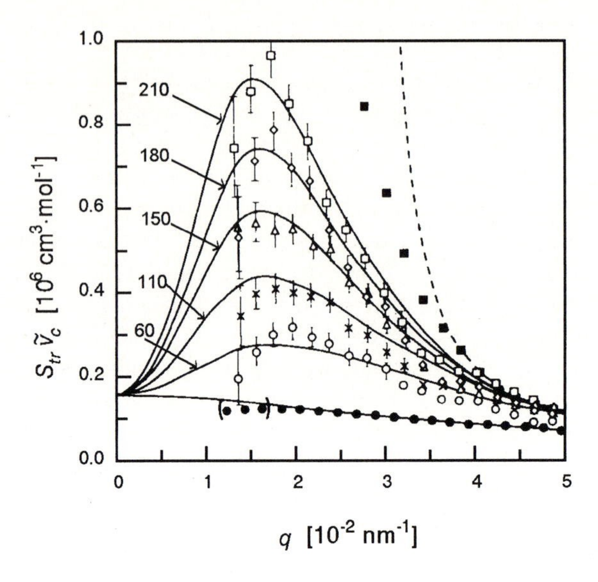

Fig. 3.27. Same system as in Fig. 3.26. Transient scattering functions $S_{tr}(q,t)$ measured after a rapid temperature change from $T_i = 130\,°C$ (one-phase region) to $T_f = 134.1\,°C$ (two-phase region). Times of evolution are indicated (in seconds)[18]

found for the scattering in forward direction. The temperature dependence of the forward scattering is shown on the right-hand side, in a plot of $S^{-1}(q \to 0)$ against $1/T$. Data indicate a divergence, and its location determines the critical temperature. Here we find $T_c = 131.8\,°C$.

When the phase boundary is crossed through the critical point, then a spinodal decomposition is initiated, and it can be followed by time dependent scattering experiments. Figure 3.27 shows the evolution of the scattering function during the first stages, subsequent to a rapid change from an initial temperature T_i two degrees below T_c, to $T_f = 134.1\,°C$, located 2.3 °C above. Beginning at zero time with the equilibrium structure factor associated with the temperature T_i in the homogeneous phase, a peak emerges and grows in intensity.

Figure 3.28 presents, as a second example, a further experiment on mixtures of polystyrene and poly(vinylmethylether), now carried out by time dependent light scattering experiments (this sample had a lower critical temperature, probabely due to differencies in behavior between normal and deuterated polystyrene). Experiments encompass a larger time range and probe the scattering at the small q's reached when using light. Again one observes the development of a peak, and it also stays at first at a constant position. Here, we can see that during the later stages it shifts to lower scattering angles.

This appearance of a peak which grows in intensity, initially at a fixed position and then shifting to lower scattering angles, can in fact be considered

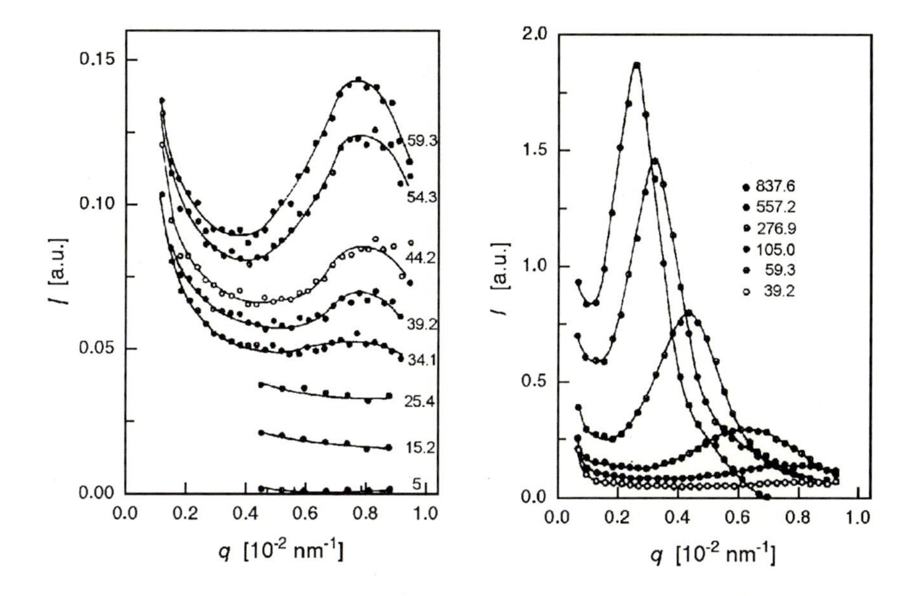

Fig. 3.28. Time dependent light scattering experiments, conducted on a (0.3:0.7)-mixture of PS ($M = 1.5 \cdot 10^5$) and PVME ($M = 4.6 \cdot 10^4$) subsequent to a rapid transfer from a temperature in the region of homogeneous states to the temperature $T_f = 101\,^\circ$C located in the two-phase region. Numbers give the time passed after the jump (in seconds). Data from Hashimoto et al. [19]

as indicative of a spinodal decomposition. One can say that the peak reflects the occurrence of wave-like modulations of the local blend composition, with a dominance of particular wave lengths. Furthermore, the intensity increase indicates a continuous amplitude growth. This, indeed, is exactly the process sketched at the bottom of Fig. 3.24.

All these findings, the steep growth of the concentration fluctuations in the homogeneous phase near to the critical point, as well as the kinetics of spinodal decomposition with its strong preference for certain wave-lengths, can be treated in a common theory. It was originally developed by Cahn, Hilliard, and Cook, in order to treat unmixing phenomena in metallic alloys and anorganic glasses, and then adjusted by de Gennes and Binder to the polymer case. Polymers actually represent systems which exhibit these phenomena in a particularly clear form and thus allow a verification of the theories. In the three subsections to follow, which concern the critical scattering as observed in the homogeneous phase, the initial stages of spinodal decomposition and the late stage kinetics, some main results will be presented.

Critical Scattering

We consider here the concentration fluctuations in the homogeneous phase and also the manner, in which these are reflected in measured scattering functions.

How can one deal with the fluctuations? At first view it might appear that the Flory-Huggins treatment does not give any help. Accounting for all microscopic states, the Flory-Huggins expression for the Gibbs free energy includes also the overall effect of all the concentration fluctuations in a mixture. The overall effect, however, is not our point of concern. We wish to grasp a *single* fluctuation state, as given by a certain distribution of the A's specified by a function $\phi(\boldsymbol{r})$, and determine its statistical weight. What we need for this purpose is a knowledge about a *constrained* Gibbs free energy, namely that associated with a single fluctuation state only.

To solve our problem we use a trick which was originally employed by Kadanoff in an analysis of the critical behavior of ferromagnets. Envisage a division of the sample volume in a large number of cubic 'blocks', with volumes v_{B} which, although being very small, still allow the use of thermodynamic laws; block sizes in the order of 10-100 nm^3 seem appropriate for this purpose. For this grained system, description of a certain fluctuation state is accomplished by giving the concentrations ϕ_i of all blocks. The (constrained) free energy of a thus characterized fluctuation state can be written down, proceeding in three steps. As we may apply the Flory-Huggins equation for each block separately, we first write a sum

$$\mathcal{G}(\{\phi_i\}) = \sum_i v_{\mathrm{B}} g(\phi_i) \tag{3.133}$$

Here, g stands for the free energy density of the mixture

$$g(\phi) = \phi g_{\mathrm{A}} + (1-\phi) g_{\mathrm{B}} + \tilde{R}T \left[\frac{\phi}{\tilde{v}_{\mathrm{A}}} \ln \phi + \frac{(1-\phi)}{\tilde{v}_{\mathrm{B}}} \ln(1-\phi) + \frac{\chi}{\tilde{v}_{\mathrm{c}}} \phi(1-\phi) \right] \tag{3.134}$$

g_{A} and g_{B} denoting the free energy densities of the one-component phases. Neighboring blocks, being in close contact, interact with each other across the interfaces and we have to inquire about the related interfacial energy. We know that it must vanish for equal concentrations and increase with the concentration difference, independent of the direction of change. The simplest expression with such properties is the quadratic term

$$\beta(\phi_i - \phi_j)^2$$

where ϕ_i, ϕ_j are the concentrations in the adjacent blocks. It includes a coefficient β which determines the strength of the interaction. We add this term to the first sum and write

$$\mathcal{G}(\{\phi_i\}) = \sum_i v_{\mathrm{B}} g(\phi_i) + \sum_{ij} \beta(\phi_i - \phi_j)^2 \tag{3.135}$$

Finally, replacing the summation by an integral we obtain

$$\mathcal{G}(\phi(\boldsymbol{r})) = \int \left(g(\phi(\boldsymbol{r})) + \beta'(\nabla\phi)^2 \right) \mathrm{d}^3\boldsymbol{r} \qquad (3.136)$$

with $\beta' := \beta v_{\mathrm{B}}^{-1/3}$. With this result we have solved our problem. Equation (3.136) describes in an approximate, empirical manner the free energy to be attributed to a given fluctuation state $\phi(\boldsymbol{r})$. It is known in the literature as 'Ginzburg-Landau functional' and is widely applied in treatments of various kinds of fluctuations.

The equation can be further simplified, if a linearization approximation is used. Clearly the state with a uniform concentration,

$$\phi(\boldsymbol{r}) = \mathrm{const} := \phi$$

has the lowest free energy, $\mathcal{G}_{\mathrm{min}}$. For considering the change in the Gibbs free energy

$$\delta\mathcal{G} := \mathcal{G} - \mathcal{G}_{\mathrm{min}}$$

as it results from a fluctuation

$$\delta\phi(\boldsymbol{r}) := \phi(\boldsymbol{r}) - \phi$$

we may use a series expansion of $g(\delta\phi)$ up to the second order

$$\delta\mathcal{G} = \int \left(\delta g(\delta\phi(\boldsymbol{r})) + \beta'(\nabla\delta\phi)^2 \right) \mathrm{d}^3\boldsymbol{r} \qquad (3.137)$$

$$= \frac{\partial g}{\partial\phi} \int \delta\phi \mathrm{d}^3\boldsymbol{r} + \frac{1}{2}\frac{\partial^2 g}{\partial\phi^2} \int (\delta\phi)^2 \mathrm{d}^3\boldsymbol{r} + \beta' \int (\nabla\delta\phi)^2 \mathrm{d}^3\boldsymbol{r} \qquad (3.138)$$

Conservation of the masses of the two species implies

$$\int \delta\phi \mathrm{d}^3\boldsymbol{r} = 0 \qquad (3.139)$$

and we only have to calculate the second derivative of g. This leads to

$$\delta\mathcal{G} = \frac{\tilde{R}T}{2}\left(\frac{1}{\tilde{v}_{\mathrm{A}}\phi} + \frac{1}{\tilde{v}_{\mathrm{B}}(1-\phi)} - \frac{2\chi}{\tilde{v}_{\mathrm{c}}} \right) \int (\delta\phi)^2 \mathrm{d}^3\boldsymbol{r} + \beta' \int (\nabla\delta\phi)^2 \mathrm{d}^3\boldsymbol{r} \qquad (3.140)$$

This is a useful result. It relates the Gibbs free energy of a given fluctuation state to two parameters only, namely the integral or mean values of $(\delta\phi)^2$ and $(\nabla\delta\phi)^2$.

Let us now turn to scattering experiments. They may generally be regarded as carrying out a Fourier analysis, in our case a Fourier-analysis of the concentration fluctuations in the mixture. We therefore represent $\delta\phi(\boldsymbol{r})$ as a sum of wave-like modulations with amplitudes $\phi_{\boldsymbol{k}}^*$

$$\delta\phi(\boldsymbol{r}) = V^{-1/2} \sum_{\boldsymbol{k}} \exp \mathrm{i}\boldsymbol{k}\boldsymbol{r} \cdot \phi_{\boldsymbol{k}}^* \qquad (3.141)$$

For a finite sample volume V, the sum includes a sequence of discrete values of $\boldsymbol{k}$ which can be selected as usually, by adopting periodic boundary conditions. When writing a Fourier series in terms of exponential functions, the amplitudes $\phi_{\boldsymbol{k}}^*$ are complex numbers. We represent them as

$$\phi_{\boldsymbol{k}}^* = \phi_{\boldsymbol{k}} \exp \mathrm{i}\varphi_{\boldsymbol{k}}$$

where $\phi_{\boldsymbol{k}}$ and $\varphi_{\boldsymbol{k}}$ denote the modulus and the phase. Since $\delta\phi(\boldsymbol{r})$ is a real quantity, we have

$$\phi_{-\boldsymbol{k}}^* = \overline{\phi_{\boldsymbol{k}}^*} \tag{3.142}$$

and therefore

$$\phi_{\boldsymbol{k}} = \phi_{-\boldsymbol{k}} \tag{3.143}$$

When we introduce the Fourier series into the integral of Eq. (3.140), we obtain

$$\int (\delta\phi)^2 \mathrm{d}^3\boldsymbol{r} = V^{-1} \sum_{\boldsymbol{k},\boldsymbol{k}'} \phi_{\boldsymbol{k}}^* \phi_{\boldsymbol{k}'}^* \int \exp \mathrm{i}(\boldsymbol{k}\boldsymbol{r} + \boldsymbol{k}'\boldsymbol{r}) \mathrm{d}^3\boldsymbol{r} \tag{3.144}$$

Since

$$\int \exp \mathrm{i}(\boldsymbol{k} + \boldsymbol{k}')\boldsymbol{r} \ \mathrm{d}^3\boldsymbol{r} = V\delta_{\boldsymbol{k},-\boldsymbol{k}'} \tag{3.145}$$

we can write

$$\int (\delta\phi)^2 \mathrm{d}^3\boldsymbol{r} = \sum_{\boldsymbol{k}} \phi_{\boldsymbol{k}}^* \phi_{-\boldsymbol{k}}^* = \sum_{\boldsymbol{k}} \phi_{\boldsymbol{k}}^2 \tag{3.146}$$

For the gradient term we obtain in similar manner

$$\int (\nabla\delta\phi)^2 \mathrm{d}^3\boldsymbol{r} = \sum_{\boldsymbol{k}} k^2 \phi_{\boldsymbol{k}}^* \phi_{-\boldsymbol{k}}^* = \sum_{\boldsymbol{k}} k^2 \phi_{\boldsymbol{k}}^2 \tag{3.147}$$

with $k := |\boldsymbol{k}|$. Introducing Eqs. (3.146), (3.147) into Eq. (3.140) we obtain

$$\delta\mathcal{G} = \frac{\tilde{R}T}{2} \sum_{\boldsymbol{k}} \left(\frac{1}{\tilde{v}_{\mathrm{A}}\phi} + \frac{1}{\tilde{v}_{\mathrm{B}}(1-\phi)} - \frac{2\chi}{\tilde{v}_{\mathrm{c}}} + \beta'' k^2 \right) \phi_{\boldsymbol{k}}^2 \tag{3.148}$$

The coupling constant $\beta'' := 2\beta'(\tilde{R}T)^{-1}$ is unknown at this point of the discussion, but later on we shall learn more about it.

As we can see, the Fourier-transformation leads to a decoupling. Whereas, in direct space, we have a short-ranged coupling between fluctuations at different positions as expressed by the gradient term in Eq. (3.140), different Fourier-amplitudes $\phi_{\boldsymbol{k}}$ contribute separately to $\delta\mathcal{G}$, thus being perfectly free. Hence, the wave-like modulations of the concentration may be regarded as the basic 'modes' of the system, which can be excited independently from each other. The general dynamics of the concentration fluctuations in a polymer mixture is described as a superposition of all these modes, each mode being characterized by a certain wave-vector.

Having an expression for the free energy increase associated with the excitation of the mode $\boldsymbol{k}$, one can calculate its mean-squared amplitude in thermal equilibrium $\langle \phi_{\boldsymbol{k}}^2 \rangle$. It follows from Boltzmann statistics as

$$\langle \phi_{\boldsymbol{k}}^2 \rangle = \int \phi_{\boldsymbol{k}}^2 \exp -\frac{\delta \mathcal{G}(\phi_{\boldsymbol{k}})}{kT} \delta\phi_{\boldsymbol{k}} \Big/ \int \exp -\frac{\delta \mathcal{G}(\phi_{\boldsymbol{k}})}{kT} \delta\phi_{\boldsymbol{k}} \tag{3.149}$$

Evaluation of the integrals yields

$$\langle \phi_{\boldsymbol{k}}^2 \rangle = \mathrm{N_L}^{-1} \left(\frac{1}{\tilde{v}_{\mathrm{A}}\phi} + \frac{1}{\tilde{v}_{\mathrm{B}}(1-\phi)} - \frac{2\chi}{\tilde{v}_{\mathrm{c}}} + \beta'' k^2 \right)^{-1} \tag{3.150}$$

The result includes a singularity, coming up if the denominator equals zero. It tells us that finite concentration fluctuations can exist only under the condition

$$\frac{1}{\tilde{v}_{\mathrm{A}}\phi} + \frac{1}{\tilde{v}_{\mathrm{B}}(1-\phi)} - \frac{2\chi}{\tilde{v}_{\mathrm{c}}} + \beta'' k^2 > 0 \tag{3.151}$$

Regarding Eq. (3.130), this is equivalent to

$$\chi_{\mathrm{sp}} - \chi + \frac{\tilde{v}_{\mathrm{c}}}{2}\beta'' k^2 > 0 \tag{3.152}$$

As we can see, in the limit $k \to 0$, the stability criterion of the Flory-Huggins theory, $\chi < \chi_{\mathrm{sp}}$, is recovered. For finite k's, the criterion becomes modified.

Next, we relate the calculated fluctuations amplitudes to the scattering function obtained in X-ray or light scattering experiments. Discussions are usually based on a scattering function which refers to the reference volume common for both species, v_{c}, or in the language of the lattice models, on a scattering function which refers to the cells of the lattice. It is denoted S_{c} and defined as

$$S_{\mathrm{c}}(\boldsymbol{q}) := \frac{1}{\mathcal{N}_{\mathrm{c}}} \langle | C(\boldsymbol{q}) |^2 \rangle \tag{3.153}$$

$C(\boldsymbol{q})$ is the scattering amplitude, and $\mathcal{N}_{\mathrm{c}}$ stands for the total number of A- and B-units in the sample. The scattering function $S_{\mathrm{c}}(\boldsymbol{q})$ can be directly related to the mean-squared amplitudes of the fluctuations $\langle \phi_{\boldsymbol{k}}^2 \rangle$. As is shown in Sect. A.4.1 in the Appendix, the relation is

$$S_{\mathrm{c}}(\boldsymbol{q}) = \frac{1}{v_{\mathrm{c}}} \langle \phi_{\boldsymbol{k}=\boldsymbol{q}}^2 \rangle \tag{3.154}$$

Making use of Eq. (3.150), we obtain the scattering function of a polymer mixture. It is given by the following equation

$$S_{\mathrm{c}}(\boldsymbol{q}) = \left(\frac{1}{N_{\mathrm{A}}\phi} + \frac{1}{N_{\mathrm{B}}(1-\phi)} - 2\chi + \tilde{v}_{\mathrm{c}}\beta'' q^2 \right)^{-1} \tag{3.155}$$

The result allows a reconsideration of the open question about the functional form of the coupling coefficient β''. Insight results from a view on the limiting properties of the scattering function for low concentrations of the polymers A and B respectively. For the discussion, it is advantageous, to change to the reciprocal of the scattering function, since this leads to a separation of the contributions of the A's and B's

$$\frac{1}{S_{\mathrm{c}}} = \frac{1}{N_{\mathrm{A}}\phi} + \frac{1}{N_{\mathrm{B}}(1-\phi)} - 2\chi + \tilde{v}_{\mathrm{c}}\beta'' q^2 \tag{3.156}$$

Let us first look at the limit $\phi \to 0$. When A is the minority species, present only in low concentration, our equation gives

$$\frac{1}{S_{\mathrm{c}}} \to \frac{1}{\phi N_{\mathrm{A}}} + \tilde{v}_{\mathrm{c}}\beta'' q^2 \tag{3.157}$$

On the other hand, for this case the exact form of S_{c} is known. Since in melts polymer chains are ideal, S_{c} is given by the Debye-structure function (Eqs. (2.60) and (2.61)), multiplied by the volume fraction ϕ in order to account for the dilution

$$S_{\mathrm{c}} = \phi N_{\mathrm{A}} S_{\mathrm{D}}(R_{\mathrm{A}}^2 q^2) \tag{3.158}$$

Using the series expansion Eq. (2.63) we may write

$$\frac{1}{S_{\mathrm{c}}} \approx \frac{1}{N_{\mathrm{A}}\phi}\left(1 + \frac{R_{\mathrm{A}}^2 q^2}{18}\right) \tag{3.159}$$

Equivalently, when choosing polymer B as the diluted species, our equation leads to

$$\frac{1}{S_{\mathrm{c}}} \to \frac{1}{(1-\phi)N_{\mathrm{B}}} + \tilde{v}_{\mathrm{c}}\beta'' q^2 \tag{3.160}$$

whereas the complete expression is

$$\frac{1}{S_{\mathrm{c}}} = \frac{1}{(1-\phi)N_{\mathrm{B}} S_{\mathrm{D}}(R_{\mathrm{B}}^2 q^2)} \tag{3.161}$$

$$\approx \frac{1}{N_{\mathrm{B}}(1-\phi)}\left(1 + \frac{R_{\mathrm{B}}^2 q^2}{18}\right) \tag{3.162}$$

Comparison of Eq. (3.157) with Eq. (3.159) and Eq. (3.160) with Eq. (3.162) gives us an explicit expression for the coupling constant β'': Equations agree if we write

$$\beta'' = \frac{1}{\tilde{v}_{\mathrm{c}}}\frac{R_{\mathrm{A}}^2}{18 N_{\mathrm{A}}\phi} + \frac{1}{\tilde{v}_{\mathrm{c}}}\frac{R_{\mathrm{B}}^2}{18 N_{\mathrm{B}}(1-\phi)} \tag{3.163}$$

If we now take β'' and insert it in Eq. (3.156) we obtain as our final result

$$\frac{1}{S_{\mathrm{c}}} = \frac{1}{\phi N_{\mathrm{A}}}\left(1 + \frac{R_{\mathrm{A}}^2 q^2}{18}\right) + \frac{1}{(1-\phi)N_{\mathrm{B}}}\left(1 + \frac{R_{\mathrm{B}}^2 q^2}{18}\right) - 2\chi \tag{3.164}$$

Is it really correct? Considering the simple Ginzburg-Landau functional, Eq. (3.136), which we chose as our starting point, this is a legitimate question and indeed, the comparisons with the known limiting behaviors for $\phi \to 0$ and $(1-\phi) \to 0$ point at limitations. Full agreement in these limits is only reached for $R_{\mathrm{A}}^2 q^2 \ll 1$, $R_{\mathrm{B}}^2 q^2 \ll 1$.

One might suspect that these limitations could be removed by an obvious extension of Eq. (3.164). It is possible to construct a scattering function which is correct for the known limits without being restricted to low q's. Evidently this is accomplished by the equation

$$\frac{1}{S_{\mathrm{c}}} = \frac{1}{\phi N_{\mathrm{A}} S_{\mathrm{D}}(R_{\mathrm{A}}^2 q^2)} + \frac{1}{(1-\phi) N_{\mathrm{B}} S_{\mathrm{D}}(R_{\mathrm{B}}^2 q^2)} - 2\chi \tag{3.165}$$

Equation (3.165) in fact represents the correct result. It can be obtained with the aid of a theoretical method superior to the Ginzburg-Landau treatment and known as the 'random phase approximation'. The interested reader finds the derivation in the Appendix, Sect. A.4.1.

Use of Eq. (3.165) enables us to make an evaluation of scattering experiments, in particular

- a determination of the Flory-Huggins parameter χ and the coil sizes $R_{\mathrm{A}}, R_{\mathrm{B}}$
- a determination of the spinodal, based on the temperature dependence of the concentration fluctuations in the homogeneous phase.

We reduce the discussion again to the case of symmetric polymer mixtures with

$$N_{\mathrm{A}} = N_{\mathrm{B}} = N$$

and apply Eq. (3.164), now in the form

$$\begin{aligned} \frac{1}{S_{\mathrm{c}}} &= \frac{1}{N}\frac{1}{\phi(1-\phi)} + \frac{q^2}{18N}\left(\frac{R_{\mathrm{A}}^2}{\phi} + \frac{R_{\mathrm{B}}^2}{1-\phi}\right) - 2\chi \\ &= \frac{1}{N}\frac{1}{\phi(1-\phi)} + \frac{1}{N\phi(1-\phi)}\frac{q^2}{18}R_\phi^2 - 2\chi \end{aligned} \tag{3.166}$$

In the last equation we have introduced a ϕ-dependent average over the coil radii, R_ϕ, defined as

$$R_\phi^2 := (1-\phi)R_{\mathrm{A}}^2 + \phi R_{\mathrm{B}}^2 \tag{3.167}$$

Applying Eq. (3.128), we may also write

$$\frac{1}{S_{\mathrm{c}}} = 2(\chi_{\mathrm{sp}} - \chi) + 2\chi_{\mathrm{sp}}\frac{R_\phi^2}{18}q^2 \tag{3.168}$$

Equation (3.168) enables us to make a determination of χ and R_ϕ for a (symmetric) polymer mixture. Figure 3.29 presents, as an example, results of small angle X-ray scattering experiments, carried out on mixtures of polystyrene and

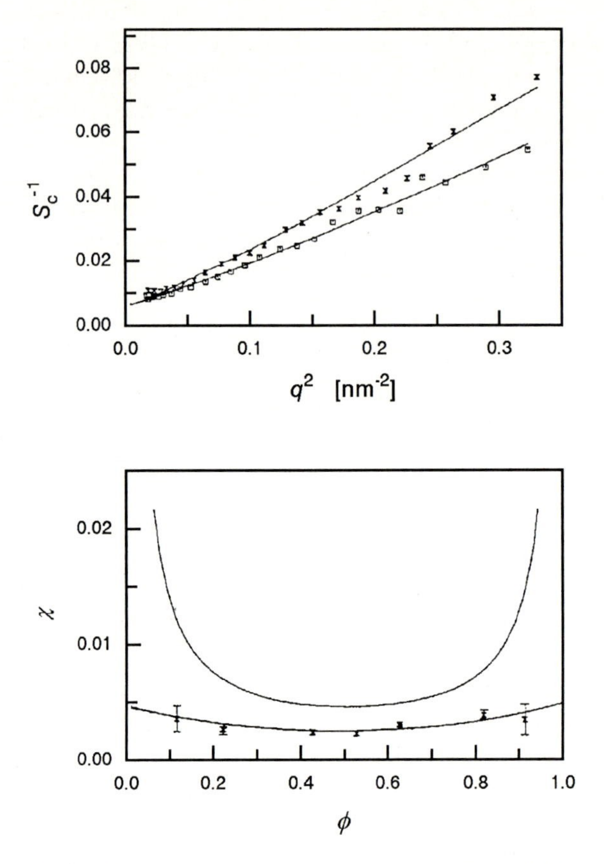

Fig. 3.29. Results of SAXS experiments on mixtures of PS and PBr_xS, both species having equal degrees of polymerization (N=430). Scattering functions for ϕ (PS) = 0.42 and 0.62 (*top*) and derived function $\chi(\phi)$ (*bottom*). The upper curve in the lower figure represents χ_{sp} [20]

partially brominated polystyrene (PBr_xS, with x=0.17). Data are represented by a plot S_c^{-1} versus q^2, as suggested by Eq. (3.168). The difference in slopes indicates a change of R_ϕ with the composition, telling us that the coil sizes of polystyrene and the partially brominated polystyrene are different (analysis of the data yielded $R(\mathrm{PS}) = 32$ Å, $R(\mathrm{PBr_xS}) = 39$ Å). The bottom part of Fig. 3.29 presents the values derived for χ, together with χ_{sp} according to Eq. (3.128). Results show that χ is not a constant, although the changes are comparatively small. Strictly speaking, the measurement yields Λ rather than χ, but the difference seems negligible.

Understanding of the microscopic origin of the observed ϕ-dependence on theoretical grounds is difficult and this is a situation where computer simulations can be quite helpful. In fact, computations for a lattice model led

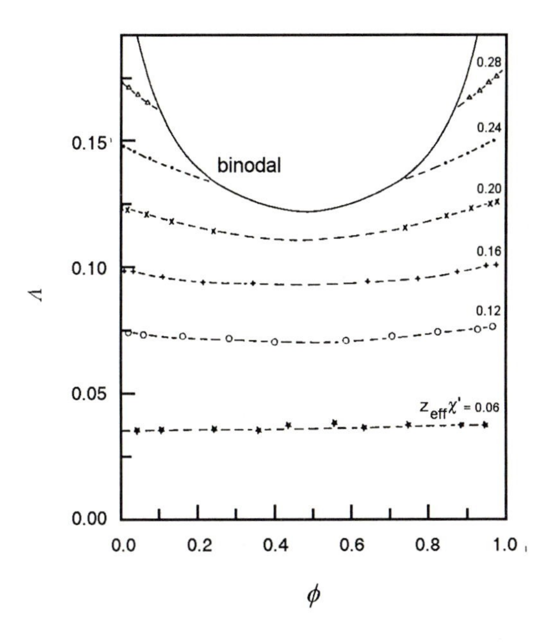

Fig. 3.30. Results obtained in a Monte-Carlo simulation for a lattice model of a polymer mixture ($N_A = N_B = 16$; simple cubic lattice, i.e. $z_{\text{eff}} = 4$; 80% of the lattice are occupied by the chains). Calculation of the function $\Lambda(\phi)$ for different values of $z_{\text{eff}}\chi'$ ($\phi := \phi_A/(\phi_A + \phi_B)$ is the relative concentration of A-chains). Calculation by Sariban and Binder [21]

to qualitatively similar results, as is demonstrated by the curves depicted in Fig. 3.30: They all exhibit the slight curvature of the experimental curves. A second result of the simulations is particularly noteworthy. Computer simulations can be used for general checks of the assumptions of the Flory-Huggins model which cannot be accomplished in an easy manner by analytical considerations. In the example, computations were carried out for a simple cubic lattice. In order to reduce the 'equilibration time' in the computer, as given by the number of steps necessary to reach the equilibrium when starting from an arbitrary configuration, 20% of the lattice sites were left empty. Calculations were carried out for different values of χ'. We discussed the predictions of the Flory-Huggins model and expect from it, for a dense system, the relation Eq. (3.83)

$$(\Lambda =)\chi = z_{\text{eff}}\chi' \tag{3.169}$$

The simulation yielded consistently lower values

$$\Lambda < z_{\text{eff}}\chi' \tag{3.170}$$

There is first a trivial reason, given by the presence of the vacancies which reduce the interaction energy, but this contributes only a factor of about 0.8.

The observed difference is definitely larger and this points at deficiencies of the mean-field approximation in the description of this model system. Obviously the number of AB-contacts is smaller than expected under the assumption of a random distribution of the chains. Closer inspection of the data indicated an enhanced number of intramolecular contacts and also some intermolecular short-range order. Hence, the simulation tells us, as a general kind of warning, that one should be careful in interpreting measured χ-parameters. There can always be perturbing effects. Shortcomings of the Flory-Huggins treatment show up in particular, if the molar masses are low. Some effects emerge only for such systems, an important one being the short-range ordering mentioned above. Short-range order effects can only arise if the distances, over which the concentration fluctuations are correlated, are larger than, or similar to the chain size. Conversely, for chains with sufficiently high degrees of polymerization, short-range order effects are ruled out; chains actually average over all local concentration fluctuations and experience the mean value of the contact energies only. In our case, both the experiment and the simulation refer to moderate or even low degrees of polymerization and the qualitative comparison appears justified.

Next, let us return once again to Fig. 3.26, showing temperature dependent measurements on mixtures of deutero-polystyrene and poly(vinylmethylether). Now, we can recognize the theoretical basis of the chosen representation, S^{-1} versus q^2, namely as corresponding to Eq. (3.165). A change in temperature with the resulting change in χ leads to a parallel shift of the curve $S^{-1}(q^2)$. The right part of the figure shows the limiting values $S^{-1}(q \to 0)$ as a function of temperature, expressing directly the T-dependence of χ according to

$$\frac{1}{S_{\mathrm{c}}(q \to 0)\tilde{v}_{\mathrm{c}}} = \frac{2(\chi_{\mathrm{sp}} - \chi)}{\tilde{v}_{\mathrm{c}}} \tag{3.171}$$

The observed straight line indicates for $\chi(T)$ a linear dependence

$$\chi_{\mathrm{sp}} - \chi \sim T^{-1} - T_{\mathrm{sp}}^{-1} \tag{3.172}$$

If we wish to account for both upper and lower miscibility gaps, we may write in linear approximation

$$\chi_{\mathrm{sp}} - \chi \sim |T - T_{\mathrm{sp}}| \tag{3.173}$$

and thus expect a temperature dependence

$$S^{-1}(0) \sim |T - T_{\mathrm{sp}}| \tag{3.174}$$

The data in Fig. 3.26 were obtained for a mixture with the critical concentration, and here the extrapolation to the point where $S_{\mathrm{c}}(0)$ diverges yielded the critical temperature. We can now also see the procedure to be used for a

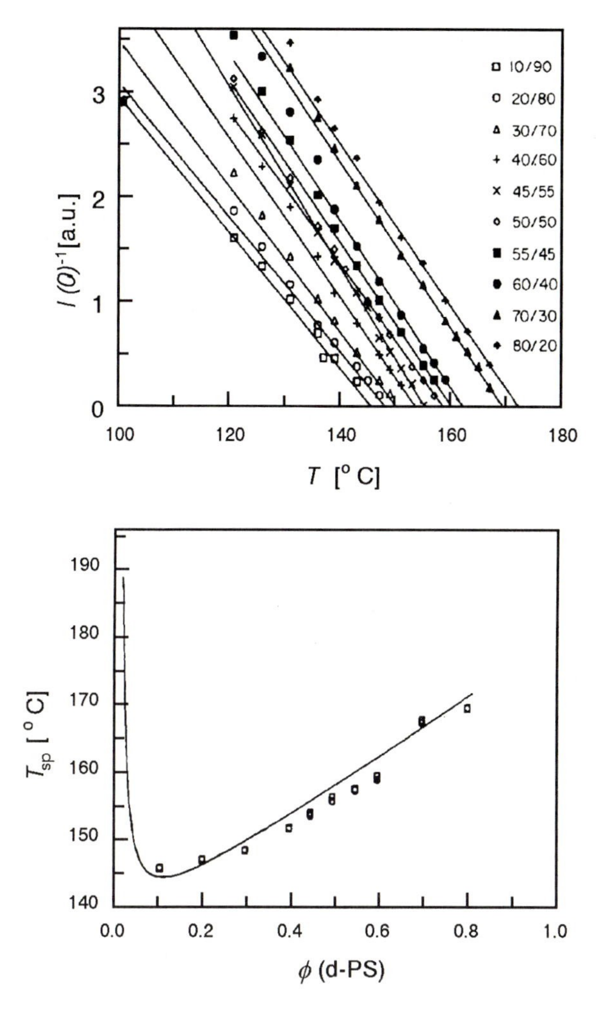

Fig. 3.31. Spinodal of a mixture of d-PS ($M = 5.93 \cdot 10^5$) and PVME ($M = 1.1 \cdot 10^6$) (*bottom*), as derived from the temperature dependence of neutron scattering intensities in forward direction (*top*). Data from Han et al. [22]

determination of the complete spinodal. One has to carry out temperature dependent measurements for a series of mixtures which cover the whole range of compositions. Extrapolations on the basis of Eq. (3.174), i.e. a continuation of the temperature dependent $S_c^{-1}(0)$ down to zero, yields the spinodal $T_{sp}(\phi)$, as represented in a (ϕ, T)-phase diagram. Figure 3.31 shows, as an example, a respective set of data which was obtained in another neutron scattering study on mixtures of deuterated polystyrene and poly(vinylmethylether). The linear relation Eq. (3.174) appears verfied, and the corresponding extrapolations then yield the spinodal depicted in the lower half.

The concentration fluctuations in a mixture are spatially correlated, with the degree of coupling decreasing with the distance. We may inquire about the 'correlation length' of the fluctuations, i.e. the maximum distance over which correlations remain essential. The answer follows from the scattering function. Rewriting Eq. (3.168), we obtain for the the small angle range a curve with Lorentzian shape

$$S_c = \frac{1}{2(\chi_{sp} - \chi)} \cdot \frac{1}{1 + \xi_\phi^2 q^2} \tag{3.175}$$

The parameter ξ_ϕ^2 is given by

$$\xi_\phi^2 = \frac{\chi_{sp} R_\phi^2}{18(\chi_{sp} - \chi)} \tag{3.176}$$

ξ_ϕ represents the correlation length, as is revealed by a Fourier-transformation of S_c. It yields the correlation function for the concentration fluctuations in direct space

$$\langle \delta\phi(0)\delta\phi(\boldsymbol{r})\rangle \sim \int \exp -\mathrm{i}\boldsymbol{qr} \cdot S_c(\boldsymbol{q})\mathrm{d}^3\boldsymbol{q} \tag{3.177}$$

(if an explanation is necessary, look at the derivation of Eq. (A.17) in the Appendix). The evaluation is straightforward and leads to

$$\langle \delta\phi(0)\delta\phi(\boldsymbol{r})\rangle \sim \frac{1}{r} \exp -\frac{r}{\xi_\phi} \tag{3.178}$$

As we see, ξ_ϕ indeed describes the spatial extension of the correlations.

In all second order phase transitions, the correlation length of the fluctuations of the order parameter diverges at the critical point. We find this behavior also in our system, when making use of Eq. (3.174). For the temperature dependence of ξ_ϕ we obtain the power law

$$\xi_\phi \sim (\chi_{sp} - \chi)^{-1/2} \sim |T - T_{sp}|^{-1/2} \tag{3.179}$$

If an experiment is conducted for the critical composition ϕ_c, then one observes the divergence of ξ_ϕ. For concentrations different from ϕ_c, the increase of ξ_ϕ stops when the binodal is reached.

Early Stage Phase Separation Kinetics

After having crossed the spinodal, either through the critical point or somewhere else by a rapid quench which passes quickly through the nucleation and growth-range, unmixing sets in by the mechanism known as spinodal decomposition. Measurements like the ones presented in Figs. 3.27 and 3.28 allow detailed investigations. The experiments yield the time dependent 'transient scattering function' which we denote $S_{tr}(q, t)$.

Theory succeeded to derive an 'equation of motion' for $S_{\mathrm{tr}}(q,t)$ which can be used for an analysis of the kinetics of structure evolution in the early stages of development. It has the following form

$$\frac{\mathrm{d}S_{\mathrm{tr}}(q,t)}{\mathrm{d}t} = -\Gamma(q)(S_{\mathrm{tr}}(q,t) - S_{\mathrm{c}}(q)) \tag{3.180}$$

S_{c} is defined by Eq. (3.165) and $\Gamma(q)$ is a rate constant, determined by

$$\Gamma(q) = 2q^2\lambda(q)S_{\mathrm{c}}^{-1}(q) \tag{3.181}$$

λ is a function which relates to the single chain dynamics in the mixture.

A derivation of this equation lies outside our scope, so that we can only consider briefly its background and some implications. First of all, note that Eq. (3.180) has the typical form of a first order relaxation equation, as it is generally used to describe irreversible processes which bring a system from an initial non-equilibrium state back to equilibrium. Therefore, if rather than crossing the spinodal, the temperature jump is carried out within the one-phase region, causing a transition of the structure into a new state with higher or lower concentration fluctuations, then the applicability of the equation is unquestionable. Indeed, Eq. (3.180) is meant to cover this 'normal' case as well. S_{c} then represents the structure factor associated with the new equilibrium state. The different factors included in the equation for the relaxation rate Γ are all conceivable. A quadratic term in q always shows up for particle flows based on diffusive motions, and these have to take place if a concentration wave is to alter its amplitude. Its background is of a twofold nature and easily seen. Firstly, according to Fick's law, flow velocities are proportional to concentration gradients and thus proportional to q. Secondly, with increasing wavelength, particles have to go over correspondingly larger distances and this produces a second factor q. Both effects together give the characteristic q^2. The origin of the factor S_{c}^{-1} is revealed by a look at Eqs. (3.148), (3.156). Equation (3.148) is formally equivalent to the energy (u) - displacement (x) relation of a harmonic oscillator

$$u = \frac{1}{2}ax^2 \tag{3.182}$$

We therefore may also address the factor in Eq. (3.148) which corresponds to a as a 'stiffness coefficient', now related to the formation of a concentration wave. Interestingly enough, exactly this stiffness coefficient shows up again in Eq. (3.156) for S_{c}^{-1}, apart from a trivial factor $\tilde{R}T/\tilde{v}_{\mathrm{c}}$. As S_{c}^{-1} is determined by this factor only, it can replace the stiffness coefficient in equations. Clearly, the latter affects the relaxation rate and therefore has to be part of any equation for Γ. Since our system shows close similarities to an overdamped harmonic oscillator, both having the same equation of motion, we can also understand the linear dependence of Γ on S_{c}^{-1}. Hence in conclusion, for temperature jumps within the one-phase region, Eq. (3.180) looks perfectly reasonable. It may appear less obvious that its validity is maintained if temperature jumps transfer the system into the two-phase region so that spinodal decomposition sets in. One could argue that, in view of the continuous character of critical phase transitions, one could expect the same kinetic equations to hold on both sides of the phase boundary, but a direct proof is certainly necessary and is indeed provided by the theoretical treatments.

A change occurs in the meaning of S_c. For temperatures in the two-phase region, S_c can no longer be identified with an equilibrium structure function. Nevertheless, its definition by Eq. (3.165) is maintained. This implies that S_c shows negative values at low q's, being positive only for high q's. We are dealing here with a 'virtual structure function', which is not a measurable quantity but defined by an extrapolation procedure. In order to obtain S_c, one has to determine the temperature dependence of χ in the homogeneous phase, introduce it into Eq. (3.165) and use this equation also for temperatures in the two-phase region.

The specific character of the spinodal decomposition can now be understood as being a consequence of the peculiar q-dependence of the rate constant Γ. Figure 3.32 presents results of a calculation applying Eq. (3.181). For q's below a critical value q_c, S_c^{-1} and therefore $\Gamma(q)$ take on negative values. A negative value of Γ indicates an amplitude growth, instead of the usual decay. The main feature in the curve is the maximum in the growth rate, $-\Gamma$, at a certain value q_{max} somewhere in the range

$$0 < q_{max} < q_c$$

Structure evolution is controlled by the concentration waves with wave vectors around q_{max}. These constitute the dominant modes of structure formation and determine the length scale of the pattern during the early stages of development. Figure 3.32 also indicates the temperature dependence of q_{max} and the largest associated growth rate. We see that the approach of the spinodal in the two-phase region is accompanied by a decrease of q_{max}. Straightforward analysis shows that the decrease obeys the power law

$$q_{max} \sim (\chi_{sp} - \chi)^{1/2} \sim |T_{sp} - T|^{1/2} \tag{3.183}$$

Simultaneously, a slowing-down of the growth rate occurs according to

$$-\Gamma(q_{max}) \sim q_{max}^2 S_c^{-1}(q_{max}) \tag{3.184}$$

Employing Eq. (3.168) we obtain

$$-\Gamma(q_{max}) \sim q_{max}^2 \left((\chi_{sp} - \chi) + \chi_{sp} \frac{R_\phi^2}{18} q_{max}^2 \right) \sim |T_{sp} - T|^2 \tag{3.185}$$

This 'critical slowing-down' also shows up on the other side of the phase boundary, when for a critical mixture the critical temperature is approached

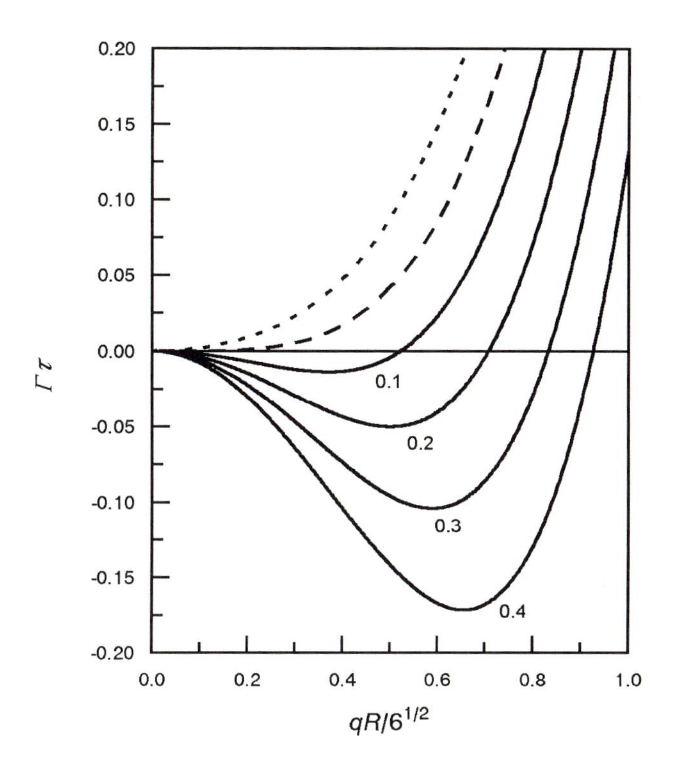

Fig. 3.32. Rate constants Γ which determine the time-dependent changes of the structure function of a symmetric polymer mixture ($R_{\mathrm{A}} = R_{\mathrm{B}} = R$) after a temperature jump from the homogeneous phase ($\chi < \chi_{\mathrm{sp}}$) into the two-phase region ($\chi > \chi_{\mathrm{sp}}$) (*continuous lines*). Curves correspond to different distances from the spinodal, $(\chi - \chi_{\mathrm{sp}})/\chi_{\mathrm{sp}} = 0.1 - 0.4$, and were obtained applying Eqs. (3.181),(3.165) ($\tau := NR^2\phi(1-\phi)/6\lambda(0)$). The *dashed line* gives the rate constants at the spinodal, the *dotted line* those associated with a temperature jump within the one-phase region to $\chi/\chi_{\mathrm{sp}} = 0.9$

from the one-phase region. The kinetic parameter of interest, to be used on both sides, is the 'collective diffusion coefficient', D_{coll}, defined as

$$D_{\mathrm{coll}} := \lim(q \to 0)\frac{\Gamma(q)}{2q^2} \tag{3.186}$$

and it is given by

$$D_{\mathrm{coll}} = \lambda(0)S_{\mathrm{c}}(0)^{-1} \tag{3.187}$$

The attribute 'collective' is used in order to distinguish this parameter from the 'self-diffusion coefficient' of the individual chains which relates to the single chain dynamics as expressed by λ only, and therefore shows no critical slowing-down. We see that D_{coll} takes on positive and negative values, crossing

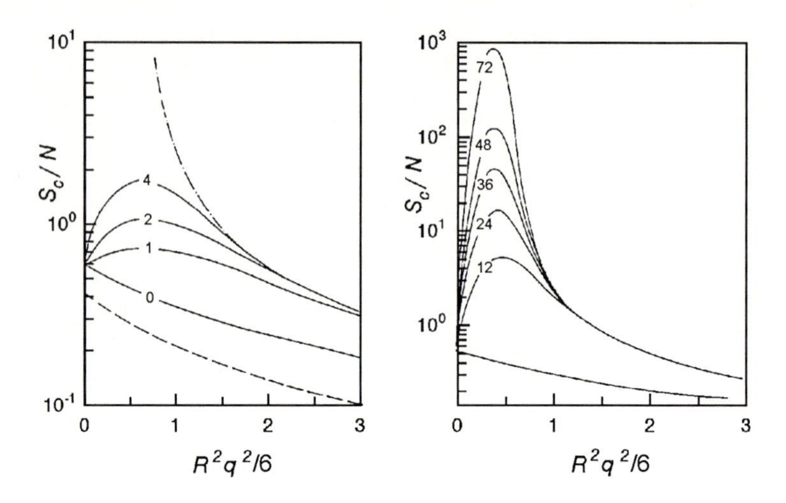

Fig. 3.33. Spinodal decomposition initiated by a jump from the one-phase region ($N\chi = 1$) to the two-phase region ($N\chi = 2.5$). Model calculation for a symmetric polymer blend ($N_A = N_B = N, R_A = R_B = R$) on the basis of Eqs. (3.189),(3.181),(3.165). The numbers represent units of time [23]

zero at the spinodal

$$D_{\text{coll}} \sim \chi_{\text{sp}} - \chi \sim \pm|T - T_{\text{sp}}| \tag{3.188}$$

Equation (3.180) can be solved exactly, and the solution is

$$S_{\text{tr}}(q,t) = S_{\text{c}}(q) + (S_{\text{tr}}(q,0) - S_{\text{c}}(q)) \cdot \exp -\Gamma(q)t \tag{3.189}$$

Figure 3.33 presents the results of model calculations performed on the basis of this equation. We find that a spinodal decomposition leads to an intensity increase for all $q's$, with a maximum at a certain q_{max}. Growing in intensity, the peak stays at a fixed position. In the long time limit we observe an exponential law

$$S_{\text{tr}}(q_{\text{max}}) \sim \exp -\Gamma t \tag{3.190}$$

As we can see, the model calculations reproduce the main features of the experimental observations during the initial stages of spinodal decompositions. In fact, the equations can be applied for a representation of experimental data and we refer here once again to the measurement presented in Fig. 3.27. Figure 3.34 shows on the left-hand side a plot according to

$$\ln \frac{S_{\text{tr}}(q,t) - S_{\text{c}}(q)}{S_{\text{tr}}(q,0) - S_{\text{c}}(q)} := \ln \frac{\Delta S(q,t)}{\Delta S(q,0)} = -\Gamma(q)t \tag{3.191}$$

The 'virtual structure function' $S_{\text{c}}(q)$ has been constructed by a linear ex-

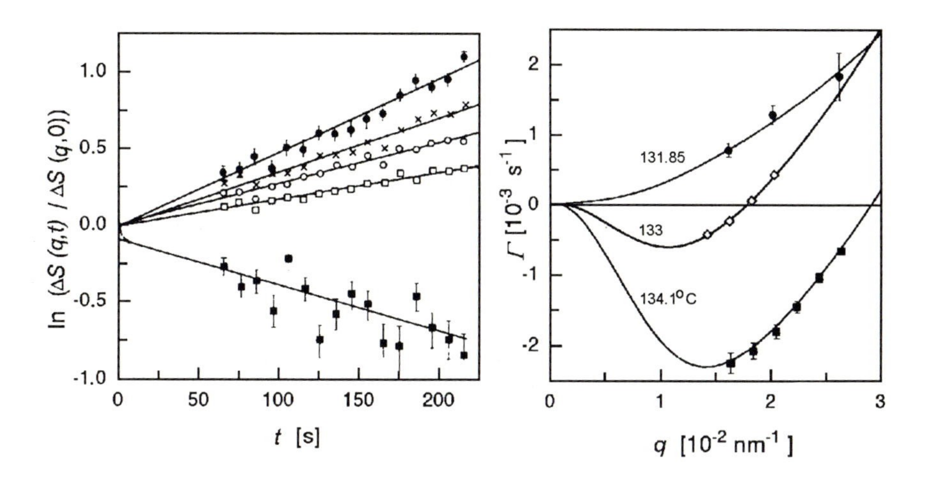

Fig. 3.34. Same system as in Figs. 3.26 and 3.27 [18]. Plot demonstrating an exponential time dependence of the transient scattering intensities at $T = 134.1\,^\circ\mathrm{C}$ for different $q's$ $(1.79 \cdot 10^{-2}; 2.2 \cdot 10^{-2}; 2.4 \cdot 10^{-2}; 2.6 \cdot 10^{-2}; 3.62 \cdot 10^{-2}\mathrm{nm}^{-1})$ (*left*). Derived rate constants $\Gamma(q)$ for growth ($\Gamma < 0$) or decay ($\Gamma > 0$), together with the results of equivalent experiments for $T = 133\,^\circ\mathrm{C}(> T_\mathrm{c} = 131.9\,^\circ\mathrm{C})$ and $T = 131.85\,^\circ\mathrm{C}(< T_\mathrm{c})$ (*right*)

trapolation of the equilibrium values in the homogeneous region shown in Fig. 3.26; the change from positive to negative values occurs for $q = 2.9 \cdot 10^{-2}$ nm. For all $q's$ we find an exponential time dependence in agreement with Eq. (3.189). The derived rate constants are given by the lowest curve on the right-hand side. One has negative values for $q < q_\mathrm{c}$ and in this range a maximum in the growth rate. The figure includes, in addition, the results of two other experiments, one conducted at $T = 133\,^\circ\mathrm{C}$, i.e. even closer to T_c, and the other at $T = 131.85\,^\circ\mathrm{C}$, which is in the one-phase region. One observes a shift of q_max towards zero for $T \to T_\mathrm{c}$, and on both sides of T_c a critical slowing down for $D_\mathrm{coll} \sim \mathrm{d}^2\Gamma(0)/\mathrm{d}^2 q$, in full agreement with the theoretical predictions.

Late Stage Kinetics

The described initial stages of spinodal decomposition constitute the entrance process, thereby setting the basic structure characteristics and the primary length- and time-scales. They represent a first part only, coming to an end when the concentration waves produce, in summary, variations $\delta\phi$ which already approach the concentrations of the two equilibrium phases. Then the exponential increase of the amplitudes cannot continue further and the kinetics must change. A first natural effect is a retardation of the growth rate, and a second is a shift of q_max towards lower values. An example for this generally

Fig. 3.35. Structure development during the late stages of spinodal decomposition observed for a PS/PBr_xS-(1:1) mixture. Micrographs were obtained during annealing at 200 °C ($< T_c = 220$ °C) for 1 min (*left*), 3 min (*center*) and 10 min (*right*) [17]

observed behavior was presented in Fig. 3.28 with the light scattering curves obtained for a polystyrene/poly(vinylmethylether) mixture. Theory has dealt with these first changes by a generalization of the linear equations valid for the initial stages and accounting for the saturation effects introduced by the bounds. Treatments are rather involved, and we cannot present them here. Interestingly enough, after this second period, there follows a third part where behavior becomes simpler again. This is the regime of the 'late stage kinetics' and, in this section, we will briefly describe some major observations.

The micrograph on the right of Fig. 3.21 was obtained during this late stage of structure evolution and represents an instructive example. The interconnected domains are set up by the two equilibrium phases. The interfaces are well-established and it can be assumed that their microscopic structure, as described by the concentration profile of the transition zone, has also reached the equilibrium form. Further observations on the same system, a mixture of polystyrene and partially brominated polystyrene, are included in Fig. 3.35. The three micrographs were obtained at somewhat earlier times. Although the observed structures are finer and therefore less well resolved, it seems clear that they are identical in general character. What we see here is a coarsening process and, importantly, the observations suggest that all these transient structures which are passed through during the late stages of unmixing are similar to each other and differ only in length scale. Checks for the suspected similarity are possible by light scattering experiments. Figure 3.36 depicts, as an example, scattering curves obtained for a mixture of polybutadiene (PB) and polyisoprene (PIP). Here a spinodal decomposition can be initiated by a temperature jump into an upper miscibility gap. Similarity implies that in a representation with reduced variables, plotting $\log(I/I_{\max})$ versus

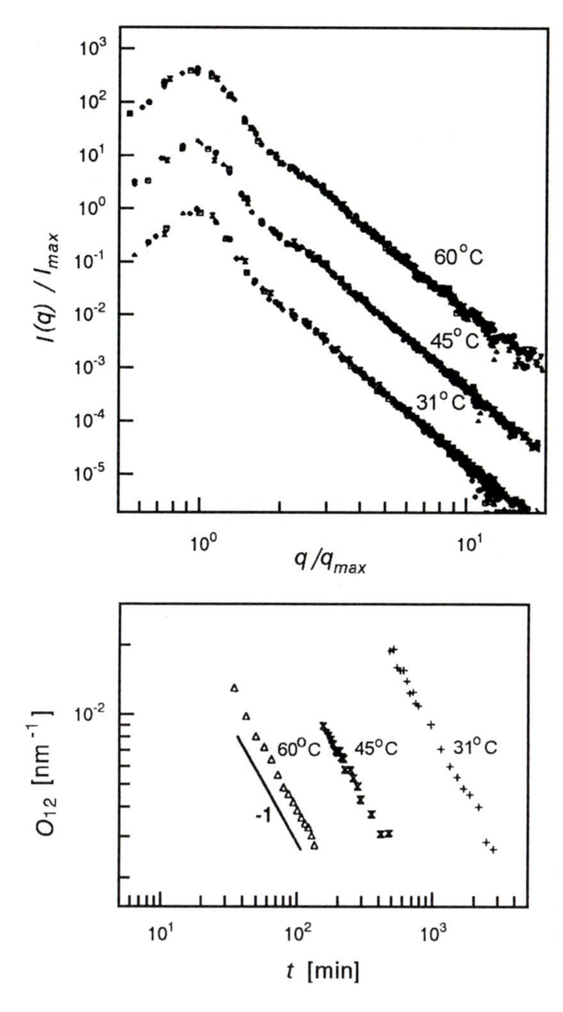

Fig. 3.36. Light scattering curves obtained for a PB ($M = 5.8 \cdot 10^5$)/PIP ($M = 1 \cdot 10^5$)-(1:1) mixture during the late stage of spinodal decomposition at the indicated temperatures (*top*; curves for 45 °C and 60 °C are shifted by constant amounts in vertical direction). Each curve contains measurements for different times and these superpose exactly. Time dependence of the interfacial area per unit volume, O_{12}, in agreement with a power law $O_{12} \sim t^{-1}$, as indicated by the straight line with slope -1 (*bottom*). Data from Takenaka and Hashimoto [24]

$\log(q/q_{\max})$, curves measured at different times must become identical. As we see, this is indeed true. We notice in addition that even the structures observed at different temperatures are similar to each other. We thus have a most simple situation which allows us to describe the kinetics of unmixing by the time dependence of just one parameter. Possible choices are either $q_{\max}^{-1}$, representing a typical length in the structure, or the interfacial area per unit volume, denoted by O_{12}. In fact, both quantities are related. Two-phase systems in general have two primary structure parameters, namely the volume fraction of one phase, ϕ, and O_{12}. As explained in Sect. A.4.2 in the Appendix, one can derive from ϕ and O_{12} a characteristic length of the structure, l_c, as

$$l_c = \frac{2\phi(1-\phi)}{O_{12}} \tag{3.192}$$

(see Eq. (A.161)). l_c and q_{max}^{-1} have equal orders of magnitude and are proportional to each other

$$l_c = \text{const} \cdot q_{max}^{-1} \tag{3.193}$$

the proportionality constant depending on the structure type. Clearly, when the formation of the equilibrium phases is completed for the first time, then ϕ is fixed and does not change any more. Hence from this point on, throughout the late stages of unmixing, one must find a strict inverse proportionality between l_c or q_{max}^{-1} and O_{12}. O_{12} can be directly derived from the scattering curve using 'Porod's law', Eq. (A.160), which states that the scattering function of a two-phase system shows generally an asymptotic behavior according to the power law

$$S(q \to \infty) \sim \frac{O_{12}}{q^4} \tag{3.194}$$

The curves in Fig. 3.36 are in agreement with this law which therefore can be employed for a determination of O_{12}. The time dependence of O_{12} is given in the lower half of Fig. 3.36. Results indicate a decrease of O_{12} inverse to t

$$O_{12} \sim t^{-1} \tag{3.195}$$

Here, we cannot discuss the theories developed for the late stage kinetics, but the physical background must be mentioned, since it is basically different from the initial stages discussed above. Whereas the kinetics in the initial stages is based on diffusive processes only, the late stages are controlled by convective flow. The driving force originates from the excess free energy of the interfaces. The natural tendency is a reduction of O_{12}, and this is achieved by a merging of smaller domains into larger ones.

The latter mechanism remains effective up to the end, however, the structure characteristics must finally change as the similarity property cannot be maintained. The very end is a macroscopic phase separation, as shown for example in Fig. 3.37, and clearly, the final structure is always of the same type independent of whether phase separation has started by spinodal decomposition or by nucleation and growth.

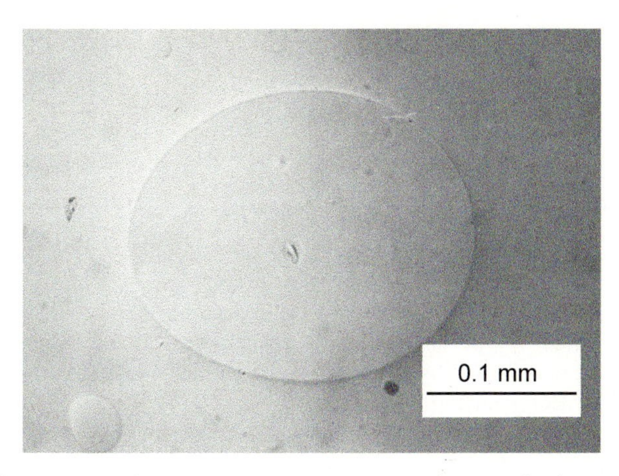

Fig. 3.37. Macroscopic domains in a two-phase PS/PBr_xS-(1:1) mixture, formed after 2 h of annealing [17]

3.3 Block Copolymers

If two different polymeric species are coupled together by chemical links, one obtains 'block copolymers'. These materials possess peculiar properties, and we will consider them in this section.

3.3.1 Phase Behavior

In the discussion of the behavior of binary polymer mixtures, we learned that, in the majority of cases, they separate into two phases. As the linkages in block copolymers inhibit such a macroscopic phase separation, one may wonder in which way these systems react under comparable conditions. Figure 3.38 gives the answer with a drawing: The A's and B's still segregate but the domains have only mesoscopic dimensions corresponding to the sizes of the single blocks. In addition, as all domains have a uniform size, they can be arranged in regular manner. As a result ordered mesoscopic lattices emerge. In the figure it is also indicated that this 'microphase-separation' leads to different classes of structures in dependence on the ratio between the degrees of polymerization of the A's and B's. For $N_{\mathrm{A}} \ll N_{\mathrm{B}}$ spherical inclusions of A in a B-matrix are formed, and they set up a body-centered cubic lattice. For larger values N_{A}, but still $N_{\mathrm{A}} < N_{\mathrm{B}}$, the A-domains have a cylindrical shape and they are arranged in a hexagonal lattice. Layered lattices form under essentially symmetrical conditions, i.e. $N_{\mathrm{A}} \approx N_{\mathrm{B}}$. Then, for $N_{\mathrm{A}} > N_{\mathrm{B}}$, the phases are inverted, and the A-blocks now constitute the matrix.

In addition to these lattices composed of spheres, cylinders and layers, under special conditions periodic structures occur where both phases are continuous. In the example presented in the figure both have the symmetry of a

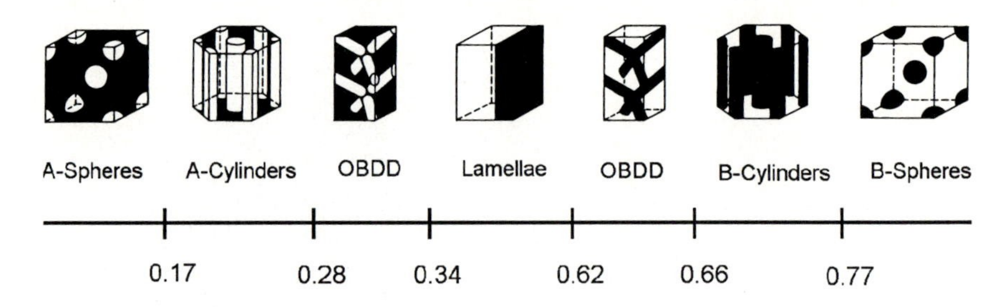

Fig. 3.38. Different classes of microphase-separated structures in block copolymers, as exemplified by polystyrene-*block*-polyisoprene. The numbers give the phase boundaries in terms of the volume fraction of the PS blocks. Figure taken from a review article by Bates and Fredrickson [25]

diamond lattice, and interpenetrate each other. These 'ordered bicontinuous double diamond (OBDD)' structures exist only in a narrow range of values N_A/N_B, between the regimes of the cylindrical and lamellar structures.

To be sure, the figure depicts the structures observed for polystyrene-*block*-polyisoprene, but these are quite typical. Spherical, cylindrical and layer-like domains are generally observed in all block copolymers. Less is known about how general special types like the OBDD lattices are. Observations are rare, since these exist in small regions only, positioned in-between the extended stability ranges of the major structures.

The majority of synthesized compounds are 'di-block copolymers' composed of one A- and one B-chain, however, tri-blocks and multiblocks, comprising an arbitrary number of A- and B-chains, can be prepared as well. One can also proceed one step further and build up multiblocks which incorporate more than two species, thus again increasing the variability. The question may arise as to if all these modifications result in novel structures. In fact, this is not the case. The findings give the impression that at least all block copolymers composed of two species exhibit qualitatively similar phase behaviors. Changes then occur for ternary systems. For the latter, the observed structures still possess periodic orders, but the lattices are much more complex. Here, we shall only be concerned with the simplest systems, the di-block copolymers.

Suitable methods for an analysis of block copolymer structures are electron microscopy and small angle X-ray scattering experiments. Figure 3.39 gives an example and presents, on the left, scattering curves obtained for a series of polystyrene-*block*-polybutadienes where both blocks had similar molecular weights. Structures belong to the layer regime and one correspondingly observes series of equidistant Bragg reflections. The right-hand side depicts micrographs obtained for the same samples in an electron microscope using ultra-thin sections of specimens where the polyisoprene blocks were stained with OsO_4. The layered structure is clearly visible and one notices an increase of the layer thicknesses with the molar masses of the blocks.

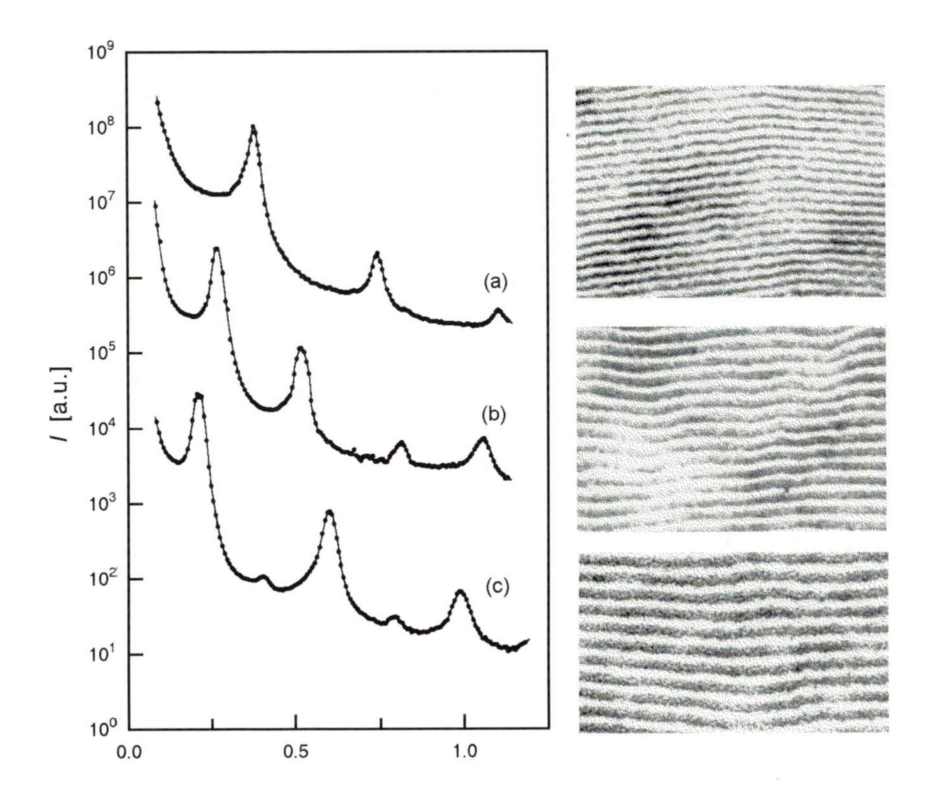

Fig. 3.39. SAXS curves measured for a series of polystyrene-*block*-polyisoprenes with different molecular weights in the microphase-separated state: $M = 2.1 \cdot 10^4$, $\phi(\mathrm{PS}) = 0.53$ (*a*); $M = 3.1 \cdot 10^4$, $\phi(\mathrm{PS}) = 0.40$ (*b*); $M = 4.9 \cdot 10^4$, $\phi(\mathrm{PS}) = 0.45$ (*c*) (*left*). Transmission electron micrographs obtained using ultra-thin sections of specimen stained with OsO_4 (*right*). Structures belong to the layer regime. Data from Hashimoto et al. [26]

In binary polymer mixtures, one finds under favorable conditions homogeneous phases . They either arise if the forces between unlike monomers are attractive or, generally, if the molar masses are sufficiently low. Block copolymers behave similarly and can also have a homogeneous phase. Actually it has a larger stability range than the corresponding binary mixture. Recall that for a symmetric mixture ($N_{\mathrm{A}} = N_{\mathrm{B}}$) the two-phase region begins at (Eq. (3.109))

$$(\chi N_{\mathrm{A}})_{\mathrm{c}} = 2$$

If, from the same A- and B-chains, a symmetric di-block copolymer is formed, the transition between the homogeneous phase and the microphase-separated

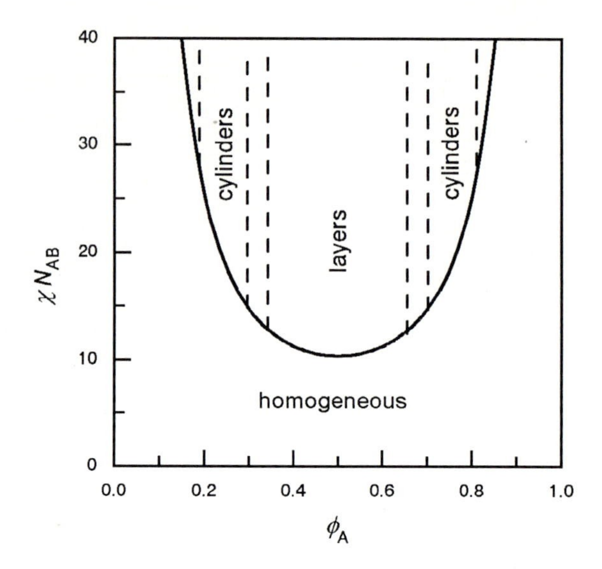

Fig. 3.40. Phase diagram of a di-block copolymer in a schematic representation. The curve describes the points of transition between the homogeneous phase and the microphase-separated states. The ordered states split into different classes. The *dashed* boundary lines between the different types must not be vertical, but may in reality be somewhat inclined

state takes place at a higher χ, namely for

$$(\chi N_{\mathrm{A}})_{\mathrm{c}} \approx 5 \tag{3.196}$$

The complete phase-diagram of a block copolymer is displayed in Fig. 3.40 in a schematic representation. Variables are the volume fraction of the A-blocks

$$\phi_{\mathrm{A}} = \frac{N_{\mathrm{A}}}{N_{\mathrm{A}} + N_{\mathrm{B}}} \tag{3.197}$$

and the product χN_{AB}, where N_{AB} describes the total degree of polymerization

$$N_{\mathrm{AB}} := N_{\mathrm{A}} + N_{\mathrm{B}}$$

The transition line which separates the homogeneous phase from the various microphase-separated structures has an appearance similar to the binodal of a polymer mixture. There is, however, a basic difference: In the block copolymer case, we are dealing with a one-component-system, rather than a binary mixture. The line therefore relates to a phase transition rather than to a miscibility gap. It should also be noticed that, in contrast to the binodal of a mixture, the transition line tells us nothing about the internal composition of the microphases. In principle, these could be mixed states, practically, however, compositions are mostly close to pure A- or B-states.

3.3.2 Layered Structures

Each of the ordered structures represents under the respective conditions the state with the lowest Gibbs free energy. Calculations of the Gibbs free energies and comparisons between the various lattices and the homogeneous phase can therefore provide an understanding of the phase diagram. In addition, they enable us to a determine the structure parameters.

Theoretical analyses were carried out by Meier and Helfand. A full presentation lies outside or possibilities but in order to gain at least an impression of the approaches, we will pick out the layered structures as an example and discuss the equilibrium conditions. The main result will be a power law which formulates the dependence of the layer thicknesses on the degree of polymerization of the blocks.

If we think about the structural changes which accompany a transition from the homogeneous phase to an ordered layer structure we find three contributions to the change in Gibbs free energy

$$\Delta g_{\mathrm{p}} = \Delta h_{\mathrm{p}} - T\Delta s_{\mathrm{p,if}} - T\Delta s_{\mathrm{p,conf}} \tag{3.198}$$

There is a change in enthalpy, a change in entropy following from the arrangement of the junction points along the interfaces, and another change in entropy resulting from altered chain conformations. We write the equation in terms of quantities which refer to one di-block polymer.

The driving force for the transitions comes from the enthalpic part. In the usual case of unfavorable AB-interactions, i.e. $\chi > 0$, there is a gain in enthalpy on unmixing. We assume a maximum gain, achieved when we have a random distribution of the monomers in the homogeneous phase and a perfect segregation in the lamellar phase. Then the enthalpy change per polymer, Δh_{p}, is given by

$$\Delta h_{\mathrm{p}} = -kT\chi N_{\mathrm{AB}}\phi_{\mathrm{A}}(1-\phi_{\mathrm{A}}) + \Delta h_{\mathrm{p,if}} \tag{3.199}$$

The first term follows directly from Eq. (3.98). The second term, $\Delta h_{\mathrm{p,if}}$, accounts for an excess enthalpy which is contributed by the interfaces. To see the background, regard that interfaces always possess a finite thickness, typically in the order of one to several nm. Within this transition layer the A's and B's remain mixed, which leads to an increase in enthalpy proportional to χ and to the number of structure units in the transition layer. Let the thickness of the transition layer be d_{t} and the interface area per polymer o_{p}, then we may write

$$\Delta h_{\mathrm{p,if}} \simeq kT\chi\frac{o_{\mathrm{p}}d_{\mathrm{t}}}{v_{\mathrm{c}}} \tag{3.200}$$

v_{c} again is the volume of the structure unit, commonly chosen for both the A- and B-chains.

The two entropic parts work in the opposite direction. There is first the loss in entropy which results from the confinement of the junction points, being localized in the transition layer. For a layered phase with layer thicknesses $d_{\rm A}$ and $d_{\rm B}$, and therefore a period

$$d_{\rm AB} = d_{\rm A} + d_{\rm B} \tag{3.201}$$

$\Delta s_{\rm p,if}$ may be estimated using a standard relation of statistical thermodynamics:

$$\Delta s_{\rm p,if} \simeq k \ln \frac{d_{\rm t}}{d_{\rm A} + d_{\rm B}} \tag{3.202}$$

The second entropic contribution, $\Delta s_{\rm p,conf}$, accounts for a decrease in entropy which follows from a change in the chain conformations. The Gaussian conformational distribution found in the homogeneous phase cannot be maintained in the microphase-separated state. Formation of a layer structure leads, for steric reasons, necessarily to a chain stretching which in turn results in a loss in entropy. For a qualitative description we employ the previous Eq. (2.93)

$$\Delta s_{\rm p,conf} \simeq -k \left(\frac{R}{R_0} \right)^2 \tag{3.203}$$

where R and R_0 now are the end-to-end distances of the block copolymer in the layered and the homogeneous phase respectively. Assuming that chain sizes and layer spacings are linearly related, by

$$R = \beta d_{\rm AB} \tag{3.204}$$

the equation converts into

$$\Delta s_{\rm p,conf} \simeq -k\beta^2 \left(\frac{d_{\rm AB}}{R_0} \right)^2 \tag{3.205}$$

We can now search for the equilibrium. First note that $o_{\rm p}$ and $d_{\rm AB}$ are related, by the obvious equation

$$o_{\rm p} d_{\rm AB} = N_{\rm AB} v_{\rm c} \tag{3.206}$$

We therefore have only one independent variable, for example $o_{\rm p}$. Using all the above expressions, we obtain for the change in the Gibbs free enthalpy

$$\frac{1}{kT}\Delta g_{\rm p} = -\chi N_{\rm AB}\phi_{\rm A}(1-\phi_{\rm A}) + \chi o_{\rm p} d_{\rm t} v_{\rm c}^{-1} + \ln \frac{d_{\rm t}}{d_{\rm AB}} + \beta^2 \left(\frac{d_{\rm AB}}{R_0} \right)^2 \tag{3.207}$$

If we neglect the slowly varying logarithmic term, we obtain for the derivative

$$\frac{1}{kT}\frac{{\rm d}\Delta g_{\rm p}}{{\rm d}o_{\rm p}} = \chi \frac{d_{\rm t}}{v_{\rm c}} - 2\beta^2 \frac{N_{\rm AB}^2 v_{\rm c}^2}{R_0^2} \frac{1}{o_{\rm p}^3} \tag{3.208}$$

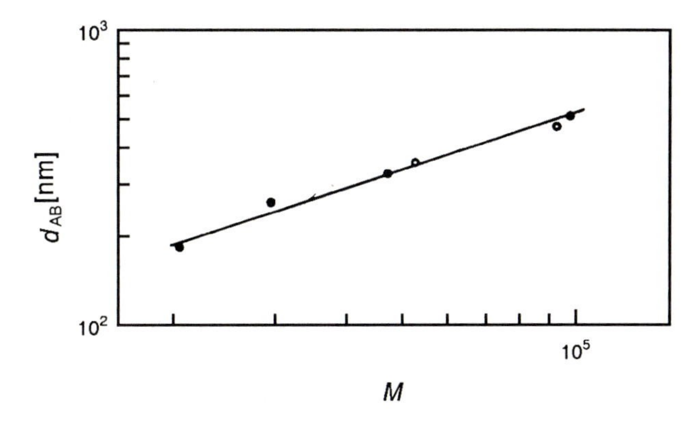

Fig. 3.41. Set of samples of Fig. 3.39. Molecular weight dependence of the layer spacing d_{AB}

The equilibrium value of o_{p} follows as

$$o_{\mathrm{p}}^3 \sim 2\frac{v_{\mathrm{c}}^3}{R_0^2 d_{\mathrm{t}}\chi} N_{\mathrm{AB}}^2 \tag{3.209}$$

With

$$R_0^2 \sim v_{\mathrm{c}}^{2/3} N_{\mathrm{AB}} \tag{3.210}$$

we find

$$o_{\mathrm{p}}^3 \sim \frac{v_{\mathrm{c}}^{7/3}}{d_{\mathrm{t}}\chi} N_{\mathrm{AB}} \tag{3.211}$$

Replacement of o_{p} by d_{AB} gives us the searched-for result

$$d_{\mathrm{AB}}^3 = \frac{N_{\mathrm{AB}}^3 v_{\mathrm{c}}^3}{o_{\mathrm{p}}^3} \sim \chi d_{\mathrm{t}} v_{\mathrm{c}}^{2/3} N_{\mathrm{AB}}^2 \tag{3.212}$$

How does this result compare with experiments? Figure 3.41 depicts the data obtained for the samples of Fig. 3.39. Indeed, the agreement is perfect. The slope of the line in the double logarithmic plot exactly equals the predicted exponent 2/3.

3.3.3 Pretransitional Phenomena

A characteristic property of polymer mixtures in the homogeneous phase is the increase of the concentration fluctuations accociated with an approaching of the point of unmixing. A similar behavior is found for the homogeneous phase of block copolymers, and a first example is given in Fig. 3.42. The figure shows scattering functions measured for a polystyrene-*block*-polyisoprene under variation of the temperature. The temperature of the transition to the

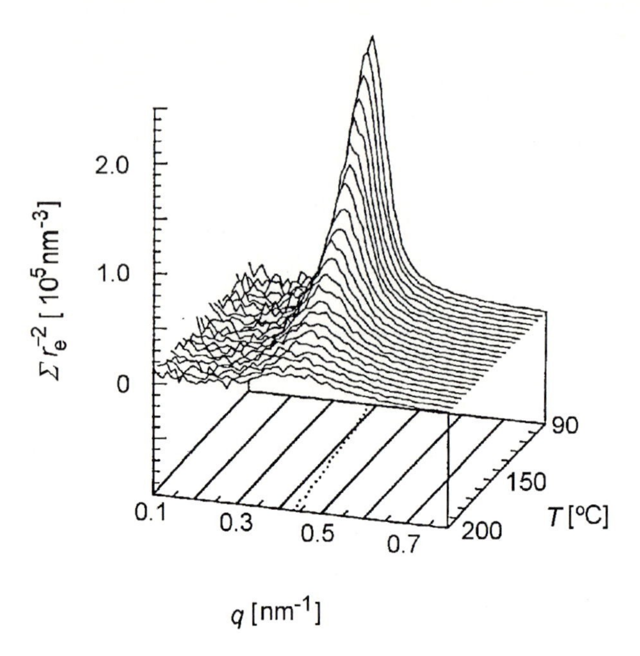

Fig. 3.42. SAXS curves measured for a polystyrene-*block*-polyisoprene ($M = 1.64 \cdot 10^4, \phi$ (PS) $= 0.22$) in the homogeneous phase. The *dotted line* on the base indicates the temperature dependence of the peak position [27]

microphase-separated state is located around 85 °C, just outside the temperature range of the plot. The curves exhibit a peak, with an intensity which strongly increases when the temperature moves towards the transition point.

The feature in common with the polymer mixtures is the intensity increase, however, we can also see a characteristic difference: The maximum of the scattering intensity and the largest increase now are found for a finite scattering vector $q_{\max}$, rather than at $q = 0$. As scattering curves display the squared amplitudes of wave-like concentration fluctuations, the observation tells us that concentration fluctuations with wavevectors in the range $|\boldsymbol{k}| \approx q_{\max}$ are always large compared to all the others and show a particularly strong increase on approaching the phase transition. What do these observations mean? Clearly, they remind on the pretransitional phenomena observed for critical phase transitions. There, the approach of the transition point is always associated with an unusual increase of certain fluctuations. Hence as it appears, not only for polymer mixtures, but also for block copolymers, one also finds properties in the homogeneous phase which have much in common with the behavior of critical systems.

The general shape of the scattering curve, showing a maximum at some $q_{\max}$ and going to zero for $q \to 0$ is conceivable. As explained in Sect. A.3.2 of the Appendix, the forward scattering, $S(q \to 0)$, always relates to the fluctuation of the number of particles in a fixed macroscopic volume. In our case, this refers to both the A's and the B's. The strict coupling between A- and B-chains in the block copolymers completely suppresses number fluctuations on length scales which are large compared to the size of the block copolymer. The limiting behavior of the scattering function, $S(q \to 0) \to 0$, reflects just this fact. On the other hand, for large q's, scattering of a block copolymer and of the corresponding polymer mixture composed of the decoupled blocks, must be identical because here only the internal correlations within the A- and B-chains are of importance. As a consequence, asymptotically the scattering law of ideal chains, $S(q) \sim 1/q^2$, shows up. Hence, one expects an increase in the scattering intensity coming down from large q's and when emanating from $q = 0$ as well. Both increases together produce a peak, located at a certain finite q_{max}.

The increase of the intensity with decreasing temperature reflects a growing tendency for associations of the junction points accompanied by some short-ranged segregation. As long as this tendency is not too strong, this could possibly occur without affecting the chain conformations, i.e. chains could still maintain Gaussian properties. If one adopts this view, then the scattering function can be calculated explicitly. Leibler and also other workers derived for the scattering function per structure unit, S_{c}, the following expression

$$\frac{1}{S_{\mathrm{c}}(q)} = \frac{1}{S_{\mathrm{c}}^0(q)} - 2\chi \tag{3.213}$$

with $S_{\mathrm{c}}^0(q)$, the scattering function in the athermal case, being determined by

$$\begin{aligned} S_{\mathrm{c}}^0(q) \cdot N_{\mathrm{AB}} S_{\mathrm{D}}(R_0^2 q^2) \quad = \quad & \phi(1-\phi) N_{\mathrm{A}} N_{\mathrm{B}} S_{\mathrm{D}}(R_{\mathrm{A}}^2 q^2) S_{\mathrm{D}}(R_{\mathrm{B}}^2 q^2) \\ & -\frac{1}{4}[N_{\mathrm{AB}} S_{\mathrm{D}}(R_0^2 q^2) - \phi N_{\mathrm{A}} S_{\mathrm{D}}(R_{\mathrm{A}}^2 q^2) \\ & -(1-\phi) N_{\mathrm{B}} S_{\mathrm{D}}(R_{\mathrm{B}}^2 q^2)]^2 \end{aligned} \tag{3.214}$$

R_0^2 denotes the mean-squared end-to-end distance of the block copolymer, given by

$$R_0^2 = R_{\mathrm{A}}^2 + R_{\mathrm{B}}^2 \tag{3.215}$$

With regard to the effect of χ, Eq. (3.213) is equivalent to Eq. (3.165). Indeed, the physical background of both equations is similar, and they are obtained in equal manner, by an application of the 'random phase approximation (RPA)'. The interested reader can find the derivation in Sect. A.4.1 in the Appendix.

Importantly, Eq. (3.213) describes the effect of χ directly. It becomes very clear if one plots the inverse scattering function. Then changes in χ result in parallel shifts of the curves only. Figure 3.43 depicts the results of model calculations for a block copolymer with a volume fraction of polystyrene blocks of $\phi = 0.22$, in correspondence to the sample of Fig. 3.42. The curves were obtained for the indicated values of the product χN_{AB}. Obviously the calcu-

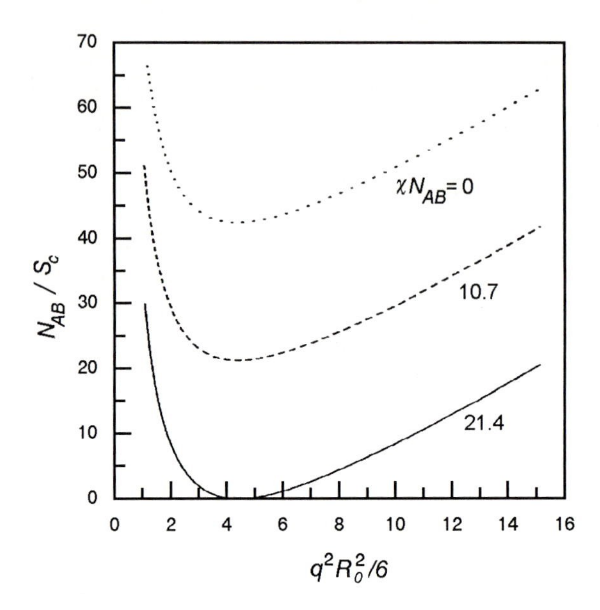

Fig. 3.43. Theoretical scattering functions of a block copolymer with $\phi = 0.22$, calculated for the indicated values of χN_{AB}

lations represent the main features correctly: They yield a peak at a certain $q_{\max}$, which grows in intensity with increasing χ, i.e. with decreasing temperature. The important result comes up for $\chi N_{\mathrm{AB}} = 21.4$. For this value we find a diverging intensity at the position of the peak, $S(q_{\max}) \to \infty$. This is exactly the signiture of a critical point. We thus realize that the RPA equation formulates a critical transition with a continuous passage from the homogeneous to the ordered phase. When dealing with critical phenomena, it is always important to see the order parameter. Here it is of peculiar nature. According to the observations it is associated with the amplitudes of the concentration waves with $|\boldsymbol{k}| = q_{\max}$.

For $\phi = 0.22$, the critical point is reached for $N_{\mathrm{AB}}\chi = 21.4$. With the aid of the RPA result, Eq. (3.214), one can calculate the critical values for all ϕ's. In particular, for a symmetric block copolymer one obtains

$$\chi N_{\mathrm{AB}} = 10.4$$

This is the lowest possible value and the one mentioned earlier in Eq. (3.196).

In polymer mixtures, one calls the curve of points in the phase diagram, where $S(q = 0)$ apparently diverges, the 'spinodal'. One can employ the same notion for block copolymers and determine this curve in an equal manner by a linear extrapolation of scattering data measured in the homogeneous phase. We denote this spinodal again $T_{\mathrm{sp}}(\phi)$.

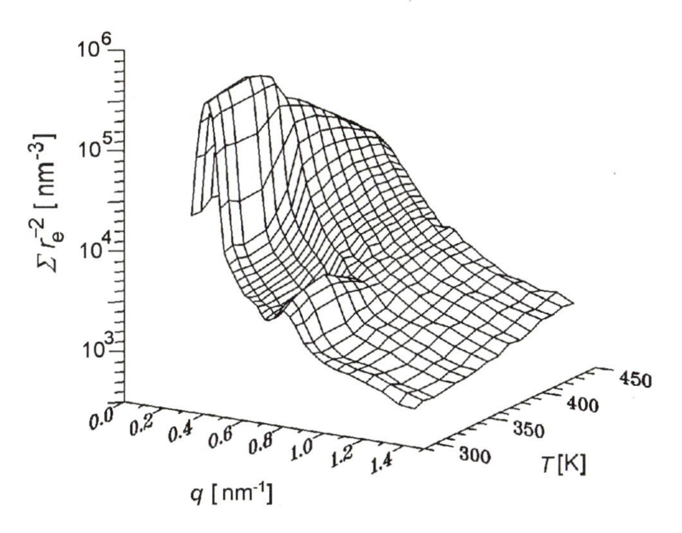

Fig. 3.44. SAXS curves measured for a polystyrene-*block*-polyisoprene ($\phi(\text{PS}) = 0.44, M = 1.64 \cdot 10^4$) in the temperature range of the microphase-separation. The transition occurs at $T_t = 362$ K. Data from Stühn et al.[28]

Regarding all these findings, one could speculate that the microphase-separation might take place as a critical phase transition in the strict sense, at least for block copolymers with the 'critical composition' associated with the lowest transition temperature. Actually, experiments which pass over the phase transition show that this is not true and they also point to other limitations of the RPA treatment. Figure 3.44 presents scattering curves obtained for a polystyrene-*block*-polyisoprene near to the critical composition ($\phi(\text{PS}) = 0.44$), in a temperature run through the transition point. As we can see, the transition is not continuous up to the end but is associated with the sudden appearance of two Bragg-reflections. Hence, although the global behavior is dominated by the steady growth of the concentration fluctuations typical for a critical behavior, there is finally a discontinuous step, which converts this transition into one of 'weakly first order'.

Of interest are also the details. Figure 3.45 depicts the temperature dependence of the inverse peak intensity $I^{-1}(q_{\text{max}})$. Equation (3.213) predicts a dependence

$$S(q_{\text{max}})^{-1} \sim \chi_{\text{sp}} - \chi \qquad (3.216)$$

or, assuming a purely enthalpic χ with $\chi \sim 1/T$ (Eq. (3.96))

$$S(q_{\text{max}})^{-1} \sim T_{\text{sp}}^{-1} - T^{-1} \qquad (3.217)$$

The findings, however, are different. We see that the data follow a linear law only for temperatures further away from the transition point and then deviate towards higher values. The transition is retarded and does not take place until a temperature 35 K below the spinodal point is reached. According

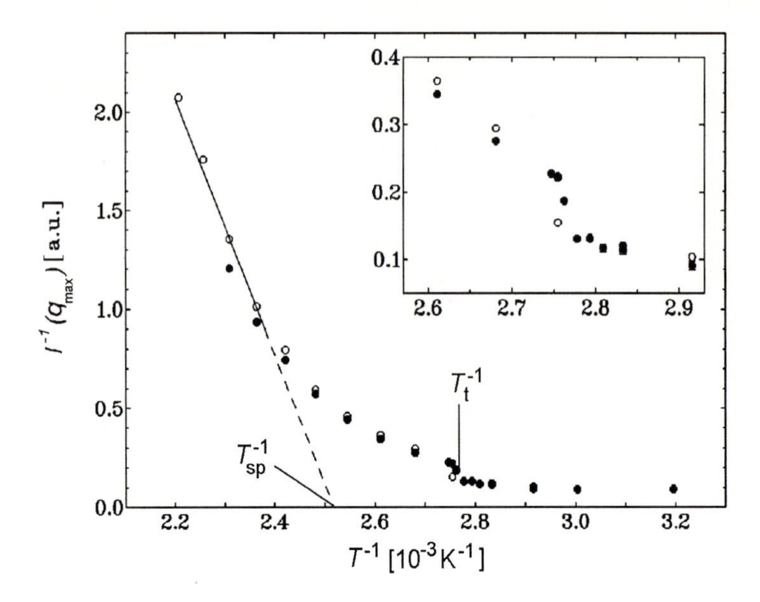

Fig. 3.45. Measurements shown in Fig. 3.44: Temperature dependence of the reciprocal peak intensity, showing deviations from the RPA predictions. The linear extrapolation determines the spinodal temperature

to theoretical explanations, which we cannot further elaborate on here, the phenomenon is due to a lowering of the Gibbs free energy, caused by the temporary short-range order associated with the fluctuations. The short-range order implies local segregations and thus a reduction of the number of AB-contacts, which in turn lowers the Gibbs free energy. We came across this effect earlier in the discussion of the causes of the energy lowering observed in computer simulations of low molar mass mixtures. Remember that there the effect exists only for low enough molar masses, since for high molecular weights a short-range ordering becomes impossible. The same prerequisite holds for block copolymers and this is also formulated by the theories.

There exists another weak point in the RPA equation. As a basic assumption, it implies that chains in the homogeneous phase maintain Gaussian statistical properties up to the transition point. The reality is different and this is not at all surprising: An increasing tendency for an association of the junction points also necessarily induces a stretching of chains, for the same steric reasons which lead in the microphase-separated state to the specific power law Eq. (3.212). This tendency is shown by the data presented in Fig. 3.44 and, even more clearly, by the results depicted in Fig. 3.42. In both cases, $q_{\max}$ shifts to smaller values with decreasing temperature, as is indicative for a chain stretching.

3.4 Further Reading

Polymer solutions:

J. des Cloizeaux, G. Jannink: *Polymers in Solution: Their Modelling and Structure*, Oxford Science Publishers, 1990

M. Doi, S.F. Edwards: *The Theory of Polymer Dynamics*, Clarendon Press, 1986

A.Y. Grosberg, A.R. Khokhlov: *Statistical Physics of Macromolecules*, AIP Press, 1994

Polymer mixtures and block copolymers:

K. Binder: *Spinodal Decomposition* in P. Haasen (Ed.): Material Science and Technology Vol.5 *Phase Transitions in Materials*, VCH Publishers, 1991

P.J. Flory: *Principles of Polymer Chemistry*, Cornell University Press, 1953

P.-G. de Gennes: *Scaling Concepts in Polymer Physics*, Cornell University Press, 1979

I. Goodman: Developments in Block Copolymers Vol.1, Applied Science Publishers, 1982

I. Goodman: Developments in Block Copolymers Vol.2, Applied Science Publishers, 1985

T. Hashimoto: *Structure Formation in Polymer Systems by Spinodal Decomposition* in R.M. Ottenbrite, L.A. Utracki, S. Inoue (Eds.): Current Topics in Polymer Science Vol.2, Hanser, 1987

D.R. Paul, S. Newman (Eds.): *Polymer Blends, Vols.1 and 2*, Academic Press, 1978

Chapter 4

Metastable Partially Crystalline States

At first it may seem questionable that polymers can set up a crystal at all but after thinking about it a little, it becomes clear how this can be accomplished: In principle, a periodic structure in three dimensions is obtained by choosing identical helical conformations for all polymers, orienting the helical axes of all chains parallel to each other, and then packing the chains laterally in a regular manner. The scheme is applicable for all polymers, provided that they have a linear architecture and a regular chemical constitution. Hence, we may conclude that polymers have the potential to crystallize. As we shall see, they do indeed form crystals, however, this occurs in a peculiar way.

For an understanding of the peculiarities, it is advantageous to begin with a look at the crystallization behavior of simpler, but related systems, namely that of 'oligomers'. Oligomers are chain molecules of low molar masses, prominent examples being the *n*-alkanes (C_nH_{2n+2}) or the perfluoro-*n*-alkanes (C_nF_{2n+2}). In contrast to polymers which always show a certain distribution in the molecular weight of the chains, oligomers usually represent 'sharp fractions' with a uniform molar mass. Oligomers readily crystallize and the crystal structures of various compounds have been determined by standard methods of X-ray crystallography. The results indicate that common principles exist in the composition of crystals, and Fig. 4.1 depicts them in a schematic drawing: Crystals are composed of stacked layers, each layer being assembled of chain molecules with identical helical conformations. The endgroups of the molecules set up and occupy the interfaces. Oligomer molecules in the melt

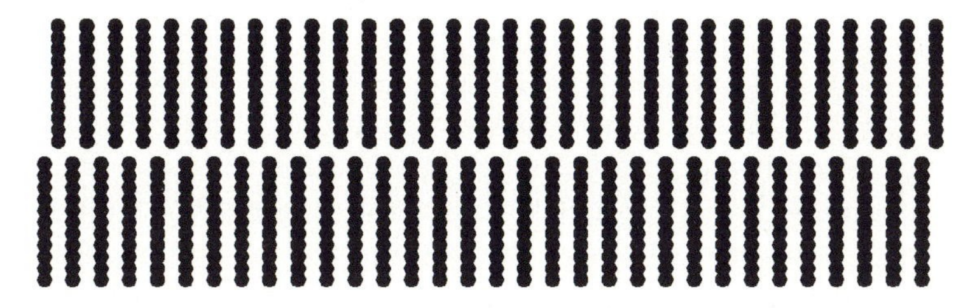

Fig. 4.1. Structure of an oligomer crystal. Schematic drawing showing two layers (for the special case of a rectangular structure, where the chains are oriented perpendicular to the interfaces)

take on coiled conformations, just like polymers. In order to form a crystal, these chains have to be straightened and separated from each other and then attached in the helical form onto the growing lateral crystal surface.

Let us now turn to the case of polymers. To start with, one cannot see any reason why the building principles found for the oligomers should not be employed for polymers in the same way, in case we have also sharp fractions. For a polymer with uniform molecular weight, the same type of crystal, composed of extended straight chains with the endgroups assembled in planar interfaces, could be formed in principle and it would again represent the equilibrium state with the lowest free energy. However, a serious problem now arises quite independently of the certainly unrealistic assumption of a uniform molecular weight: Starting from a melt of coiled, mutually interpenetrating macromolecules it is just impossible to reach this ideal crystalline state from purely kinetical reasons. The required complete disentangling would need a too long time as it is associated with an extremely high entropic activation barrier. What happens instead? The way in which polymer systems react on these conditions is that cooling a melt below the equilibrium melting point produces structures which are only *in part* crystalline. One observes layer-like crystallites which are separated by disordered regions, thus setting up a lamellar two-phase structure. That the crystallites formed have the shape of layers is not surprising if one considers the principles governing the crystallization of oligomers. There, the formation of the interfaces can be regarded as a natural way to deal with the endgroups, which cannot be incorporated into the single layers. Similarly, polymer crystallization requires the entanglements present in the melt be dealt with and got rid of, as the large majority of them cannot be resolved and eliminated within the given time. Adopting this view, we can address the basic mechanism leading to the formation of two-phase structures in crystallizable polymer systems as a separation process: Crystallization occurs together with a preceding unmixing, whereby sequences which can be stretched and incorporated into a growing crystal are separated from chain parts near to entanglements which can only be removed and shifted into the amorphous regions. To be sure, not only entanglements constitute the non-crystallizable chain parts, but endgroups, chemical pertubations like short chain branches, or specific local conformations which oppose a transformation into the helical form, as well. They all become accumulated in the amorphous parts of a partially crystalline polymer.

In the previous chapter, we discussed liquid polymer systems. These exist in specific states selected by the laws of equilibrium thermodynamics. The rules which control structure formation during crystallization are different and this is a most important point to be noticed: Structure formation here is governed by kinetical criteria rather than by equilibrium thermodynamics. What does this mean? Indeed, we encounter here a new criterion: The structure which develops at a given temperature is that with the maximum growth rate rather than that with the lowest free energy. As a consequence, treatment of the crystallization behavior of polymers requires kinetical considerations.

Thermodynamics is still necessary for the description of the driving forces, but it now constitutes only a partial aspect of the problem and time comes into play as a decisive variable.

4.1 Structure Characteristics

Being kinetically controlled, structures of partially crystalline samples are always strongly affected by the processing and show a memory of the thermal history, i.e. temperatures and times of crystallization, cooling rates, etc. A first requirement for the analysis are measurements enabling a characterization of the evolving structures. In this section, we will deal with the main observations and some of the applied techniques.

Partially crystalline polymers exhibit on different length scales different characteristic features. Let us proceed from low to high resolutions, and begin with the structural features in the μm - mm range, as observed in an optical microscope. Figure 4.2 gives a typical example. It shows optical micrographs obtained for a sample of poly(ethyleneoxide)(PEO) which has been cooled from the melt to a temperature where crystallization occurs. We observe spherical objects, so-called 'spherulites', which appear as small nuclei somewhere in the view-field and then grow in size. Correspondingly, the process ist addressed as a crystallization by 'nucleation and growth of spherulites'. Closer inspection shows that the spherulites grow with a constant rate up to the point when they touch each other. For two spherulites which were nucleated at the same time, the area of contact ist planar, if starting times were different, the boundary is bent. Finally, the whole volume is covered by bound spherulites. Their final sizes depend on the nucleation density and can vary over a large range, from several μm up to to some cm.

Fig. 4.2. Growing spherulites observed during the crystallization of PEO in an optical microscope (polarized light, crossed nicols)

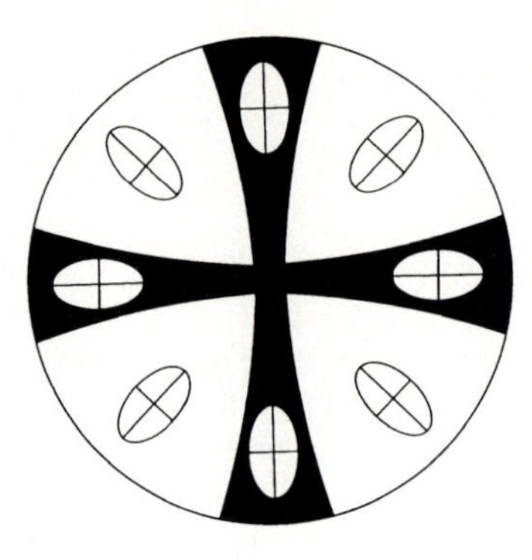

Fig. 4.3. Ordering of the indicatrices in a spherulite and the resulting Maltese cross extinction pattern, as observed between cross nicols. The orientation of the Maltese cross coincides with the directions of polarizer and analyzer

Spherulites are optically anisotropic objects. As in the example of Fig. 4.2, one observes between crossed nicols in many cases a 'Maltese cross'. Appearance of a Maltese cross is indicative of a specific arrangement of the optical indicatrices, and Fig. 4.3 represents the type of order. The basic property is a systematic variation of the orientation of the indicatrices. As indicated in the drawing, one of the axes is directed at each point along the radius vector. The cause of the birefringence is obvious: It originates from the optical anisotropy of the stretched polymer chains in the crystallites. Findings tell us directly that the chains in the crystallites must be oriented either parallel or perpendicular to the spherulite radius. Indeed, closer analysis based on a determination of the sign of the birefringence shows that the chain orientation is always *perpendicular* to the radius vector.

The optical observations cannot resolve the crystalline-amorphous structure. Observation of single crystallites requires methods which provide an analysis in the 10-nm range. Electron microscopy is particularly suited for this purpose. Figure 4.4 shows as an example the surface of a partially crystalline polyethylene, as it becomes reproduced in the electron microscope when using a carbon film replica technique. The picture of the surface resembles a landscape with many terrasses. These obviously result from cuts through stacks of laterally extended, slighty curved lamellae which have thicknesses in the order of 10 nm.

More insight into the internal structure of the lamellae follows from electron microscopic studies on ultra-thin slices if these are stained by OsO_4. Figure 4.5 shows a typical picture. Being rejected by the crystallites, the staining agent only enters the disordered regions. The image is due to the different absorption of the electron beam which is high in the Os-containing amorphous regions and low for the crystallites. The white lines therefore correspond to the crystallites which are layer-like and separated by amorphous regions given by the dark parts. More accurately, we observe those crystallites

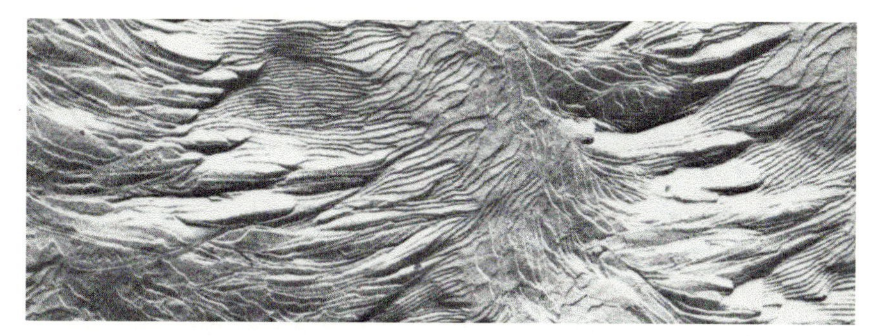

Fig. 4.4. Electron micrograph of a carbon film replica of a surface of PE. Picture obtained by Eppe and Fischer [29]

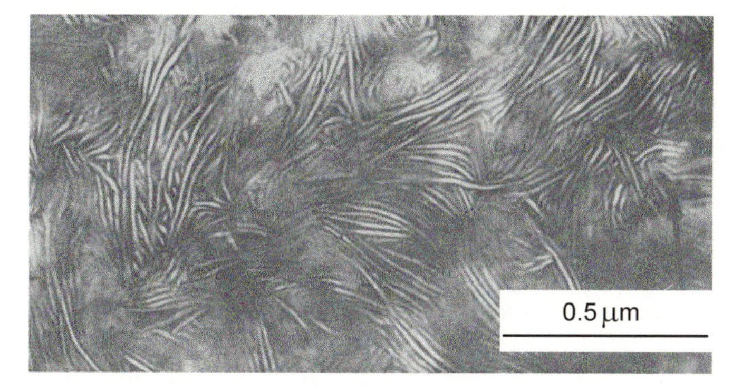

Fig. 4.5. Ultra-thin slice of a PE sample stained with OsO_4. Electron micrograph obtained by Kanig [30]

which happened to be oriented with their surface perpendicular to the slice surfaces so that the electron can pass through with minor absorption. The two micrographs are quite typical and indeed exemplify the basic structural principle emerging in the morphology of partially crystalline polymers: These are built up as a two-phase structure, being composed of layer-like crystallites which are separated by disordered regions.

If we now regard both, the essentially planar structure in the 10-nm range and the isotropic spherulites observed in the μm range, one may wonder, how these two can fit together. Figure 4.6 shows, in which way this is accomplished, and how the stacks of layers set up and fill the space of the spherulites. The left-hand side depicts an electron micrograph of the center of a spherulite of isotactic polystyrene at an early stage of development. The center is sheaf-like and formed by an aggregate of layers. On further growth, these become curved, and finally a stable spherical growth surface is established. The right-hand side of Fig. 4.6 shows a principle which has to be obeyed during further

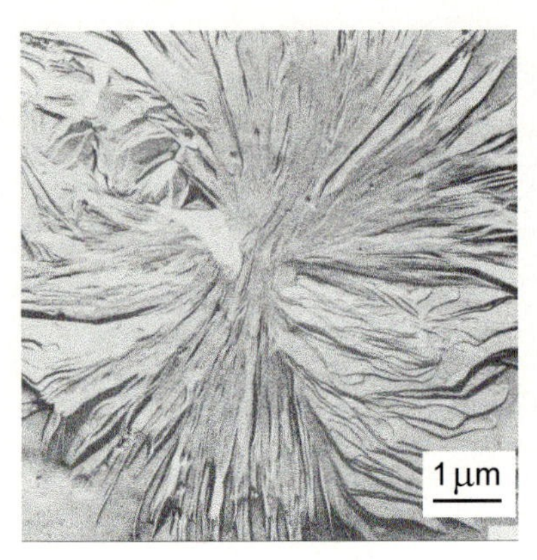

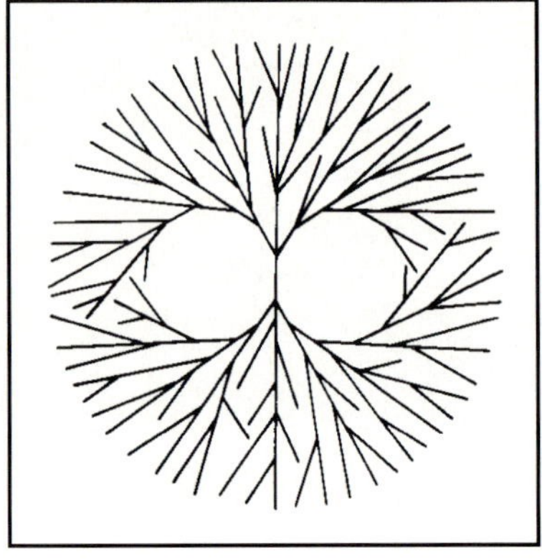

Fig. 4.6. Central sheaf-like part observed at an early stage of development of a spherulite of it-PS. Electron micrograph obtained by Vaughan and Bassett [31] (*left*). Schematic drawing showing the branching and splaying microstructures in the fully developed spherulite (*right*)

growth: In order to keep the increasing surface of the spherulite filled with layers, branching and splaying is a necessary requirement. The orientational distribution of the crystallites within a developed spherulite, that is to say the internal 'texture', is well determined: The surface normal of the crystalline layers is always directed perpendicular to the radius vector. As we deduced from the birefringence properties of the spherulites, the same holds for the direction of the crystalline chains. Chain direction and surface normal must not be identical, but the enclosed angle is usually small.

Quite often, 'banded spherulites' are observed. Figure 4.7 presents such a case, as given by a sample of polyethylene. As shown by the optical micrograph on the left, one finds here, in addition to the Maltese cross, light extinctions along circles in a periodic manner. The observation indicates a regular rotation of the chain direction and the layer normal about the radius vector, and this is fully confirmed by electron microscopic investigations, for example the micrograph shown on the right-hand side. Both experiments tell us that, on length scales of some μm, the crystallites are twisted and that this occurs strictly periodically. Really astonishing is the coherence of this texture throughout the whole spherulite, where the orientations of all crystallites are well-determined and exactly correlated. The mechanism leading to this peculiar texture is still under discussion and not yet clarified.

The first parameter to be quoted in the description of the structure of a partially crystalline polymeric solid is the degree of crystallization, shortly the 'crystallinity'. We have two slightly different choices for its definition, and the selection depends on the method of determination. Consider for example a measurement of the density of a sample. For a two-phase structure the total

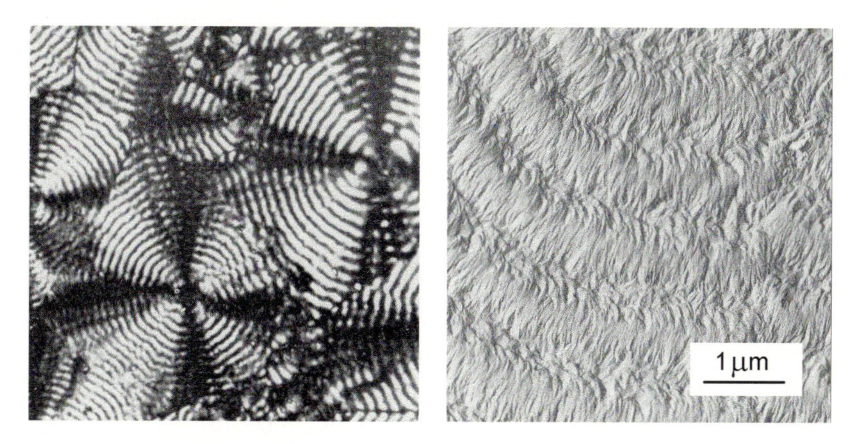

Fig. 4.7. Banded spherulites of PE. Optical micrographs showing a regular sequence of concentric rings (polarized light, crossed nicols; *left*) and electron micrographs of a surface which cuts through a spherulite (obtained by Vaughan and Bassett [31]; *right*)

mass is given by

$$V\rho = V_{\mathrm{a}}\rho_{\mathrm{a}} + V_{\mathrm{c}}\rho_{\mathrm{c}} \tag{4.1}$$

Here, V_{a} and V_{c} denote the volumes occupied by the amorphous and the crystalline parts and V is the total volume; ρ_{a}, ρ_{c} and ρ give the respective densities. As one possibility, the crystallinity can be identified with the volume fraction of the crystalline material. We denote it ϕ_{c} and write

$$\phi_{\mathrm{c}} := \frac{V_{\mathrm{c}}}{V} \tag{4.2}$$

Using Eq. (4.1) ϕ_{c} follows as

$$\phi_{\mathrm{c}} = \frac{\rho - \rho_{\mathrm{a}}}{\rho_{\mathrm{c}} - \rho_{\mathrm{a}}} \tag{4.3}$$

To apply Eq. (4.3) one has to know the densities of the crystalline and the amorphous phase. There is no problem with ρ_{c}, as this can be derived from the lattice constants measured by X-ray diffraction. The knowledge of ρ_{a} is less certain. Usually it is obtained by an extrapolation of the values measured in the melt, with the assumption of a constant expansion coefficient.

A second convenient method for a determination of the crystallinity is a measurement of the heat of fusion, $\Delta\mathcal{H}_{\mathrm{f}}$, of a sample. It leads us to the other possibility, namely to use the weight fraction of the crystalline material rather than the volume fraction. The 'crystallinity by weight', ϕ'_{c}, follows from $\Delta\mathcal{H}_{\mathrm{f}}$ by

$$\phi'_{\mathrm{c}} = \frac{\Delta\mathcal{H}_{\mathrm{f}}}{\Delta\mathcal{H}_{\mathrm{f}}(\phi'_{\mathrm{c}} = 1)} \tag{4.4}$$

for obvious reasons. The prerequisite for an application of this equation is a knowledge of the heat of fusion of a fully crystalline sample. Since such a sample normally cannot be prepared, extrapolation methods have to be used to determine this value. If a homogeneous series of oligomers is available, then one can measure their heats of fusion and subsequently carry out an extrapolation to the limit of infinite molecular weights, either empirically, or better, on the basis of a theoretical formula. Another feasible procedure is a combination of calorimetric and density measurements, i.e. a measurement of $\Delta\mathcal{H}_{\mathrm{f}}$ as a function of the density for samples with different crystallinity. These can often be prepared by a variation of the crystallization temperature, or by changes in the branching or comonomer content of samples. One can then try to extrapolate a series of data to $\rho = \rho_{\mathrm{c}}$. It must be mentioned that Eq. (4.4) is not exact, because it neglects the effect of the crystallite surfaces. As will be discussed later in this chapter, the surface free energy results in a decrease in the melting points and the heats of fusion. In principle, these effects can be accounted for but in practice, they are mostly ignored.

As a general remark, crystallinity values should not be regarded as quantities of high accuracy, for various reasons. ϕ_{c} or ϕ'_{c} are introduced assuming a two-phase structure with well-defined properties of the single phases. Actually this is not strictly true as properties of both the crystallites and the amorphous regions may vary between different samples, or even within a sample. Therefore, when dealing with crystallinities, one should always be aware of possible slight variations.

Representing a bulk property, the crystallinity tells us nothing about the characteristic lengths of the partially crystalline structure. Rough first values can be taken from the electron micrographs but, in order to obtain accurate data, one has to use X-ray scattering experiments. As typical length scales of partially crystalline structures are in the order of 10 nm the associated scattering curves are found in the small angle range. Figure 4.8 shows as an example some scattering functions measured for a sample of polyethylene. Data were obtained for a series of different temperatures, beginning at the temperature of primary crystallization coming from the melt, $T = 125\,^{\circ}\mathrm{C}$, and then continuing down to $31\,^{\circ}\mathrm{C}$ during a stepwise cooling. The general shape of the curves corresponds to the morphological features appearing in the electron micrographs. We have stacks of layer-like crystallites which show some periodicity, although not in the sense of a strict long-range order. This quasi-periodicity, called 'long spacing', becomes reflected in the peaks. By applying Braggs' law one can derive from the peak position, $q_{\max}$, an approximate value for the long spacing d_{ac}

$$d_{\mathrm{ac}} \approx 2\pi/q_{\max} \tag{4.5}$$

The temperature dependent changes in the intensity and in the shape of the scattering curves in the figure are due to both changes in the electron density difference between the crystallites and the amorphous regions and continuous modifications in the structure, as will be discussed later in this chapter.

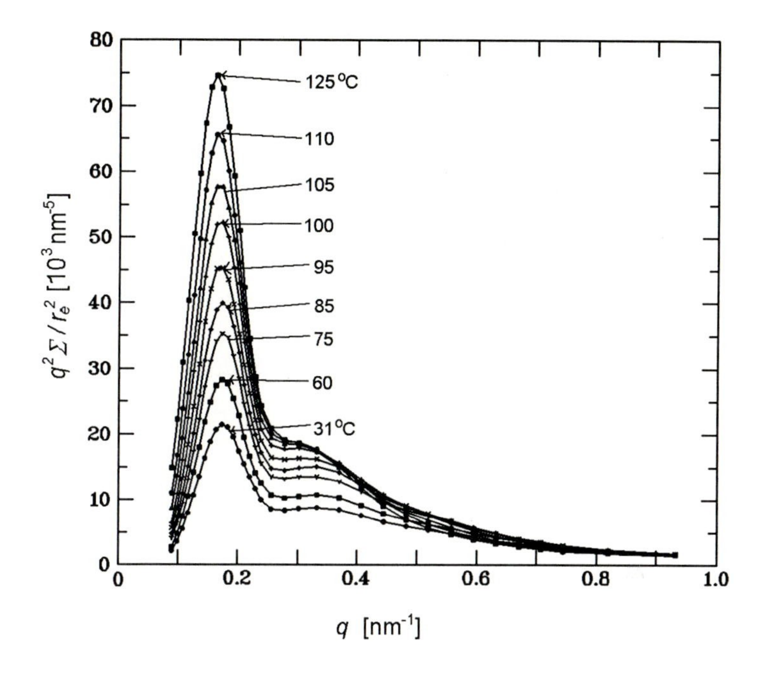

Fig. 4.8. SAXS curves measured for a sample of PE ('Lupolen 6011') after completion of the primary crystallization at 125 °C, and at the indicated temperatures during a subsequent cooling [32],[33]

It is possible to evaluate the small angle X-ray scattering curves in greater detail. In the resolution range of the experiments, the weak bending of the crystalline lamellae is unobservable, and the crystalline-amorphous structure then equals an isotropic distribution of stacks of planar crystalline and amorphous layers. As a consequence, small angle X-ray scattering curves can be related to the internal stack structure only, and the latter is fully characterized by the electron density distribution along the stack normal, $\rho_e(z)$. More specifically, the scattering cross-section per unit volume, $\Sigma(q)$, and the one-dimensional correlation function of the electron density fluctuations about the mean value

$$K(z) := \langle (\rho_e(0) - \langle \rho_e \rangle) \cdot (\rho_e(z) - \langle \rho_e \rangle) \rangle \qquad (4.6)$$

are related by Fourier-transformations. As explained in detail in Sect. A.4.2 in the Appendix, when emanating from a measured scattering function $\Sigma(q)$ the Fourier transformation giving the correlation function has the explicit form (Eq. (A.134))

$$K(z) = \frac{1}{r_e^2} \cdot \frac{1}{(2\pi)^3} \int_{q=0}^{\infty} 4\pi q^2 \Sigma(q) \cos qz \cdot dq \qquad (4.7)$$

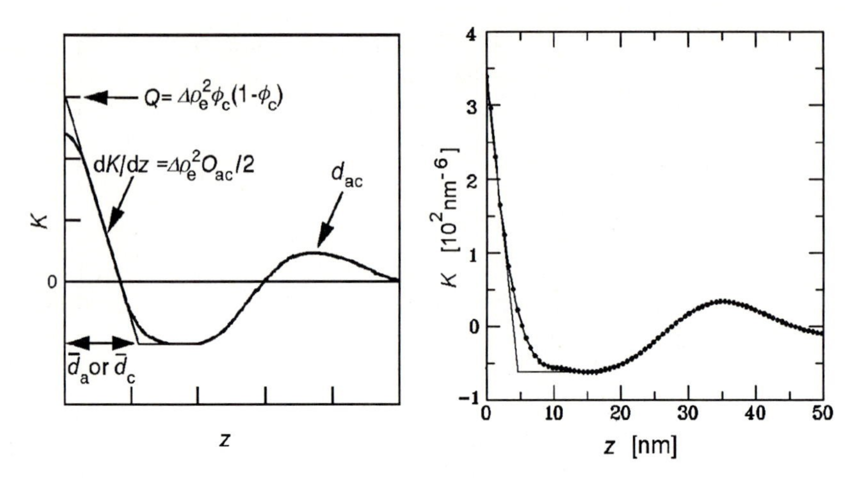

Fig. 4.9. Schematic drawing showing the basic properties of the one-dimensional electron density correlation function $K(z)$ associated with a stack of crystalline and amorphous layers (*left*). $K(z)$ derived from the scattering curve for 31 °C in Fig. 4.8 (*right*). Analysis gives $\phi_c = 0.85$, $O_{ac} = 0.065\ \mathrm{nm}^{-1}$, $\bar{d}_a = 4.6$ nm, $\rho_{e,c} - \rho_{e,a} = 52\ \mathrm{nm}^{-3}$, $d_{ac} = 34$ nm

$K(z)$ possesses peculiar properties which can be directly used for a structure characterization. These are summarized in Fig. 4.9 on the left-hand side in a schematic drawing, which represents the general shape of $K(z)$. In the correlation function there are included, in addition to the crystallinity ϕ_c, the following structure parameters:

- the specific inner surface O_{ac}, i.e. the area of interface which separates crystalline and amorphous regions, taken per unit volume
- the electron density difference between the crystalline and amorphous parts, $\Delta\rho_e = \rho_{e,c} - \rho_{e,a}$
- the long spacing d_{ac}, now being defined unambiguously as the most probable distance between the centers of adjacent crystallites
- Depending on the crystallinity, either the mean thickness $\bar{d}_c$ of the crystallites (for $\phi_c < 0.5$), or the mean thickness of the disordered regions $\bar{d}_a$ (for $\phi_c > 0.5$) shows up in $K(z)$. $\bar{d}_c$ and $\bar{d}_a$ are related to O_{ac} and ϕ_c by

$$\phi_c = \frac{O_{ac}}{2}\bar{d}_c \tag{4.8}$$

and

$$(1 - \phi_c) = \frac{O_{ac}}{2}\bar{d}_a \tag{4.9}$$

On the right of Fig. 4.9, the correlation function deduced from the scattering curve measured at 31 °C is given as an example. The derived parameters, given in the figure caption, have typical values.

Evaluation of the small angle X-ray scattering curves assumes a two-phase model. This cannot be strictly valid as the interfaces are certainly not as sharp as, for example, for a low molar mass crystal in contact with its melt at the equilibrium melting point. However, for the limited resolving power of small angle X-ray scattering experiments, it is difficult to find the transition range. More sensitive are techniques which probe mobilities and the conformational statistics of the chain molecules, like Raman- and NMR spectroscopy, and using them, there is in fact clear evidence for a transition zone between the crystallites and the disordered regions. We will first discuss here a Raman spectroscopic experiment carried out on polyethylene. At the end of this chapter there will be presented an NMR experiment which also demonstrates the existence of transition zones.

Clearly the Raman-spectrum of any polymer must be sensitive to the chain conformation since the latter determines the vibrational properties. In particular, one expects characteristic differencies between a crystalline sequence with a unique regular helical conformation and the wide distribution of different conformations typical for a melt. For polyethylene, there is a spectrum typical for the crystalline *all-trans* sequences and another spectrum which is specific for the melt, and we may wonder how the spectrum of a partially crystalline sample looks like. Figure 4.10 presents Raman-spectra obtained for these three states of order. Part (a) shows the spectrum of a sample of 'extended chain' polyethylene which by use of a special crystallization technique reached a crystallinity close to 100%, part (b) depicts the spectrum of the melt and part (c) the spectrum of a partially crystalline sample. The spectrum of the crystal shows sharp bands because, here, highly restrictive selection rules apply. For example, in order to be Raman active, i.e. to produce a non-vanishing fluctuation of the total polarizability, all monomers have to move in-phase and this holds only for a minor part of the vibrational normal modes of a helical chain. This contrasts with the disordered liquid state, where, due to lack of symmetry, selection rules are much less restrictive so that the Raman-spectrum gets a more diffuse appearance. On first view, the spectrum (c) of the partially crystalline sample may look like a superposition of the elementary spectra (a) and (b), however, detailed analysis shows deviations which are significant. Take for example the spectral range between 1400 and 1500 cm^{-1}. It is assigned to the CH_2-'scissor' (or -'bending') vibrations, which mainly vary the H-C-H valence angles. The crystal structure of polyethylene is orthorhombic, with two C_2H_4-groups in a unit cell (Fig. 4.11 shows the detailed structure). As a result crystalline bands become split in doublets, corresponding to the frequency difference between in-phase and anti-phase vibrations of the two groups. For the CH_2-scissor vibrations, the interaction and thus the splitting is particularly strong, leading to the two sharp lines located at 1416 and 1440 cm^{-1}. We can now observe that the intensity ratio of the two lines in the spectrum of the fully crystalline sample (a) is clearly different from that in the partially crystalline state (c). The findings can be interpreted, as being indicative for the occurrence of regions in the partially

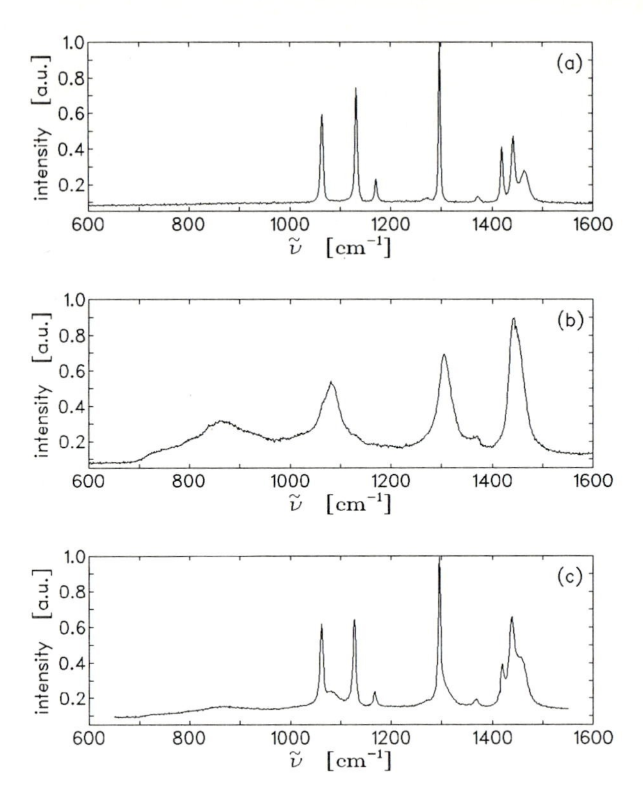

Fig. 4.10. Raman-spectra measured for different states of order of PE: extended chain sample with a crystallinity close to 100% (*a*); melt (*b*); partially crystalline sample (*c*) [34]

crystalline sample where chain sequences are still in the stretched *all-trans* conformation but not packed in an orthorhombic unit cell. Then the splitting disappears, and only the band at 1440 cm^{-1} remains (as can be derived from the spectrum of a triclinic modification with only one chain per unit cell, sometimes observed for *n*-alkanes). Superposition results in the observed enhancement of the 1440 cm^{-1} band. One encounters also a second kind of misfit. As it turns out, a fitting is possible either in the 1000-1550 cm^{-1} range or in the spectral range from 650 to 1000 cm^{-1}, but not simultaneously. Hence, we can see that the Raman-spectrum of a partially crystalline sample includes more contributions than just one corresponding to the orthorombic crystalline state and a second one corresponding to the pure melt and it appears that these additional contributions originate from regions which, although being disordered, still include an enhanced fraction of *all-trans* sequences.

It seems to be a reasonable assumption to associate these regions with the transition from the crystallites to the amorphous layers as one cause for the existence of an intermediate zone is easily seen. Just consider the number

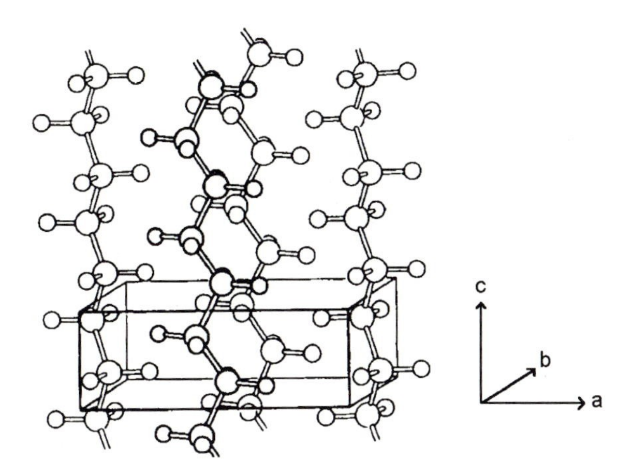

Fig. 4.11. Unit cell of PE crystallites: orthorhombic symmetry, $a = 7.42$ Å, $b = 4.95$ Å, $c = 2.55$ Å. Each cell is occupied by two C_2H_4-groups. Structure determination by Bunn [35]

of chains which pass through a unit area oriented parallel to the crystallite surface, that is to say, the 'chain flux'. Within the crystal, where all chains pass through essentially perpendicularly, the flux is necessarily larger than in the isotropic amorphous region, where crossings are mostly oblique. Straightforward geometrical considerations give a ratio of 2 in the case of orthogonal crystallites when neglecting a possible density difference between the two phases. Consequently, the chain flux has to be reduced on passing from the crystallite into the disordered layer and one may anticipate that this does not occur abruptly but is distributed over a certain range accompanied by a continuous change in the fraction of *trans* bonds.

A decomposition of a given spectrum in three parts can be accomplished without difficulties and yields the respective fractions. In the case of the example given in Fig. 4.10 (c), which refers to a sample with ultra high molecular weight ($M > 10^6$), the following typical values were obtained:

- orthorhombic crystalline phase 66%
- transition zones 13%
- melt-like amorphous phase 21%.

We end this section with some brief remarks on the internal microscopic structure of the polymer crystallites. Earlier, we mentioned the building principle valid for polymer crystallites: These are composed of polymer chains in that helical conformation which corresponds to the minimum of the intramolecular energy. The lattices are built up by a regular packing of these helices. Figure 4.11 presents one example showing the structure of the unit cell of polyethylene. It includes 2 C_2H_4-groups which are oriented with their C-C-planes roughly perpendicular to each other.

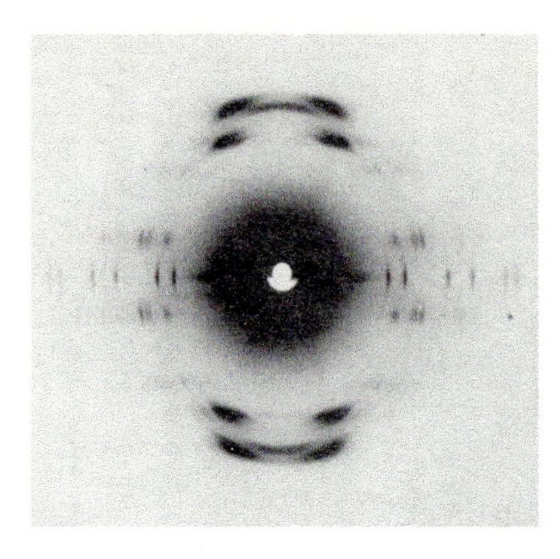

Fig. 4.12. Fiber diagram obtained for a sample of PTFE with uniaxial orientation in an X-ray scattering experiment at 15 °C [36]

Structure analysis for polymer crystallites, i.e. the determination of the edge lengths and angles of the unit cell and the positions of all atoms therein, is carried out, as usually, by X-ray scattering experiments. The conventional straightforward methods of X-ray structure analysis, however, cannot be applied since these require single crystals, which are unavailable for polymers. As the best choice under the given conditions, analysis can be based on the evaluation of scattering patterns obtained for fibers. In fibers or oriented films, crystallites can exhibit a high degree of orientation, with an essentially uniform direction for one of the crystallographic axes. Figure 4.12 shows, as an example, the X-ray scattering pattern obtained for an uniaxially oriented film of poly(tetrafluoroethylene). Bragg-reflections in such a 'fiber-diagram' are generally arranged along 'layer-lines' which are oriented perpendicular to the drawing direction. The latter usually coincides with the chain direction in the crystallites. To carry out a structure analysis, all reflections have to be assigned to lattice planes, i.e. associated with the respective Miller-indices. Although the assignment is facilitated when using fiber diagrams rather than an isotropic scattering pattern, it is usually not straightforwardly accomplished and then requires trial-and-error methods.

Even if one solves the indexing problem and then proceeds with the analysis by an evaluation of measured reflection intensities, one cannot expect to achieve an accuracy in the crystal structure data which would be comparable to those of low molar mass compounds. This is not only a result of the lack of single crystals, but represents also a principal property: In small crystallites, as they are found in partially crystalline polymers, lattice constants can be affected by their size. In many cases crystallites are not only limited in chain direction by the finite thickness of the crystalline lamellae but also laterally since polymer crystallites are often composed of mosaic blocks. Existence of these blocks is indicated in electron microscopic investigations on

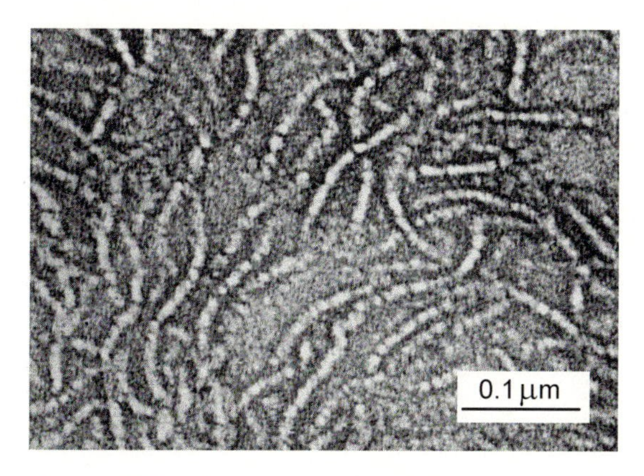

Fig. 4.13. Sample of LDPE , having a crystallinity $\phi_c \approx 0.5$. EM micrographs of a stained ultra-thin section, obtained by Michler [37]. Crystallites are composed of mosaic blocks

stained samples when the stain enters not only the disordered regions, but also the mosaic block boundaries. Figure 4.13 presents as an example electron micrographs obtained for a sample of 'low density'-polyethylene ('LDPE'). The block structure of the crystallites is clearly apparent, and as we see, the lateral extension of the mosaic blocks is comparable to the crystallite thickness.

4.2 Primary Crystallization

Partially crystalline structures in polymer systems form as a result of the kinetics of the crystallization process, and an understanding therefore requires an analysis of the prevalent mechanisms. One meets different conditions, and we begin with briefly addressing them.

The usual and therefore most important case is that encountered when an isotropic melt is cooled to a temperature below the melting point. Then crystallization occurs by nucleation and growth of spherulites, as was shown in the example of Fig. 4.2. In Sect. 4.2.1 we shall discuss this process in detail and consider how it controls the evolving structure.

Special cases exist, where crystallization proceeds in different manner. This can occur if crystallization is induced in oriented amorphous fibers, either starting from the melt during a spinning process, or from a glassy state. Glassy oriented fibers can be prepared by 'cold-drawing' of polymers which crystallize very slowly, so that crystallization can be completely suppressed by a rapid quench. For these oriented samples an alternative crystallization mode is sometimes observed, which is reminescent of the process of spinodal decomposition. The transition from the disordered phase to the better ordered partially crystalline phase proceeds here in a continuous way by passing through a continuous sequence of states rather than building up a two-phase structure from the very beginning. Studies of this process so far are rare and we shall consider them only briefly in Sect. 4.2.2.

It is a general phenomenon in polymers that, after completion of the primary crystallization at the first fixed temperature, crystallization does not come to an end but continues upon further cooling. The temperature range where crystallization occurs and, consequently, also the range of melting during a subsequent heating are always broad. There are two different processes which can contribute to this 'secondary crystallization', the 'insertion mode' and the 'surface crystallization', and they will be discussed in Sect. 4.3.

4.2.1 Spherulite Nucleation and Growth

Figure 4.14 presents crystallization isotherms of polyethylene obtained in time dependent measurements of the density during a spherulite nucleation and growth process. Isotherms are given in terms of the crystallinity, $\phi_{\mathrm{c}}(t)$, deduced from the measured densities by application of Eq. (4.3). Experiments were conducted at different temperatures and we can see that all curves have similar sigmoidal shapes and differ only in time scale. As we notice, decreasing the temperature results in a rapid increase of the crystallization rate. One observes first an induction period required for the formation of the nuclei, which is then followed by a period of accelerated crystallization during which the spherulites grow in radius. When the spherulites begin to touch each other, crystallization rates slow down again. After completion of the spherulitic crystallization, i.e. a complete filling of the volume by spherulites, the crystallinity still increases but does this very slowly.

Let us briefly address the different periods and begin with the nucleation step. In principle, crystallization can start at the equilibrium melting point T_{f}^{∞}, i.e. the temperature where the chemical potentials of a monomer in the melt, $g_{\mathrm{m}}^{\mathrm{a}}$, and in a perfect infinite crystal composed of extended chains, $g_{\mathrm{m}}^{\mathrm{c}}$, are equal

$$g_{\mathrm{m}}^{\mathrm{a}}(T_{\mathrm{f}}^{\infty}) = g_{\mathrm{m}}^{\mathrm{c}}(T_{\mathrm{f}}^{\infty}) \tag{4.10}$$

In practice, however, because nucleation has to take place at first and represents an activated process, a sufficient supercooling below T_{f}^{∞} is a necessary requirement for starting the crystallization process within an acceptable time. If nuclei form in a purified melt, nucleation rates are given by Eqs. (3.125) and (3.126). These include, as a dominant factor, the driving force for the crystallization process, now being given by the chemical potential difference

$$\Delta g_{\mathrm{m}} = g_{\mathrm{m}}^{\mathrm{a}} - g_{\mathrm{m}}^{\mathrm{c}} \tag{4.11}$$

Δg_{m} increases with the supercooling. In linear approximation, for low supercoolings, we may write

$$\Delta g_{\mathrm{m}} \approx \Delta s_{\mathrm{m}}^{\mathrm{f}}(T_{\mathrm{f}}^{\infty} - T) = \frac{\Delta h_{\mathrm{m}}^{\mathrm{f}}}{T_{\mathrm{f}}^{\infty}}(T_{\mathrm{f}}^{\infty} - T) \tag{4.12}$$

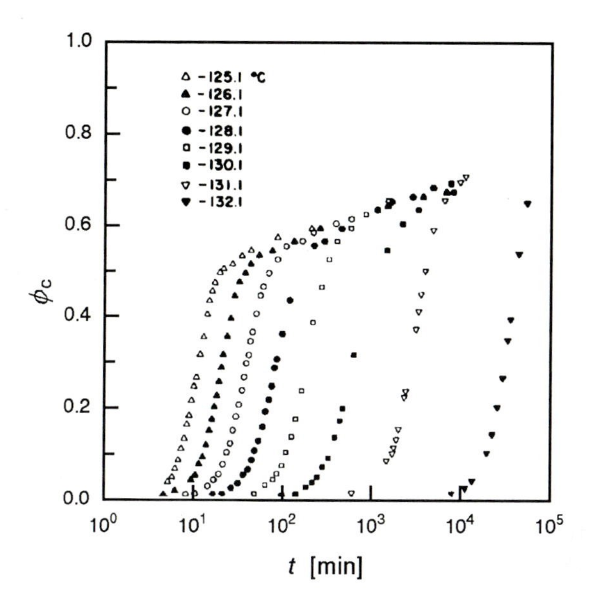

Fig. 4.14. Kinetics of crystallization associated with spherulite nucleation and growth, observed for PE ($M = 2.85 \cdot 10^5$) in time dependent density measurements at the indicated temperatures. Dilatometric data of Ergoz et al.[38]

Here, $\Delta s_{\rm m}^{\rm f}$ and $\Delta h_{\rm m}^{\rm f}$ denote the entropy and the heat of fusion per monomer respectively. Using this linear relation, one obtains for the temperature dependence of the nucleation rate the expression

$$\nu_{\rm nuc} \sim \exp - \frac{\mathrm{const} \cdot (T_{\rm f}^{\infty})^2 \sigma^3 v_{\rm m}^2}{kT \Delta g_{\rm m}^2} = \exp - \frac{\mathrm{const} \cdot \sigma^3 v_{\rm m}^2}{kT (\Delta h_{\rm m}^{\rm f})^2 (T_{\rm f}^{\infty} - T)^2} \tag{4.13}$$

σ designates the excess free energy per unit area of the surface of the nucleus and $v_{\rm m}$ is the monomer volume; the numerical constant depends on the geometrical form (spherical, sheaf-like, rectangular, etc.) of the nucleus. The result predicts a rapid change of the nucleation rate with T, as it is observed. Actually, the equation can only be applied in rare cases. It refers to really pure polymers, since only then must nuclei form in the described manner somewhere in the homogeneous melt. Under practical conditions, nucleation mostly starts on the surface of low molar mass additives, which come into the sample either uncontrolled, or deliberately as 'nucleating' agents, the latter representing a procedure of considerable technical importance. The necessary supercooling is greatly diminished by these additives as a result of a strong reduction of the interfacial free energy σ.

Usually spherulites grow with a constant radial growth rate. The crystallization rate, as given by $\mathrm{d}\rho/\mathrm{d}t$, is then proportional to the total area of free, i.e. non-touching spherulite surfaces. For a growth rate u and a fixed number of nuclei this area, O_{sph}, and thus $\mathrm{d}\rho/\mathrm{d}t$ first increase according to

$$\frac{\mathrm{d}\rho}{\mathrm{d}t} \sim O_{\mathrm{sph}} \sim (ut)^2 \tag{4.14}$$

After exceeding a maximum, both decrease again due to the increasing areas of contact between adjacent spherulites. Reflecting this kinetics, crystallization isotherms $\phi_{\mathrm{c}}(t)$ possess the typical sigmoidal shape.

As it turns out, this time dependence is often well represented by the 'Avrami equation'

$$\phi_{\mathrm{c}}(t) \sim 1 - \exp -(zt)^{\beta} \tag{4.15}$$

It was originally obtained by statistical geometrical considerations when dealing with the problem of how a sample volume gets covered by growing objects of different shapes. Thereby it is assumed that these start at random points, either all at once or at random times. It is the objective of the Avrami treatment to relate the 'Avrami exponent' β and the rate coefficient z to the particular shape of the particles, their growth rate, and the time distribution of the nucleation events, in order to enable in reverse a corresponding data evaluation. Applying it for an analysis of crystallization isotherms of polymers, however, only rarely gives the expected results, for various reasons which we cannot elaborate on here. Notwithstanding, as qualitative conclusions can sometimes be drawn, use of the Avrami equation is popular. At least, it can be employed as a means for rational data representations.

The very slow on-going increase in the crystallinity with long times indicates that the structure which has formed does not correspond to the thermodynamic equilibrium and demonstrates, at the same time, that it is rather difficult to approach further the perfect fully crystalline state. A decrease in the Gibbs free energy can only be achieved by a crystal thickening, which requires highly cooperative motions and these processes are slow. There is also a prerequisite, and this is the presence of an active mechanism of segmental diffusion within the crystallites. In polyethylene such a mechanism exists at elevated temperatures and is being known as the 'α-process', and it enables the observed long time changes. According to Fig. 4.14, the changes are described by a logarithmic time dependence

$$\frac{\mathrm{d}}{\mathrm{d}t}\phi_{\mathrm{c}} \sim \log t \tag{4.16}$$

which implies a progressive decrease in the rate of change with increasing time.

Kinetics of Lateral Crystal Growth

Let us now turn to the most important mechanism, namely that controlling the growth of a spherulite. Figure 4.15 sketches the situation given for an individual crystallite which has its growth face incorporated into the surface

Fig. 4.15. Growth of a crystallite in a polymer melt. The growth face, being part of the spherulite surface, moves in the direction indicated by the *arrow*

of a spherulite. Growth occurs in the lateral direction only by an attachment of chain sequences which are straightened over a length corresponding to the crystallite thickness d_c. Growth in the chain direction is largely suppressed by the presence of folds and entanglements at the crystallite surface. Hence, and this is a main property of polymer crystallization, growth of crystallites is essentially restricted to two dimensions. Spherulite growth results from the simultaneous two-dimensional crystallization processes of the constituing crystallites.

Growth rates u vary strongly with temperature. Studies were performed on numerous systems, and Figs. 4.16, 4.17 present two examples. Figure 4.16 shows the temperature dependencies of the spherulite growth rates of isotactic polystyrene, polyamide 6 and poly(tetramethyl-p-silpheylene siloxane). We observe a maximum for all three polymers, with rapid decreases on both sides, towards the equilibrium melting points at high temperatures and the glass transitions at low temperatures. The drop of u is steep on both sides , with temperature changes being linearly related to a change in the logarithm of the growth rate, $\mathrm{d}\log u \sim \mathrm{d}T$. Figure 4.17 gives in addition growth rates observed for a sample of polyethylene ($M = 2.66{\cdot}10^5$) on the high temperature side.

The reason for the steep decrease on the low temperature side is obvious as being due to the slowing down of the segmental mobility in the melt when approaching the glass transition. We will concern ourselves with this slowing down in later chapters, in particular in Sects. 5.3.2 and 6.3.1, and here only give a result. During the necessary reorganization, chain segments experience frictional forces emanating from the contacting adjacent segments, and these increase progressively on cooling. The increase is well described by

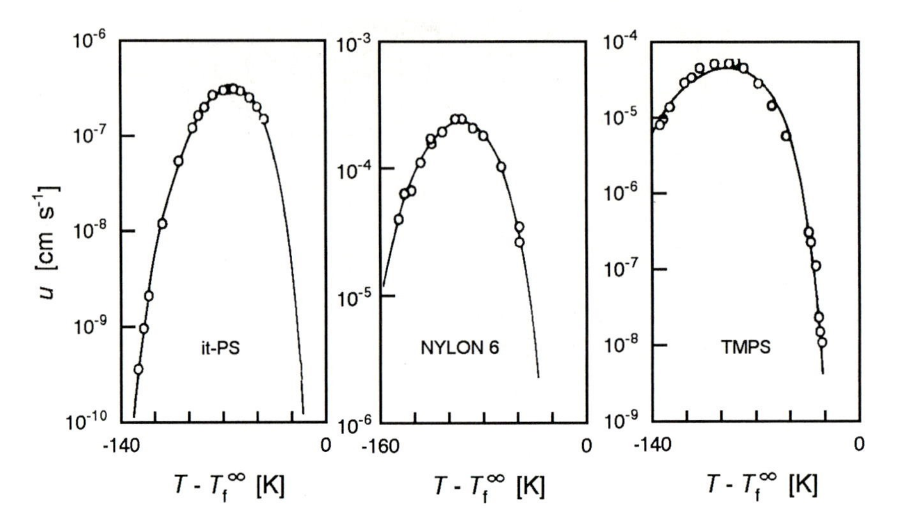

Fig. 4.16. Temperature dependence of the radial growth rate u of spherulites in isotactic polystyrene (*left*), polyamid 6 (*center*) and poly(tetramethyl-*p*-silpheylene siloxane) (*right*). Data from different authors taken from [39]

an empirical equation, the 'Vogel-Fulcher law', as

$$\zeta \sim \exp \frac{T_A}{T - T_V} \tag{4.17}$$

ζ is the 'frictional force coefficient', and its inverse determines the segmental mobility. The Vogel-Fulcher equation includes two parameters. T_V, the 'Vogel-temperature', is usually located $30 - 70\,°C$ below the glass transition temperature, and T_A is an 'activation temperature', with typical values in the order of 1000-2000 K. The low temperature sides of the growth rate data in Fig. 4.16 are mainly determined by the respective Vogel-Fulcher dependencies of the mobility, being given by

$$\log u \sim \log\ \zeta^{-1} \sim -\frac{T_A}{T - T_V} + \text{const} \tag{4.18}$$

Figure 4.17, showing on the left the high temperature growth rate data of a polyethylene sample as they were measured, presents this data on the right in a peculiar form. First, in order to extract the effect of the supercooling, the temperature dependence of the mobility is eliminated by a multiplication with function $\zeta(T)$ as given by the Vogel-Fulcher law. Secondly, the crystallization temperature T is replaced by the inverse of the supercooling. We can see that a linear relation is then obtained and this suggests that we can represent the dependence of u on the supercooling by the following expression

$$u \sim \exp -\frac{T_A}{T - T_V} \cdot \exp -\frac{B_0}{T_f^\infty - T} \tag{4.19}$$

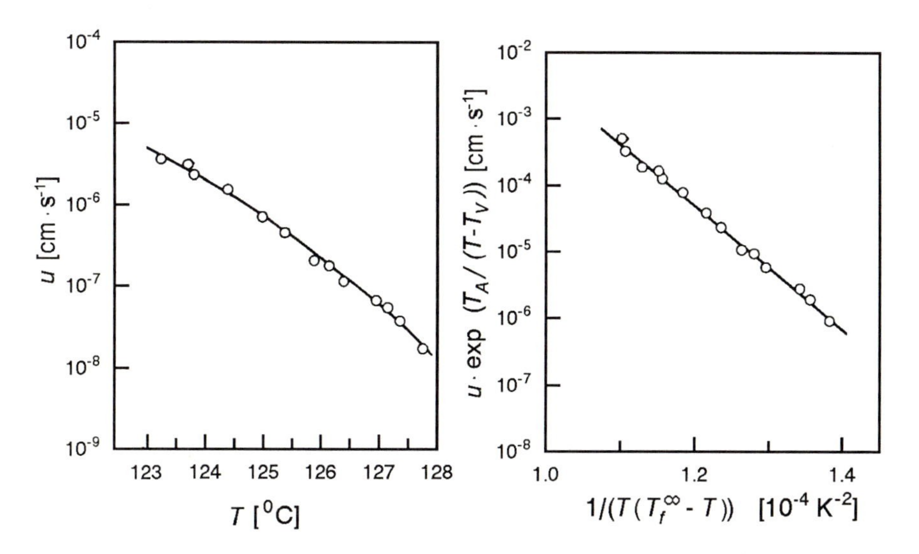

Fig. 4.17. Temperature dependence of the rate of spherulite growth, u, observed in a sample of PE ($M = 2.66 \cdot 10^5$) (*left*). Plot indicating a linear relation between the logarithm of the reduced growth rate $\exp(T_A/(T - T_V)) \cdot u$ (setting $T_A = 750$ K, $T_V = 203$ K) and the inverse supercooling (*right*). Data from Hoffmann et al. [40]

Hence, we have similar functional dependencies on both the low temperature and the high temperature side, but for quite different physical reasons. We have considered one example here, but, in fact, it is representative. It has turned out in experiments on various crystallizable polymer systems that Eq. (4.19) does indeed provide a generally satisfactory description of growth rate data.

One may wonder if the drastic changes in the growth rate near to the equilibrium melting point are accompanied by changes in the structures. Figure 4.18 shows that this is actually the case. Crystallite thicknesses were determined for a sample of polyethylene under variation of the crystallization temperature. The finding is that d_c exhibits pronounced changes with T, with the general trend that increasing the supercooling $T_f^\infty - T$ results in a decrease of d_c. More explicitly, data analysis indicates that the functional dependence is well described by the equation

$$d_c(T) = \frac{B_1}{T_f^\infty - T} + B_2 \tag{4.20}$$

Hence, crystallite thicknesses and supercooling are inversely related.

The qualitative features thus seem clear. We see that on approaching T_f^∞, the crystallizing chains become progressively extended and that this is accompanied by a drastic decrease in the crystallization rate. From this behavior

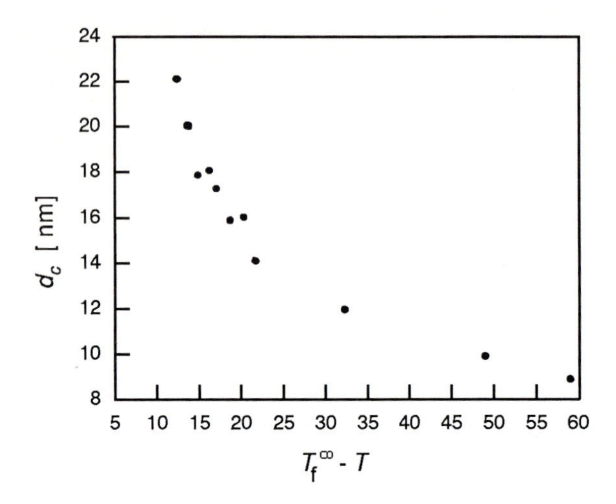

Fig. 4.18. Dependence of the initial thickness of growing crystals on the supercooling, as observed for a sample of PE. Data from Barham et al.[41]

a direct consequence arises: Growth of extended chain crystals, which should form in principle at temperatures close to T_f^∞, is practically ruled out. To transfer a melt completely in the crystalline state appears impossible.

Understanding of the mechanism of spherulite growth and the formation of the partially crystalline structure relies essentially on an understanding of the characteristic temperature dependencies formulated by Eqs. (4.19) and (4.20). We now must consider their origin. It is important, to state clearly at first the selection rule which determines the thickness of the growing crystallites. As was pointed out in the introduction to the chapter, structure formation in partially crystalline polymers is kinetically controlled. This implies, more specifically, that the crystallites which form at a given temperature are those with the maximum growth rate. Figure 4.19 depicts a nice and highly instructive experiment which demonstrates this principle. At the top we can see an optical micrograph of crystallites of poly(ethyleneoxide) as they grow in the melt. Their molecular weight is $M = 6000$ and thus rather low. For these short chains, a complete disentangling and straightening is possible and crystallization, when conducted close to the equilibrium melting point, in fact results in the stable extended chain form. At the bottom of Fig. 4.19, the temperature dependence of the growth rate is plotted as obtained by a direct measurement under the microscope. Crystallization sets in at slightly above 63 °C, and the first, immediately following part of the curve is associated with the formation of these extended chain crystals. One observes the expected increase in the growth rate with the supercooling. Importantly, at 59.5 °C there is a break and it is caused by a change in the crystallite structure. Crystals which form in this second part of the curve are composed of once-folded chains rather than

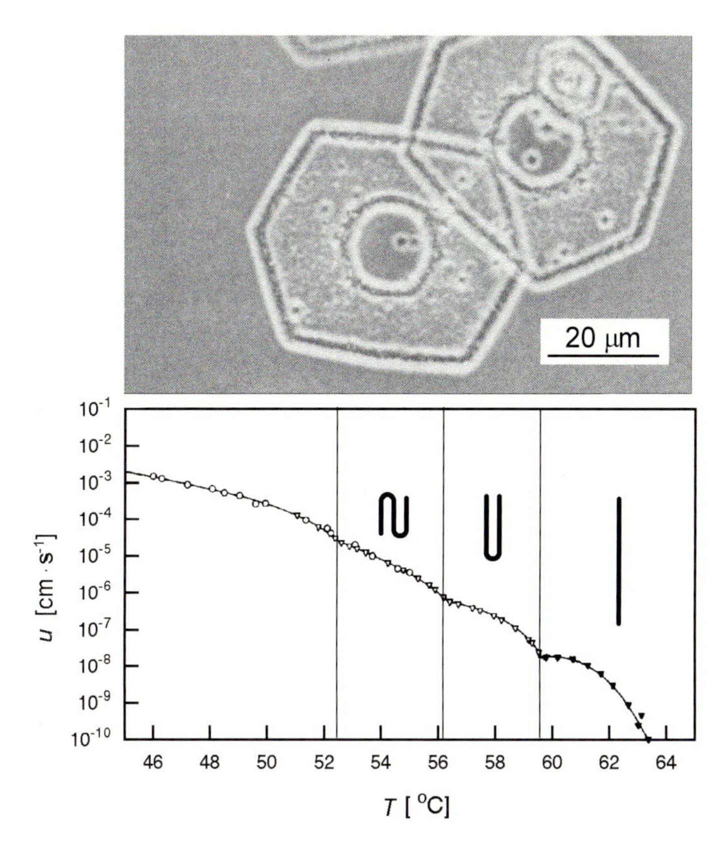

Fig. 4.19. Growth of crystals of PEO ($M = 6000$) in a melt. Picture obtained with an optical microscope using a phase contrast technique (*top*). Growth rate dependence on temperature, exhibiting different branches which are associated with the formation of crystals of extended chains, once- and twice-folded chains and crystals with a continuously decreasing layer thickness (*bottom*). Experimental results obtained by Kovacs et al. [42]

extended molecules. These crystals have a lower melting point and a higher Gibbs free energy than the thermodynamically stable extended chain form. However, as indicated by the curve shape, they obviously possess a higher growth rate, and as we see, it is this form, which is preferred by the crystallizing system. On further cooling, the growth rate curve shows a second break, associated with a transition to crystallites composed of twice-folded chains, followed by a third break, and finally ends up in a continuous part, where we find the typical polymer behavior, $\mathrm{d}\log u \sim \mathrm{d}T$. Within this last, continuous part, the layer thickness changes continuously rather than in discrete steps, agreeing now with the polymer growth behavior described above. We have here an experiment which tells us very clearly what to do in an analysis of the crystallization process in polymer melts: In order to derive the characteristic functional dependencies $d_c(T)$ and $u(T)$, we have to determine that crystallite thickness which results in the maximum growth rate at the chosen crystallization temperature.

Generally, crystal growth rates are determined by the thermodynamic driving force and relevant relaxation times. In order to formulate the driving force when dealing with crystallites of finite thickness, one has first to be concerned with their melting point. For Gibbs free energies $g_{\mathrm{m}}^{\mathrm{a}}$ and $g_{\mathrm{m}}^{\mathrm{c}}$ of a monomer in the melt and in an infinite perfect crystal respectively, and an excess free energy $\sigma_{\mathrm{m,e}}$ to be attributed to the two monomeric units located at the surfaces, the equilibrium condition at the melting point of crystallites composed of sequences of n^* monomers, $T_{\mathrm{f}}(n^*)$, becomes

$$g_{\mathrm{m}}^{\mathrm{c}} n^* + 2\sigma_{\mathrm{m,e}} = g_{\mathrm{m}}^{\mathrm{a}} n^* \tag{4.21}$$

Combination of this equation with the linear approximation for the difference $g_{\mathrm{m}}^{\mathrm{a}}$ - $g_{\mathrm{m}}^{\mathrm{c}}$, Eq. (4.12), yields

$$T_{\mathrm{f}}(n^*) = T_{\mathrm{f}}^{\infty} - \frac{2\sigma_{\mathrm{m,e}} T_{\mathrm{f}}^{\infty}}{\Delta h_{\mathrm{m}}^{\mathrm{f}}} \frac{1}{n^*} \tag{4.22}$$

Equation (4.22) is known as 'Thompson's rule' and describes the melting point depression resulting from the finite size of the crystallites in one direction. A prerequisite for this equation must be recognized. It is only valid under the given conditions, where crystallization and melting occurs on the lateral crystal face only, through the attachment or removal of a complete sequence. When formulating Eq. (4.21), this is implicitly assumed.

Clearly, crystals growing at a given crystallization temperature T have to be thicker than the crystals which are at this temperature in equilibrium with the melt. This implies for n, i.e. the thickness of a growing crystal, and n^*, i.e. the thickness of the equilibrium crystal, the relation

$$n > n^* = \frac{2\sigma_{\mathrm{m,e}} T_{\mathrm{f}}^{\infty}}{\Delta h_{\mathrm{m}}^{\mathrm{f}}} \frac{1}{T_{\mathrm{f}}^{\infty} - T} \tag{4.23}$$

In other words, there must always be a non-vanishing 'excess length' δn

$$n = n^*(T) + \delta n \tag{4.24}$$

δn determines the thermodynamic driving force and therefore the growth rate.

Now we write down a general expression for the velocity of the growth front, u. For the sake of simplicity we consider only the case of a weak thermodynamic driving force, δg, for which a linear relation $u \sim \delta g$ can be assumed. Then we can simply write

$$u = b\tau^{-1} \frac{\delta g}{kT} \tag{4.25}$$

δg describes the driving force per attached sequence of n monomers which, as pointed out, is proportional to the excess length δn

$$\delta g = \Delta g_{\mathrm{m}} \delta n \tag{4.26}$$

The parameter b denotes the monomer diameter. Of greatest importance is the rate constant τ^{-1}. It has to be included in the equation for dimensional reasons, however, it also has a well-defined physical meaning. Consider the equilibrium situation, $\delta g = 0$, when the growth front is stationary. We are dealing here with a dynamic equilibrium rather than with static conditions. Sequences on the crystal face are not fixed but become attached and detached. In stationary state the rates of attachment and detachment are exactly balanced. τ^{-1} refers to this state and describes the 'equilibrium exchange rate per site'. If the temperature is slightly reduced, starting from equilibrium conditions, the attachment rate becomes larger than the detachment rate and the growth face moves forward towards the melt. Conversely, if the temperature is slightly increased, the detachment rate becomes greater than the attachment rate and the growth front moves backwards into the crystal. In both cases, the basic time scale is set by the equilibrium rate constant, τ^{-1}, and it therefore has to be part of the kinetic equation. Equation (4.25) here was constructed by simply considering the relevant parameters and their dimensions. Although it may look plausible, a better foundation is certainly needed. What is necessary is a proper analysis of the balance of the chain fluxes in the two directions, crystallite to melt and melt to crystallite. We shall exemplify it in a model treatment supplemented at the end of this section.

In an important first step, we can demonstrate that the experimental observations are indicative of an exponential dependence of τ^{-1} on the sequence length n

$$\tau^{-1}(n) = \tau_0^{-1} \exp -\mu n \tag{4.27}$$

The rate constant τ_0^{-1} relates to the segmental mobility. It is proportional to ζ^{-1} and thus accounts for the effect of the glass transition. If we now take Eqs. (4.27) and (4.26) and introduce them into Eq. (4.25), we obtain the dependence of the growth rate u on δn:

$$u(\delta n) = \frac{b}{\tau_0} \frac{\Delta g_{\mathrm{m}}}{kT} \exp -\mu n^* \cdot \delta n \cdot \exp -\mu \delta n \tag{4.28}$$

Figure 4.20 shows this dependence in a schematic drawing. Crystals grow for $n > n^*$ and melt from the surface for $n < n^*$. The growth rate has its maximum at

$$n_{\max} = n^* + \frac{1}{\mu} \tag{4.29}$$

where it becomes

$$u = \frac{b \Delta g_{\mathrm{m}}(T)}{\mathrm{e} \mu \tau_0 kT} \exp -\mu n^* \tag{4.30}$$

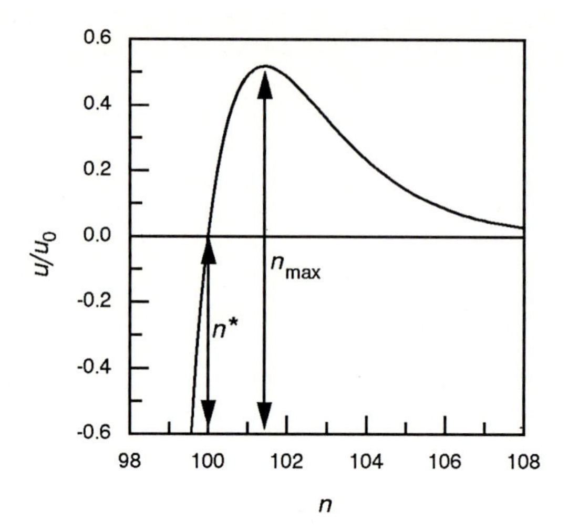

Fig. 4.20. Dependence of the growth rate on the sequence length, as described by Eq.(4.28) (calculated for $n^* = 100,\ \mu = 0.71$)

For a fixed crystallization temperature T, the large majority of crystallites grows with the thickness and maximum growth rate as given by Eqs. (4.29) and (4.30). We write for the crystal thickness

$$d_{\mathrm{c}} = a_{\mathrm{m}} n_{\mathrm{max}} \tag{4.31}$$

where a_{m} is the length per monomer. If we now take n^* from Eq. (4.23) and introduce it into these equations, we obtain the functional dependencies of d_{c} and u on the supercooling $T_{\mathrm{f}}^{\infty} - T$, as

$$d_{\mathrm{c}} = \frac{a_{\mathrm{m}} 2\sigma_{\mathrm{m,e}} T_{\mathrm{f}}^{\infty}}{\Delta h_{\mathrm{m}}^{\mathrm{f}}} \cdot \frac{1}{T_{\mathrm{f}}^{\infty} - T} + \frac{a_{\mathrm{m}}}{\mu} := \frac{B_1}{T_{\mathrm{f}}^{\infty} - T} + B_2 \tag{4.32}$$

and

$$u = \frac{1}{\mathrm{e}\mu} \frac{b}{\tau_0} \frac{\Delta h_{\mathrm{m}}^{\mathrm{f}}}{k T_{\mathrm{f}}^{\infty}} \frac{T_{\mathrm{f}}^{\infty} - T}{T} \cdot \exp - \frac{2\mu\sigma_{\mathrm{m,e}} T_{\mathrm{f}}^{\infty}}{(T_{\mathrm{f}}^{\infty} - T)\Delta h_{\mathrm{m}}^{\mathrm{f}}} \tag{4.33}$$

$$\sim \frac{b}{\tau_0} \frac{T_{\mathrm{f}}^{\infty} - T}{T} \exp - \frac{B_0}{T_{\mathrm{f}}^{\infty} - T} \tag{4.34}$$

These two equations do indeed reproduce the experimental results as represented by the Eqs. (4.20) and (4.19) (apart from the additional factor $(T_{\mathrm{f}}^{\infty} - T)/T$, which brings in another weak temperature dependence).

So far the analysis seems clear, but it is, of course, uncomplete. Actually, we are led to the crucial point for understanding the crystallization process, and this is the question about the cause of the exponential dependence of the equilibrium exchange rate τ^{-1} on n. A simple first answer is at hand. Before a sequence, which lies coiled in the melt, can be incorporated into the growing crystal, it has to be stretched and all intervening entanglements with other chains have to be removed. This is associated with a decrease in entropy proportional to the sequence length, $\Delta\mathcal{S} \sim -n$, which occurs, in thermal equilibrium, with a frequency proportional to

$$\exp \frac{\Delta S}{k} \sim \exp -(\text{const} \cdot n)$$

in correspondence to Eq. (4.27). There is an equivalent view which may be adopted as well. One could state that only a fraction

$$\phi = (\phi_0 < 1)^n = \exp -(\text{const} \cdot n)$$

of the sequences of n monomers in the melt shows the proper straight helical conformation required for an attachment and incorporation into the growing crystal. The conditions thus arising are very similar to the crystallization in a multi-component mixture where only one species is able to crystallize. The different species here are the different rotational isomers and only the sequences with a straight helical conformation build up the crystal. With increasing sequence length they become more and more diluted and the crystallization rate decreases dramatically.

In addition to the large effect following from this basic factor, the crystallization rate may also be influenced by the structure of the crystal growth face. There are two limiting cases. For growth faces which are atomically smooth and crystallographically well-defined, 'layer growth' is prevalent. In this mode, complete layers become attached onto the crystal face and each layer has first to be nucleated on the smooth surface. In this case, growth rates are controlled by this process of 'secondary nucleation'. It becomes the rate determining step since layer completion occurs very rapidly once a nucleus is formed. Hoffmann and Lauritzen showed in early work that the rate of secondary nucleation is given by an equation with the same exponential form as Eq. (4.27). The reason is easy to see. Just like the primary nucleation, secondary nucleation is also an activated process. Since the activation energy $\Delta\mathcal{G}_\mathrm{b}$ is proportional to the size of the nucleus which must possess one length equal to the thickness of the crystallite, the nucleation rate becomes

$$\nu_\mathrm{nuc} \sim \exp -\Delta\mathcal{G}_\mathrm{b}/kT \sim \exp -(\text{const} \cdot n) \tag{4.35}$$

and therefore does indeed agree in form with Eq. (4.27). Hence, if layer growth prevails, two exponential factors of different origin contribute by multiplication. A different situation exists if the growth faces are rough and not well defined. Then another mode of growth is found, usually addressed in the literature as 'normal growth'. In the attachment of sequences on a rough surface, obviously no nucleation step is necessary, since there are always 'niches' available. In this case Eq. (4.27) is related to the entropic factor only.

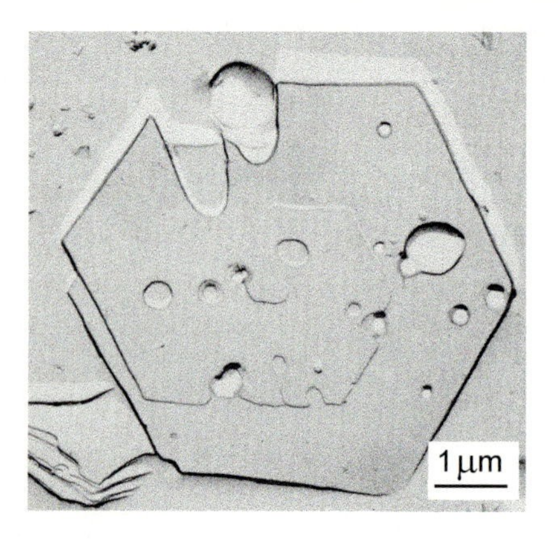

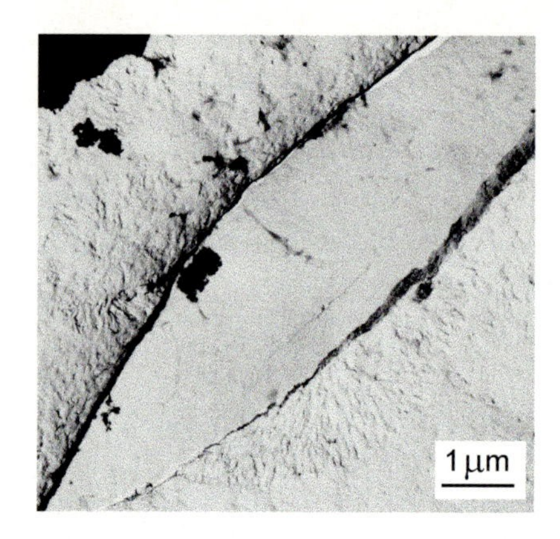

Fig. 4.21. Faceted crystallite of it-PS grown at 228 °C (*left*) and a PE crystallite with elliptically curved contours, grown at 130 °C (*right*). Electron micrographs obtained by Vaughan and Bassett [31]

It appears that both layer growth and normal growth exist in polymer crystallization, together with various transition states. The signature for the prevalence of layer growth is a faceted crystal with a clear habit related to well-defined lattice planes. Normal growth, on the other hand, is indicated, if curved habits with rounded contours are observed. In general, high temperatures promote a surface roughening and therefore normal growth, and lower temperatures result in a smoothing of crystal faces, thus favoring layer growth. Figure 4.21 presents two examples, which appear clear in their signature, both being observed in electron microscopic studies. On the left, we see a faceted crystal of isotactic polystyrene, on the right a curved crystallite of polyethylene which was formed at a high growth temperature.

At the end of this section we come back once again to the kinetical equation, Eq. (4.25), and consider for further clarification a much simplified, but explicit model. We analyse, in more detail, the movement of the lateral growth front of a crystallite which is composed of stretched sequences of n monomers, and introduce for the description two rate coefficients, j_- and j_+. These give the detachment- and the attachment-rate respectively, referring to one site at the lateral growth face to be occupied by a sequence.

At first we consider the case of a roughened growth face. Here, the energy change associated with the transition of a sequence from the crystallite surface into the melt equals the heat of fusion, apart from the endgroup correction. The heat of fusion, referred to one monomer, may be split into two parts

$$\Delta h_{\mathrm{m}}^{\mathrm{f}} = \Delta h_{\mathrm{m}}^{\mathrm{inter}} + \Delta h_{\mathrm{m}}^{\mathrm{conf}} \tag{4.36}$$

The first contribution concerns the change in the intermolecular interaction energy experienced by the stretched sequence on a transfer into the melt and the second part gives the intramolecular energy change associated with the transformation into a coiled state. Obviously, only $\Delta h_{\mathrm{m}}^{\mathrm{inter}}$ is included in the detachment rate and, accordingly, we formulate for j_- the following equation

$$j_- = j_0 \exp -\frac{1}{kT}(\Delta h_{\mathrm{m}}^{\mathrm{inter}} n - 2\sigma_{\mathrm{m,e}}) \tag{4.37}$$

The equation says that a sequence at the crystal face has to collect sufficient thermal energy to overcome the crystal binding energy; the term $-2\sigma_{\mathrm{m,e}}$ accounts for the reduced binding energy of the two endgroups.

For the attachment rate, j_+, we write simply

$$j_+ = j_0 \phi_{\mathrm{H}}(n) \tag{4.38}$$

Here, $\phi_{\mathrm{H}}(n)$ stands for the probability that a sequence of n monomers in front of the growth face has a straight helical conformation. The formulation implies the main assumption of this model, namely, that only such sequences have the potential to crystallize.

We also introduced the parameter j_0. This is a rate constant which relates to the mobility of the straight sequences. Evidently, j_0 has to be included in both equations equally.

$\phi_{\mathrm{H}}(n)$ can be calculated by an application of the rotational isomeric state model described in Sect. 2.4. We have generally

$$\phi_{\mathrm{H}}(n) = \frac{1}{Z_n} \tag{4.39}$$

where Z_n denotes the partition function of a sequence of n monomers. The latter is explicitly given by Eq. (2.137)

$$Z_n = A_{11}(A^{-1})_{11}\lambda_1^{n-1} \tag{4.40}$$

and we write shortly

$$Z_n = \beta_1^{-1}\lambda_1^n \tag{4.41}$$

The velocity of the growth face depends on both the crystallization temperature T and n and is determined by the difference between the rates of attachment and detachment

$$u(T,n) = b(j_+ - j_-) \tag{4.42}$$

$$= bj_0\left[\beta_1\lambda_1^{-n} - \exp -\frac{1}{kT}(n\Delta h_{\mathrm{m}}^{\mathrm{inter}} - 2\sigma_{\mathrm{m,e}})\right] \tag{4.43}$$

The conformational free energy of a coiled sequence of n monomers, $f_{\mathrm{m}}^{\mathrm{conf}}$, is given by

$$Z_n = \exp\frac{nf_{\mathrm{m}}^{\mathrm{conf}}}{kT} = \beta_1^{-1}\lambda_1^n \tag{4.44}$$

and by insertion we obtain

$$u = bj_0\beta_1\lambda_1^{-n}\left[1 - \exp -\frac{1}{kT}n(\Delta h_{\mathrm{m}}^{\mathrm{inter}} + f_{\mathrm{m}}^{\mathrm{conf}} - 2\sigma_{\mathrm{m,e}})\right] \tag{4.45}$$

$\Delta h_{\mathrm{m}}^{\mathrm{inter}}$ and $f_{\mathrm{m}}^{\mathrm{conf}}$ taken together equal the change Δg_{m} of the chemical potential

$$\Delta g_{\mathrm{m}} = \Delta h_{\mathrm{m}}^{\mathrm{inter}} + f_{\mathrm{m}}^{\mathrm{conf}} \tag{4.46}$$

and this leads us to

$$u = bj_0\beta_1\lambda_1^{-n}\left[1 - \exp -\frac{1}{kT}(n\Delta g_{\mathrm{m}} - 2\sigma_{\mathrm{m,e}})\right] \tag{4.47}$$

Writing

$$n\Delta g_{\mathrm{m}} - 2\sigma_{\mathrm{m,e}} = (n^* + \delta n)\Delta g_{\mathrm{m}} - 2\sigma_{\mathrm{m,e}} = \Delta g_{\mathrm{m}}\delta n := \delta g \tag{4.48}$$

we obtain our final result

$$u = bj_0\beta_1\lambda_1^{-n}\left(1 - \exp -\frac{\delta g}{kT}\right) \tag{4.49}$$

For $\delta g/kT \ll 1$, in linear approximation, it converts into

$$u = bj_0\beta_1\lambda_1^{-n}\frac{\delta g}{kT} \tag{4.50}$$

With this result Eq. (4.25) is recovered. Comparison gives

$$\tau^{-1} = j_0\beta_1\lambda_1^{-n} = j_0\beta_1 \exp -n\ln\lambda_1 \tag{4.51}$$

Hence, we find

$$\mu = \ln\lambda_1 \tag{4.52}$$
$$\tau_0^{-1} = j_0\beta_1 \tag{4.53}$$

Our model is very simple indeed and only meant to provide some basic understanding rather than a solid ground for quantitative descriptions of experimental data. One main shortcoming is the complete neglect of all transition states, i.e. conformational states, where the chain is partially attached to the crystal face. Our assumption, that only a single one out of the multitude of the conformational states of the sequence is potentially able to crystallize, is certainly too rigid to be realistic. In addition, the assumed ideal two-phase structure is probably not given since one might expect that a transition zone is built up in front of the growth face, wherein conformations deviate from the distribution in an isotropic melt. Accounting for these effects is difficult, simple solutions do not exist, and a presentation of the various approaches lies outside the scope of this book.

A final comment refers to the effect of the structure of the growth face, in the case where it is smooth rather than rough. The effect can be included into the rate coefficient j_0. If layer growth prevails, the time scale of the dynamics in the stationary state is also controlled by the rate of secondary nucleation

$$j_0 \sim \nu_{\mathrm{nuc}} \sim \exp -\frac{\Delta\mathcal{G}_{\mathrm{b}}}{kT} \tag{4.54}$$

The activation barrier $\Delta\mathcal{G}_b$ is due to the free lateral surfaces on both sides of the rod-like nucleus which extends approximately over n^* monomers, thus becoming

$$\Delta\mathcal{G}_b = n^* 2\sigma_m \tag{4.55}$$

Note that the associated excess free energy differs from $\sigma_{m,e}$. We obtain

$$j_0 \sim \nu_{nuc} \sim \exp -\frac{4\sigma_m\sigma_{m,e}T_f^\infty}{k\Delta h_m^f T(T_f^\infty - T)} \tag{4.56}$$

Experimental growth rate data are frequently evaluated on the basis of this factor alone, with neglect of the 'dilution term' $\exp -n\ln\lambda_1$ which was not included in earlier treatments, where the focus was on secondary nucleation exclusively. Results are then represented by plots of $\log[u\cdot\exp(T_A/(T-T_V))]$ versus $1/(T(T_f^\infty - T))$, as in the example on the right-hand side of Fig. 4.17, and the slope is interpreted as yielding the quantity $4\sigma_m\sigma_{m,e}T_f^\infty/(k\Delta h_m^f)$.

4.2.2 Spinodal Mode

If crystallization is induced in a melt or glass which is not isotropic, but oriented by a drawing process, a mechanism different from nucleation and growth has been observed in a number of cases. Studies by X-ray scattering and using an optical microscope indicate a continuous transition from the amorphous to the partially crystalline state. While changing X-ray scattering curves show that density fluctuations arise in the bulk and grow continuously in amplitude, the sample remains optically homogeneous during the initial stages, contrasting with the heterogeneous appearance of probes where crystallization proceeds by the growth of spherulites. The observations suggest the occurrence of a second, special mode of crystallization.

A first example is given in Fig. 4.22, dealing with the crystallization behavior observed for an oriented glassy fiber of poly(ethylene terephthalate) (PET). Crystallization was initiated by annealing at temperatures slightly above the glass transition. The left-hand side of the figure shows the evolution of the structure factor $S(q)$ with time, as observed in a small angle X-ray scattering experiment. We notice that for scattering vectors $q < q_c = 0.95\ \mathrm{nm}^{-1}$ intensities increase with time; for $q > q_c$ they decrease. As indicated by the plot on the right, kinetics follow for $q < q_c$ an exponential law, $S(q,t) \sim \exp\Gamma(q)t$, with positive rate constants Γ.

The observed scattering curves reflect the continuous development of density variations in the initially homogeneous sample. The experiment shows that long wavelength fluctuations have the tendency to grow and short wavelength fluctuations to decay. The structure which emerges is controlled by the waves with maximum growth rates which are found here around $q = 0.5\ \mathrm{nm}^{-1}$. As a consequence, the structure exhibits a quasi-periodicity with a long spacing $d_{ac} \approx 12$ nm.

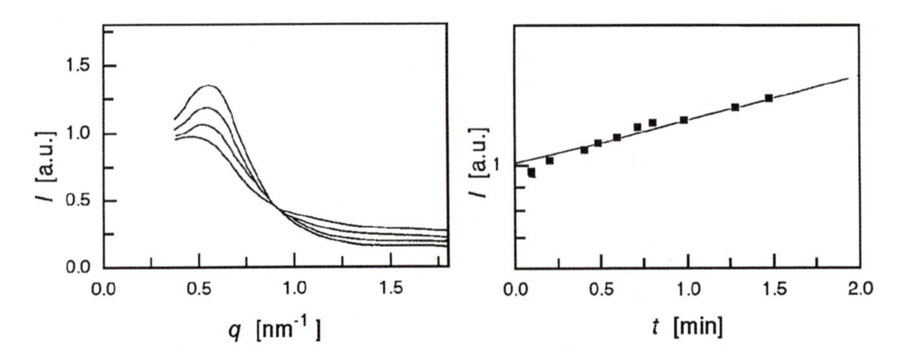

Fig. 4.22. Crystallization of a PET fiber preoriented by drawing, starting from the amorphous glassy state. SAXS curves measured after 0.1, 0.3, 0.8 and 1.5 min of annealing at 95 °C (*left*). Time dependence of the scattering intensity at the maximum $q = 0.55\ \mathrm{nm}^{-1}$ (*right*) [43]

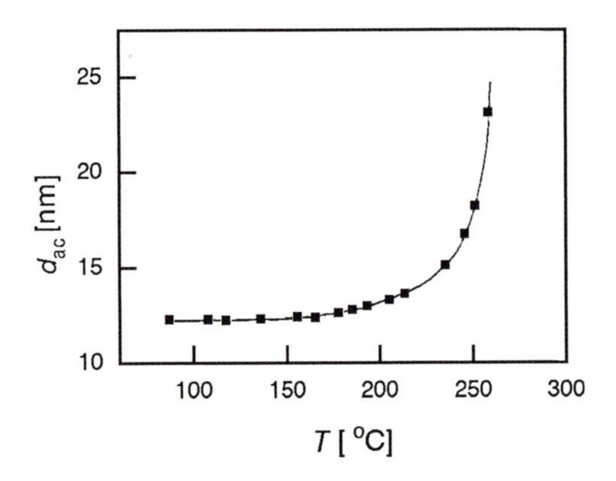

Fig. 4.23. Crystallization of an oriented PET fiber. Long spacing in dependence on the crystallization temperature [43]

The location of the growth rate maximum and thus of the long spacing d_{ac} change with the crystallization temperature. As demonstrated by Fig. 4.23, d_{ac} shifts to larger values with increasing temperature and its dependence can be approximately described as

$$d_{\mathrm{ac}}(T) - d_{\mathrm{ac},0} \sim (T_{\mathrm{c}} - T)^{-1/2} \tag{4.57}$$

The temperature T_{c} is located near to the equilibrium melting point of poly(ethylene terephthalate); $d_{\mathrm{ac},0}$ is a limiting value of the long spacing, reached at the lowest temperatures.

The second example concerns the crystallization of poly(vinylidene fluoride) (PVDF) during a melt spinning process. Extrusion of the melt through a

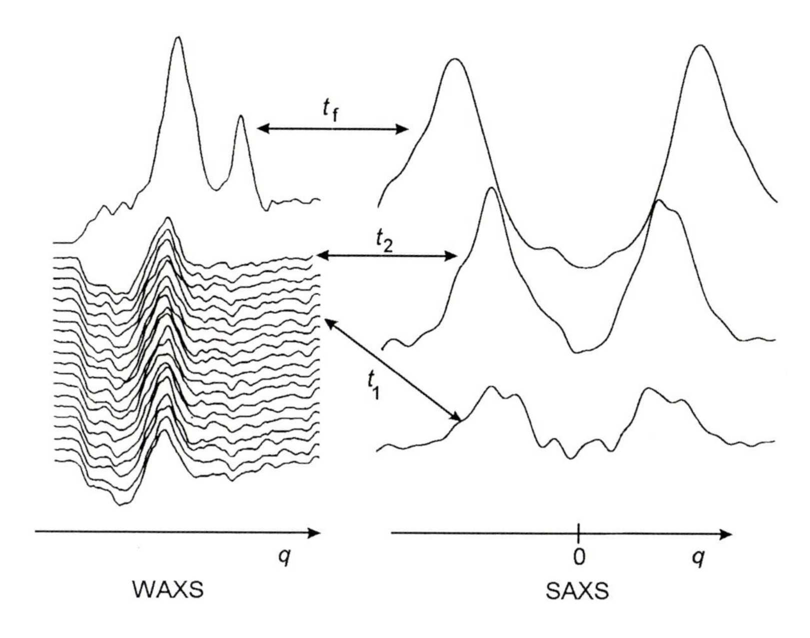

Fig. 4.24. WAXS- and SAXS patterns measured for a melt-spun tape of PVDF at different distances from the extruder exit. Arrows indicate equal locations of two curves, corresponding to times of structure evolution t_1, t_2 and t_f. Work carried out by Cakmak et al [44]

slit-like die yielded a tape which was drawn by a take-up motor with speeds in the order of 10 m min^{-1}. After leaving the extruder, the melt cools rapidly and the tape solidifies by crystallization. In the experiment the crystallization process was followed by simultaneous small angle and wide angle X-ray scattering experiments conducted at different distances from the extruder exit. Figure 4.24 presents some of the scattering curves obtained, WAXS curves on the left-hand side and SAXS curves on the right, both being registered along the direction of drawing. The correspondencies are indicated by arrows, and we notice that the two sets of curves exhibit strikingly different time dependencies: While the WAXS patterns maintain their original liquid-like appearance up to a location which corresponds to a time t_2, a SAXS peak occurs already at the earlier time t_1 and then grows in amplitude. The upper curves belong to the final state of the tape (t_f), and now there is clear evidence for a crystalline phase in the WAXS curves. Hence, according to the small angle X-ray scattering evidence, we again have a continous development of density variations, with the highest growth rates around a certain q. Interestingly enough, these are in the initial stages not accompanied by the formation of crystals. They rather precede the crystallization somewhat.

Clearly, the kinetics of structure formation observed in these two experiments remind one of the process of spinodal decomposition discussed in the last chapter when dealing with phase separation mechanisms in polymer mixtures. To consider a possible role of this mechanism also in the field of polymer crystallization looks appealing. Of course, regarding the chemical constitution, one is dealing with a one-component system. However, the monomers in a polymer chain can vary greatly in their mobility and therefore in their potential to crystallize. For example, chain parts contributing to an entanglement essentially act as cross-links which cannot be incorporated into a crystal. Other unfavorable conformations which are preserved for longer times are isolated hair-pin folds or chain torsions. All these 'conformational defects' are not fixed at certain monomers, but can move along the chain. Considering these properties, one might envisage a process of 'defect clustering' which continuously builds up a varying density distribution, comprising regions with a lowered defect concentration and other regions, where the defect concentration is enhanced. At the end there have formed crystalline regions free of defects and amorphous regions where all the conformational defects are accumulated.

The idea, to have a 'spinodal mode' of crystallization is attractive, and support comes from the experiments described, but it must be said that the evidence is presently restricted to a few examples and not yet really confirmed and well established. In any case, the spinodal mode is certainly the exception, spherulite nucleation and growth being the rule. In a sense, the two processes have something in common in that there has to occur in both cases an unmixing of crystallizable and non-crystallizable chain parts. The difference is that the unmixing in one case occurs locally at the surface of the growing spherulite and, in the other case, the spinodal mode, in the whole sample at once. As we learned in the previous chapter, the basic feature in the spinodal mechanism is the intrinsic instability of the local structure with regard to a continuous unmixing in small steps. The necessary prerequisite for that is now a high mobility of the defects along the chains so that they can be shifted and assembled without any particular activation. A preorientation of chains as it is found in fibers may well promote a high axial mobility, thus setting the basis for the spinodal mode indicated by the experiments.

4.3 Secondary Crystallization

One might expect at first that the formation of the partially crystalline structure is essentially completed when the crystallization at the first chosen temperature is finished. In this case, nothing more would happen on cooling to room temperature. Reheating the sample would produce a narrow melting peak located a couple of degrees above the previous crystallization temperature just at the stability limit of the formed crystallites. Observations on polymer samples contradict this expectation. Figure 4.25 shows the crystallization and melting curves in a measurement of the specific heat of a sample

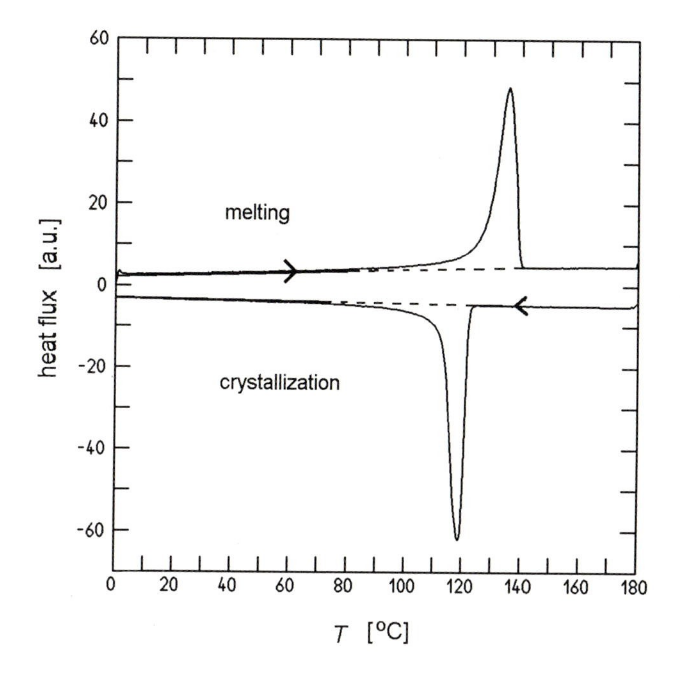

Fig. 4.25. DSC curves measured for a sample of PE ('Lupolen 6011') during a cooling- and heating-cycle ($\mathrm{d}T/\mathrm{d}t = 10$ K min^{-1})

of polyethylene using a differential scanning calorimeter (DSC). We see that the crystallization process extends from the onset at $T = 120\,°\mathrm{C}$ all the way down to about 80 °C. In correspondence to that, the melting curve also shows a broad melting range rather than a sharp melting peak. The maximum of the crystallization curve is displaced to lower temperatures in comparison to the maximum of the melting curve thus showing the necessary supercooling.

Figure 4.26 depicts the results of a Raman-spectroscopic experiment on another sample of polyethylene, which is again indicative of a continuing secondary crystallization process. Spectra were measured in the ranges of the CC-stretching and CH_2-twisting vibrations for different temperatures during cooling. Pronounced changes are observed and the results of the data evaluation are given at the bottom of the figure. One notices a continuous decrease in the melt-like fraction, an increase in the crystalline fraction, and an essentially constant value for the fraction of methylene units in the transition zone.

Especially broad crystallization and melting ranges are observed for polymers which include a small amount of non-crystallizable co-units, like short-chain branches or chemically different monomers. Figure 4.27 presents, as an example, two melting curves of a commercial low density polyethylene which has about 3% of short chain branches. The melting process here was followed

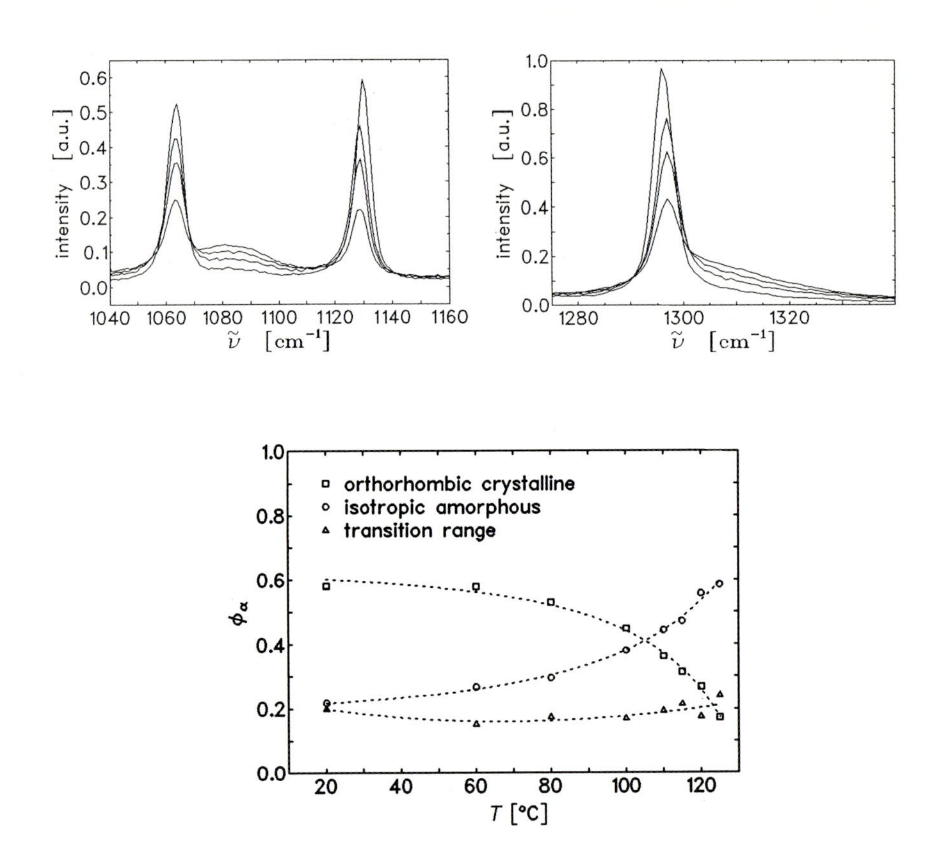

Fig. 4.26. Raman-spectra measured for a sample of PE ($M \geq 10^6$) in the frequency ranges of the CC-stretching vibrations (*top left*) and the CH_2-twisting vibrations (*top right*). Spectra were registered at 126 °C, after completion of isothermal crystallization, and then successively at 115 °C, 100 °C and 25 °C. Fractions of methylene units in the orthorhombic-crystalline, amorphous and intermediate phase, as derived from a decomposition of the spectra (*bottom*) [34]

again by DSC and in a parallel experiment also by a measurement of the thermal expansion coefficient, $\beta(T)$, using a dilatometer. This particular sample had first been crystallized by a step-wise cooling, keeping it for extended times at a series of discrete temperatures. The melting curve is very broad and exhibits a fine-structure which obviously keeps a memory of the special thermal history.

More insight into the nature of these subsequent crystallization processes comes from small angle X-ray scattering experiments. In Fig. 4.8, we presented a series of small angle X-ray scattering curves measured for a sample of polyethylene at the end of isothermal crystallization at 125 °C, and at several discrete temperatures passed through during the subsequent cooling. This

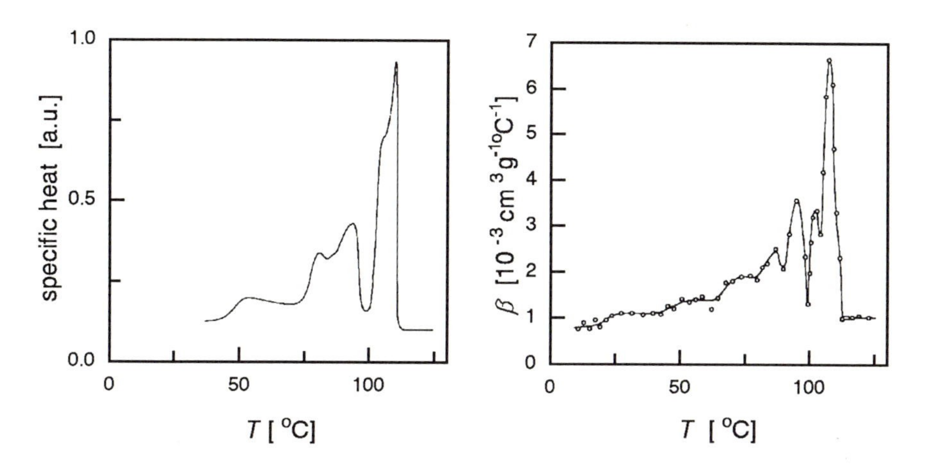

Fig. 4.27. Melting curves of a sample of low density PE ('Lupolen 1800S') crystallized by stepwise cooling (melt at 115 °C → 100 °C → 75 °C → 50 °C → 25 °C), obtained in measurements of the specific heat using a differential calorimeter (*left*) and of the expansion coefficient β, registered in a dilatometer (*right*) [45]

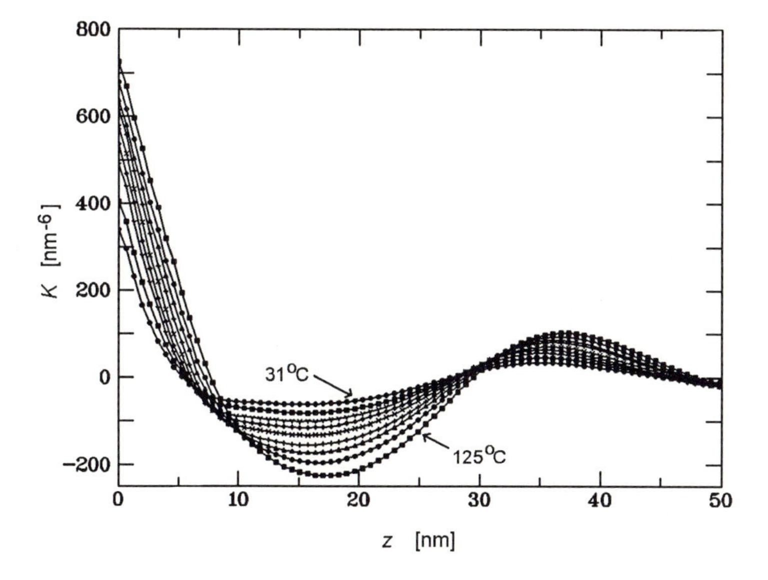

Fig. 4.28. Same sample as Fig. 4.8. Electron density correlation functions $K(z)$ derived from the scattering curves [33]

is now complemented in Fig. 4.28 by the derived electron density correlation functions. Pronounced changes are observed, and the analysis of the shapes of $K(z)$ tells us, how the partially crystalline structure changes during cooling. First insights are provided by a look at the 'self-correlation triangle' with a

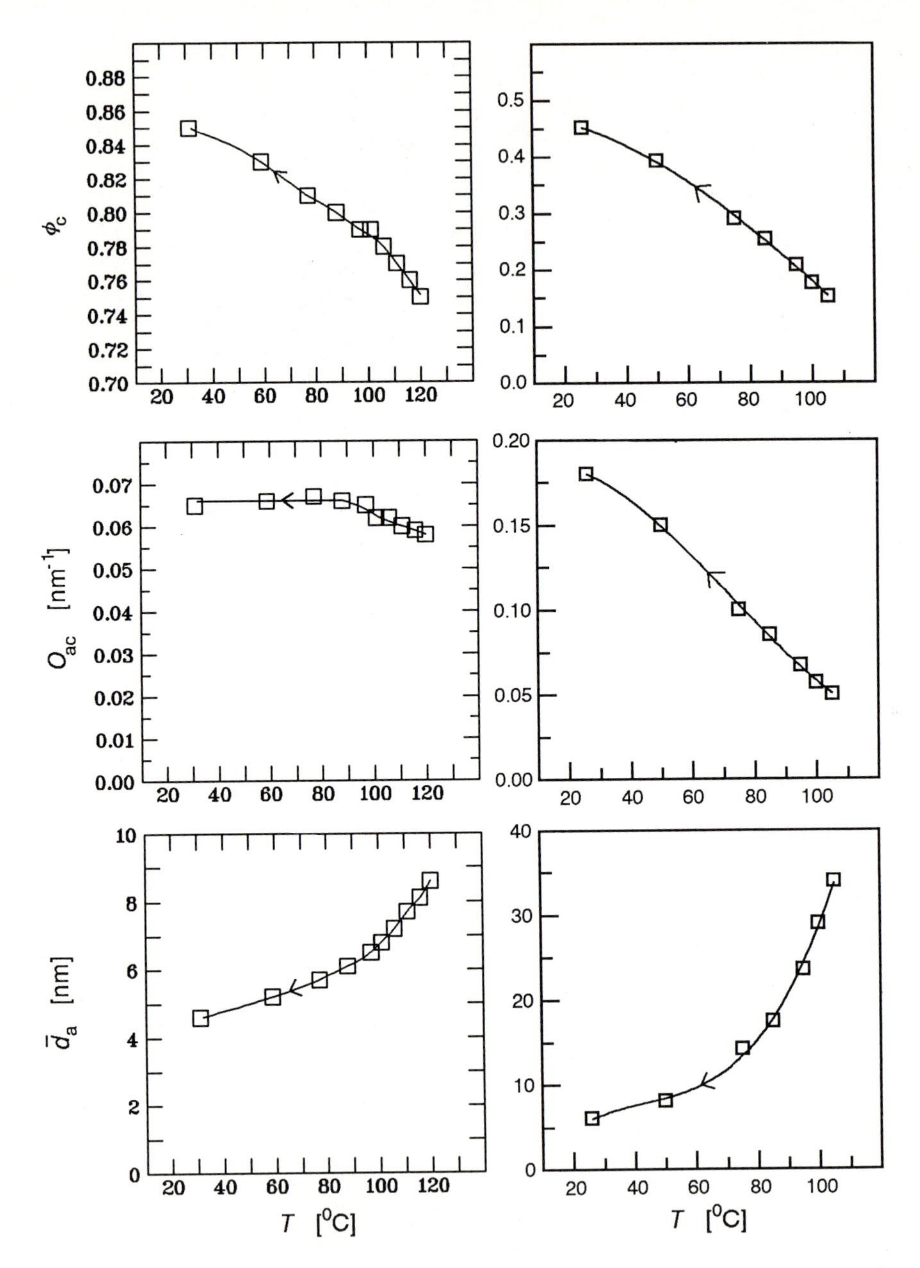

Fig. 4.29. Changes of the crystallinity ϕ_c, the interface area per unit volume O_{ac} and the mean thickness of the amorphous layers $\bar{d}_a$ during cooling of samples of linear PE (*left column*) and branched PE (*right column*), as derived from SAXS experiments and complementary density measurements [32], [46]

maximum at the origin. As indicated in the schematic drawing on the left of Fig. 4.9, the width of its base gives the average thickness of the amorphous layers. Obviously this thickness decreases continuously during cooling. The finding indicates first of all that the regions between the crystallites which remained amorphous after the primary isothermal crystallization still have the potential to crystallize and partially realize this on cooling. For this additional crystallization a higher thermodynamic driving force is needed, and it develops at the lower temperatures. The correlation functions $K(z)$ may be evaluated quantitatively and Fig. 4.29 presents in the left column of plots the derived structure parameters. We can see that the crystallinity, reaching 75% after the primary crystallization, increases further during cooling by another 10%, and that this is associated with a decrease in the thickness of the amorphous layers to half of its original value. The interface area O_{ac} is much less affected by the structure change and remains constant below 100 °C. The column of plots on the right-hand side collects the results of an analogous SAXS experiment, now conducted on the sample of Fig. 4.27, a low density polyethylene with 3% of short chain branches. Interestingly enough, we find a different behavior. The primary crystallization process yields only a crystallinity of 15% which then goes up by another 35% during cooling. Hence, secondary crystallization provides the larger part. Furthermore, in contrast to linear polyethylene, one observes for the branched species a large change in O_{ac} which is even larger than for ϕ_c.

These are clear results, and they enable structural interpretations to be made. The observations in the second case obviously indicate the formation of additional crystallites during cooling. These become successively inserted into the original stack built up during the primary isothermal crystallization. On the other hand, the mechanism which dominates the behavior in linear polyethylene is a continuous shift of the interface towards the amorphous regions corresponding to a 'surface crystallization'-process. Both processes have in common that they reduce the thickness of the amorphous layers but they accomplish it in different ways. Further observations also showed that both processes are largely reversible. They both emanate from the primary structure built up during the first crystallization and leave this structure, as given by the positions of the primary crystallites, unchanged as long as the temperature of the primary crystallization is not exceeded. The two last sections of this chapter concern these two processes, the 'insertion mode' of secondary crystallization and the surface crystallization and melting process, and provide qualitative explanations of their physical origins.

4.3.1 Insertion Mode

Using an electron microscope, it is possible to have a look at the structure which exists at the end of the primary isothermal crystallization at elevated temperature, and to compare it with the final state reached after cooling the sample down to room temperature. Figure 4.30 presents two micrographs

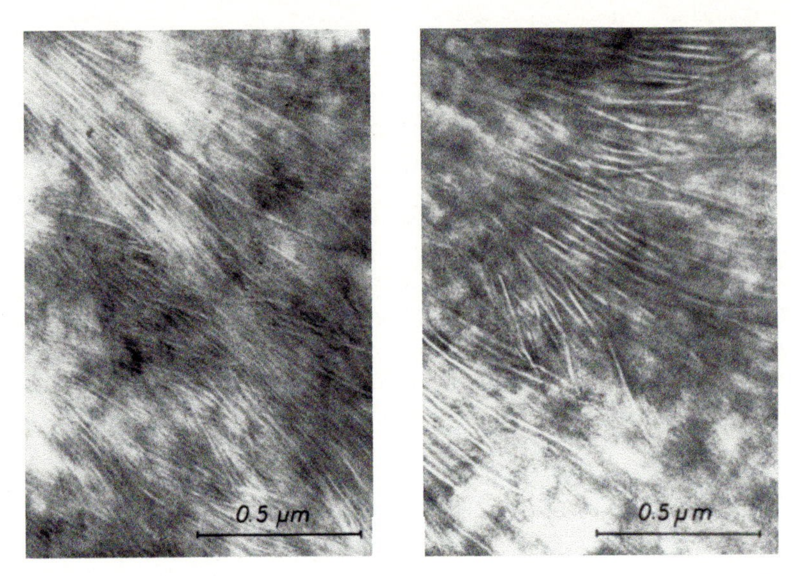

Fig. 4.30. Electron micrographs showing the structures of a sample of LDPE after isothermal crystallization at 100 °C (*right*) and a subsequent cooling to room temperature (*left*). Images were obtained on thin sections stained with OsO_4 [46]

showing the structure of a low density polyethylene sample at 100 °C and at ambient temperature. The images were obtained for thin sections where the crystallites show up as white lines, just as in Fig. 4.5. The difference in the morphologies for the two temperatures is evident. The partially crystalline structure at 100 °C is composed of crystallites with large lateral extensions and a mostly uniform thickness. At room temperature this uniformity is lost, and as it appears, many thin lamellae have formed in-between the primary crystallites.

The observations suggest a peculiar mechanism of the structural development and it is shown schematically in Fig. 4.31. On the left-hand side, what happens during the lateral growth of a single crystallite is depicted. This is necessarily associated with transport to the surface of all non-crystallizable chain parts such as short chain branches, end groups, and also the entanglements. As a consequence, a zone is created which has an enhanced concentration (c_B) of non-crystallizable units. This zone sets up together with the crystallite a region with thickness d_{min}, which cannot be entered by any other growing crystalline lamella. As a result, after completion of the primary crystallization at T_0, distances between the centers of adjacent crystallites will vary between d_{min} as the lower and $d_{max} \approx 2d_{min}$ as an upper limit, the latter following from the fact that if a distance is larger than $2d_{min}$, another crystallite will grow in-between. Cooling the sample then leads to the formation of additional crystallites by insertion into the primary stack in the manner sketched on the right-hand side. The new crystallites which become inserted

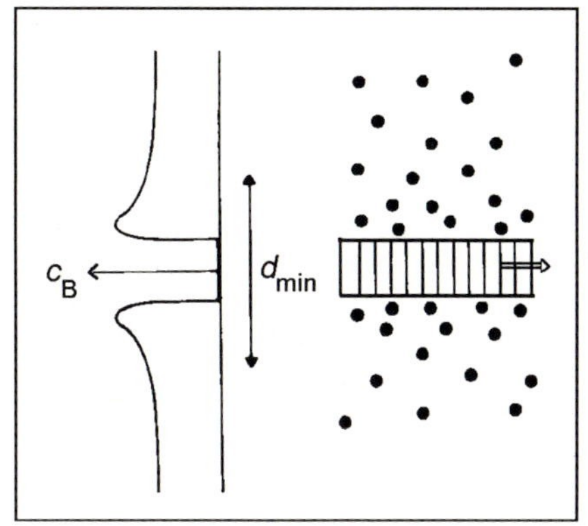

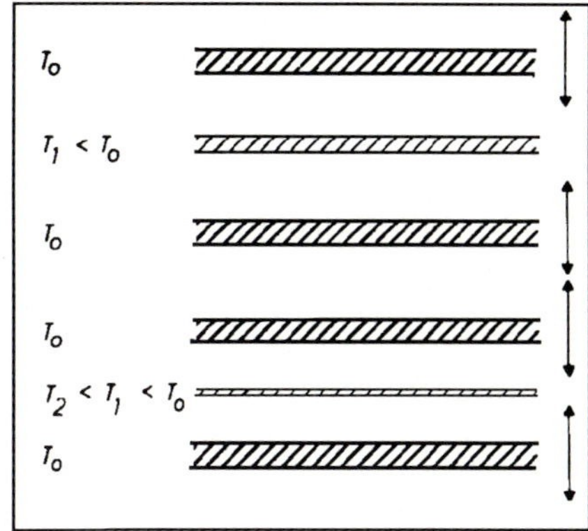

Fig. 4.31. Consecutive building up of a stack of crystallites during cooling via the insertion mechanism. Formation of a crystallite generates a zone of thickness d_{min} which cannot be entered by other crystallites (*left*). Primary crystallization at T_0 produces a stack of crystallites with varying distances d_{ac} in the range $d_{\mathrm{min}} < d_{\mathrm{ac}} < 2d_{\mathrm{min}}$. On cooling to T_1 and T_2 further lamellae become successively inserted, their thicknesses decreasing with temperature (*right*)

at the first temperature, T_1, appear in the thickest amorphous regions. Then on cooling to the next temperature, T_2, crystallization enters the next thinner ones, and the process continues in this manner on further cooling. The variable which controls the temperature at which a crystallite forms within a given amorphous layer is the concentration of non-crystallizable units therein. Their presence leads to a melting point depression, similar to the case when a non-crystallizable low molar mass solute is dissolved in a melt of some crystallizing compound. As described by Raoults' law, the depression increases with the concentration. Clearly, crystals which form at lower temperatures are generally thinner and can also include more internal disorder.

It is important to see the central point in the model and this is the assumed non-uniform distribution of the non-crystallizable units in the primary stack, being higher in the thinner and lower in the thicker amorphous layers. The necessary prerequisite for this is a suppression of any transport of non-crystallizable units through the crystallites so that the difference in concentrations is maintained and an equilibration prevented. For branched polyethylene, this condition is obviously fulfilled, which appears reasonable, considering the size of the short-chain branches.

Further experiments indicate that the additional crystallites which form during cooling do not need a separate nucleation. Figure 4.32 shows the difference in crystallization kinetics between the primary isothermal crystallization and the subsequent crystallization steps. The time dependence of the density for the primary crystallization shows the common sigmoidal shape, beginning with a nucleation period. In contrast, the curves for the subsequent crystal-

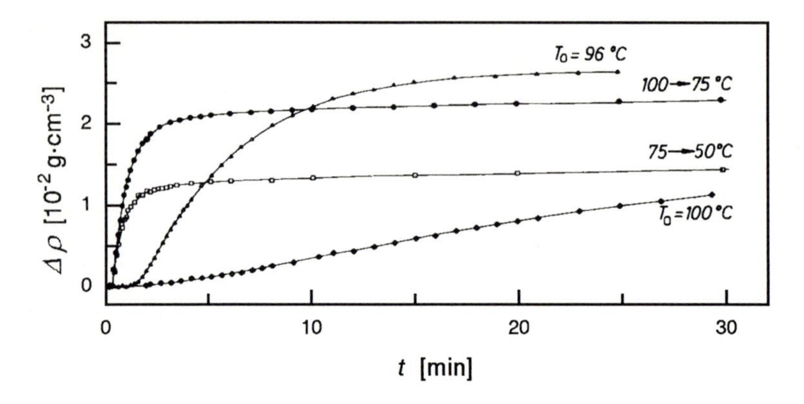

Fig. 4.32. Crystallization isotherms observed in density measurements on a sample of LDPE, using a dilatometer. Note the difference in shape between the primary crystallizations at 100 °C or 96 °C, and the subsequent crystallizations on stepwise cooling, from 100 °C to 75 °C and 50 °C [45]

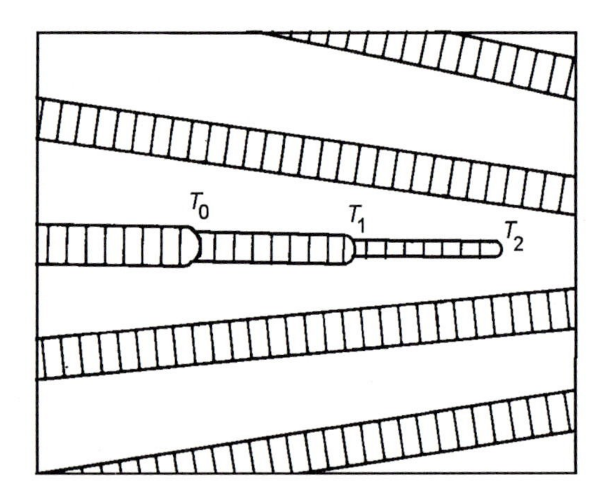

Fig. 4.33. Growth of a lamella in-between two tilted crystallites. At the primary temperature T_0 growth stops at first and then proceeds continuously upon cooling to T_1 and T_2, accompanied by a crystallite thinning

lization steps exhibit no nucleation period at all. The maximum crystallization rate is observed at the very beginning and then it decreases continuously. How can these observations be interpreted? A possible explanation is schematically indicated in Fig. 4.33. Crystallites are often slightly tilted with respect to each other. In this situation, there arises a continuous increase, i.e. a gradient in the concentration of non-crystallizable units. The consequences are obvious:

A lamella which is growing at a given temperature, say T_0, in-between two tilted crystallites stops at first at a certain point, and then continues on further cooling to T_1 and T_2. Necessarily, from step to step, the thickness of the crystallite becomes successively smaller. Indeed, closer inspection of electron micrographs corroborates this view, and the absence of a renewed nucleation is thus fully conceivable.

4.3.2 Surface Crystallization and Melting

By indicating a continuous shift of the interfaces, the results of the small angle X-ray scattering experiments on linear polyethylene given in the left column of Fig. 4.29 suggest the occurrence of a surface crystallization and melting process. Further evidence comes from the Raman-spectroscopic findings depicted in Fig. 4.26. There we see that in spite of pronounced changes in the crystallinity the fraction of material in the transition zones is largely constant, which is indicative for an unchanged interfacial area, just as is expected for a surface melting process. Figure 4.34 shows a sketch of this peculiar mode of crystallization. A change of temperature is accompanied by a continuous shift of the interface, on cooling towards the amorphous regions and on heating towards the crystallites.

That this process exists at all tells us something specific about the structure of the amorphous regions, namely that they must be different from a normal polymer melt. The behavior actually suggests that there is a local thermodynamic equilibrium between crystallites and adjacent amorphous layers. The physical background is easily seen when one envisages the chain structure in the amorphous regions. Rather than having free chains, one finds only loops which are fixed with their ends in the crystallites. Furthermore, the amorphous layers include trapped entanglements in a concentration which is enhanced compared to the melt. As a consequence, the mean chemical potential of the units differs from that in a pure melt, and one can expect that it varies with the thickness of the disordered layer. To see the direction of change, one should realize that the numbers of entanglements and points of chain entrance into the crystallites are constant. The restrictions for the chain motion therefore become released if the layer thickness increases and this implies a decrease in the chemical potential. Under these conditions, for each

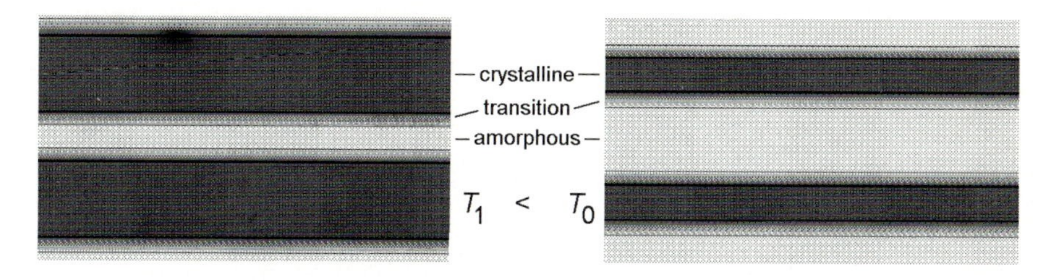

Fig. 4.34. Schematics of the process of surface crystallization and melting

change in temperature, a new local equilibrium between crystallites and disordered regions becomes established and this means a surface crystallization or melting.

Another point is important: The envisaged process can only be realized if the continuously renewed equilibration is kinetically possible. A shift of the interface requires a rearrangement of the chains in the crystallites and this can only be accomplished if chains are perfect, i.e. do not include co-units or branches which cannot enter the crystallites, and in particular, if chains possess sufficient mobility to facilitate a longitudinal transport. For linear polyethylene both conditions are fulfilled, the necessary longitudinal mobility being provided by the action of the 'α-process'. Direct evidence comes from a nuclear magnetic resonance (NMR) experiment.

Since we are considering an NMR experiment for the first time in this book, some introductory remarks are needed in order to at least roughly explain the result of the measurements. To begin with a general statement, NMR spectroscopy provides information on the local state of order and the molecular dynamics in a solid or liquid and this is probed by spin carrying nuclei. Protons (^{1}H), deuterium (^{2}H) and the carbon isotope ^{13}C (the major carbon isotope, ^{12}C, is spinless) are particularly suitable for studies. There is a large variety of NMR experiments, distinguished by the way the spins are excited and the manner in which the resulting magnetization is further modified and probed.

We will briefly discuss here an NMR experiment on polyethylene based on ^{13}C, and it is based on the observation that the resonance frequency of ^{13}C, being dependent on the local surroundings, differs between the crystalline and the amorphous phase. In terms of the ppm-units used in NMR spectrocopy (spectral shifts are expressed in 'parts per million', referring to the resonance frequency of a standard substance), the resonance of the crystallites is found at 32.5 ppm and the signal of the amorphous parts at 30.5 ppm. One has to mention here that this small shift can only be observed if the much larger shifts due to the magnetic interactions between all the spins in the sample are completely removed and this can be accomplished by a technical procedure, namely a rapid rotation of the sample about the 'magic angle' $\theta = 54.7°$ (θ is the angle enclosed by the magnetic field and the rotation axis). The experiment considered uses this technique and therefore is called a 'magic angle spinning ^{13}C two-dimensional exchange experiment'. It nicely demonstrates the capabilities of NMR spectroscopy, and it is possible to explain in broad outlines its information content, without the need to refer to the technique applied.

Figure 4.35 shows the results of measurements at two different temperatures, $T = 363$ K (left-hand side) and $T = 373$ K (right-hand side). Functions which depend on two variables, $p(\omega_1, \omega_2)$, are represented in two-dimensional plots with level-line plots as inserts. To each curve, there belongs a certain 'mixing time' t_{m}. The functions $p(\omega_1, \omega_2)$ have a well-defined meaning: They represent the probability that a ^{13}C-nucleus, which was at a position with a

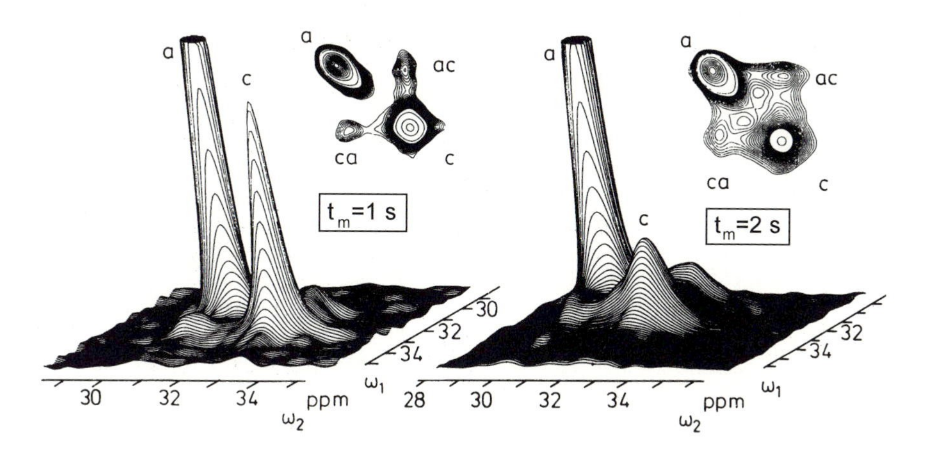

Fig. 4.35. MAS ^{13}C 2d exchange spectra measured for a sample of PE ($M = 4 \cdot 10^6$) at 363 K (*left*) and 373 K (*right*). Work reported by Schmidt-Rohr and Spiess [47]

resonance frequency ω_1 at zero time, changes within a time t_m to a position with a resonance frequency ω_2. We observe in the figures four peaks, two large ones, designated 'a' and 'c', located on the diagonal $\omega_1 = \omega_2$ and two smaller ones, 'ac' and 'ca', found in off-diagonal positions. Their assignments are obvious. Contributions to the diagonal peaks originate from those ^{13}C-nuclei which either remained within the crystalline phase ($\omega_1 = \omega_2 = 32.5$ ppm) or within the amorphous phase ($\omega_1 = \omega_2 = 30.5$ ppm). Importantly, the two off-diagonal peaks provide direct evidence for a move of monomers, either from the crystalline to the amorphous phase ($\omega_1 = 32.5, \omega_2 = 30.5$), or in reverse direction ($\omega_1 = 30.5, \omega_2 = 32.5$). Comparison of the two results in Fig. 4.35 indicates that the amount of exchange increases with time and temperature.

The experiment tells us definitely that a motional mechanism is active which produces the exchange. A longitudinal transport of monomers through the crystallites is required. It looks reasonable to assume that the transport is mediated by diffusing conformational defects, such as a local chain twist by 180°. Based on such a model, a large set of data obtained under variation of t_m and the temperature was evaluated and it was possible to derive the rate with which a crystalline sequence becomes displaced over the length of one CH_2-unit. Figure 4.36 shows the temperature dependence of this jump rate and, as can be seen, it obeys an Arrhenius-law with an activation energy $\tilde{A} = 105\ \mathrm{kJ \cdot mol^{-1}}$.

A second observation in Fig. 4.35 is also noteworthy. We see non-vanishing values of $p\ (\omega_1, \omega_2)$ also away from the peaks and in particular between the peak 'c' and the off-diagonal peaks 'ac' and 'ca'. This is just the range where contributions of the transition zones are expected, when units are passing through during a change from the crystallite into the amorphous phase. One

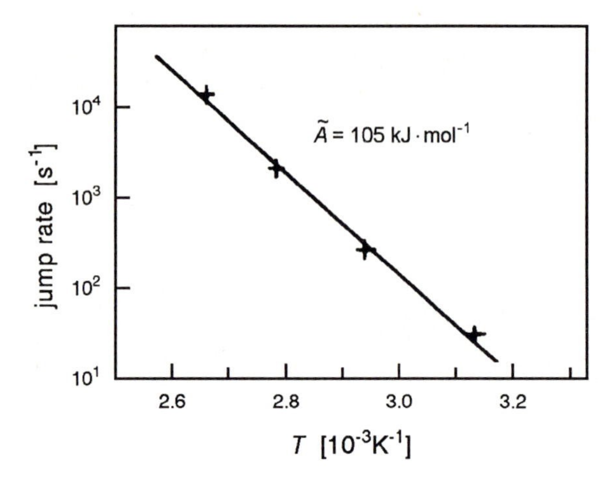

Fig. 4.36. Rate of jumps over one CH_2-unit performed by the crystalline sequences in PE. Result of an NMR experiment by Schmidt-Rohr and Spieß[47]

can also check for the contributions of the transition zones in-between the two peaks on the diagonal, corresponding to units which stay there for a time t_m. These contributions are small indeed, which indicates that the monomers remain only for a short time in this region, then diffusing away, either into the crystallite or towards the center of the amorphous zones.

Hence, the small angle X-ray scattering and NMR experiments, considered together, tell us that linear polyethylene shows a surface crystallization and melting process and that the prerequisite for its occurence, a mechanism of longitudinal transport enabling continuous reajustments of the chains, is fulfilled. One may wonder if surface crystallization and melting represents a widespread phenomenon, similar conditions as to polyethylene being also found for many other polymers. It appears today that this is not the case. Indeed, positive evidence so far is restricted to polyethylene and there is, for example, a clearly negative result for polypropylene where longitudinal transport is in fact lacking. There are other promising candidates, for example poly(oxymethylene), where NMR spectroscopy has detected helical jumps but, so far, these have not been checked for surface melting.

Surface crystallization and melting being the exception, the insertion mode is the rule and is indeed mainly responsible for the generally observed secondary crystallization. As it does not require a mobile crystalline phase, it can always occur.

This chapter would not be complete without a warning. It is natural that a textbook based on a lecture course has a personal touch, expressed in the selection of the topics and the weights attibuted to the various arguments and concepts in the literature. For this chapter this is even more applicable than for the others. Readers who already have some knowledge in the field perhaps miss a notion which plays a major role in many discussions of polymer crystallization, namely the 'chain folding' by 'adjacent reentry'. In fact, I consider this mechanism as representing a secondary feature only, rather than as a primary phenomenon and see a danger if it is put in the foreground of the considerations. There is the danger, at least for newcomers in the field, to compare it to the process of 'refolding' of a protein, taking place during the transformation from the denaturalized coiled state into the biologically active uniquely structured equilibrium state. Indeed, while the latter *is* a primary process, in synthetic polymers a comparable intrinsic mechanism which would enforce a regular chain folding, does not exist. Chain folding by adjacent reentry, as it surely takes place with a certain probability when crystallites form, only represents an efficient means to accomplish the reduction in chain flux which must accompany the transition from a crystallite into the adjacent amorphous regions. If layer growth is prevalent, the sharp folds may become equally oriented and, as a result, a limited degree of order arises on the crystallite surface, effective enough to induce the sometimes observed surface sectorizations and epitaxial interactions between crystallites. In the view expressed in this chapter the primary step in polymer crystallization is the unmixing between crystallizable and non-crystallizable chain parts, since this necessarily has to precede the formation and growth of crystallites. This view is not at all novel and has been put forward among others by two of the pioneers of polymer science, namely Flory, when postulating the 'switchboard model', and Fischer, when suggesting the 'solidification model' of polymer crystallization. For reasons which are difficult to analyse, it remained a minority opinion. In the majority of discussions in the literature the emphasis is on the chain folding concept as introduced by Keller and in theories of polymer crystallization based upon it, by Hoffmann, to cite only two prominent authors. With my warning I wish to draw the attention of readers to this situation. The people who are new to the field should also check more conventional presentations which are included in the literature recommended for reading.

4.4 Further Reading

P. Barham: *Crystallization and Morphology of Semicrystalline Polymers* in R.W. Cahn, P. Haasen, E.J. Kramer, E.L. Thomas (Eds.): Materials Science and Technology Vol.12 *Structure and Properties of Polymers*, VCH Publishers, 1993

E.W. Fischer *Investigation of the Crystallization Process of Polymers by Means of Neutron Scattering* in L.A. Kleintjens, P.J. Lemstra (Eds.): *Integration of Fundamental Polymer Science and Technology*, Elsevier, 1886

U.W. Gedde: *Polymer Physics*, Chapman & Hall, 1995.

J.D. Hoffmann, G.T. Davis, J.I. Lauritzen: *The Rate of Crystallization of Linear Polymers with Chain Folding* in N.B. Hannay (Ed.): Treatise in Solid State Chemistry Vol.3, Plenum Press, 1976

A. Keller: *Chain-folded Crystallisation of Polymers From Discovery to Present Day: A Personalised Journey* in R.G.Chambers, J.E.Enderby, A.Keller, A.R.Lang, J.W.Steeds (Eds.): *Sir Charles Frank, An Eightieth Birthday Tribute*, Adam Hilger, 1991

L. Mandelkern: *The Crystalline State* in J.E. Mark, A. Eisenberg W.W. Graessley, L. Mandelkern, J. L. Koenig: *Physical Properties of Polymers*, Am.Chem.Soc., 1984

B. Wunderlich: *Macromolecular Physics* Vols.1,2,3, Academic Press, 1973

Chapter 5

Mechanical and Dielectric Response

In the large majority of present day uses of polymeric materials, the focus is on their mechanical performance. Properties are of a peculiar nature since polymer melts are different from low molar mass liquids and polymer solids differ from conventional crystalline solids. While the latter usually represent perfectly elastic bodies and low molar mass liquids develop viscous forces only, bulk polymers combine elastic and viscous properties in both the fluid and the solid state. Therefore they are generally addressed as 'viscoelastic' and, in fact, polymers are the main representatives of this special class of materials.

Viscoelastic behavior does not just mean a superposition of independent viscous and elastic forces, but it includes in addition a new phenomenon known as 'anelasticity', where both become coupled. It becomes apparent in the observation that part of the deformation, although being reversible, requires a certain time to become established when a load is applied.

The contributions of perfect elasticity, anelasticity, and viscous flow to the total mechanical response of a sample possess different weights for different polymers and, in particular, they greatly vary with temperature. This strong temperature dependence represents another characteristic property of polymeric materials and contrasts with the much less sensitive behavior of metals or ceramics. As a consequence of the changes, the temperature range for a certain application of a polymer is limited. The most important limitation results from the 'glass transition', where the elasticity and strength shown by a glassy solid get lost and the polymer becomes melt-like or, if it is cross-linked, turns into a rubber. In addition, there are other 'transitions' in the sense of further, usually weaker changes in the mechanical properties occurring within a narrow temperature range, and they sometimes induce undesired effects. It is clear that, for the use of a polymeric compound, one requires a good knowledge of all these processes. Since this pattern is complex, analysis necessitates special measures in both the experimental methods of characterization and the theoretical descriptions.

Different fields are concerned and they all need their own approaches

- the properties under moderate loads, where deformations and velocities of viscous flow remain small

- the case of large reversible deformations realized in rubbers and the rheological properties of polymer melts at higher strain rates, both representing non-linear behavior
- and finally, of special importance for applications, yielding and fracture.

We shall treat the first topic in this and the next chapter and large deformations, non-linear flow and the 'ultimate properties' yield and break subsequently, in chapters 7 and 8.

In electrical applications, polymers are mostly used as isolators. Since it is then important to be informed about possible electric losses, one needs to know their dielectric properties in dependence on frequency and temperature. As we shall see, description of the response of dielectric materials to applied time dependent electric fields is formally equivalent to the treatment of time dependent mechanical responses. Therefore, we shall discuss both together in one chapter.

5.1 Response Functions

If a mechanical or an electric field is applied to a polymer sample and remains sufficiently small, then the reaction, as given by the deformation and the polarization respectively, can be described by linear equations. We shall deal first with the linear viscoelasticity, which can be specified by various mechanical response functions, and then with the linear dielectric behavior, as characterized by the time- or frequency dependent dielectric function.

5.1.1 Viscoelasticity

A direct simple method to study the viscoelastic properties of a given sample is the *creep experiment*. It is carried out by instantaneously applying a constant force, which is then followed by a measurement of the resulting deformation as a function of time. Figure 5.1 indicates schematically a possible result, referring to the case where an uniaxial tensile load is applied, which then leads to an elongation ΔL_z. In general, it will be found that the 'creep curve' represents a superposition of three contributions

- a perfectly elastic, i.e. instantaneous response
- a retarded elastic deformation, i.e. an anelastic part, and
- viscous flow.

The first two contributions are reversible, the last one is irreversible.

It is of interest to determine separately the reversible and irreversible parts and this can be accomplished in an easy manner by removing the load from the sample and monitoring the subsequent 'recovery'-process. As indicated in Fig. 5.1, this leads first to an immediate shortening, which is then followed by a retarded further length reduction; only the irreversible part, caused by the viscous flow, remains.

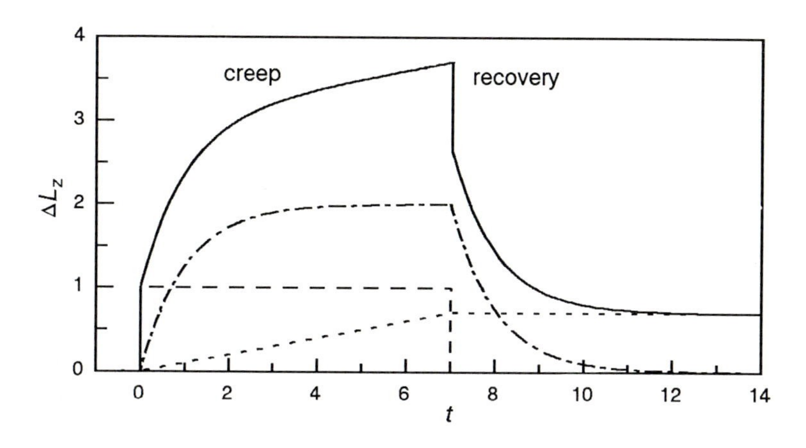

Fig. 5.1. Creep curve of a polymer sample under tension (schematic). The elongation ΔL_z induced by a constant force applied at zero time is set up by a superposition of an instantaneous elastic response (*dashed line*), a retarded anelastic part (*dash-dot line*) and viscous flow (*dotted line*). An irreversible elongation is retained after an unloading and the completion of the recovery process

If the force applied is sufficiently small, then one finds that the creep curve, $\Delta L_z(t)$, becomes proportional to the force. This suggests the use, for a description of the response in this 'linear viscoelastic range', of the ratio between the time dependent elongation and the force. More specifically, for a sample under tension one introduces the 'tensile creep compliance', $D(t)$, defined as

$$D(t) := \frac{e_{zz}(t)}{\sigma_{zz}^0} \tag{5.1}$$

Here $e_{zz}(t)$ denotes the time dependent longitudinal strain

$$e_{zz}(t) := \frac{\Delta L_z}{L_z} \tag{5.2}$$

where L_z is the original sample length; σ_{zz}^0 stands for the constant tensile stress applied at zero time (as usual, the first subscript indicates the normal vector of the face acted upon and the second gives the direction of the stress component). It is important to recognize that $D(t)$ provides in principle a complete characterization of the tensile properties of a given sample. Practical measurements are, of course, limited because registration cannot start before a certain minimum and extend over a certain maximum time.

A second method in mechanical tests is the *stress relaxation experiment.* Here, a certain constant strain is instantaneously imposed on a sample and the stress induced by this procedure is measured as a function of time. Figure 5.2 shows schematically for an uniaxially deformed sample the possible shape of a stress relaxation curve. The tensile stress has its maximum directly after

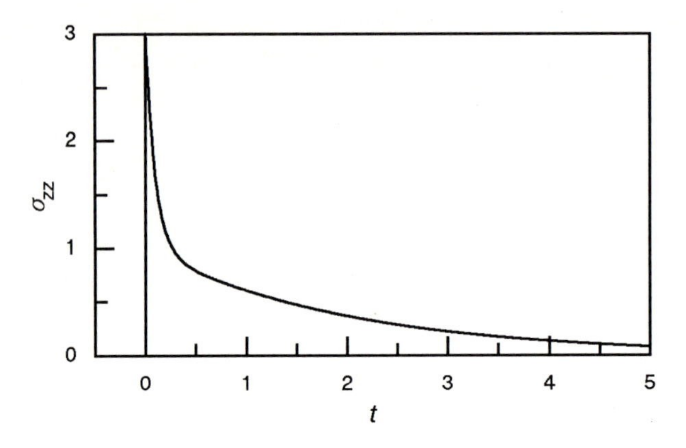

Fig. 5.2. Stress relaxation curve (schematic)

the deformation act, and then it decays. Anelastic components produce a first downward step. If the sample can flow, the stress will further decrease and finally vanish completely. The result of such an experiment can be described by the 'time dependent tensile modulus' $E(t)$, defined as

$$E(t) := \frac{\sigma_{zz}(t)}{e_{zz}^0} \tag{5.3}$$

whereby e_{zz}^0 denotes the imposed longitudinal strain.

The third method in use are *dynamic-mechanical experiments*. In these measurements, samples are exposed to a periodically varying stress field, for example to a tensile stress

$$\sigma_{zz}(t) = \sigma_{zz}^0 \exp \mathrm{i}\omega t \tag{5.4}$$

The resulting time dependent longitudinal strain, $e_{zz}(t)$, is indicated in Fig. 5.3. It varies with the frequency of the stress, but shows in general a phase-lag. Therefore we write for the strain

$$e_{zz}(t) = e_{zz}^0 \exp -\mathrm{i}\varphi \cdot \exp \mathrm{i}\omega t \tag{5.5}$$

A full description of the relation between $\sigma_{zz}(t)$ and $e_{zz}(t)$ is provided by the complex 'dynamic tensile compliance' D^*, defined as

$$D^*(\omega) := \frac{e_{zz}(t)}{\sigma_{zz}(t)} = \frac{e_{zz}^0 \exp -\mathrm{i}\varphi}{\sigma_{zz}^0} = D' - \mathrm{i}D'' \tag{5.6}$$

We write the dynamic compliance as a function, $D^*(\omega)$, because it varies generally with the frequency. To completely characterize the viscoelastic tensile properties of a given sample, one does indeed require to know the complete functional dependence.

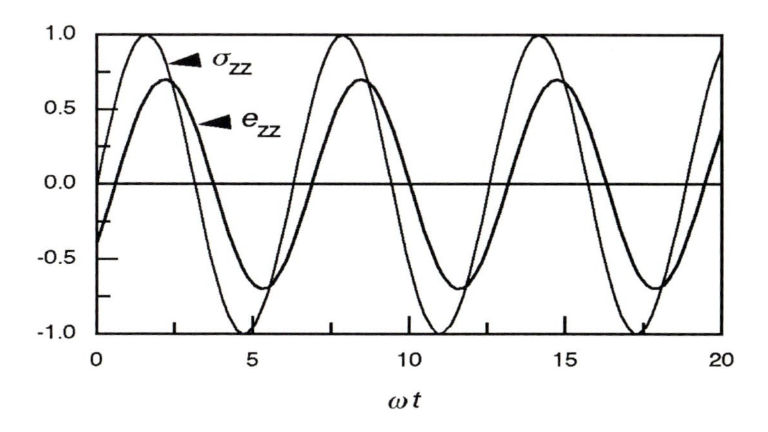

Fig. 5.3. Time dependence of stress σ_{zz} and strain e_{zz} in a dynamic-mechanical experiment (schematic)

As an alternative, one can employ as well the complex 'dynamic tensile modulus' $E^*(\omega)$, defined as

$$E^*(\omega) := \frac{\sigma_{zz}(t)}{e_{zz}(t)} = \frac{1}{D^*(\omega)} = E' + \mathrm{i}E'' \tag{5.7}$$

The different choices for the sign in front of the imaginary parts D'' and E'' represent a convention and result in positive values for both.

Analogous experiments can be carried out for other kinds of mechanical loading. Of particular importance are measurements for 'simple shear', which determine the relation between the shear strain e_{zx}, giving the displacement along x per unit distance normal to the shear plane z =const, and the shear stress σ_{zx}, as given by the force per unit area, acting on the shear plane along x. Shear properties of samples are described by

- the 'shear compliance'

$$J(t) := \frac{e_{zx}(t)}{\sigma^0_{zx}} \tag{5.8}$$

- the 'time dependent shear modulus'

$$G(t) := \frac{\sigma_{zx}(t)}{e^0_{zx}} \tag{5.9}$$

- the 'dynamic shear compliance'

$$J^*(\omega) := \frac{e^0_{zx}}{\sigma^0_{zx}} \cdot \exp -\mathrm{i}\varphi \tag{5.10}$$

• the 'dynamic shear modulus'

$$G^*(\omega) := \frac{\sigma_{zx}^0}{e_{zx}^0} \cdot \exp i\varphi \tag{5.11}$$

As already mentioned, a dynamic-mechanical measurement at one frequency does not possess the same information content as creep or stress relaxation experiments, and in order to achieve equivalence, these measurements have to be carried out under variation of ω over a sufficiently large range. Commercially available mechanical spectrometers usually scan a range of 3-4 orders of magnitude. As this is still limited, one might suspect at first that the information content is reduced compared to the time dependent experiments. However, as it turns out, if measurements are combined with temperature variations, one can also achieve a satisfactory overall characterization. We shall come back to this point for an explanation below.

Having introduced three different methods for a characterization of viscoelastic properties, one may wonder if the results are interrelated. This is indeed the case but before entering into this matter, we will take a look at dielectric measurements.

5.1.2 Orientational Polarization

Application of an electric field $\boldsymbol{E}$ on a non-conducting sample leads to a polarization $\boldsymbol{P}$. In electrostatics this process is described by the linear relation

$$\boldsymbol{P} = \epsilon_0(\epsilon - 1)\boldsymbol{E} \tag{5.12}$$

Employing the dielectric displacement vector $\boldsymbol{D}$, one can write equivalently

$$\boldsymbol{D} := \epsilon_0\boldsymbol{E} + \boldsymbol{P} = \epsilon_0\epsilon\boldsymbol{E} \tag{5.13}$$

In both equations the response is given by the dielectric function ϵ.

If the experiment is carried out with a constant field $\boldsymbol{E}_0$ being switched on at zero time, one finds in general a time dependent polarization $\boldsymbol{P}(t)$, set up of an instantaneous part $\boldsymbol{P}_{\mathrm{u}}$ and a retarded part $\boldsymbol{P}_{\mathrm{or}}$

$$\boldsymbol{P}(t) = \boldsymbol{P}_{\mathrm{u}} + \boldsymbol{P}_{\mathrm{or}}(t) \tag{5.14}$$

One describes this result with the aid of a time dependent dielectric function $\epsilon(t)$, as

$$\boldsymbol{P}(t) = \epsilon_0(\epsilon(t) - 1) \cdot \boldsymbol{E}_0 \tag{5.15}$$

or, split up into the two contributions, by

$$\boldsymbol{P}(t) = \epsilon_0(\epsilon_{\mathrm{u}} - 1)\boldsymbol{E}_0 + \epsilon_0\Delta\epsilon(t)\boldsymbol{E}_0 \tag{5.16}$$

The immediately reacting part, $\boldsymbol{P}_{\mathrm{u}}$, is due to the shift of the electron clouds and the deformation of the molecular skeletons which occur within times corresponding to frequencies in the UV- and IR-range respectively. The retarded part, $\boldsymbol{P}_{\mathrm{or}}$, arises for polar molecular fluids and originates from the orientation of the permanent dipoles.

More common than using dc-fields is for the characterization of dielectric properties of samples the use of ac-fields. On applying a sinusoidally varying field, represented in complex notation as

$$\boldsymbol{E}(t) = \boldsymbol{E}_0 \exp \mathrm{i}\omega t \tag{5.17}$$

there results, in general, a time dependent polarization

$$\boldsymbol{P}(t) = \boldsymbol{P}_0 \exp -\mathrm{i}\varphi \cdot \exp \mathrm{i}\omega t \tag{5.18}$$

The angle φ denotes a possible phase-lag. For a description of the relation between the polarization and the field, one can again choose, as in the analogous case of the dynamic-mechanical experiment, their complex ratio, known as 'complex dielectric susceptibility'. Rather than the latter, we will use the closely related 'complex dielectric function' $\epsilon^*(\omega)$, defined as

$$\epsilon^*(\omega) := \frac{\boldsymbol{D}(t)}{\boldsymbol{E}(t)} = \epsilon_0 + \frac{P_0}{E_0} \exp -\mathrm{i}\varphi \tag{5.19}$$

$\epsilon^*(\omega)$ splits up in a real and an imaginary part, whereby one conventionally chooses a negative sign in front of ϵ''

$$\epsilon^*(\omega) = \epsilon'(\omega) - \mathrm{i}\epsilon''(\omega) \tag{5.20}$$

5.1.3 General Relationships

It may have already been noticed that all the described experiments correspond to a common basic scheme: There is a force or field, represented here by the stress or the electric field, which leads to a 'displacement', as given by the strain or the polarization. In all the cases considered, the force and the resulting displacement are related by a linear equation. Hence, we dealt throughout with '*linear responses*'. Clearly, many other effects exist which represent linear responses too. There are the reactions on still other kinds of mechanical loading but also on the applications of other fields, as for example, a magnetic field $\boldsymbol{B}$ which induces a magnetization $\boldsymbol{M}$.

There is a second characteristic property which all cases have in common: One always deals with a pair of energy conjugated variables, that is to say, the 'displacement' caused by the 'field' results in work. More specifically, if a 'field' ψ gives rise to a 'displacement' $\mathrm{d}x$, then the work per unit volume is

$$\frac{\mathrm{d}\mathcal{W}}{V} = \psi \mathrm{d}x \tag{5.21}$$

This holds for the tensile load where

$$\frac{\mathrm{d}\mathcal{W}}{V} = \sigma_{zz}\mathrm{d}e_{zz} = D\sigma_{zz}\mathrm{d}\sigma_{zz} \tag{5.22}$$

for the applied shear stress where

$$\frac{\mathrm{d}\mathcal{W}}{V} = \sigma_{zx}\mathrm{d}e_{zx} = J\sigma_{zx}\mathrm{d}\sigma_{zx} \tag{5.23}$$

and also for the dielectric experiment, when we identify $\mathrm{d}x$ with the change of the dielectric displacement vector

$$\frac{\mathrm{d}\mathcal{W}}{V} = \boldsymbol{E}\mathrm{d}\boldsymbol{D} \tag{5.24}$$

So far, we have discussed the response of systems only for forces with special time dependencies. The creep compliance describes the reaction on a force which is switched on at zero time and then remains constant, the dynamic compliance specifies the response on a sinusoidally varying stress. What happens in the general case, when an arbitrary time dependent force $\psi(t)$ is applied? There is a specific function which enables us to deal with this general situation, sometimes called the 'primary response function'. It is introduced by considering the effect of an infinitely short pulse, as represented by

$$\psi(t) = \psi_0\delta(t) \tag{5.25}$$

where $\delta(t)$ is the delta-function. The primary response function, denoted $\mu(t)$, describes the time dependent displacement $x(t)$ caused by this pulse, as

$$x(t) = \psi_0\mu(t) \tag{5.26}$$

It is instructive to look at some typical examples as sketched in Fig. 5.4. A damped harmonic oscillator reacts to a pulse by starting an oscillation with exponentially decaying amplitudes and this is shown in part (a). The effect on a perfectly viscous body is quite different since it just becomes plastically deformed and then maintains the new shape (b). Part (c) shows the reaction of a perfectly elastic sample, i.e. a Hookean solid, which is only deformed during the short time of the pulse. Finally, part (d) represents the reaction of an overdamped oscillator or 'relaxator' which exhibits an exponential decay.

The primary response function indeed enables us generally to formulate the displacement which results from an arbitrary time dependent force. It is given by

$$x(t) = \int_{-\infty}^{t} \mu(t-t')\psi(t')\mathrm{d}t' \tag{5.27}$$

The physical background of Eq. (5.27) is easily seen. The integral relation just follows from the two basic properties of linear systems, namely the causality principle and the validity of the superposition principle:

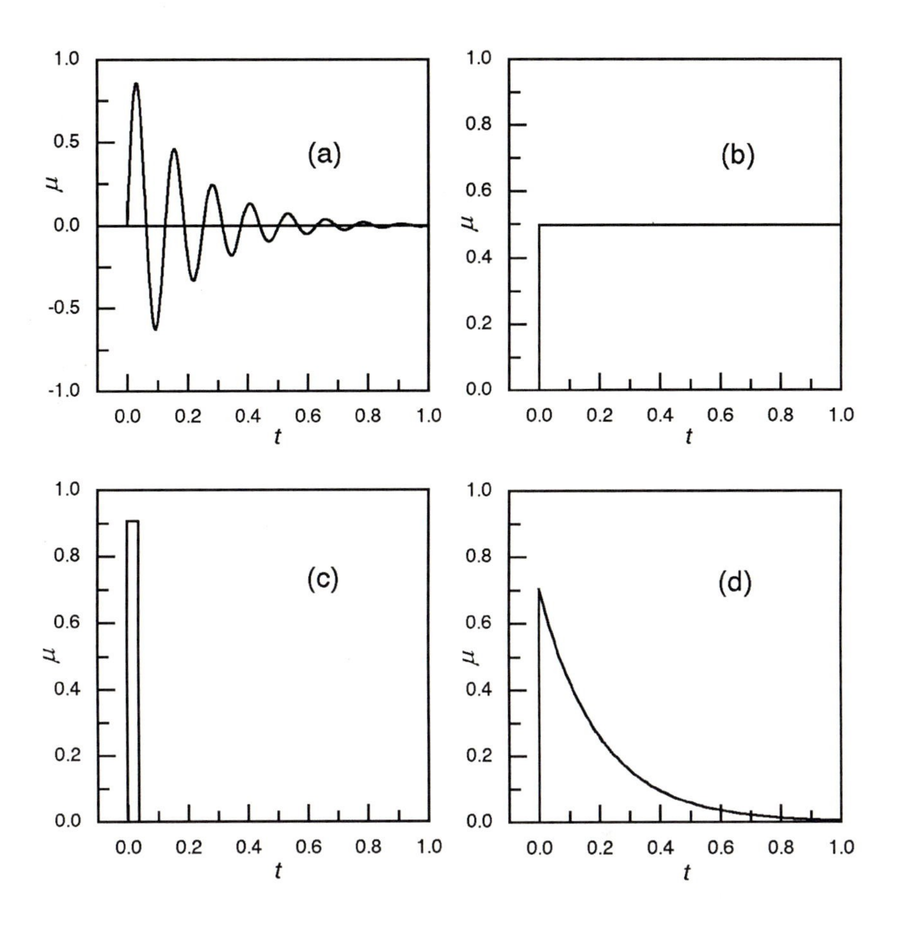

Fig. 5.4. Primary response function of a damped harmonic oscillator (a), a perfectly viscous body (b), a Hookean solid (c), a simple relaxatory system (d)

- Causality requires that the displacement at a given time can only depend on the forces in the past and this finds its expression in the limits chosen for the integral.
- Employing the superposition principle, an arbitrary time dependent force can first be divided into a sequence of pulses with adjusted heights, and then the total reaction can be represented as a sum over the responses to all the pulses. The integral expresses exactly this procedure.

Emanating from Eq. (5.27), one can now derive the interrelations between the various response functions. We formulate them in terms of the general variables x and ψ and consider, as a first example, the displacement $x(t)$ resulting from a force with amplitude ψ_0 which is switched on at zero time.

Equation (5.27) yields directly

$$x(t) = \int_0^t \mu(t-t')\psi_0 \mathrm{d}t' \tag{5.28}$$

As for the creep experiment, the result can also generally be described with the aid of the ratio

$$\frac{x(t)}{\psi_0} := \alpha(t) \tag{5.29}$$

The name used for this function $\alpha(t)$ is 'time dependent susceptibility'. We obtain

$$\alpha(t) = \int_0^t \mu(t-t')\mathrm{d}t' = \int_0^t \mu(t'')\mathrm{d}t'' \tag{5.30}$$

Taking the derivative on both sides gives the relation between $\mu(t)$ and $\alpha(t)$

$$\mu(t) = \frac{\mathrm{d}\alpha}{\mathrm{d}t}(t) \tag{5.31}$$

Secondly, consider the stress relaxation experiment. In terms of the general variables it is to be described as

$$x_0 = \int_0^t \mu(t-t')\psi(t')\mathrm{d}t' \tag{5.32}$$

where x_0 denotes the imposed displacement. We introduce a 'general time dependent modulus' $\mathrm{a}(t)$ as

$$\mathrm{a}(t) := \frac{\psi(t)}{x_0} \tag{5.33}$$

$\mathrm{a}(t)$ and $\mu(t)$ are related by the integral equation

$$1 = \int_0^t \mu(t-t')\mathrm{a}(t')\mathrm{d}t' \tag{5.34}$$

If we eliminate $\mu(t)$ from this equation with the aid of Eq. (5.31) we find the interrelation between the time dependent susceptibility and the time dependent modulus

$$1 = \int_0^t \frac{\mathrm{d}\alpha}{\mathrm{d}t}(t-t')\mathrm{a}(t')\mathrm{d}t' \tag{5.35}$$

An important equation is obtained, if we take Eq. (5.31), introduce it in Eq. (5.27)

$$x(t) = -\int_{-\infty}^t \frac{\mathrm{d}\alpha}{\mathrm{d}t'}(t-t')\psi(t')\mathrm{d}t' \tag{5.36}$$

and then carry out a partial integration, leading to

$$x(t) = \int_{-\infty}^t \alpha(t-t')d\psi(t') \tag{5.37}$$

Rather than beginning with Eq. (5.27), in the polymer literature, Eq. (5.37) is often used as the basis. It is then understood as the mathematical formulation of the 'Boltzmann superposition principle' which states that each loading step makes an independent contribution to the final deformation and that the latter then follows by the simple addition. This is exactly the physical meaning of the integral. The contributions are described in the integral. As we can see, the after-effect of each step is a creep curve, starting with the step and being weighted by the step height.

Rather than considering the displacement $x(t)$ in dependence on the forces in the past, $\psi(t' < t)$, one can ask conversely for the functional dependence of the force $\psi(t)$ on the previous displacements $x(t' < t)$. The solution is obvious: We just have to exchange the susceptibility $\alpha(t - t')$ against the generalized time dependent modulus $\mathrm{a}(t - t')$ and represent $\psi(t)$ as a sum of relaxation curves:

$$\psi(t) = \int_{-\infty}^{t} \mathrm{a}(t - t')\mathrm{d}x(t') \tag{5.38}$$

This is the alternative form of the Boltzmann superposition principle and it is of importance in particular in rheological treatments.

Let us finally come to the dynamic experiments, where an oscillatory force

$$\psi(t) = \psi_0 \exp \mathrm{i}\omega t \tag{5.39}$$

is applied. It results in a displacement, to be written in general as

$$x(t) = x_0 \exp -\mathrm{i}\varphi \cdot \exp \mathrm{i}\omega t \tag{5.40}$$

Employing Eq. (5.27), this relation corresponds to

$$x_0 \exp -\mathrm{i}\varphi \exp \mathrm{i}\omega t = \int_{-\infty}^{t} \mu(t - t')\psi_0 \exp \mathrm{i}\omega t' \mathrm{d}t' \tag{5.41}$$

The complex amplitude ratio

$$\frac{x_0 \exp -\mathrm{i}\varphi}{\psi_0} := \alpha^*(\omega) \tag{5.42}$$

describes the result of a dynamic experiment. $\alpha^*(\omega)$ is called 'general dynamic susceptibility' and we obtain for it

$$\alpha^*(\omega) = \int_{-\infty}^{t} \mu(t - t') \exp -\mathrm{i}\omega(t - t')\ \mathrm{d}t' \tag{5.43}$$

With the substitution

$$t - t' := t''$$

we can write

$$\alpha^*(\omega) = \int_{0}^{\infty} \mu(t'') \exp -\mathrm{i}\omega t'' \mathrm{d}t'' = \int_{-\infty}^{\infty} \mu(t'') \exp -\mathrm{i}\omega t'' \mathrm{d}t'' \tag{5.44}$$

taking into account that $\mu(t'' < 0) = 0$. As we can see, the primary response function, $\mu(t)$, and the dynamic susceptibility, $\alpha^*(\omega)$, correspond to a pair of Fourier transforms.

The dynamic susceptibility is in general a complex quantity

$$\alpha^*(\omega) = \alpha'(\omega) - \mathrm{i}\alpha''(\omega) \tag{5.45}$$

Actually, the real part $\alpha'(\omega)$ and the imaginary part $\alpha''(\omega)$ are not independent, but related by the 'Kramers-Kronig dispersion relations'. These have the following forms

$$\alpha'(\omega_0) = \frac{1}{\pi}\mathrm{P}\int_{-\infty}^{\infty} \frac{\alpha''(\omega)}{\omega - \omega_0}\mathrm{d}\omega \tag{5.46}$$

$$\alpha''(\omega_0) = -\frac{1}{\pi}\mathrm{P}\int_{-\infty}^{\infty} \frac{\alpha'(\omega)}{\omega - \omega_0}\mathrm{d}\omega \tag{5.47}$$

The relations include a special type of integral, the 'Cauchy integral', which eliminates the singularity at $\omega = \omega_0$. In the case of Eq. (5.46) it is defined by

$$\mathrm{P}\int_{-\infty}^{\infty} \frac{\alpha''(\omega)}{\omega - \omega_0}\mathrm{d}\omega := \lim_{\delta\to 0}\left[\int_{-\infty}^{\omega_0-\delta} \frac{\alpha''}{\omega - \omega_0}\mathrm{d}\omega + \int_{\omega_0+\delta}^{\infty} \frac{\alpha''}{\omega - \omega_0}\mathrm{d}\omega\right] \tag{5.48}$$

and for Eq. (5.47) equivalently. According to this expression, ω_0 is approached in a synchronized manner from both sides. Then the positive and negative divergent values compensate each other and the singularity does not emerge. A derivation of the Kramers-Kronig relations can be found in many of the textbooks of statistical mechanics. Here, we leave it with one remark concerning their physical origin: The relations can be regarded as a consequence of the causality principle since the derivation makes use of one condition only, namely that $\mu(t'')$ vanishes for $t'' < 0$.

It is important that we now discuss the work which results from the movement under the action of the force. A most useful and simple result is obtained for the dynamic experiments. Here the force

$$\psi(t) = \psi_0 \exp \mathrm{i}\omega t \tag{5.49}$$

produces a displacement

$$x(t) = \alpha^*(\omega)\psi(t) = (\alpha' - \mathrm{i}\alpha'')\psi(t) \tag{5.50}$$

For a calculation of the work one has to use the true values $\psi(t)$ and $x(t)$, which follow from the complex notation by an extraction of the real part. We therefore write

$$\psi(t) = \psi_0 \cos\omega t \tag{5.51}$$

and

$$x(t) = \alpha'\psi_0 \cos\omega t + \alpha''\psi_0 \sin\omega t \tag{5.52}$$

The power is given by

$$\frac{1}{V}\frac{\mathrm{d}\mathcal{W}}{\mathrm{d}t} = \psi \cdot \frac{\mathrm{d}x}{\mathrm{d}t} = -\frac{\psi_0^2}{2}\omega\alpha' \sin 2\omega t + \psi_0^2 \omega\alpha'' \cos^2 \omega t \tag{5.53}$$

There are two contributions, one proportional to α' and the other proportional to α''. These two contributions just represent two different aspects of the work. The first contribution varies periodically between positive and negative values, which indicates an energy exchange between the driving part and the driven system. Obviously this first term is associated with an energy which during one half-period is stored in the driven system and during the successive half-period then is completely returned.

The second contribution is qualitatively different, as it yields a non-vanishing positive value in the time average:

$$\frac{1}{V}\overline{\frac{\mathrm{d}\mathcal{W}}{\mathrm{d}t}} = \frac{1}{2}\psi_0^2 \omega\alpha''(\omega) \tag{5.54}$$

Hence, work is expended on the driven system. If the sample is kept under isothermal conditions, the internal energy $\mathcal{E}$ does not change

$$\overline{\frac{\mathrm{d}\mathcal{E}}{\mathrm{d}t}} = 0 \tag{5.55}$$

Since generally, according to the first law of thermodynamics, we have

$$\mathrm{d}\mathcal{E} = \mathrm{d}\mathcal{W} + \mathrm{d}\mathcal{Q} \tag{5.56}$$

we find

$$\overline{\frac{\mathrm{d}\mathcal{W}}{\mathrm{d}t}} = -\overline{\frac{\mathrm{d}\mathcal{Q}}{\mathrm{d}t}} \tag{5.57}$$

This means that the power is 'dissipated', i.e. returned by the system in the form of heat.

We see that the susceptibility separates the elastic part and the viscous dissipative part of the work expended on the system. They show up in the real part $\alpha'(\omega)$ and the imaginary part $\alpha''(\omega)$ respectively. It is important to note that the two parts are not independent even if they represent quite different physical properties. In fact, they are related by the Kramers-Kronig relations, Eqs. (5.46) and (5.47).

We finish this section with the mention of another frequently used quantity, known as the 'loss tangent'. It is defined as

$$\tan\delta(\omega) := \frac{\alpha''(\omega)}{\alpha'(\omega)} \tag{5.58}$$

According to the definition, $\tan\delta(\omega)$ describes the ratio between the dissipated and the reversibly exchanged work.

5.2 Relaxatory Modes

Orientational polarization, as it is found in polar liquids, provides a good example for explaining the physical background of reversible retarded responses. First consider the natural state without a field. Here we have no polarization, and this arises from distributing the orientations of the polar units in the sample isotropically. If now an electric field is applied, the orientational distribution changes. Since dipole orientations in the field direction are preferred, the distribution function becomes anisotropic. As a consequence we find a non-vanishing value for the orientational polarization $\boldsymbol{P}_{\mathrm{or}}$.

For systems, where the coupling between the polar units is so weak that they can reorient largely independent from each other, $\boldsymbol{P}_{\mathrm{or}}$ can be calculated in simple manner. It then just emerges from the competition between the interaction energy of the dipoles with the electric field and the kinetic energy of the molecular rotation. The calculation is carried out in many textbooks of physical chemistry and the result reads, in an approximate form,

$$\boldsymbol{P}_{\mathrm{or}} \simeq c_{\mathrm{m}} \frac{|\boldsymbol{p}|^2 |\boldsymbol{E}|}{3kT} \tag{5.59}$$

$\boldsymbol{p}$ is the dipole moment of the reorienting units, and c_{m} gives their number density.

Establishment of the new equilibrium subsequent to a sudden application of an electric field requires a finite time. To see the origin of the retardation, envisage the rotational dynamics in the fluid. Owing to the strongly varying intermolecular forces, the dipole carrying units cannot rotate freely, but rather show a statistical kind of motion. For independent units, this motion may be described as a 'rotational diffusion', that is to say, it equals a succession of uncorrelated angular steps. The diffusive motion leads to a complete reorientation within a certain time, say τ. In fact, τ is the only parameter required to characterize completely the state of rotational motion in a system of independent units. Therefore, it sets the time scale generally for all changes in the orientational distribution function. Hence, in particular, it also determines the time needed to attain the new equilibrium set by a changed electric field.

What is the microscopic background of the retarded mechanical responses? As we have learned, one can envisage a polymeric fluid as an ensemble of macromolecules which change between the various conformational states. The populations of the different states are determined by the laws of Boltzmann statistics. If a mechanical field is now applied, a change in the population numbers is induced. For example, consider a rubber to which a tensile stress is imposed. Clearly, now preferred are all conformations which are accompanied by an extension along the direction of stress. The repopulation of the conformational states and the resulting increase in the sample length require a finite time, which must correspond to the time scale of the conformational transitions.

Compared to the dielectric response first considered, the mechanical reaction is more complex. Dielectric relaxation in a system of independent polar units originates from their individual reorientational motions. In mechanical relaxation of a rubber we find a different situation. Here, we are dealing with transitions between the different conformations of a chain and not with individual movements of single groups. Rather than having one process only, in this case, a large number of different 'modes' exist, and these may vary over a wide range in the characteristic times. As a consequence, the sample's response showing up in the time dependent change of its length subsequent to the application of the tensile stress, cannot be associated with a single time constant only, but is of a complex nature.

There is also a simple situation equivalent to the dipole reorientations in mechanical behavior. In glassy polymers, large scale conformational changes are inhibited but there remains the possibility of localized conformational transitions. These can be observed, for example, for polymers with side-groups. In the next section, an example will be presented where the side-groups of a polymer chain possess just two conformational states. Application of stress here leads to a change in the respective occupation numbers and the redistribution occurs within a time as given by the rate of jumps between the two states. Macroscopically, a detectable change in the shape of the loaded sample results which can be related to a single characteristic time only.

In all the examples discussed, we are concerned with the passage from a non-equilibrium situation, created by the sudden imposition of an external field, to the new equilibrium. The change is accompanied and driven by a decrease in the free energy. Using mechanistic terms one could say that a system which at first, when having an enhanced free energy, is 'strained', 'relaxes' while going to the equilibrium. Correspondingly, all these retarded transitions into a new equilibrium are generally addressed as 'relaxation processes'. The name includes even more, namely the underlying microscopic motions as well. The notion 'relaxation' thus has a broad meaning in the literature, and is not at all restricted to the first introduced stress relaxation experiment which just represents one special case.

5.2.1 Single-Time Relaxation Process

As we have seen, the time dependence of a macroscopic relaxation process always reflects the underlying microscopic dynamics. We may now proceed and look for kinetical equations which correctly describe the time dependence of the observed retarded responses.

There is an obvious choice for the simple case when only a single characteristic time is included. It goes back to Debye, who proposed it in a famous work on the dielectric properties of polar liquids, based on a statistical mechanical theory. We formulate the equation for the above mentioned simple mechanical relaxation process, associated with transitions between two conformational states only, and consider a creep experiment under shear stress.

The equation has the following form

$$\frac{\mathrm{d}e_{zx}}{\mathrm{d}t} = -\frac{1}{\tau}(e_{zx}(t) - \Delta J \sigma_{zx}^0) \tag{5.60}$$

It represents a linear differential equation of the first order, implying the assumption that, for a system in non-equilibrium, relaxation takes place with a rate which increases linearly with the distance from the equilibrium state. This is not a specific kind of expression devised to deal exclusively with our problem. Equivalent equations are broadly used in thermodynamics to describe the kinetics of all sorts of irreversible processes. Importantly, the equation includes one time constant only, the 'relaxation time' τ.

A further parameter, ΔJ, determines here the equilibrium value of the anelastic contribution to the shear strain following from an applied stress σ_{zx}^0. The equilibrium value is given by the product $\Delta J \sigma_{zx}^0$. In accordance with the physical meaning, ΔJ is called 'relaxation strength'. This is, in fact, a general name also used for the analogous parameters in other relaxation processes. They all have in common that they determine the magnitude of the effect, as given for example by a contribution to the strain, the stress or the polarization.

The solution of the relaxation equation for the creep experiment, i.e. a step-like application of a stress σ_{zx}^0 at zero time, can be written down directly. It is given by

$$e_{zx}(t) = \Delta J \sigma_{zx}^0 \left(1 - \exp -\frac{t}{\tau}\right) \tag{5.61}$$

This is indeed a correct representation of the creep curve observed for a single-time relaxation process. The physical properties of the process are included in the two parameters ΔJ and τ. The latter agrees with the transition rates between the two conformational states. It is more difficult to predict and thus to interpret the relaxation strength. As is intuitivally clear, the prerequisite for an alteration of the population numbers under an applied shear stress is a change in the 'shape' of the active unit, as only then a change in the energy can arise. To formulate this interaction energy for a unit as represented by a whole monomer or only the side-group can, however, be difficult, much more difficult than the case considered above of the interaction of a dipole with an electric field. It is therefore not surprising that calculations of mechanical relaxation strengths on microscopic grounds are rare and quantitative interpretations of measured values an exception.

The 'relaxation equation' (5.60) is not restricted in use to step-like changes in the external conditions, but also holds, if the equilibrium value $\Delta J \sigma_{zx}$ is not a constant and changes with time. In particular, it can be employed to treat dynamic-mechanical experiments. As is clear, applying an oscillatory shear stress

$$\sigma_{zx}(t) = \sigma_{zx}^0 \exp \mathrm{i}\omega t \tag{5.62}$$

means to impose on the sample an oscillatory variation for the equilibrium

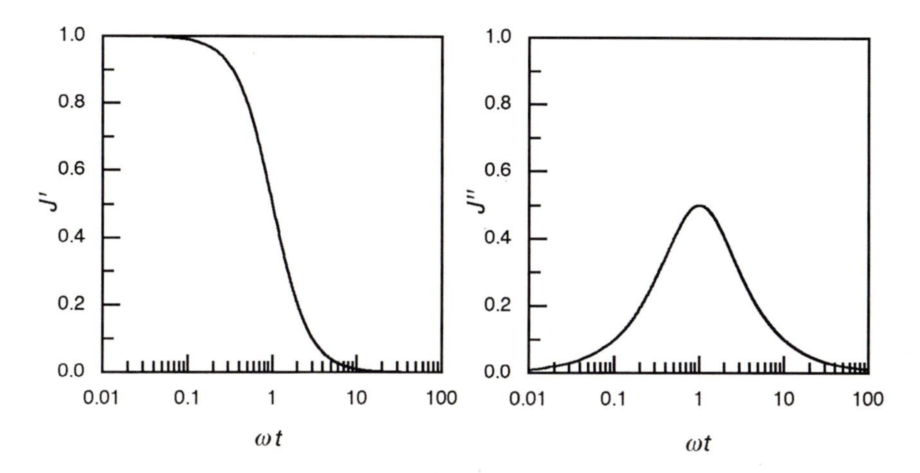

Fig. 5.5. Real part (*left*) and imaginary part (*right*) of the dynamic compliance associated with a mechanical Debye-process

strain, $\Delta J\sigma_{zx}(t)$. If we introduce it into the relaxation equation we obtain

$$\frac{\mathrm{d}e_{zx}}{\mathrm{d}t} = -\frac{1}{\tau}(e_{zx}(t) - \Delta J\sigma_{zx}^0 \exp \mathrm{i}\omega t) \tag{5.63}$$

An oscillatory stress results in an oscillatory strain, as given by Eq. (5.10)

$$e_{zx}(t) = \sigma_{zx}^0 J^*(\omega) \exp \mathrm{i}\omega t \tag{5.64}$$

If we take this solution, introduce it into Eq. (5.63) and take the common exponential factor off on both sides, we obtain the dynamic compliance. It is given by

$$J^*(\omega) = \frac{\Delta J}{1 + \mathrm{i}\omega\tau} \tag{5.65}$$

The dynamic compliance of the single-time relaxation process, in the literature also addressed as 'Debye-process', thus has a simple form, being a function of the product $\omega\tau$ and ΔJ only. Separation into the real and the imaginary part yields

$$J^*(\omega) = J' - \mathrm{i}J'' = \frac{\Delta J}{1 + \omega^2\tau^2} - \mathrm{i}\frac{\Delta J\omega\tau}{1 + \omega^2\tau^2} \tag{5.66}$$

Figure 5.5 depicts J' and J'' using a logarithmic scale for the variable $\omega\tau$. The use of a logarithmic scale is not only convenient, regarding that experiments usually cover several orders of magnitude, but also has the advantage that the curves then exhibit characteristic symmetries. The imaginary part which describes the loss forms a symmetric bell-shaped curve with maximum at $\omega\tau = 1$. If J'' is written in the form

$$J'' = \frac{\Delta J}{10^{-\log\omega\tau} + 10^{\log\omega\tau}} \tag{5.67}$$

the symmetry is obvious.

The real part shows a steep decrease in the range of the loss maximum and its physical cause is easily revealed. If the mechanical field applied has a frequency which is small compared to the transition rates in the system, establishment of thermal equilibrium is rapid compared to the period of the field and the system can always remain in equilibrium. Hence, we encounter quasistatic conditions and observe the full relaxation strength. At the other limit, when the frequencies of the field applied are large compared to the transition rates, equilibrium cannot be established and the system reacts to the average strain only, which is zero. The crossover from one to the other regime just occurs in the range $\omega\tau \simeq 1$.

The maximum in J'' and the steepest descent of J' are located at the same frequency. Furthermore, the area under the loss curve and the relaxation strength ΔJ are proportional to each other. Integration gives the following relation

$$\int_{-\infty}^{\infty} J'' \mathrm{d} \log \omega\tau = \frac{\pi}{2 \ln 10} \Delta J \tag{5.68}$$

In fact, these properties are not specific to the Debye-process, but have a deeper basis which extends their validity. According to the Kramers-Kronig relations, J' and J'' are mutually dependent and closer inspection of the equations reveals that it is impossible, in principle, to have a loss without a simultaneous change in J'. Both effects are coupled, the reason being, as mentioned above, the validity of the causality principle.

The loss curve has a characteristic width, the total width at half height amounting to 1.2 decades. Compared to the loss at the resonance frequency of an oscillating system, the loss curve of the Debye-process is much broader. A halfwidth of 1.2 decades in fact presents the lower limit for all loss curves found in relaxing systems. Loss curves which are narrower are therefore indicative of the presence of oscillatory contributions, or more generally speaking, indicate effects of moments of inertia.

A simple check, if a measured dynamic compliance or a dielectric function agrees with a Debye-process, is provided by the 'Cole-Cole plot'. Let us illustrate it with a dielectric single-time relaxation process. If we choose for the dipolar polarization an expression analogous to Eq. (5.66) and take also into account the instantaneous electronic polarization with a dielectric constant ϵ_{u}, the dielectric function $\epsilon^*(\omega)$ shows the form

$$\epsilon^* = \epsilon_{\mathrm{u}} + \frac{\Delta\epsilon}{1+\omega^2\tau^2} - \mathrm{i}\frac{\omega\tau\Delta\epsilon}{1+\omega^2\tau^2} \tag{5.69}$$

When for all values of $\omega\tau$ the associated pairs $\epsilon'(\omega\tau)$ and $\epsilon''(\omega\tau)$ are plotted in a plane as shown in Fig. 5.6, a semi-circle is obtained. The circle begins for $\omega\tau = 0$ at $\epsilon' = \epsilon_{\mathrm{u}} + \Delta\epsilon$, and ends for $\omega\tau \to \infty$ at $\epsilon' = \epsilon_{\mathrm{u}}$. The proof for the circular form of the Cole-Cole plot is straightforward, as we can write

$$\left[\epsilon' - (\epsilon_{\mathrm{u}} + \frac{\Delta\epsilon}{2})\right]^2 + (\epsilon'')^2 \quad = \quad \left[\frac{2\epsilon' - 2\epsilon_{\mathrm{u}} - \Delta\epsilon}{2}\right]^2 + (\epsilon'')^2$$

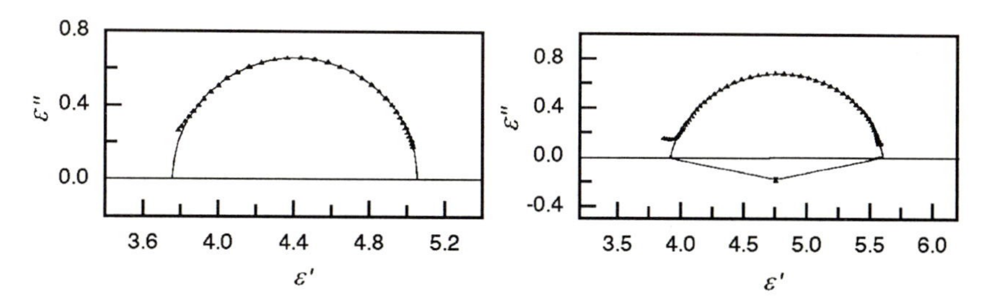

Fig. 5.6. Cole-Cole plots of dielectric data obtained for a dipole carrying rod-like molecule of low molar mass (*left*) and a polysiloxane which has these molecules attached as side-groups (*right*) [48]

$$= \left[\frac{2\Delta\epsilon - \Delta\epsilon(1+\omega^2\tau^2)}{2(1+\omega^2\tau^2)}\right]^2 + \frac{\Delta\epsilon^2\omega^2\tau^2}{(1+\omega^2\tau^2)^2}$$

$$= \frac{\Delta\epsilon^2(1+\omega^2\tau^2)^2}{4(1+\omega^2\tau^2)^2} = \frac{\Delta\epsilon^2}{4} \qquad (5.70)$$

Figure 5.6 presents, as examples, results of dielectric studies on certain rod-like molecules, shortly designated as 'C_6C_3', which carry a longitudinal electric dipole. The Cole-Cole plot on the left-hand side represents measurements in the liquid phase of this low molar mass compound. Points are arranged along a semi-circle, as is indicative for a Debye-process. The data given on the right-hand side were obtained for a polymer, which has the C_6C_3-rods attached as side-groups onto a polysiloxane backbone. Obviously, the coupling modifies the rotational kinetics so that it no longer equals a Debye-process. Here a satisfactory representation of data is achieved by use of an empirical function, with the form

$$\epsilon^*(\omega) = \frac{\Delta\epsilon}{(1+\mathrm{i}\omega\tau)^{1-\beta}} \qquad (5.71)$$

It is known in the literature as 'Cole-Cole-function' , and includes with β an additional parameter. The larger the value of β, the more pronounced are the deviations from a single-time behavior. In the example, we have $\beta = 0.08$.

5.2.2 Retardation and Relaxation Time Spectra

Having established the properties of the single-time relaxation process, we have now also a means to represent a more complex behavior. This can be accomplished by applying the superposition principle, which must always hold in systems controlled by linear equations. Considering shear properties again, we write for a dynamic compliance $J^*(\omega)$ with general shape a sum of Debye-processes with relaxation times τ_l and relaxation strengths ΔJ_l

$$J^*(\omega) = J_\mathrm{u} + \sum_l \frac{\Delta J_l}{1+\mathrm{i}\omega\tau_l} \qquad (5.72)$$

Often it is more appropriate to employ a representation in integral form

$$J^*(\omega) = J_\mathrm{u} + \int \frac{1}{1+\mathrm{i}\omega\tau} \mathrm{L}_J(\log\tau)\mathrm{d}\log\tau \tag{5.73}$$

The characteristic function in this integral which specifies the relaxation properties of the system is $\mathrm{L}_J(\log\tau)$, being called the 'retardation time spectrum of the shear compliance J'.

The identical function can be used in order to describe the result of a creep experiment on the system. One has just to substitute the dynamic compliance of the Debye-process by the associated elementary creep function, as given by Eq. (5.61). This leads to

$$J(t) = J_\mathrm{u}\Theta(t) + \int \left(1 - \exp -\frac{t}{\tau}\right) \mathrm{L}_J(\log\tau)\mathrm{d}\log\tau \tag{5.74}$$

The immediate reaction with amplitude J_u here is represented by the heavyside function $\Theta(t)$, which is unity for $t > 0$ and zero for $t < 0$.

The superposition approach may first look as a purely formal one, but a physical basis also exists, valid for many of the cases of interest. We are mostly dealing with the kinetics of transitions between the different conformational states. Motions include both local changes and cooperative movements of many monomers up to the full length of a chain. It appears reasonable to assume that the resulting total dynamics can be described as a superposition of a large number of independent 'relaxatory modes', each one representing a single-time relaxation process. In fact, theories like the Rouse- or the reptation model, to be discussed in the next chapter, lend support to this mode picture which therefore may well be regarded as a notion suitable for general considerations.

Alternative to the shear compliances, $J(t)$ and $J^*(\omega)$, one can also use for the description of the properties under shear the shear moduli, $G(t)$ and $G^*(\omega)$. As we shall find, this can drastically change the values of the relaxation times. Let us first consider a single Debye-process, now in combination with a superposed perfectly elastic part, and calculate the associated dynamic modulus. We have

$$J^*(\omega) = J_\mathrm{u} + \frac{\Delta J}{1+\mathrm{i}\omega\tau} \tag{5.75}$$

and therefore

$$\begin{aligned} G^*(\omega) = \frac{1}{J^*(\omega)} &= \frac{1+\mathrm{i}\omega\tau}{J_\mathrm{u}(1+\mathrm{i}\omega\tau)+\Delta J} \\ &= \frac{1}{J_\mathrm{u}} \frac{J_\mathrm{u} + \mathrm{i}\omega\tau J_\mathrm{u} + \Delta J - \Delta J}{J_\mathrm{u} + \Delta J + \mathrm{i}\omega\tau J_\mathrm{u}} \\ &= \frac{1}{J_\mathrm{u}} - \frac{\Delta J}{J_\mathrm{u}J_\mathrm{r}} \frac{1}{1+\mathrm{i}\omega\hat{\tau}} \end{aligned} \tag{5.76}$$

Here J_r is defined as

$$J_\mathrm{r} := J_\mathrm{u} + \Delta J \tag{5.77}$$

and $\hat{\tau}$ denotes another time constant, defined as

$$\hat{\tau} := \tau \frac{J_{\rm u}}{J_{\rm r}} \tag{5.78}$$

The subscripts 'r' and 'u' stand for the attributes 'relaxed' and 'unrelaxed' respectively. We can also introduce the 'relaxed' and 'unrelaxed' limiting values of the shear modulus, by

$$G_{\rm r} := \frac{1}{J_{\rm r}} \tag{5.79}$$

$$G_{\rm u} := \frac{1}{J_{\rm u}} \tag{5.80}$$

and the change of G, giving the relaxation strength, by

$$\Delta G := G_{\rm u} - G_{\rm r} \tag{5.81}$$

Using these parameters, the dynamic modulus obtains the simple form

$$G^*(\omega) = G_{\rm u} - \frac{\Delta G}{1 + {\rm i}\omega\hat{\tau}} \tag{5.82}$$

The important point in this result is that, compared to the dynamic compliance, the characteristic time has changed. This change from τ to $\hat{\tau}$ can be quite large. For example, the relaxation processes which are responsible for the glass transition transfer a polymer sample from the glassy to the rubbery state, which means a change in the compliance by four orders of magnitude. This large change then shows up correspondingly in the ratio between τ and $\hat{\tau}$.

Rather than representing the viscoelastic properties of a given sample in the form of Eq. (5.73), i.e. by a superposition of Debye-processes which are specified by ΔJ_l and τ_l, one can perform an analogous procedure based on single-time relaxation processes specified by ΔG_l and $\hat{\tau}_l$. We then write in the integral form

$$G^*(\omega) = G_{\rm u} - \int \frac{1}{1 + {\rm i}\omega\hat{\tau}} {\rm H}_G(\log\hat{\tau}) {\rm d}\log\hat{\tau} \tag{5.83}$$

It now includes the characteristic function ${\rm H}_G(\log\hat{\tau})$ which obviously differs from ${\rm L}_J(\log\tau)$. ${\rm H}_G(\log\hat{\tau})$ is called 'relaxation time spectrum of G'.

The corresponding expression for the time dependent modulus is

$$G(t) = G_{\rm r} + \int \exp -\frac{t}{\hat{\tau}} {\rm H}_G(\log\hat{\tau}) {\rm d}\log\hat{\tau} \tag{5.84}$$

In order to show that this is true, we have to prove that the time dependent modulus for the Debye-process does indeed equal the exponential function $\exp -(t/\hat{\tau})$. For the proof, we calculate first the primary response function, $\mu(t)$, by use of Eq. (5.31):

$$\mu(t) = \frac{{\rm d}J}{{\rm d}t} = \frac{\Delta J}{\tau} \exp -\frac{t}{\tau} + J_{\rm u}\delta(t) \tag{5.85}$$

Then we apply Eq. (5.34)

$$1 = \frac{\Delta J}{\tau} \int_0^t \exp -\frac{t-t'}{\tau} \cdot G(t')\mathrm{d}t' + J_\mathrm{u} \int_0^t \delta(t-t')G(t')\mathrm{d}t' \tag{5.86}$$

This leads to

$$\exp\frac{t}{\tau} = \frac{\Delta J}{\tau} \int_0^t \exp\frac{t'}{\tau} \cdot G(t')\mathrm{d}t' + J_\mathrm{u} \exp\frac{t}{\tau} \cdot G(t) \tag{5.87}$$

Taking on both sides the time derivative and dividing by $\tau^{-1} \cdot \exp(t/\tau)$ yields

$$1 = \Delta J \cdot G(t) + J_\mathrm{u} G(t) + \tau J_\mathrm{u} \frac{\mathrm{d}G}{\mathrm{d}t} \tag{5.88}$$

Differentiating for a second time gives the equation

$$0 = J_\mathrm{r} \frac{\mathrm{d}G}{\mathrm{d}t} + \tau J_\mathrm{u} \frac{\mathrm{d}^2 G}{\mathrm{d}t^2} \tag{5.89}$$

It is solved by

$$G(t) = G_\mathrm{r} + \Delta G \exp -\frac{t}{\hat{\tau}} \tag{5.90}$$

for

$$0 = -\frac{J_\mathrm{r}}{\hat{\tau}} + \tau J_\mathrm{u} \frac{1}{(\hat{\tau})^2} \tag{5.91}$$

or

$$\hat{\tau} = \tau \frac{J_\mathrm{u}}{J_\mathrm{r}} \tag{5.92}$$

in agreement with Eq. (5.78). Equations (5.90) and (5.92) confirm that Eq. (5.84) is correct.

Relaxation or retardation time spectra like $\mathrm{H}_G(\log\hat{\tau})$ or $\mathrm{L}_J(\log\tau)$ can always be used for a representation of measured data although its derivation from experimentally obtained modul- or compliance-functions can be difficult. The 'inversion' of one of the integral equations, Eqs. (5.73), (5.74), (5.83) or (5.84), belongs to a class of problems which are called 'ill-posed'. Here small fluctuations in the data, which cannot be avoided, become greatly magnified by the mathematical solution algorithm, thus leading to large variations in the derived quantities. Modern mathematical procedures can provide help, by allowing us to include in the solution any additional knowledge about the spectral functions. If they can be applied, situations improve and calculated spectra may then possess a satisfactory accuracy.

In this section, the focus was on the difference in the characteristic times observed in measurements of compliances or moduli respectively. In the explanation, we had to use two different symbols, the notation τ for the 'retardation time' and $\hat{\tau}$ for the 'relaxation time'. In what follows we shall not perstist in this differenciation and write τ generally, for all kinds of characteristic times observed in experiments.

5.3 Specific Relaxation Processes and Flow Behavior

After the introduction of the various interrelated response functions and basic concepts like the Debye-process and the derived spectral representations we come now in the second part of this chapter to the description and discussion of actual polymer behavior. In fact, relaxation processes play a dominant role and result in a complex pattern of temperature and frequency dependent properties.

We already have a general picture suitable for considerations. As has been repeatedly emphasized, dealing with fluid polymers means dealing with an ensemble of chains which can exist in a manifold of different conformational states. Thermal equilibrium is a dynamical situation where chains change between these states activated by thermal energies. The microscopic dynamics shows up in the macroscopic experiments. Relaxation rates observed in certain mechanical or dielectrical measurements equal the rates of transitions within a certain group of conformations.

The rates of conformational transitions of a chain encompass an enormously wide range. Local rearrangements which include only a few adjacent monomers are usually rapid and take place with rates similar to those in ordinary liquids. Conformational changes of more extended sequences require much longer times. In particular, relaxatory modes which are associated with the chain as a whole show a pronounced dependence on the molecular weight, as relaxation then has to propagate over larger and larger distances. Flow behavior is governed by these sluggish modes and therefore sets up the 'terminal region', i.e. the long time end, of the spectrum of relaxations.

The rates of the relaxatory modes in a sample do not cover the whole spectral range homogeneously, but usually one observes a separation into several zones where relaxation rates are accumulated. Each zone belongs to a group of processes with similar roots. It has become a convention to designate these different groups by Greek letters, α, β and γ, and to use the symbol α for the process with the lowest transition rates showing up at the highest temperature. On the other hand, the symbol γ is used for the processes observed at the low temperature end, and that means those with the highest transition rates.

In the remaining part of this chapter, we discuss the properties of some major groups of relaxation processes in polymers as there are

- local processes, to be observed in the glassy state
- cooperative processes in longer chain sequences which provide the basis for the elasticity of rubbers and the viscoelasticity of polymer melts
- chain diffusion, which controls the flow behavior
- specific processes in partially crystalline states, associated with coupled motions of sequences in the crystallites and the amorphous regions

5.3.1 Local Processes

Figure 5.7 shows the results of a dynamic shear experiment carried out on poly(cyclohexyl methacrylate) (PCHMA) in the glassy state. One observes a relaxation process which produces a loss maximum just in the frequency range of the mechanical spectrometer. With increasing temperature the position of the loss maximum shifts to higher values.

Considering the chemical constitution of PCHMA, there is an obvious assignment for this 'γ-process': It reflects the flip-motion between the 'chair'- and the 'boat'-conformation of the cyclohexane side-group. Since this process changes the shape of the side-group, it couples to the applied shear field. The assignment is corroborated by the observation that this process shows up whenever a cyclohexyl group is attached to a polymer chain. For all investigated samples the relaxation rates were similar, as to be expected for a mode with local character.

Figure 5.8 shows the temperature dependence of the relaxation rate in an Arrhenius-plot. The data were obtained in several experiments on polyacrylates and poly(methylacrylates) with pendant cyclohexyl groups. The linearity of the plot is indicative of an activated process, the relaxation time being given by the Arrhenius law

$$\tau \sim \exp \frac{\tilde{A}}{\tilde{R}T} \tag{5.93}$$

The relaxation rate τ^{-1} equals the rate of transitions between the two conformational states. The observed activation energy, $\tilde{A} = 47$ kJ $\cdot$mol^{-1}, therefore has to be identified with the height of the energy barrier which has to be passed over during a change.

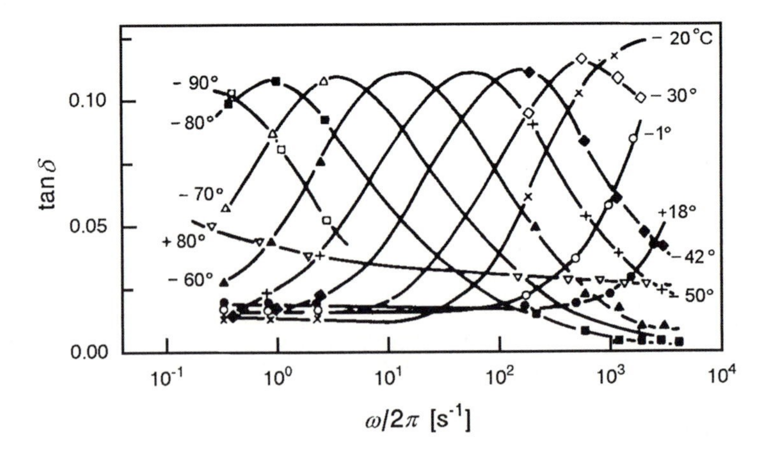

Fig. 5.7. Frequency dependence of the mechanical loss tangent measured for PCHMA at the indicated temperatures (after Heijboer [49])

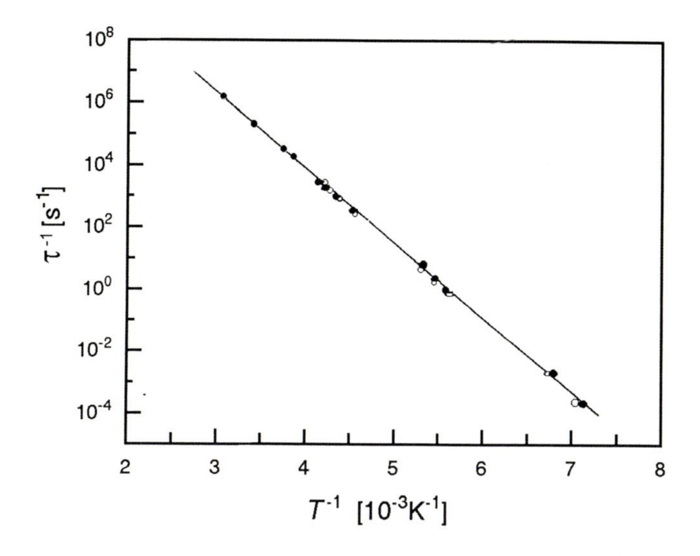

Fig. 5.8. Temperature dependence of the relaxation rates of the γ-process in polyacrylates (*open symbols*) and poly(methacrylates) (*filled symbols*) with pendant cyclohexyl groups. Data from Heijboer [50]

A remarkable fact to be noted in Fig. 5.7 is the constancy in the peak amplitude and the shape of the loss curves when varying the temperature. This behavior, in combination with the regular temperature shift according to Arrhenius' law, opens the way for an alternative experimental procedure. Rather than carrying out frequency dependent measurements at one temperature, loss curves may also be registered by temperature dependent measurements at constant frequency. Figure 5.9 presents such measurements, and as seen, they provide equivalent information. The relationship between the relaxation rate and the temperature follows equally from both measurements by a registration of the loss maxima.

In the combination of frequency- and temperature dependent measurements, one can even go one step further, thereby establishing an important general procedure. For groups of relaxation processes which encompass a broader time range, it often happens that the experimentally limited frequency range of the experimental device is not large enough to include the curves completely. Measurements carried out at a sequence of different temperatures can provide the missing information. As indicated by our example, different parts of the loss curve are placed into the accessible frequency window on changing the temperature. This property can now be used to set up the complete loss curve by a synthesis. The sections obtained at the different temperatures can be coupled together by carrying out appropriate shifts along the log ω-axis, thus ending up in one continuous curve.

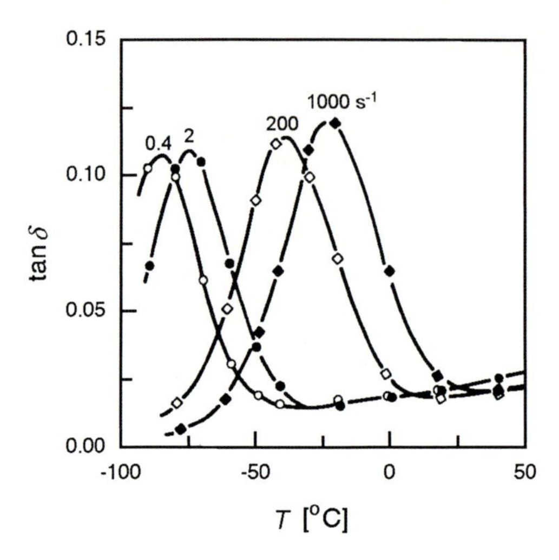

Fig. 5.9. Temperature dependent measurements of the loss tangent of the γ-process of PCHMA for several fixed frequencies $\omega/2\pi$ (After Heijboer [49])

What is applied here is known in the literature as the 'time-temperature superposition principle'. The result of the synthesis is called a 'master-curve'. For a thermally activated Debye-process, the basis of the principle is easily seen. According to Eq. (5.65), the dynamic compliance and the dynamic modulus here are functions of the product $\omega\tau$, or equivalently, of $\log \omega\tau$. If we also use Eq. (5.93), we may then represent the compliance as a function of a sum of terms

$$J^*(\log \omega\tau) = J^*\left(\log \omega + \log \tau_0 + \frac{\tilde{A}}{\tilde{R}T} \log \mathrm{e} \right) \tag{5.94}$$

The expression tells us that there are two ways of achieving a change in J^*, namely either by a shift in $\log \omega$, or by a shift in T^{-1}. The effects of frequency and temperature thus appear as 'superposed', and Eq. (5.94) informs us about the correspondencies.

As a prerequisite for the construction of a master-curve, the shape of the loss curve must remain constant under temperature variations. For the system under discussion, this is obviously fulfilled. Measured curves coincide after appropriate shifts along the $\log \omega$-axis, as is shown in Fig. 5.10 for the real and imaginary part of the dynamic shear modulus. The example represents an ideal case, and here there is also no need for a synthesis of the curves from parts. In many other cases, however, construction of the master-curve is the only means to explore a group of relaxation processes in total. Even if one is not sure if curve shapes are really temperature independent, construction of a master-curve remains useful as it can always provide a rough overall view, good for qualitative purposes.

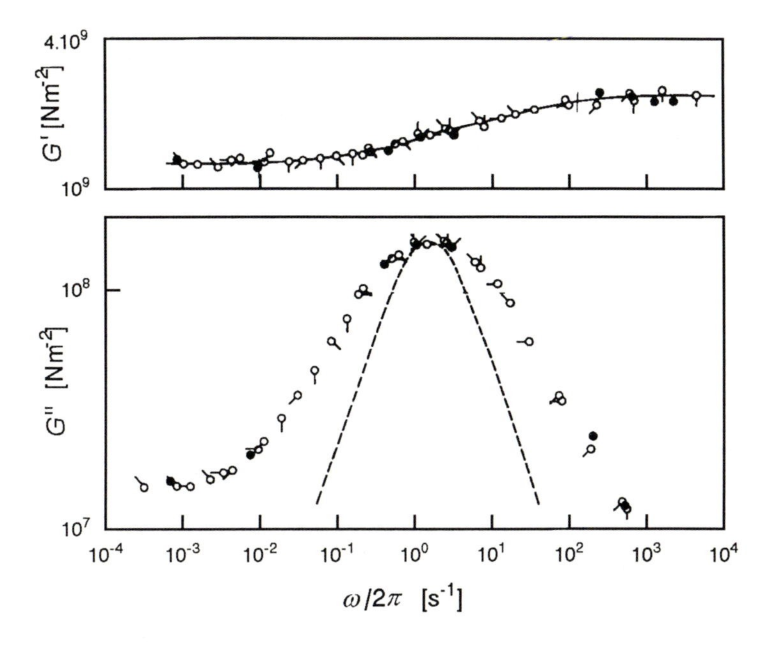

Fig. 5.10. Real and imaginary part of the dynamic shear modulus in the range of the γ-process of PCHMA, synthesized as a master-curve using measurements at various temperatures. Curves represent the viscoelastic behavior at $-80\,°C$. The dashed curve indicates a perfect Debye-process. Data from Heijboer [50]

Figure 5.10 shows also a comparison with the Debye-process. We notice that the γ-process of the cyclohexyl groups does not agree with a single-time relaxation process, but exhibits some broadening. This may be caused by a coupling between adjacent side-groups, as a conformational change in one side-group may well affect the neighbors. More specifically, the jump rate could depend on the conformations of the neighbors, which then would lead to a distribution of relaxation times, as is indicated by the broadened loss spectrum.

5.3.2 Glass-Rubber Transition and Melt Flow

Figure 5.11 presents creep curves, registered for a sample of polystyrene under shear-stress at various temperatures between $-268\,°C$ and $296.5\,°C$. We observe a creep compliance which encompasses the enormously broad range of nine orders of magnitude. At the lowest temperatures, the mechanical properties are those of a glass. At the other limit, the high temperature end, the behavior is dominated by viscous flow as indicated by the characteristic linear increase of J with time. The transition from the solid-like to the liquid-like behavior occurs continuously, and most importantly, obviously in a system-

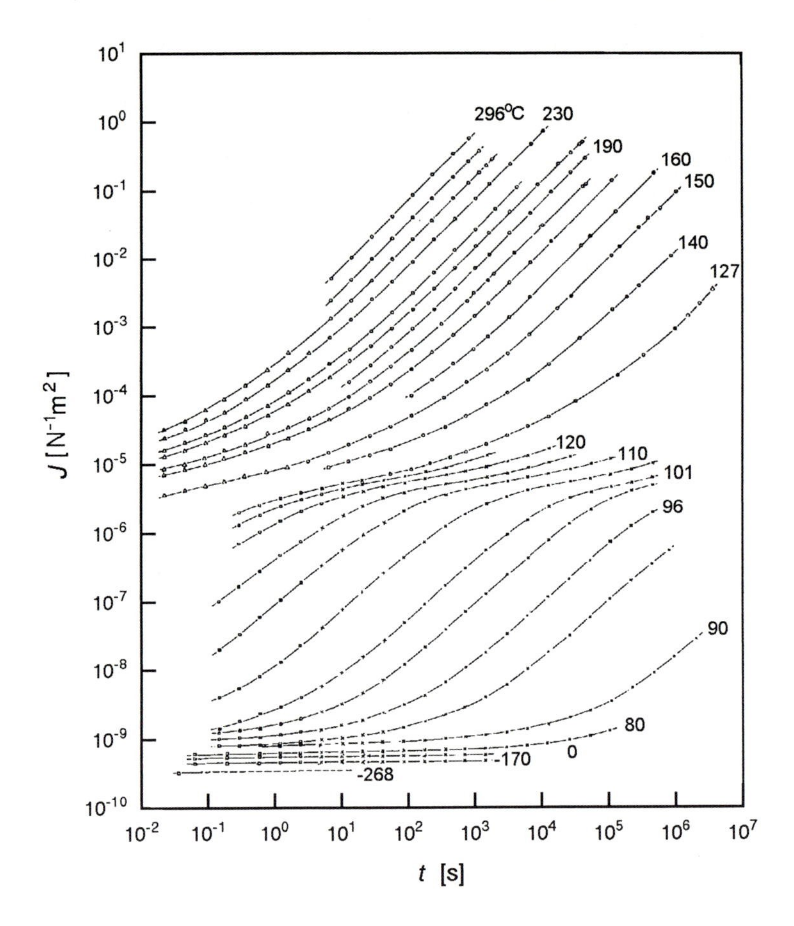

Fig. 5.11. Creep compliance of PS ($\overline{M}_{\mathrm{w}} = 3.85 \cdot 10^5$), as measured at the indicated temperatures. Data from Schwarzl [51]

atic manner. Indeed, the way curves change with temperature indicates that again time-temperature superposition is obeyed. Temperature variations result in shifts of the creep compliance along the $\log t$-axis, apparently without essential modifications in shape. The consequence is the same as for the just discussed local processes: On varying the temperature, different parts of $J(t)$ show up in the time-window of the experiment, and they can be reassembled to form a master-curve. Applying this procedure yields the overall creep curve and it evidently has a shape as is indicated schematically in Fig. 5.12. We can estimate the encompassed total time range by roughly summing up the time ranges of the sections included and we find an enormous extension of about 20 orders of magnitude.

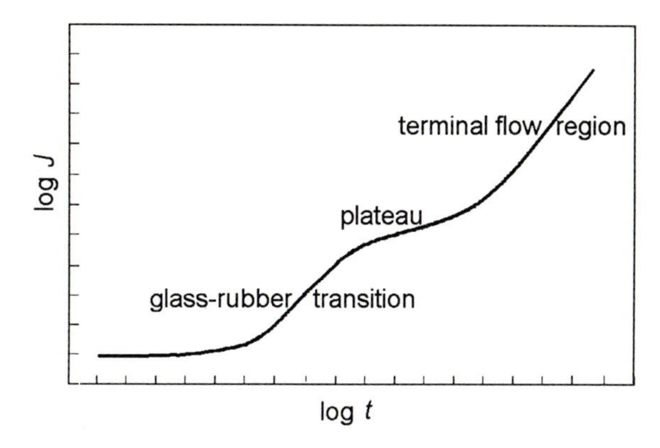

Fig. 5.12. General shape of the complete creep curve of PS, as suggested by the appearance of the different parts shown in Fig. 5.11

$J(t)$ has a characteristic shape composed of several parts. Subsequent to the glassy range with a solid-like compliance in the order of 10^{-9} $\mathrm{N^{-1}m^2}$, an additional anelastic deformation emerges and eventually leads to a shear compliance in the order of 10^{-5} $\mathrm{N^{-1}m^2}$. The latter value is typical for a rubber. For a certain time a plateau is maintained but then there finally follows a steady linear increase of J, as is indicative for viscous flow. The displayed creep curve of polystyrene is really not a peculiar one and may be regarded as representative for all amorphous, i.e. noncrystalline polymers. One always finds these four parts

- a glassy region
- the glass-rubber transition, often also called the 'α-process'
- a rubber-elastic plateau
- the terminal flow range.

These are the basic ingredients determining the mechanical properties of amorphous polymers and we discuss them now in a brief overview.

A most important conclusion can be drawn immediately and it concerns the nature of the main part, the glass-rubber transition. As we find a systematic shift of the time range of the transition with temperature, it is obvious that we are dealing here with a purely kinetical phenomenon rather than with a structural transition like the melting process or a solid-solid phase change. Curves demonstrate that whether a sample reacts like a glass or a rubber is just a question of time. Temperature enters only indirectly, in that it determines the characteristic time which separates glassy from rubbery behavior.

In chapter 7, we will discuss the properties of rubbers. These are networks, composed of chemically cross-linked macromolecules. Owing to the weak restoring forces, application of stress here induces a deformation which is very large compared to solids. The observation of a plateau in the creep compliance at a height comparable to the compliance of rubbers indicates that a polymer melt actually resembles a temporary network. This behavior expresses a major property specific for polymeric liquids: These include chain entanglements, i.e. constraints for the motion arising from the chain connectivity, which act like cross-links. Different from true cross-links of chemical nature, entanglements are only effective for a limited time during which they are able to suppress flow. This time becomes apparent in the creep-curve as the end of the plateau region.

Subsequent to the plateau, flow sets in. As is intuitively clear, the time needed for the chain disentangling increases with the molecular weight and this shows up in a corresponding broadening of the plateau. Results of dynamic-mechanical experiments on polystyrene, presented below in Fig. 5.15, exemplify the behavior. The data indicate also a lower limit: When decreasing the molecular weight one reaches a point, where the plateau vanishes. Then the glass-rubber transition and the terminal flow region merge together. Absence of the plateau means the absence of an entanglement network. The observation tells us that entanglement effects only exist above a certain minimum molecular weight. For each polymer one finds a characteristic value, known as the 'critical molecular weight at the entanglement limit', usually denoted M_{c}.

The measurements at high temperatures in Fig. 5.11 indicate a viscous flow with a constant creep rate, determined by a viscosity η_0

$$\frac{\mathrm{d}J}{\mathrm{d}t} \sim \frac{1}{\eta_0} \tag{5.95}$$

As the flow velocity relates to the disentangling time, this also holds for the melt viscosity. Indeed, η_0 and the disentangling time for entangled melts show the same dependence on the molecular weight. Figure 5.13 collects the results of viscosity measurements for various polymers. As should be noted, a power law behavior

$$\eta_0 \sim M^{\nu} \tag{5.96}$$

is generally observed. One finds two regions, with different values of the exponent ν and a cross-over at the entanglement limit M_{c}. For molecular weights below M_{c} one has $\nu = 1$, above M_{c} one observes $\nu \approx 3.2 - 3.6$.

Importantly, as is also shown by Fig. 5.15, the two parts of the mechanical response separated by the rubber-elastic plateau differ in their molecular weight dependence. In contrast to the terminal flow region, the glass-rubber transition remains largely unaffected by the molecular weight. The findings teach us that chain equilibration in reaction to an applied field takes place as a two-step process with a finite delay time in between. In the first step equilibration by relaxatory modes only includes chain sequences up to a certain length which is determined by the distance between the entanglements. As this distance is independent of M, this holds likewise for the characteristic time of this first step. Further relaxation is postponed until a chain extricates itself from the 'tube' formed by the other surrounding molecules and this process is of course strongly affected by the molecular weight.

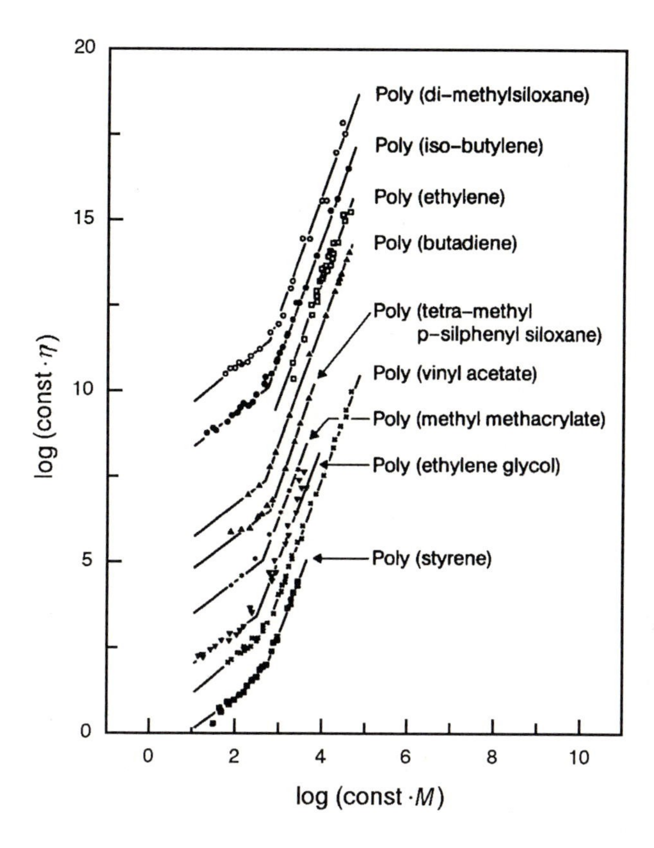

Fig. 5.13. Molecular weight dependence of the viscosity as observed for the indicated polymers. For better comparison curves are suitably shifted in horizontal and vertical direction. Data from Berry and Fox [52]

As explained in the first part of this chapter, the viscoelastic properties of polymers may also be studied by stress relaxation experiments or dynamic mechanical measurements. Since all response functions are interrelated, the mentioned ingredients of the mechanical behavior of amorphous polymers must show up in the other experiments as well. To give an example, Fig. 5.14 displays the time dependent tensile modulus registered for polyisobutylene (PIB). Measurements were again conducted for a series of temperatures. As expected, data show the glass-rubber transition (for temperatures in the range 190 – 220 K), followed by a plateau (around 230 K) and finally the onset of flow. The right-hand side presents the composite master-curve, set up by shifting the partial curves as indicated by the arrows. The amounts of shift along the $\log t$-axis are displayed in the insert. In the construction of the master-curve the time dependent modulus obtained at 298 K was kept fixed, while all other curves were displaced. The shift factor, denoted $\log a_T$,

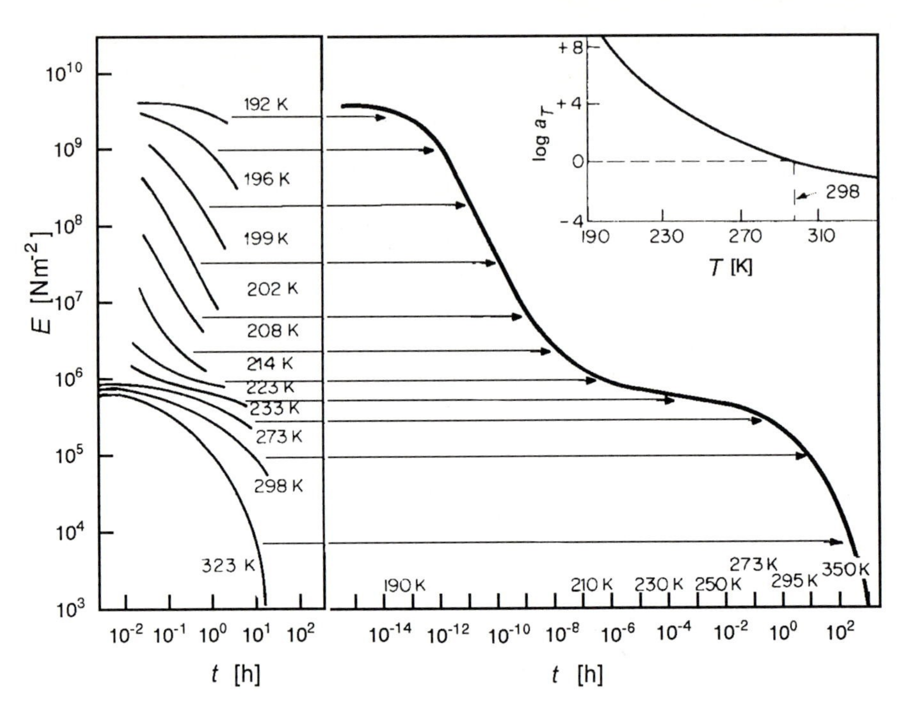

Fig. 5.14. Time dependent tensile modulus of PIB. Measurements at the indicated temperatures (*left*) and master-curve, constructed for a reference temperature $T =$ 298 K (*right*). The insert displays the applied shifts. Data from Castiff and Tobolsky [53]

is zero at this reference temperature. The result represents the complete time dependent shear modulus at the reference temperature. Comparable to the creep compliance in Fig. 5.12, this tensile modulus again encompasses a huge range of about 20 orders of magnitude in time.

Regarding the large number of conformational changes which must take place if a rubber is to be extended, the glass-rubber transition cannot equal a single-time relaxation process and this is shown by the curve shapes. To describe $E(t)$, empirical equations exist which often provide good data fits. A first one is concerned with the beginning of the transition range. It is known as the 'Kohlrausch-Williams-Watts (KWW)' function and has the form of a 'stretched exponential'

$$E(t) \sim \exp -(\frac{t}{\tau})^{\beta} \tag{5.97}$$

The KWW function employs two parameters: τ sets the time scale and β determines the extension in time of the decay process. For values $\beta < 1$ a broadening results, as is always observed for the glass-rubber transition. Typical values are in the order $\beta \simeq 0.5$. The KWW function holds only at the beginning, i.e. in the short-time range of the glass-rubber transition.

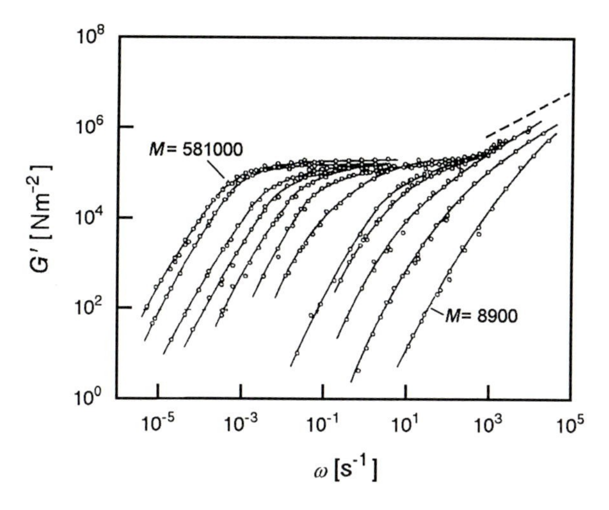

Fig. 5.15. Storage shear moduli measured for a series of fractions of PS with different molecular weights in the range $M = 8.9 \cdot 10^3$ to $M = 5.81 \cdot 10^5$. The *dashed line* in the upper right corner indicates the slope corresponding to the power law Eq. (6.81) derived for the Rouse-model of the glass-transition. Data from Onogi et al.[54]

Subsequently, there often follows a power law

$$E(t) \sim t^{-\nu} \tag{5.98}$$

Experimentally it is indicated by a linear range in the center, when using a log-log plot. Typical values of the exponent are $\nu \simeq 0.5$.

Figure 5.15 presents, as a third example, results of dynamic-mechanical measurements. They were obtained for a series of monodisperse polystyrenes, i.e. fractions with sharp molecular weights. The curves depict the frequency dependence of the storage shear modulus, $G'(\omega)$. As we note, the order of appearance of the viscous flow and the α-process is reversed when compared to the time dependent measurements. The flow-dominated long-time behavior emerges first at low frequencies, whereas an investigation of the rubber-glass transition requires measurements at the high frequency end. The plateau appears in between. Its width varies systematically with the molecular weight, as already mentioned and discussed. There is no plateau at all for the sample with the lowest molar mass ($M = 8.9 \cdot 10^3$), but after its first appearance, it widens progressively with further increasing molecular weight.

Low Frequency Properties of Polymer Melts

Also of interest, in Fig. 5.15, is the finding that the shapes of curves in the terminal region remain similar to each other for all molecular weights. More specifically, within the limit of low frequencies, a constant slope emerges, indicating a power law $G'(\omega) \sim \omega^2$. It is possible to explain this asymptotic behavior and to relate it to the properties of flowing polymer melts.

For a Newtonian low molar mass liquid, knowledge of the viscosity is fully sufficient for the calculation of flow patterns. Is this also true for polymeric liquids? The answer is no under all possible circumstances. Simple situations are encountered for example in dynamical tests within the limit of low frequencies or for slow steady state shears and even in these cases, one has to include one more material parameter in the description. This is the 'recoverable shear compliance', usually denoted J_{e}^0, and it specifies the amount of recoil observed in a creep recovery experiment subsequent to the unloading. J_{e}^0 relates to the elastic and anelastic parts in the deformation and has to be accounted for in all calculations. Experiments show that, at first, for $M < M_{\mathrm{c}}$, J_{e}^0 increases linearly with the molecular weight and then reaches a constant value which essentially agrees with the plateau value of the shear compliance.

At higher strain rates even more complications arise. There the viscosity is no longer constant and shows a decrease with increasing rate, commonly called 'shear-thinning'. We will discuss this effect and related phenomena in chapter 7, when dealing with non-linear behavior. In this section, the focus is on the limiting properties at low shear rates, as expressed by the 'zero shear rate viscosity', η_0, and the recoverable shear compliance at zero shear rate, J_{e}^0.

Our concern is to find out how the characteristic material parameters η_0 and J_{e}^0 are included in the various response functions. To begin with, consider a perfectly viscous system in a dynamic-mechanical experiment. Here the dynamic shear compliance is given by

$$J^* = -\mathrm{i}\frac{1}{\eta_0\omega} \tag{5.99}$$

This is seen when introducing the time dependencies

$$\begin{aligned} \sigma_{zx} &= \sigma_{zx}^0 \exp \mathrm{i}\omega t \\ e_{zx} &= J^* \sigma_{zx}^0 \exp \mathrm{i}\omega t \end{aligned}$$

into the basic equation for Newtonion liquids

$$\sigma_{zx} = \eta_0 \frac{\mathrm{d}e_{zx}}{\mathrm{d}t} \tag{5.100}$$

which results in

$$\sigma_{zx}^0 \exp \mathrm{i}\omega t = \eta_0 \mathrm{i}\omega J^* \sigma_{zx}^0 \exp \mathrm{i}\omega t \tag{5.101}$$

In a polymer melt, the viscous properties of Newtonian liquids combine with elastic forces. The latter ones contribute a real part to the dynamic shear compliance, to be identified with J_{e}^0

$$J'(\omega \to 0) := J_{\mathrm{e}}^0 \tag{5.102}$$

Combining Eqs. (5.99) and (5.102) gives the dynamic shear compliance of polymeric fluids in the limit of low frequencies

$$J^*(\omega \to 0) = J_{\mathrm{e}}^0 - \mathrm{i}\frac{1}{\eta_0\omega} \tag{5.103}$$

As we see, η_0 and J_{e}^0 show up directly and separately, in the limiting behavior of J' and J''.

The dynamic shear modulus follows as

$$\begin{aligned} G^*(\omega \to 0) &= \frac{1}{J^*(\omega \to 0)} = \frac{\eta_0\omega}{\eta_0\omega J_{\mathrm{e}}^0 - \mathrm{i}} \\ &= \frac{\eta_0^2\omega^2 J_{\mathrm{e}}^0 + \mathrm{i}\eta_0\omega}{(\eta_0\omega J_{\mathrm{e}}^0)^2 + 1} \end{aligned} \tag{5.104}$$

giving

$$G'(\omega \to 0) = J_{\mathrm{e}}^0\eta_0^2\omega^2 \tag{5.105}$$

in agreement with Fig. 5.15, and

$$G''(\omega \to 0) = \eta_0\omega \tag{5.106}$$

We thus find characteristic power laws also for the storage and the loss modulus which again include J_{e}^0 and η_0 in a well-defined way.

One may wonder if η_0 and J_{e}^0 can also be deduced from the time dependent response functions, as for example from $G(t)$. Indeed, direct relationships exist, expressed by the two equations

$$\eta_0 = \int_0^\infty G(t)\mathrm{d}t \tag{5.107}$$

and

$$J_{\mathrm{e}}^0\eta_0^2 = \int_0^\infty G(t)t\mathrm{d}t \tag{5.108}$$

The first relation follows immediately from Boltzmann's superposition principle in the form of Eq. (5.38) when applied to the case of a deformation with constant shear rate $\dot{e}_{zx}$. We have

$$(\mathrm{d}x\hat{=})\mathrm{d}e_{zx} = \dot{e}_{zx}\mathrm{d}t \tag{5.109}$$

and thus

$$(\psi\hat{=})\ \sigma_{zx} = \dot{e}_{zx}\int_{t'=-\infty}^{t} G(t-t')\mathrm{d}t' = \dot{e}_{zx}\int_{t''=0}^{\infty} G(t'')\mathrm{d}t'' \tag{5.110}$$

Since per definition

$$\sigma_{zx} := \eta_0 \dot{e}_{zx}$$

we find

$$\eta_0 = \int_{t=0}^{\infty} G(t)\mathrm{d}t$$

To derive the second equation, we consider a dynamic-mechanical experiment and treat it again on the basis of Boltzmann's superposition principle, writing

$$\sigma_{zx} = \int_{t'=-\infty}^{t} G(t-t')\dot{e}_{zx}(t')\mathrm{d}t' \tag{5.111}$$

Introducing

$$e_{zx}(t) = e_{zx}^0 \exp \mathrm{i}\omega t \tag{5.112}$$

and

$$\sigma_{zx}(t) = G^* e_{zx}(t) \tag{5.113}$$

we obtain

$$G^* = \int_{t''=0}^{\infty} G(t'')\mathrm{i}\omega \exp -\mathrm{i}\omega t'' \mathrm{d}t'' \tag{5.114}$$

setting $t'' := t - t'$. In the limit $\omega \to 0$ we can use a series expansion

$$G^*(\omega \to 0) = \int_{t''=0}^{\infty} G(t'')(\mathrm{i}\omega + \omega^2 t'' + \ldots)\mathrm{d}t'' \tag{5.115}$$

giving

$$G'(\omega \to 0) = \omega^2 \int_{t=0}^{\infty} G(t)t\mathrm{d}t \tag{5.116}$$

Comparison with Eq. (5.105) yields Eq. (5.108).

Combination of Eqs. (5.107) and (5.108) can be used for estimating the average time of stress decay subsequent to a sudden shear deformation of a melt. We may introduce this time, denoted $\bar{\tau}$, as

$$\bar{\tau} := \frac{\int_{t=0}^{\infty} G(t)t\mathrm{d}t}{\int_{t=0}^{\infty} G(t)\mathrm{d}t} \tag{5.117}$$

and then obtain simply

$$\bar{\tau} = J_{\mathrm{e}}^0 \eta_0 \tag{5.118}$$

Equation (5.118) for the mean viscoelastic relaxation time may be applied for both non-entangled and entangled melts and yields different results for the two cases. For non-entangled melts, i.e. $M < M_c$, we have $J_e^0 \sim M$ and $\eta_0 \sim M$, hence

$$\bar{\tau} \sim M^2 \tag{5.119}$$

For molecular weights above the entanglement limit, i.e. $M > M_c$, one finds J_e^0=const and $\eta_0 \sim M^{3.4}$, therefore

$$\bar{\tau} \sim \eta_0 \sim M^{3.4} . \tag{5.120}$$

Vogel-Fulcher Law and WLF Equation

We turn now to another important point and consider the temperature dependence. Recall that the data indicate the validity of time-temperature or frequency-temperature superposition. This has an important implication: The findings show that the processes comprising the terminal flow region and the glass-rubber transition change with temperature in the same manner. Particularly suited for the description of this common temperature dependence is the shift parameter $\log a_T$. We introduced it in connection with the construction of the master-curves but it has also a well-defined physical meaning. This becomes revealed when we look at the equations valid in the terminal range, Eqs. (5.105) and (5.106). It should be noted that ω and η_0 enter into the expressions for the dynamic modulus and the dynamic compliance not separately, but only as a product. As temperature affects just η_0, we conclude that a_T and η_0 must be proportional quantities. The exact relationship follows when taking into account that shift parameters always relate to a certain reference temperature. Let this reference temperature be T_0. Then a_T is given by

$$a_T = \frac{\eta_0(T)}{\eta_0(T_0)} \tag{5.121}$$

With the aid of a_T we can express response functions at any temperature in terms of the respective response function at T_0. Explicitly, for the dynamical shear modulus, the following relation holds

$$G^*(T, \omega) = G^*(T_0, a_T\omega) \tag{5.122}$$

or for a logarithmic frequency scale

$$G^*(T, \log\omega) = G^*(T_0, \log\omega + \log a_T) \tag{5.123}$$

In correspondence to this, we write for the time dependent shear modulus

$$G(T, t) = G(T_0, \frac{t}{a_T}) \tag{5.124}$$

or

$$G(T, \log t) = G(T_0, \log t - \log a_T) \tag{5.125}$$

The uniform temperature dependence implies a joint rescaling of the relaxation times of all modes in both the glass-transition range and the terminal flow region, and one may wonder how this might arise. One should be aware that these modes vary greatly in their spatial extensions, which begin with the length of a Kuhn segment and go up to the size of the whole chain, and vary also in character, as they include intramolecular motions as well as diffusive movements of the whole chain, and nevertheless, all modes behave uniformly. There seems to be only one possible conclusion: The temperature dependence must be a property of the individual segments. Since all modes are based on the motion of segments, their mobility affects each mode alike. There is a notion which suitably expresses this property and this is the 'segmental frictional coefficient'. We will introduce it in the next chapter, in the treatment of microscopic dynamics. For the moment it is sufficient to say that frictional forces exist which act in an identical manner on all the segments. They control uniformly the kinetics of all the relaxatory modes of the chains. The common temperature dependence of all relaxatory modes in the α-transition range and the terminal zone, and thus of the viscosity, just reflects that of the segmental frictional force.

Equation (5.121) relates a_T to the temperature dependence of the viscosity. Numerous experiments were carried out to measure this function. They led to a specific result. As it turns out, for the majority of polymer systems, $\eta_0(T)$ is well represented by an empirical equation known as the 'Vogel-Fulcher law'. It has the form

$$\eta_0(T) = B \exp \frac{T_{\mathrm{A}}}{T - T_{\mathrm{V}}} \tag{5.126}$$

In addition to the prefector B two parameters are included, namely the 'activation temperature' T_{A} and the 'Vogel temperature' T_{V}. The introduction of the latter makes up the difference to Arrhenius' law.

The function $\eta_0(T)$, as formulated by the Vogel-Fulcher law, includes a singularity at $T = .T_V$. However, whether the viscosity really diverges if T approaches T_{V} cannot be checked by any experiment. Measurements of viscosities always come to an end about 50 K above T_{V}, because η_0 then is already very large, reaching values in the order of 10^{13} poise. Notwithstanding the fact that the point of divergence is out of reach, validity of the Vogel-Fulcher equation is well established since effects of a finite Vogel temperature are clearly observable also in the range of accessible temperatures. There, the function $\eta_0(T)$ exhibits a characteristic curvature which distinguishes it from Arrhenius behavior. Figure 5.16 depicts, as an example, results obtained for polyisobutylene (PIB). An increase to high values of η_0 is observed at low temperatures and it can be described by a Vogel-Fulcher function, as given by the continuous line. The figure also includes the temperature dependence of the characteristic time τ_α of the glass-rubber transition. It is given by

$$\tau_\alpha = \tau_0 \exp \frac{T_{\mathrm{A}}}{T - T_{\mathrm{V}}} \tag{5.127}$$

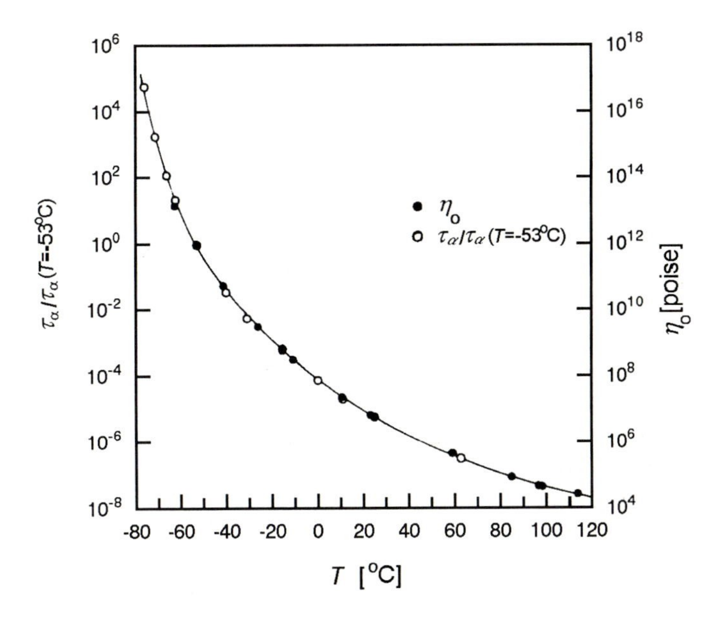

Fig. 5.16. Temperature dependencies of the viscosity η_0 of PIB (*open symbols, right axis*) and of the relaxation time of the α-process τ_α (*filled symbols, left axis*). Both correspond to a Vogel-Fulcher function (*continuous line*). Data from Plazek et al.[55]

with the same values for $T_{\rm A}$ and $T_{\rm V}$.

Having an equation for the temperature dependence of the viscosity, we may also formulate the shift factor $\log a_T$. Equations (5.126) and (5.121) yield

$$\begin{aligned} \log a_T &= \log \mathrm{e} \cdot T_{\rm A} \left(\frac{1}{T - T_{\rm V}} - \frac{1}{T_0 - T_{\rm V}} \right) \\ &= \log \mathrm{e} \cdot (-T_{\rm A}) \frac{T - T_0}{(T_0 - T_{\rm V})(T - T_{\rm V})} \\ &= \log \mathrm{e} \cdot \frac{(-T_{\rm A})}{T_0 - T_{\rm V}} \cdot \frac{T - T_0}{T - T_0 + (T_0 - T_{\rm V})} \end{aligned} \tag{5.128}$$

This is usually expressed as

$$\log a_T = -C_1 \frac{T - T_0}{T - T_0 + C_2} \tag{5.129}$$

introducing two paramters, C_1 and C_2, defined as

$$C_1 := \log \mathrm{e} \cdot \frac{T_{\rm A}}{T_0 - T_{\rm V}} \quad \text{and} \quad C_2 := T_0 - T_{\rm V} \tag{5.130}$$

Equation (5.129) was postulated by Williams, Landel and Ferry and is well-known in the literature under the short name 'WLF equation'.

When master-curves are constructed one chooses in most cases the 'glass transition temperature' $T_{\rm g}$ as reference temperature. $T_{\rm g}$ is obtained by a standard calorimetric or volumetric measurement, as explained in one of the following sections. It is found that, for this choice of T_0, the parameters C_1 and C_2 of the WFL equation have values which are bound to certain ranges, namely

$$C_1 = 14 - 18$$

$$C_2 = 30 - 70 \text{ K}$$

The values of C_2 indicate that $T_{\rm V}$ is located $30 - 70$ K below $T_{\rm g}$.

Dielectric α-Process and Normal Mode

The two groups of relaxatory modes which lead in mechanical relaxation experiments to the α-transition and the final viscous flow also emerge in the dielectric response. Figure 5.17 presents, as a first example, the frequency dependencies of the real and imaginary part of the dielectric constant, obtained for poly(vinylacetate) (PVA) at the indicated temperatures. One observes a strong relaxation process.

Figure 5.18 displays the temperature dependence of the relaxation rates, as derived from the maxima of the loss curves. For a comparison it also includes the temperature dependencies of the loss maxima of the mechanical α-process, as observed in measurements of either $J''(\omega)$ or $G''(\omega)$. As we can see, the dielectric relaxation rates are located intermediately between the rates obtained in the mechanical experiments and importantly, all three temperature dependencies are similar, the rates differing only by constant factors. The assignment of this dielectric relaxation process is therefore obvious: It originates from the same group of processes as the mechanical α-process and thus is to be addressed as the 'dielectric α-process'.

There are other polymers which show in addition the chain disentangling associated with the flow transition. An example is given by *cis*-polyisoprene (PIP). Figure 5.19 depicts the dielectric loss ϵ'' in a three-dimensional representation of the functional dependence on frequency and temperature. Two relaxation processes show up. The one with the higher frequency again represents the α-process, the other is called the 'normal mode', for reasons to be seen in a moment.

To learn more about the two processes, it is instructive to check for the molecular weight dependencies. In fact, one finds here a characteristic difference. The results of studies on a set of samples with different molecular weights are displayed in Fig. 5.20. We observe that the α-process is molecular weight independent, whereas the normal mode shows quite pronounced changes. Figure 5.21 depicts these changes, in a plot of the relaxation time τ of the normal mode in dependence on the molecular weight. Interestingly enough, a power law is found

$$\tau \sim M^{\nu}$$

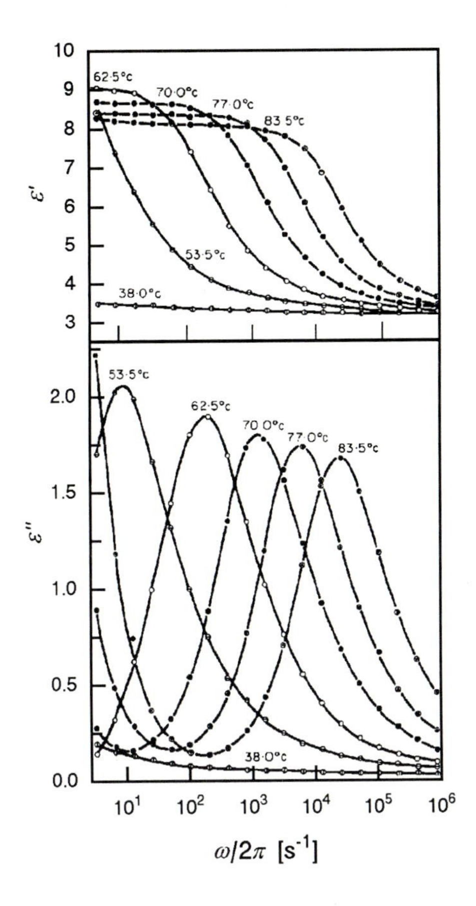

Fig. 5.17. Dielectric α-process in PVA. Data from Ishida et al. [56]

with two different values for the exponent

$$\nu = 3.7 \quad \text{for} \quad M > 10^4$$

and

$$\nu = 2 \quad \text{for} \quad M < 10^4$$

The cross-over from one to the other regime shows up as a sharp bend in the curve.

We have already met this particular molecular weight dependence, in Eqs. (5.119) and 5.120), when formulating the average viscoelastic relaxation time $\bar{\tau}$ of polymer melts. Roughly speaking, $\bar{\tau}$ gives the time required by a chain for a complete conformational reorganisation. This also implies a full reorientation of the end-to-end distance vector of the chain. It is exactly this motion which shows up in the dielectric normal mode.

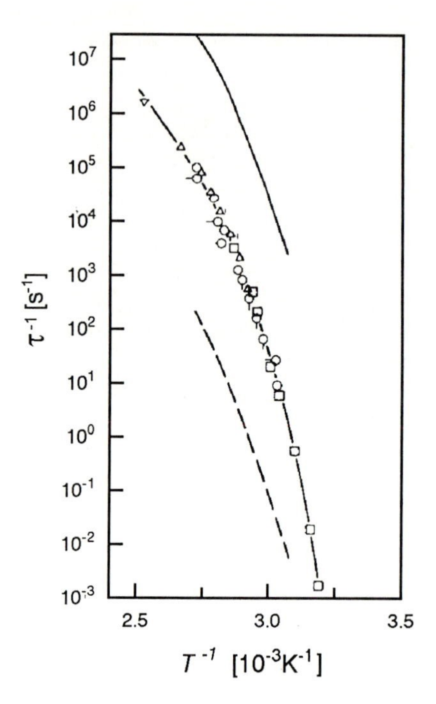

Fig. 5.18. Frequency-temperature locations of the dielectric loss maxima (*open symbols*) of PVA, compared to the maxima of G'' (*continuous line*) and J'' (*broken line*) observed in mechanical experiments. Collection of data published in [57]

The question arises why *cis*-polyisoprene, different from poly(vinyl-acetate), shows in its dielectric spectrum the chain reorientation. The reason becomes clear when we look at the chemical constitution of polyisoprene, and focus in particular on the associated dipole moments. Figure 5.22 displays the chemical structure. The main point is that isoprene monomers are polar units which possess a longitudinal component $p_{||}$ of the dipole moment, which always points in the same direction along the chain. As a consequence, the longitudinal components of the dipoles of all monomers become added up along the contour, giving a sum which is proportional to the end-to-end distance vector $\boldsymbol{R}$. In the dielectric spectrum the kinetics of this total dipole of the chain is observable, hence also the chain reorientation as described by the time dependence $\boldsymbol{R}(t)$.

The peculiar name 'normal mode' needs a comment. As will be explained in detail in the next chapter, chain dynamics in melts may be described with the aid of two theoretical models known as the 'Rouse-model' and the 'reptation model'. In the frameworks of these treatments chain kinetics is represented as a superposition of statistically independent relaxatory 'normal modes'. As

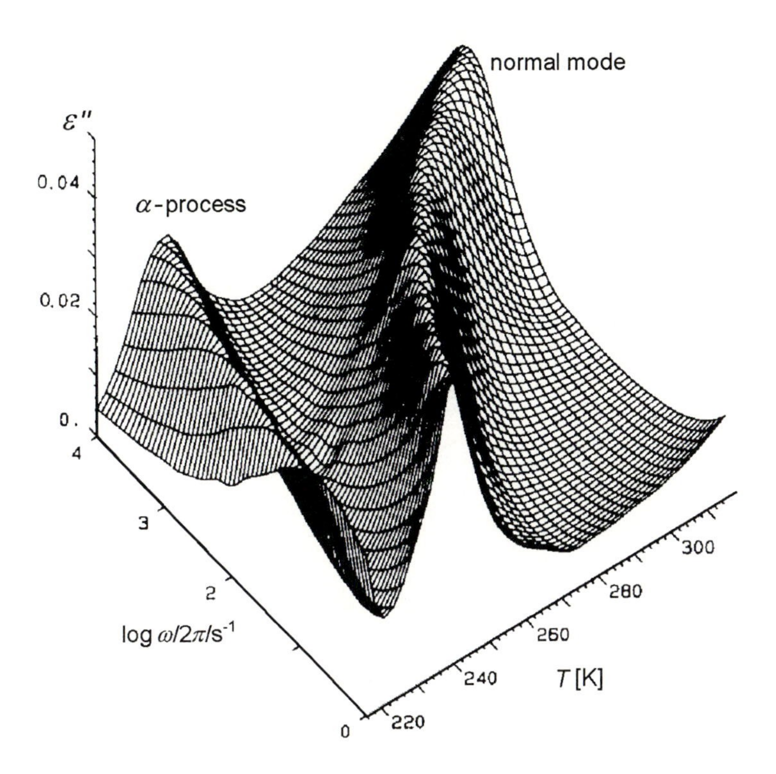

Fig. 5.19. Frequency- and temperature dependence of the dielectric loss in *cis*-PIP ($M = 1.2 \cdot 10^4$), indicating the activity of two groups of relaxatory modes. Spectra obtained by Boese and Kremer [58]

it turns out, the dielectric normal mode is associated with the mode with the longest relaxation time. For non-entangled melts this is the lowest order Rouse-mode, for entangled melts the lowest order reptation mode.

In addition to the longitudinal component of the dipole per monomer, there is also a transverse part. As the reorientation of the transverse component requires only local changes in the conformation, it can take place much more rapidly than the spatially extended normal mode. Hence, a qualitativ change in the kinetics occurs and indeed, it is this movement which shows up in the α-process. It is important to notice that both the α-process and the normal mode obey the Vogel-Fulcher law, in full analogy to the common behavior of the α-process and the terminal relaxation in mechanics.

It is possible to write down approximate expressions for the relaxation strengths $\Delta\epsilon$ of the two processes. As a chain may be described as a sequence of freely jointed segments, we can just make use of Eqs. (5.16) and (5.59), and introduce for the α-process and the normal mode the transverse and the longitudinal component of the dipole moment respectively. The relaxation

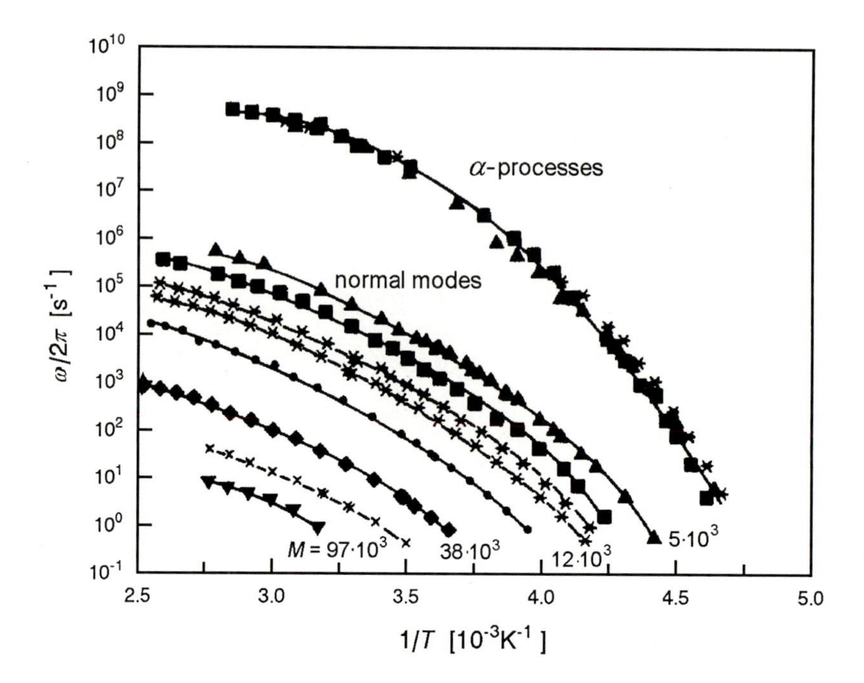

Fig. 5.20. Temperature dependence of the relaxation rates of the dielectric α-process and the normal mode, observed for samples of *cis*-PIP with different molecular weights (four values are indicated). The solid lines are fits based on the WLF equation. Data from Boese and Kremer [58]

strength of the α-process then follows as

$$\epsilon_0 \Delta\epsilon_\alpha \simeq c_{\mathrm{s}} \frac{\langle (p_\perp^{\mathrm{s}})^2 \rangle}{3kT} \tag{5.131}$$

Here, $p_\perp^{\mathrm{s}}$ is the transverse dipole moment per segment, and c_{s} gives the number density of segments. The brackets indicate an averaging over all rotational isomeric states of one segment. The relaxation strength of the normal mode follows equivalently by introduction of the mean longitudinal dipole moment per segment

$$\epsilon_0 \Delta\epsilon_{\mathrm{nm}} \simeq c_{\mathrm{s}} \frac{\langle (p_\parallel^{\mathrm{s}})^2 \rangle}{3kT} \tag{5.132}$$

Neither the α-process nor the normal mode equal a single-time relaxation process. A good representation of data is often achieved by use of the empirical 'Havriliak-Nagami equation' which has the form

$$\epsilon^* - \epsilon_{\mathrm{u}} = \frac{\Delta\epsilon}{(1 + (\mathrm{i}\omega\tau)^{\beta_1})^{\beta_2}} \tag{5.133}$$

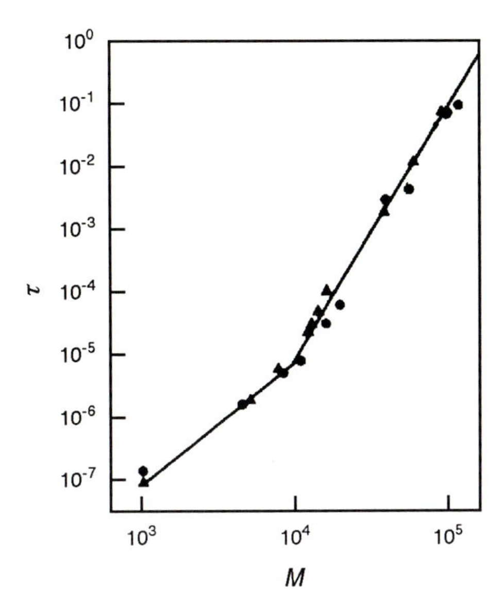

Fig. 5.21. Molecular weight dependence of the relaxation time of the dielectric normal mode in *cis*-PIP. Data from Boese and Kremer [58]

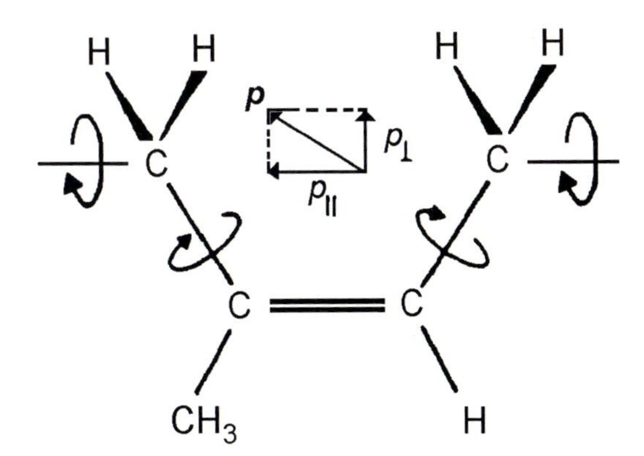

Fig. 5.22. Stereochemical constitution of a monomer unit of *cis*-PIP. The electric dipole moment, split into a longitudinal and a transverse component, is indicated

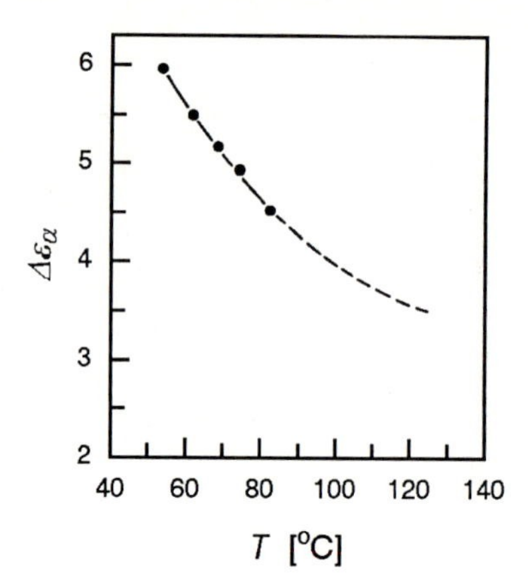

Fig. 5.23. Temperature dependence of the relaxation strength $\Delta\epsilon_\alpha$ of the dielectric α-process in PVA. Data from Ishida et al.[56]

This function is a formal generalization of the single-time relaxation function, achieved by an inclusion of two additional parameters, β_1 and β_2 (for $\beta_1 = 1$ it equals the Cole-Cole function mentioned earlier). These determine the asymptotic behavior, β_1 on the low frequency side, since

$$\beta_1 \approx \frac{\mathrm{d}\log\epsilon''}{\mathrm{d}\log\omega} \qquad \text{for } \omega\tau \ll 1 \tag{5.134}$$

and the product $\beta_1\beta_2$ on the high frequency side, since

$$\beta_1\beta_2 = -\frac{\mathrm{d}\log\epsilon''}{\mathrm{d}\log\omega} \qquad \text{for } \omega\tau \gg 1 \tag{5.135}$$

Obviously β_2 determines the curve asymmetry. It is needed for the data representation because the observed curves $\epsilon''(\omega)$ generally exhibit a larger broadening on the high frequency side. Typical values are $\beta_1 \simeq 0.5$, $\beta_2 \simeq 0.7$ for the α-process, $\beta_1 \simeq 1$, $\beta_2 = 0.4$ for the normal mode.

Equations (5.131) and (5.133) together provide a description of the dielectric α-transition, with the assumption that dipoles of different segments reorient independently. In fact, this is only true at larger distances from the glass transition temperature. On approaching T_{g} deviations show up. Figure 5.23 shows the temperature dependence of the relaxation strength of poly(vinylacetate) and one observes a pronounced increase. The behavior indicates increasing correlations between the motions of the transverse dipoles, not only along one chain, but possibly also between adjacent segments on different chains.

5.3.3 Glass Transition Temperature

The mechanical experiments clearly demonstrate that the transition from the glassy to the liquid state is a purely kinetical phenomenon. Whether the compliance of a sample is small as in a glass, or large as for a rubber, depends only on the measuring time or the applied frequency. The reasons were discussed above. Rubber elasticity originates from the activity of the 'α-modes', a major group of relaxation processes in polymer fluids. Establishment of the deformation subsequent to the application of a load requires a certain time, given by the time scale of the α-modes. If the load varies too rapidly, the deformation cannot follow and the sample reacts like a glass. We also discussed the effect of temperature and found, as a main property of the α-modes, that relaxation times change according to the Vogel-Fulcher law. The progressive increase of the relaxation times on cooling implied by this law finally leads to a 'freezing' of the α-modes within a comparatively small temperature range. If they are frozen, we have a glass.

We thus find glass-like reactions for both, sufficiently high frequencies and sufficiently low temperatures, but are the two situations really comparable? The answer is, yes and no, depending on the point of view. Yes, because both situations have in common that the α-modes cannot equilibrate. No, if we consider the thermodynamic state of order. In the first case we deal with a system in thermal equilibrium and study its reaction on perturbations, in the second situation, however, the system has become non-ergodic, i.e. thermal equilibrium is only partially established. A major part of the internal degrees of freedom, as represented by the α-modes, cannot equilibrate. The temperature, where the transition from a liquid equilibrium state to a non-ergodic, i.e. only partially equilibrated state takes place, is called the 'glass transition temperature', with the general designation T_{g}.

How can T_{g} be determined? In principle this can be achieved in various ways, however, two of the methods are of special importance and used in the majority of cases. These are temperature dependent measurements of the expansion coefficient or the heat capacity of a sample, carried out during heating or cooling runs. They need only small amounts of material, and standard equipment is commercially available.

Figures 5.24 and 5.25 present as examples the results of a volumetric and a calorimetric measurement on poly(vinylacetate). The glass transition has a characteristic signature which shows up in the curves. As we can see, the transition is associated with steps in the expansion coefficient $\mathrm{d}\rho^{-1}/\mathrm{d}T$ and the heat capacity $\mathrm{d}\mathcal{H}/\mathrm{d}T$, i.e. changes in the slope of the functions $\rho^{-1}(T)$ and $\mathcal{H}(T)$. The transition extends over a finite temperature range with typical widths in the order of 10 degrees. The calorimetric experiment also exhibits another characteristic feature. One can see that the location of the step depends on the heating rate $\dot{T}$, showing a shift to higher temperatures on increasing the rate.

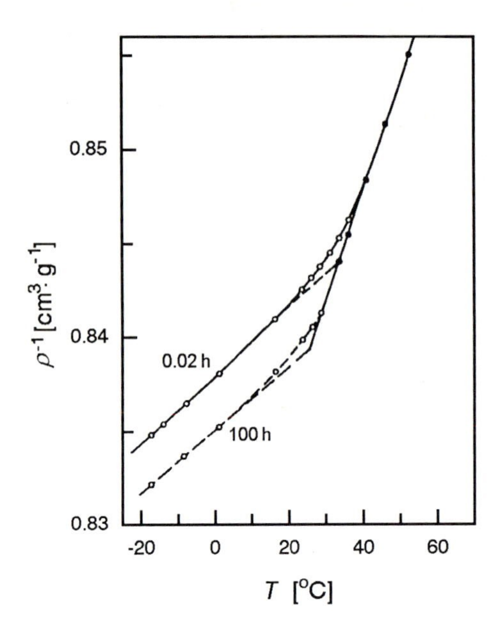

Fig. 5.24. Temperature dependence of the specific volume of PVA, measured during heating. Dilatometric results obtained after a quench to $-20°$C, followed by 0.02 or 100 h of storage. Data from Kovacs [59]

In view of the broadening of the step and the rate effects, it does not seem appropriate to introduce a sharply defined T_g. For the practical use as material parameter and for comparisons it is sufficient to conduct the measurements with a standard heating or cooling rate ($|\dot{T}| = 10^{-1} - 1\ \mathrm{Ks}^{-1}$) and to pick out some temperature near the center of the step, for example that associated with the maximum slope. The so-obtained values of T_g have a tolerance of some degrees but this must be accepted regarding the physical nature of the phenomenon.

The cause for the occurrence of the steps in the heat capacity and the expansion coefficient is easily seen. Cooling a sample below T_g results in a freezing of the α-modes. The observations tell us that the α-modes affect not only the shape of a sample, but also its volume and its enthalpy. This is not at all surprising. If segments move, they produce in their neighborhoods an additional volume. In the literature, this is often called a 'free volume' to stress that is is not occupied by the hard cores of the monomers. The free volume increases with temperature because motions intensify, that is to say the jump rates increase and, more importantly, a growing number of conformational states becomes populated and not all of them allow a dense chain packing. Therefore, when on crossing T_g from low temperatures the α-modes become active, beginning slowly and then steadily increasing in intensity, correspond-

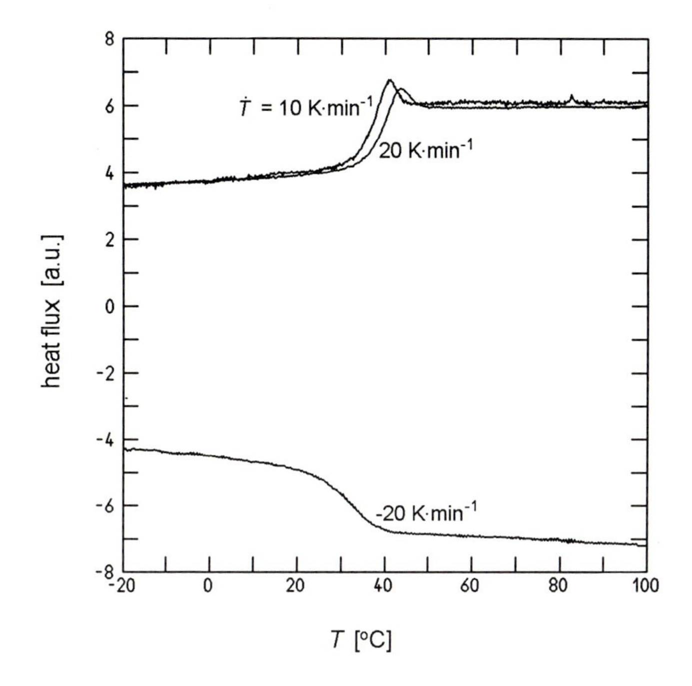

Fig. 5.25. Heat capacity of PVA, as measured in a differential calorimeter during heating (with 2 different heating rates) and cooling.

ingly a growing additional free volume arises. Thermal expansion in the glass is due to the anharmonicity of vibrational motions, as in crystalline solids. As we can see, the α-modes contribute another, even larger part to the expansion coefficient and it comes into effect at T_g.

That a corresponding behavior is found for the enthalpy and the heat capacity is conceivable. As the free volume incorporates energy, changes in the volume and in the enthalpy are interrelated and this results in simultaneous steps in the expansion coefficient and the heat capacity.

Being in a non-equilibrium state, liquids below T_g have a tendency further to change the structure in the direction towards the equilibrium. A slow decrease in volume and enthalpy is often observed. Figure 5.24 also exemplifies this behavior. Prolonged storage of the sample of poly(vinylacetate) below T_g for 100 h resulted in a shrinkage in volume. Note that, as a consequence, T_g, as measured during the subsequent heating, is shifted to lower values. Formally, this may be associated with the shift of the point of intersection of the two lines which represent the glassy and liquid state. Physically, it is caused by a change in the microstructure. 'Ageing' is the technical term used in general for these processes and they can produce problems since often the mechanical properties deteriorate.

We saw that the position of the step in the expansion coefficient or the heat capacity, observed during a cooling run, depends on the cooling rate. It is possible to analyse this dependence in more detail and to derive a criterion for the location of the glass transition. To begin with, recall once again what happens when cooling a polymeric liquid. In the fluid state, all degrees of freedom equilibriate rapidly so that thermal equilibrium is always maintained. Conditions change on approaching the glass transition since, here, the relaxation times of the α-modes reach values which are too high to further allow for a continuous equilibriation. Vibrations and local modes still react immediately to programmed temperature changes but the sluggish α-modes lag behind more and more. Finally, after having crossed the transition range, the energy exchange between the instantaneously reacting modes and the α-modes stops completely.

In an analysis, one has to consider the time dependence of relaxation processes under non-isothermal conditions as imposed during a cooling or heating run. Observations suggest that we represent the sample volume as a sum of two contributions

$$V(T) = V_{\mathrm{u}}(T) + \Delta V_{\alpha}(T) \tag{5.136}$$

The first part V_{u} describes the volume of a hypothetical system without α-modes, being determined by the hard cores of the molecules, the anharmonicity of the vibrations and possible effects of local relaxation processes. The second term ΔV_{α} accounts for the free volume produced by the α-modes. An analogous description is suggested for the enthalpy and we formulate correspondingly

$$\mathcal{H}(T) = \mathcal{H}_{\mathrm{u}}(T) + \Delta\mathcal{H}_{\alpha}(T) \tag{5.137}$$

Since $V_{\mathrm{u}}(T)$ and $\mathcal{H}_{\mathrm{u}}(T)$ pass continuously through the glass transition without any peculiar effects, we may reduce the discussion to the contributions of the α-modes. To keep the equations simple, we disregard the multi-mode character of the α-process and assume a single relaxation time, τ_{α}, with a temperature dependence following the Vogel-Fulcher law. Then kinetics can be analysed by an application of the relaxation equation

$$\frac{\mathrm{d}\Delta\mathcal{H}_{\alpha}}{\mathrm{d}t} = -\tau_{\alpha}^{-1}(T)(\Delta\mathcal{H}_{\alpha} - \Delta\mathcal{H}_{\alpha,\mathrm{eq}}(T)) \tag{5.138}$$

which is formulated here for the calorimetric experiment. According to this equation, the change from a non-equilibrium value of the enthalpy associated with the α-modes, $\Delta\mathcal{H}_{\alpha}$, to the equilibrium value, $\Delta\mathcal{H}_{\alpha,\mathrm{eq}}$, occurs with a temperature dependent rate τ_{α}^{-1}.

The experimental heat capacity, c_{α}, is given by

$$c_{\alpha} := \frac{\mathrm{d}\Delta\mathcal{H}_{\alpha}}{\mathrm{d}T} \tag{5.139}$$

while the equilibrium value, $c_{\alpha,\text{eq}}$, is represented by

$$c_{\alpha,\text{eq}} := \frac{\mathrm{d}\Delta\mathcal{H}_{\alpha,\text{eq}}}{\mathrm{d}T} \tag{5.140}$$

We consider a temperature program based on a constant cooling rate $\dot{T} < 0$

$$T(t) = T(0) + \dot{T}t \tag{5.141}$$

Under these conditions the relaxation equation converts into the relation

$$c_\alpha(t)\dot{T} = -\tau_\alpha^{-1}(T)(\Delta\mathcal{H}_\alpha(t) - (\Delta\mathcal{H}_{\alpha,\text{eq}}(0) + \dot{T}c_{\alpha,\text{eq}} \cdot t)) \tag{5.142}$$

Differentiation on both sides yields

$$\begin{aligned} \dot{T}\frac{\mathrm{d}c_\alpha}{\mathrm{d}t} = & - \tau_\alpha^{-1}(T)\dot{T}(c_\alpha - c_{\alpha,\text{eq}}) \\ & - \frac{\mathrm{d}\tau_\alpha^{-1}}{\mathrm{d}T} \cdot \dot{T} \cdot (\Delta\mathcal{H}_\alpha(t) - (\Delta\mathcal{H}_{\alpha,\text{eq}}(0) + \dot{T}c_{\alpha,\text{eq}} \cdot t)) \end{aligned} \tag{5.143}$$

assuming a constant value for $c_{\alpha,\text{eq}}$. Combination of both differential equations, followed by a division by $\dot{T}$ on both sides, results in

$$\frac{\mathrm{d}c_\alpha}{\mathrm{d}t} = -\tau_\alpha^{-1}(T)(c_\alpha - c_{\alpha,\text{eq}}) + c_\alpha \frac{\mathrm{d}\ln\tau_\alpha^{-1}}{\mathrm{d}T} \cdot \dot{T} \tag{5.144}$$

We find here two terms which determine the quantity of interest, $\mathrm{d}c_\alpha/\mathrm{d}t$. At temperatures far above T_g, the first term dominates. Here, the relaxation rates are high and change only slowly with temperature. Within this limit, Eq. (5.144) may be replaced by

$$\frac{\mathrm{d}c_\alpha}{\mathrm{d}t} \approx -\tau_\alpha^{-1}(T)(c_\alpha - c_{\alpha,\text{eq}}) \tag{5.145}$$

To see the consequences, just consider a cooling run starting at a temperature T. Beginning with $c_\alpha = 0$, the equilibrium value $c_{\alpha,\text{eq}}$ will be reached within a time $\tau_\alpha(T)$. From thereon, c_α remains constant and equal to $c_{\alpha,\text{eq}}$.

In the other limit, for temperatures around and below T_g, the relaxation rates are very low and their changes, as described by the Vogel-Fulcher law, are large. Hence, the second term dominates and Eq. (5.144) can be approximated by

$$\frac{\mathrm{d}c_\alpha}{\mathrm{d}t} \approx c_\alpha \frac{\mathrm{d}\ln\tau_\alpha^{-1}}{\mathrm{d}T}\dot{T} \tag{5.146}$$

or

$$\frac{\mathrm{d}\ln c_\alpha}{\mathrm{d}T} \approx \frac{\mathrm{d}\ln\tau_\alpha^{-1}}{\mathrm{d}T} \tag{5.147}$$

The solution is

$$c_\alpha \sim \tau_\alpha^{-1} \tag{5.148}$$

or, using the Vogel-Fulcher law Eq. (5.127),

$$c_\alpha(T) \sim (\tau_0)^{-1} \exp - \frac{T_\mathrm{A}}{T - T_\mathrm{V}} \tag{5.149}$$

We see that in this range, c_α converges rapidly to zero.

The cross-over from one to the other regime and thus the glass transition takes place when both terms have the same order of magnitude. Equating the two terms yields

$$\tau_\alpha^{-1} \simeq \frac{\mathrm{d} \ln \tau_\alpha^{-1}}{\mathrm{d}T} \dot{T} \tag{5.150}$$

or

$$\frac{\mathrm{d} \ln \tau_\alpha^{-1}}{\mathrm{d}t} \tau_\alpha \simeq 1 \tag{5.151}$$

Equation (5.151) formulates a criterion for T_g. It states that T_g is reached during a cooling run when the relative change of the relaxation rate τ_α^{-1} within a time in the order of τ_α is no longer negligible.

It looks interesting to calculate the relaxation time at T_g, as it follows from these equations. Using the Vogel-Fulcher law gives

$$\frac{T_\mathrm{A}}{(T_\mathrm{g} - T_\mathrm{V})^2} \cdot \dot{T} \cdot \tau_\alpha(T_\mathrm{g}) \simeq 1 \tag{5.152}$$

We have typically $T_\mathrm{A} \simeq 2000$ K, $T_\mathrm{g} - T_\mathrm{V} \simeq 50$ K and therefore

$$\frac{T_\mathrm{A}}{(T_\mathrm{g} - T_\mathrm{V})^2} \simeq 1 \ \mathrm{K}^{-1} \tag{5.153}$$

For a cooling rate $\dot{T} \simeq 10^{-2}$ Ks^{-1} we thus arrive at

$$\tau_\alpha(T_\mathrm{g}) \simeq 10^2 \ \mathrm{s} \tag{5.154}$$

Hence, a calorimetric measurement and also a volumetric experiment produce the step at a temperature where the relaxation time of the α-modes is in the order of minutes.

Equation (5.152) correctly describes the effect of the heating or cooling rates. If $|\dot{T}|$ is increased, the step occurs at an even shorter relaxation time τ_α, and therefore at higher temperatures, which is in accordance with the experiment. Figure 5.26 presents, for illustration and further confirmation, the results of numerical solutions of Eq. (5.144), as obtained for three different cooling rates, choosing typical values for the other parameters. The shifts largely agree with the experimentally observed behavior.

In the literature, one also finds another definition for T_g, which refers to the viscosity. T_g is determined by the condition

$$\eta_0(T_\mathrm{g}) = 10^{13} \ \mathrm{poise} = 10^{12} \ \mathrm{Nm^{-2}s} \tag{5.155}$$

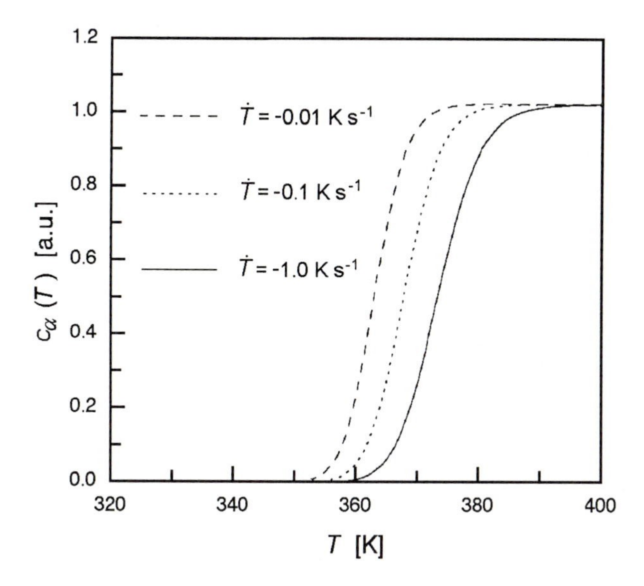

Fig. 5.26. Model calculation on the basis of Eq. (5.144), simulating heat capacity measurements during cooling runs with the indicated rates ($T_{\mathrm{A}} = 2000$ K, $T_{\mathrm{V}} = 300$ K, $\tau_\alpha(\infty) = 10^{-11}$ s, $T(t=0) = 400$ K)[60]

Since viscosity measurements at such high values are cumbersome they are only rarely performed. Instead, one measures the viscosity away from the glass transition and determines T_{g} by an extrapolation based on the Vogel-Fulcher law. The values obtained in such measurements are similar to those derived from the calorimetric or volumetric experiments.

To show some typical values, Table 5.1 collects the T_{g}'s of several amorphous polymers. One finds two major groups, first the natural and synthetic rubbers, together with poly(dimethyl siloxane), which have T_{g}'s far below ambient temperature, and secondly polymers which are glassy at room temperature. The latter mostly have T_{g}'s located around 100 °C.

A concern of practical importance is a knowledge of relations between the chemical constitutions of polymers and their respective glass transition temperatures. A good microscopic understanding is lacking, but there are a number of empirical rules based on the assumption that different molecular units give separate contributions to T_{g}. Also quite useful are 'mixing rules' which describe the T_{g}-values of blends as a function of the T_{g}'s of the components in pure states and their volume fractions. As it turns out, in many cases a good representation is achieved by the 'Fox-Flory'-equation, given by

$$\frac{1}{T_{\mathrm{g}}} = \frac{\phi_{\mathrm{A}}}{T_{\mathrm{g}}^{\mathrm{A}}} + \frac{\phi_{\mathrm{B}}}{T_{\mathrm{g}}^{\mathrm{B}}} \tag{5.156}$$

Table 5.1. Glass transition temperatures of some common polymers

Polymer	T_g [K]
poly(dimethyl siloxane)	146
polybutadiene (*cis*)	164
polyisoprene (*cis*) (natural rubber)	200
polyisobutylen	200
poly(vinyl methylether)	242
poly(α-methyl styrene)	293
poly(vinyl acetate)	305
polystyrene	373
poly(methyl methacrylate) (atactic)	378
poly(acrylic acid)	379
poly(acrylonitrile)	398
polycarbonate	418

5.3.4 Relaxation in Partially Crystalline Systems

The relaxation behavior of partially crystalline systems is complex and different from amorphous polymers. Observations give the general impression that, in comparison to amorphous systems, partially crystalline samples are much less uniform in behavior. Many of the systems exhibit peculiarities and these can dominate the viscoelastic properties. This is not the place to explore this large field in the necessary depth, which would mean we would have to discuss separately the mechanical behavior of polyethylene, poly(ethylene terephthalate), polypropylene, it-polystyrene, poly(tetrafluoroethylene) etc. What can be done for illustration is to pick out one instructive example and we select polyethylene.

We begin with a look at the results of a temperature dependent measurement of the storage shear modulus and the mechanical loss tangent shown in Fig. 5.27. Data were obtained with the help of a torsion pendulum working with a frequency of about 1 s^{-1}. The figure includes data obtained for two commercial samples, a linear polyethylene (LPE) with high crystallinity and a low density polyethylene (LDPE) with short-chain branches. Two relaxation processes show up for LPE, one at low temperatures, designated as γ-process, and another at high temperatures, called the 'α-process'. For LDPE a third process emerges in addition, the 'β-process' characterized by a loss maximum at around $-20\,°C$.

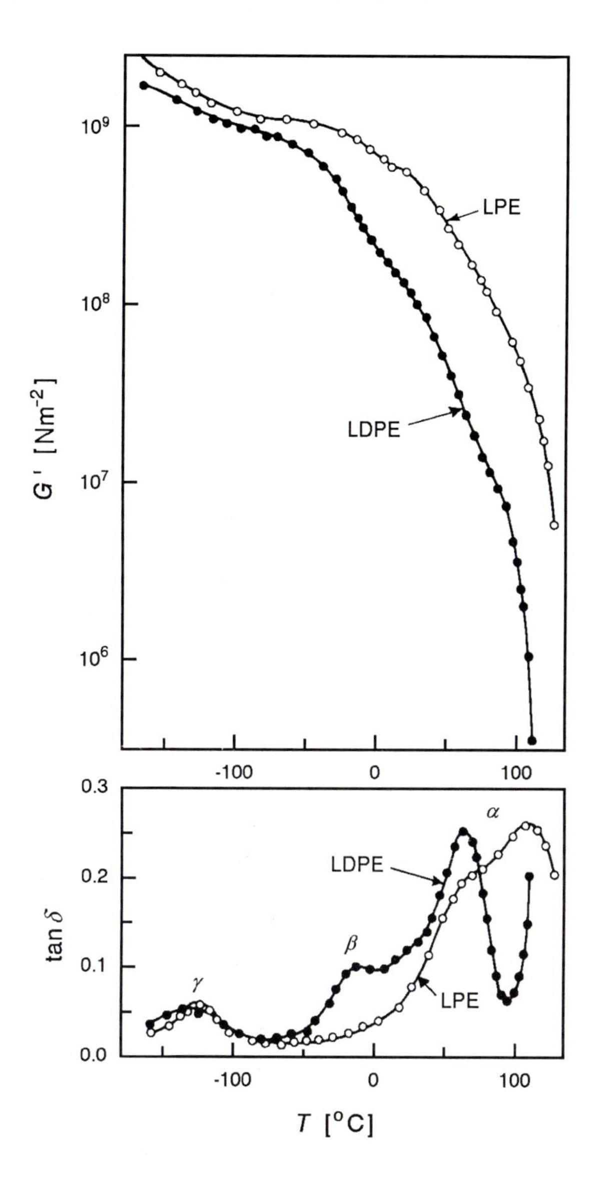

Fig. 5.27. Temperature dependence of the storage shear modulus (*top*) and the loss tangent (*bottom*) of linear (LPE) and branched polyethylene (LDPE). Data obtained by Flocke [61], using a torsion pendulum with frequencies in the order of 1 s^{-1}

The change in behavior from the highly crystalline LPE to samples with lower crystallinity like the LDPE occurs continuously, as is demonstrated by the measurement shown in Fig. 5.28. Here chlorination was employed for a controlled reduction of the crystallinity. We observe that decreasing the

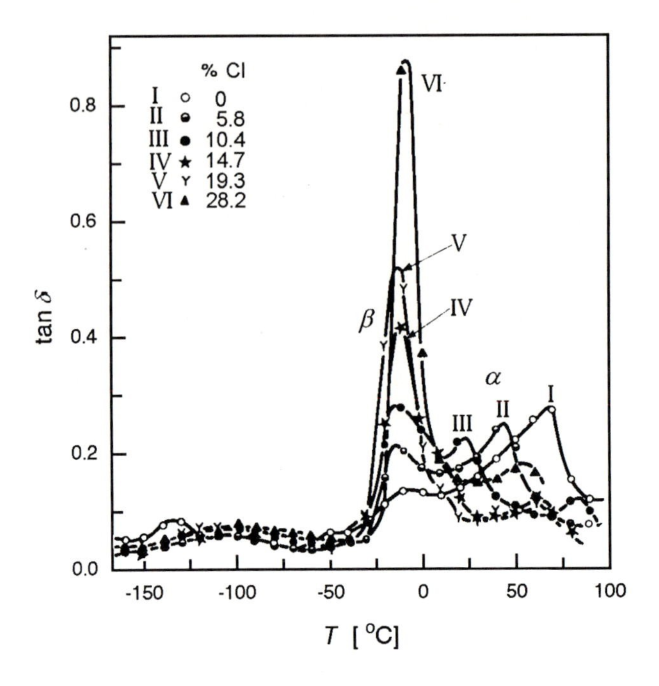

Fig. 5.28. Temperature dependence of the loss tangent of chlorinated PE, obtained for a series of different samples with chlorine contents between 0 and 28.2 %. Results of a torsion pendulum measurement by Schmieder and Wolf [62]

crystallinity results in a strong increase of the loss signal related to the β-process.

The relaxation behavior of amorphous polymers was dominated by two processes, the glass-rubber transition and the terminal flow region, which are both characterized by a WLF temperature dependence. For polyethylene, one cannot expect a flow transition because flow is suppressed by the crystallites in the sample. The interesting fact is that for linear polyethylene, i.e. polyethylene with high crystallinity, there is no WLF-controlled process at all. The numerous measurements in the literature provide clear evidence that the two processes observed in LPE, α and γ, are both based on activated mechanisms obeying the Arrhenius law. The process which does show a WLF behavior is the β-process. The disappearance of the β-process in polyethylene samples with high crystallinity tells us that the state of order and the molecular dynamics in the disordered regions, which still make up a non-negligible part of the sample volume (typically 20%), differs qualitatively from that in an amorphous polymer. On the other hand, increasing the volume fraction of disordered material evidently changes the mobility in these regions towards

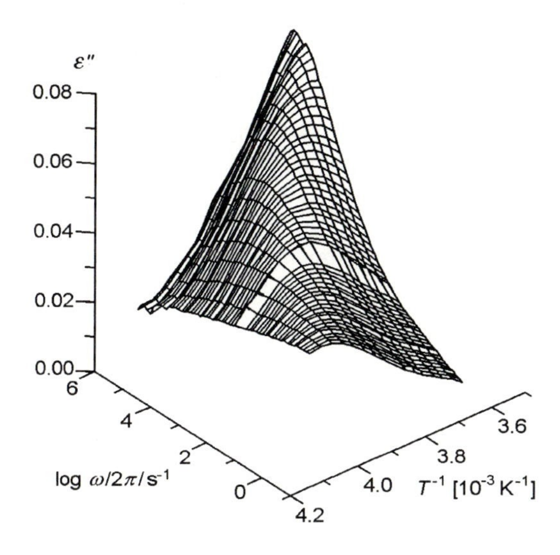

Fig. 5.29. Frequency- and temperature dependence of the dielectric loss associated with the β-process in poly(ethylene-*co*-vinylacetate) with 17% vinylacetate units [63]

the usual properties of a melt. The β-process observed for polyethylene with low crystallinity indeed corresponds to the glass-rubber transition in amorphous polymers.

The use of the Greek letters in agreement with the convention to choose the symbols α, β, γ etc. for a designation of the processes in the sequence they show up from high temperatures downwards, produces some confusion in the meaning of the term 'α-process'. While for amorphous polymers, it is identical with the glass-rubber transition, this is not the case for polyethylene. Here, if the glass-rubber transition occurs at all, it is denoted as the 'β-process'.

The mobility in the disordered regions of polyethylene shows great variations. The β-process, as observed in dielectric measurements, is unusually broad and changes its shape with temperature. Figure 5.29 presents results obtained for a sample of polyethylene which included 17% of vinylacetate-groups as co-units in the chains. These co-units are rejected from the crystallites and accumulate in the amorphous regions. As the groups carry a dipole moment, their dynamics shows up in the dielectric spectrum. The difference in behavior between the glass transition in an amorphous polymer and the β-process in a partially crystalline system such as polyethylene becomes very clear when comparing Fig. 5.29 with Fig. 5.17, which shows the α-process for poly(vinylacetate). For the partially crystalline system, loss curves are much broader and also exhibit a pronounced temperature dependence. The observations are indicative of large variations in the segmental mobility in the disordered regions, obviously caused by the restrictions and limitations imposed by the crystallites and the trapped entanglements. The temperature dependence is partly due to the continuous change in crystallinity which we discussed in Sect. 4.3. This, however, is not the whole effect since the structural changes practically end at about 0 °C, while the progressive broadening goes on. The β-process has a cooperative character which implies that segmental motions are correlated up to a certain distance ξ_β. Consequently, hindrances such as crystallite surfaces become effective over a similar range. The observed increase in the variance of the mobilities on approaching T_g may thus be understood as being caused by an increase in ξ_β. Since it appears that, close to T_g, the whole amorphous layer is affected, one may furthermore conclude that the correlation length then becomes similar to the layer thickness, i.e. is in the order of some nanometers.

The largest changes in the mechanical properties of polyethylenes with moderate to high crystallinity are caused by the α-process. Figures 5.30 and 5.31 present results of frequency dependent measurements of the tensile modulus of a sample of branched polyethylene conducted at different temperatures between 26 °C and 95 °C. As can be seen, the loss tangent shows a systematic shift to higher frequencies. The temperature dependence of the loss maxima depicted at the bottom of Fig. 5.30 is indicative of an activated process with an activation energy $\tilde{A} = 104\ \mathrm{kJ \cdot mol^{-1}}$. Figure 5.31 shows that the storage tensile modulus decreases dramatically by about two orders of magnitude. To a larger extent, the decrease is caused by the continuous melting (compare for example the right column of curves in Fig. 4.29 which were obtained for a similar sample), the effect of the α-process, however, is also comparatively strong, as can be seen for example in the curve for 85 °C.

In the search for the origin of the α-process different observations must be included in the considerations. First, remember the results of the NMR experiment presented in Sect. 4.3.2. Here, a longitudinal chain transport through the crystallites was clearly indicated. The chain motion seems to be accomplished by a 180°-twist defect which is created at a crystal surface and then moves through the crystallite to the other side. As a result all monomers of a crystalline sequence are rotated by 180° and shifted over the length of one CH_2-unit. This screw-motion alone, however, cannot set up the α-process, since it is mechanically inactive. As the crystals remain unchanged, both internally and in their external shape, there is no coupling with a stress field. On the other hand, the high value of the relaxation strength of the α-process strongly suggests a location of the associated deformation mode in the weak parts of the structure. These are the amorphous regions, which can be sheared quite easily.

How can these different facts be reconciled and brought together in one common picture? The answer is: The α-process in polyethylene has a composite nature. The mechanical relaxation indeed originates from an additional shear of the amorphous regions. The prerequisite for this shear, however, is a chain transport through the crystallites. By this intracrystalline motion the

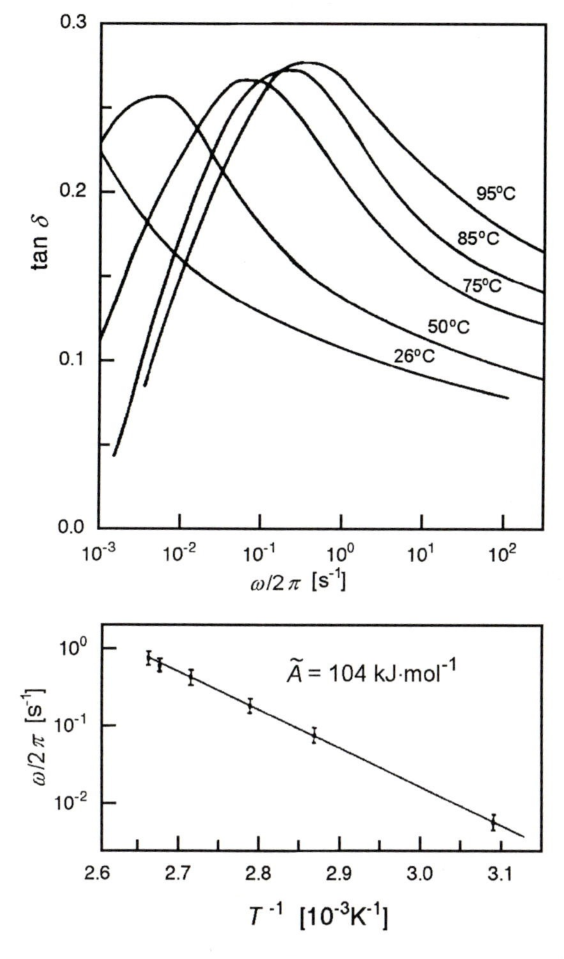

Fig. 5.30. Mechanical α-process in LDPE ('Lupolen 1800S'). Loss tangent (*top*) and temperature dependence of the frequency of the loss peak (*bottom*) [64]

strict pinning of the amorphous sequences at the crystallite surfaces is removed, which facilitates a reorganisation in the amorphous regions and gives rise to a further stress relaxation. Hence, in the α-process, a relaxation mode in the crystallites and another located in the amorphous zones become combined. As indicated by the broadening of the loss curves, the α-process is based on a larger group of relaxatory modes. We furthermore notice that a temperature increase leaves the shape of the loss curves essentially unchanged and we may conclude that all modes employ the same elementary process. The experiments suggest identifying this elementary process with the longitudinal shifts of the crystalline sequences since we find nearly identical activation energies in the NMR experiment ($\tilde{A} = 105 \text{ kJ} \cdot \text{mol}^{-1}$, Fig. 4.36) and the dy-

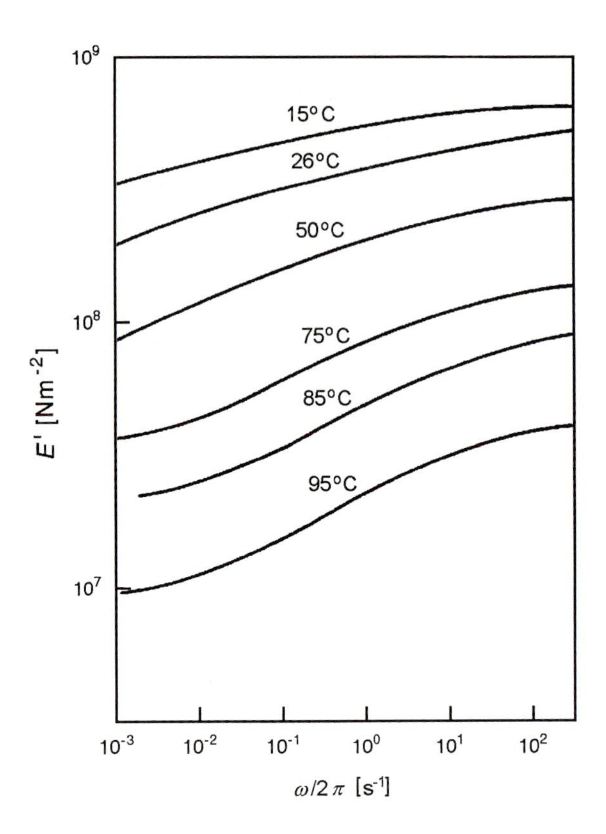

Fig. 5.31. Same system as Fig. 5.30. Frequency dependent storage tensile modulus [64]

namic mechanical measurements ($\tilde{A} = 104$ kJ $\cdot$ mol^{-1}, Fig. 5.30). Of interest is a comparison of the rate of elementary steps with the mechanical relaxation rate. We find a difference of about four orders of magnitude, telling us that the reorganization of the chains in the amorphous regions is a complicated procedure which requires a large number of elementary steps.

We can now see more clearly the principal effect of the crystallites. While in an amorphous polymer the transition from the glassy state, where motions remain localized, to a melt with large scale segmental mobility is accomplished by just one group of relaxatory modes within a limited temperature range, in partially crystalline systems the same process becomes extended over a much larger temperature range and obtains a more continuous character. This holds in particular for LPE, but is also true for LDPE. In the latter systems, the β-process provides only a first step in the introduction of segmental motion into the amorphous regions. If it gave rise to a transition into a truly melt-like

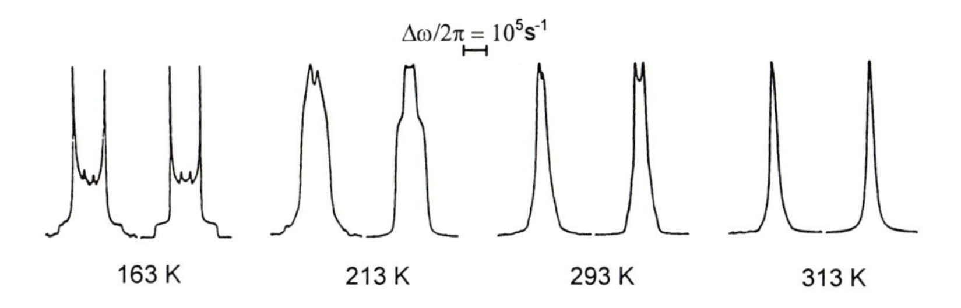

Fig. 5.32. Observed (*left*) and calculated (*right*) pairs of ^{2}H NMR spectra of the mobile deuterons in the amorphous regions of LPE at different temperatures. Work of Spiess et al.[65]

state, the mobility in the crystallites could add nothing more to the molecular dynamics and no mechanical α-process would arise.

NMR spectroscopy is particularly suitable for revealing these continuous mobility changes. We have a look at an experiment conducted on a sample of LPE. Again it is not possible to provide here all the explanations required for a thorough understanding. We must restrict ourselves to some brief statements, needed as a minimum for addressing the main results. We will consider a measurement on a deuterated system since these kinds of experiments are most appropriate for studies of polymer dynamics. Deuterons (^{2}H) possess twice the spin of protons (I = 1 compared to I = 1/2), but the magnetic moment is much smaller so that effects of dipole-dipole interactions are comparatively weak. The peculiar property of ^{2}H is a large electric quadrupole moment and this interacts with the local electric field gradient tensor. Usually one is dealing with CH-bonds, where the electrons do indeed produce a strong field gradient at the location of the proton or deuteron. As a result of the interaction, the transitions $I_z = 1 \to 0$ and $I_z = 0 \to -1$ become associated with different energy changes. The two lines may be represented as

$$\omega = \omega_0 \pm \omega_Q \tag{5.157}$$

where ω_0 stands for the Larmor-frequency of an isolated ^{2}H, and ω_Q represents the quadrupole splitting. ω_Q depends only on the angle θ enclosed by the magnetic field and the CH-bond direction which sets the orientation of the uniaxially symmetric field gradient tensor. The splitting is rather large and, for the usual magnetic fields amounts to values in the order $\omega_Q/2\pi \simeq 10^5\ \mathrm{s}^{-1}$. Deuterons thus represent spin labels which are almost exclusively governed by the local interaction with the electric field gradient.

Under static conditions the line shape of a ^{2}H NMR absorption spectrum reflects directly the orientational distribution of the CH-bonds in the sample. Molecular dynamics modifies the spectral shape in a well-defined way. Only the reorientational motion of individual CH-bonds is probed by the experiment which leads generally to a narrowing of the line. A simple situation arises in the 'rapid exchange limit' when the reorientation rate τ_{r}^{-1} becomes large compared to the inverse of the spectral width

$$\tau_{\mathrm{r}}^{-1} \gg \omega_{\mathrm{Q}}^{-1} \tag{5.158}$$

Then the quadrupole coupling $\pm\, \omega_{\mathrm{Q}}(\theta)$ is replaced by its time average $\overline{\omega}_{\mathrm{Q}}^{t}$, which reduces the amount of splitting in most cases. In particular, for an isotropic reorientational motion one expects

$$\overline{\omega}_{\mathrm{Q}}^{t} = 0 \tag{5.159}$$

i.e. a complete vanishing of the splitting. If reorientations include only a limited angular range, only a partial averaging of the quadrupole coupling occurs. Specific motions lead to specific spectra; the basic equations enable precise predictions to be made and thus a discrimination between different motional processes.

Figure 5.32 depicts results of a study on linear polyethylene which makes use of this capability of ^{2}H resonance experiments. We can see spectra of deuterons in the amorphous phase of a perdeuterated sample, as measured at different temperatures between 163 K and 313 K (they can be discriminated and separated from the deuterons in the crystallites due to different 'spin-lattice relaxation' times). The experimental spectra are always compared to the result of a model calculation adjusted to the data. The model spectrum computed for comparison with the spectrum at 163 K represents a superposition of a rigid phase (70%) and deuterons incorporated in mobile 'kink-defects' (30%). Kink-defects correspond to a sequence *gauche*$^{-}$*-trans-gauche*$^{+}$ or *gauche*$^{+}$*-trans-gauche*$^{-}$ in an otherwise stretched part of the chain. The transition between the two conformers is mechanically active, as the chain contour becomes locally modified. Therefore, it is generally regarded as a mechanism which could be responsible for the γ-process. The NMR experiments corroborate this assignment. In fact, analysis of the subsequent NMR spectra led to the conclusion that the state of motion based on the presence of mobile kink-defects pertains up to 200 K, whereby the mobility increases continuously as a result of a growing defect concentration. Above 200 K the situation changes. The spectra measured at 213 K and 293 K can only be accounted for when assuming, for a fraction of the deuterons, a reorientational motion which covers a larger angular range. Conformational considerations showed that these more extended motions can be accomplished by correlated rotations of 5 subsequent bonds in a chain, which contrasts with the 3-bond motion of a kink-defect. Finally, interpretation of the spectrum at 313 K requires that one assumes an even larger correlation range along the chain, including 7 subsequent bonds. Figure 5.33 summarizes these results in a schematical drawing.

The picture which emerges from these studies is thus again that of an amorphous phase strongly affected by the limitations for the motion imposed by constraints. These constraints, originating from the adjacent crystallites

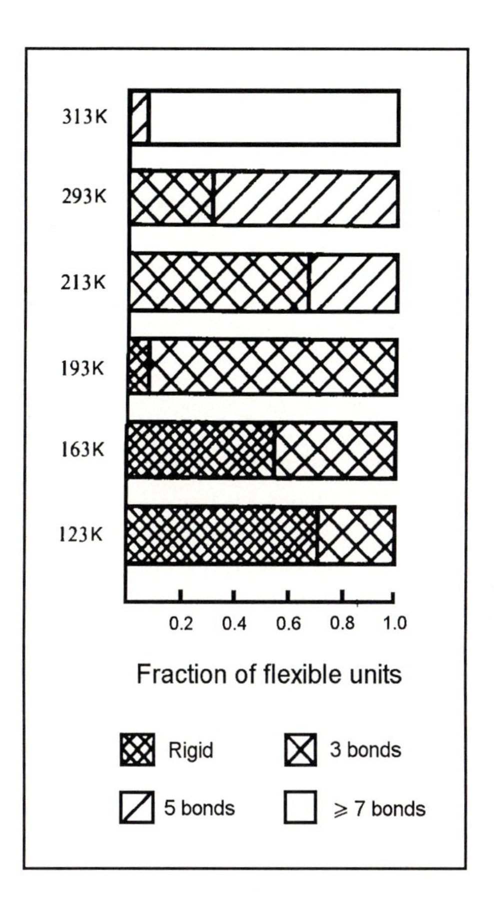

Fig. 5.33. Fractions of flexible units with different lengths in the amorphous regions of LPE derived from the analysis of the ^{2}H NMR line shapes displayed in Fig. 5.32 [65]

and the trapped entanglements, are long lived and relax only gradually upon heating. They control the number of conformations which are available for a unit in the chain and this number slowly increases with temperature in correspondence with the weakening of the constraints. In fact, this continuous change in mobility differs drastically from the behavior of an amorphous polymer like polystyrene when observed in an analogous ^{2}H resonance experiment. Here the chain motion becomes essentially isotropic within the comparatively narrow temperature range of the α-process, and no indications for a restriction of the motion are left.

At the end of this section, let us return once again to the mechanical data of Figs. 5.30 and 5.31. While the loss-tangent essentially obeys frequency-temperature superposition, with

$$\tan\delta(T,\log\omega) \approx \tan\delta\left(T_0, \log\omega - \frac{\tilde{A}}{R}\left(\frac{1}{T_0} - \frac{1}{T}\right)\right) \tag{5.160}$$

this is not at all true for the tensile storage modulus $E'(\omega)$. Both observations are formally reconciled if one assumes for $E^* = E' + \mathrm{i}E''$ a dependence on T and ω of the form

$$E^*(T,\log\omega) = \lambda(T)E^*\left(T_0, \log\omega - \frac{\tilde{A}}{R}\left(\frac{1}{T_0} - \frac{1}{T}\right)\right) \tag{5.161}$$

The new parameter $\lambda(T)$ is meant to account for the structure changes in the sample caused by the continuous crystallite melting. Being introduced as an isolated factor, it affects E^*, but is eliminated in the loss tangent. What can be said about the dependence of λ on the changing structure parameters? In fact, not much in detail, because the functional dependence is far from being simple. We encounter here the important and difficult problem of how to calculate the modulus or compliance of a two-phase system. In our case the mechanical properties of the two phases, i.e. the crystallites and the amorphous regions after the β-transition, differ greatly. Computation of the sample modulus implies a determination of the distribution of stress and strain through the different regions in the sample and this distribution is inhomogeneous. There is no general analytical solution to the problem and also no particular solution for the given situation where stacks of lamellar crystallites are incorporated in spherulites. As the only simple procedure available, one can calculate an upper and a lower bound for the modulus. The bounds are strictly valid for the static mechanical properties of perfectly elastic composite materials but may be used also for polymers, provided that the loss is negligible, i.e. stress and strain are in phase. The upper bound, known as the 'Reuss'-limit, is reached in the limiting case of a homogeneous distribution of the strain, when one finds

$$G = \phi_c G_c + (1 - \phi_c) G_a \tag{5.162}$$

Here we refer to the shear modulus G, being expressed in terms of the modulus of the crystallites, G_c, and the modulus of the amorphous regions, G_a. The lower bound, often called the 'Voigt'-limit, is attained for a homogeneous distribution of the stress, where we can write

$$J = \phi_c J_c + (1 - \phi_c) J_a \tag{5.163}$$

Equation (5.163) expresses the shear compliance J as a weighted average of the shear compliances J_c and J_a of the crystalline and amorphous regions. Realizations of the two limiting cases can be easily envisaged. Just consider the

Fig. 5.34. Shear deformations of a stack of crystallites (*black*) separated by amorphous layers (*gray*), representing the limiting cases of a uniform strain (*left*) and a uniform stress (*right*)

different mechanical responses of the stack of crystallites shown in Fig. 5.34 to shear stresses oriented along the basal and the lateral faces of the crystallites respectively. In the first case we have a uniform stress, in the second case a uniform strain, and the shear compliance and modulus then follow by Eqs. (5.162) and (5.163).

The experimental results, as expressed by $\lambda(T)$, are located somewhere in-between the two limiting cases of uniform stress and uniform strain. The moduli at low temperatures tend to be close to the upper limit, whereas those at elevated temperatures approach the lower bound. The observed drastic decrease of E with temperature is due to this change.

A last remark concerns the factorized dependence, Eq. (5.161), of E^* on a structure dependent parameter, λ, and temperature or frequency, as suggested by the constant amplitude of the loss tangent. The behavior may be regarded as evidence that both the α-process and the preceding deformation rely mainly on a shearing of the amorphous parts, the crystallites responding as essentially stiff objects just by rotations. Under these conditions, storage and dissipation of energy both remain restricted to the amorphous regions. As a consequence the loss tangent, giving the ratio between the dissipated and stored work, coincides with that of the amorphous parts, and thus is not affected by changes in the crystallinity.

5.4 Further Reading

E.-J. Donth *Relaxation and Thermodynamics in Polymers: Glass Transition*, Akademie Verlag, 1992

J.D. Ferry: *Viscoelastic Properties of Polymers*, John Wiley & Sons, 1970

N.G. McCrum, B.E. Read, G. Williams: *Anelastic and Dielectric Effects in Polymeric Solids*, John Wiley & Sons, 1967

D.J. Meier (Ed.): *Molecular Basis of Transitions and Relaxations*, Gordon and Breach, 1978

F.R. Schwarzl: *Polymermechanik*, Springer, 1990
I.M. Ward: *Mechanical Properties of Solid Polymers*, John Wiley & Sons, 1971
G. Williams: *Dielectric Properties of Polymers* in R.W. Cahn, P. Haasen, E.J. Kramer, E:L. Thomas (Eds.): Materials Science and Technology Vol.12 *Structure and Properties of Polymers*, VCH Publishers, 1993

Chapter 6

Microscopic Dynamical Models

The previous chapter provided an overview of the characteristics of the mechanical and dielectric behavior of polymer systems. We discussed the material properties as they are described by the various response functions and were in particular concerned with the pronounced effects of temperature. These macroscopic properties have a microscopic basis. So far, we have addressed this basis only in qualitative terms. This chapter deals with microscopic dynamical models which yield a quantitative description for some of the observed macroscopic properties.

Before we come to these models, we will first introduce a basic law of statistical thermodynamics which we require for the subsequent treatments and this is the 'fluctuation-dissipation theorem'. We learned in the previous chapter that the relaxation times showing up in time- or frequency dependent response functions equal certain characteristic times of the molecular dynamics in thermal equilibrium. This is true in the range of linear responses, where interactions with applied fields are always weak compared to the internal interaction potentials and therefore leave the times of motion unchanged. The fluctuation-dissipation theorem concerns this situation and describes explicitly the relation between the microscopic dynamics in thermal equilibration and macroscopic response functions.

6.1 The Fluctuation-Dissipation Theorem

Imagine that we select within a sample a subsystem contained in a volume v, which is small but still macroscopic in the sense that statistical thermodynamics can be applied. If we could measure the properties of this subsystem we would observe time dependent fluctuations, for example in the shape of the volume, i.e. the local strain, the internal energy, the total dipole moment, or the local stress. The fluctuation-dissipation theorem relates these spontaneous, thermally driven fluctuations to the response functions of the system. We formulate the relationship for two cases of interest, the fluctuations of the dipole moments in a polar sample and the fluctuations of stress in a melt.

The total dipole moment of a subsystem

$$\boldsymbol{p}_v = \sum_i \boldsymbol{p}_i \tag{6.1}$$

is the result of the superposition of the group dipole moments $\boldsymbol{p}_i$ contained in v. Together with the $\boldsymbol{p}_i$'s the total dipole moment varies in time, and a characterization can be achieved with the aid of correlation functions. The simplest, but also most important one is the second order time correlation function

$$\langle \boldsymbol{p}_v(0) \cdot \boldsymbol{p}_v(t) \rangle$$

It describes the correlation between the results of two measurements of $\boldsymbol{p}_v$, carried out at t' and $t' + t$, whereby t' is arbitrary since systems in thermal equilibrium are homogeneous in time. Fluctuations occur independently along x, y and z. The correlation function for one of the components, denoted p_v, is therefore

$$\langle p_v(0) p_v(t) \rangle = \frac{\langle \boldsymbol{p}_v(0) \cdot \boldsymbol{p}_v(t) \rangle}{3} \tag{6.2}$$

In the previous chapter we dealt with the dielectric response. Application of an electric field produces a polarization. If the field is imposed at zero time the polarization develops as described by Eq. (5.16)

$$P = \epsilon_0 \Delta\epsilon(t) E_0 \tag{6.3}$$

finally leading to the equilibrium value

$$P = \epsilon_0 \Delta\epsilon(\infty) E_0 \tag{6.4}$$

If the electric field then is switched off, the polarization returns back to zero, in a time dependent process described by

$$P = \epsilon_0 (\Delta\epsilon(\infty) - \Delta\epsilon(t)) E_0 \tag{6.5}$$

t here gives the time elapsed since the moment of switching off.

We now can formulate the fluctuation-dissipation theorem. It relates the correlation function of the fluctuations of the component of the total dipole moment along the field direction, p_v, to the decay function of the polarization, by

$$\langle p_v(0) p_v(t) \rangle = vkT\epsilon_0 (\Delta\epsilon(\infty) - \Delta\epsilon(t)) \tag{6.6}$$

The left-hand side involves a correlation function associated with the spontaneous fluctuations in thermal equilibrium, as they arise from the molecular dynamics. The response function on the right-hand side incorporates the reaction of the sample to the imposition of an external field. The fluctuation-dissipation theorem states that linear responses of macroscopic systems are related to and can, indeed, be calculated from equilibrium fluctuations. More specifically, the equation states that the regression of spontaneous fluctuations, as they occur in subsystems of mesoscopic size, follows the same law as the relaxation subsequent to an external perturbation of a macroscopic system. This important principle was first proposed as a hypothesis by Onsager and later, in 1951, proved by Callen and Welton.

As a second example, we formulate the theorem with regard to the stress fluctuations in quiescent melts, focusing on the shear component σ_{zx}. In this case it has the form

$$\langle \sigma_{zx}(0)\sigma_{zx}(t)\rangle = kT\frac{G(t)}{v} \tag{6.7}$$

The background is the same as in the first example: The equation states that the regression of the fluctuations of the local shear stress obeys the same law as the macroscopic stress relaxation.

The limiting values for $t = 0$ of Eqs. (6.6) and (6.7) provide the variances of the fluctuating variables. We obtain the equations

$$\langle p_v^2\rangle = v\epsilon_0\Delta\epsilon(\infty)kT \tag{6.8}$$

and

$$\langle \sigma_{zx}^2\rangle = kT\frac{G(0)}{v} \tag{6.9}$$

which tell us that all fluctuations increase proportional to T. When comparing the two expressions we also recognize a characteristic difference. The total dipole moment represents an extensive variable, $p_v \sim v$, and here the variance is proportional to the size of the subsystem. On the other hand, for the local stress, an intensive variable, the variance decreases inversely to the size of the chosen subsystem. The latter behavior needs a comment since the question may arise about the value to be attributed to an intensive variable in a subsystem. The answer is that σ_{zx}, as used above, and intensive variables in general, have to be identified with the spatial average in v. This average varies between different subsystems in a sample, or for one given subsystem it varies with time. Equation (6.9) states that the variations decrease if the averaging volume is increased. This, indeed, expresses a basic requirement of thermodynamics. In the thermodynamic limit, i.e. for systems of infinite size, intensive variables must become sharp.

The name 'fluctuation-dissipation theorem' actually refers to another form of the relationship, namely that holding if times are replaced by frequencies. Let us switch in the explanation to general terms and choose the symbol X_v for the extensive variable and, as earlier, the notation ψ for the field. For the linear response, we write in equivalence to Eq. (5.29)

$$X_v(t) = v\alpha(t)\psi_0 \qquad (t \geq 0) \tag{6.10}$$

with $\alpha(t)$ as general time dependent susceptibility. The general form of Eq. (6.6) is

$$\langle X_v(0)X_v(t)\rangle = vkT(\alpha(\infty) - \alpha(t)) \tag{6.11}$$

Here, we exclude from the considerations any viscous flow so that existence of the limit $\alpha(t \to \infty)$ is ensured.

Rather than characterizing the dynamics of a fluctuating thermodynamic variable by the time dependent correlation function $\langle X_v(0)X_v(t)\rangle$, one can also describe it by the spectral density $\langle X_v(\omega)^2\rangle$. We utilize here another famous theorem of statistical physics, known as the 'Wiener-Chinchin theorem'. It states that these two functions represent a pair of Fourier transforms

$$\langle X_v(0)X_v(t)\rangle = \frac{1}{2\pi}\int_{-\infty}^{\infty} \langle X_v(\omega)^2\rangle \exp \mathrm{i}\omega t \mathrm{d}\omega \tag{6.12}$$

and

$$\langle X_v(\omega)^2\rangle = \int_{-\infty}^{\infty} \langle X_v(0)X_v(t)\rangle \exp -\mathrm{i}\omega t \mathrm{d}t \tag{6.13}$$

The steady state susceptibility, $\alpha(t \to \infty)$, agrees with the limiting value at zero frequency of the dynamic susceptibility

$$\alpha(t \to \infty) = \alpha'(\omega = 0) \tag{6.14}$$

If we apply Eqs. (6.12) and (6.14) in Eq. (6.11), setting $t = 0$, we obtain

$$\frac{1}{2\pi}\int_{-\infty}^{\infty} \langle X_v(\omega)^2\rangle \mathrm{d}\omega = vkT\alpha'(\omega = 0) \tag{6.15}$$

or, employing the the Kramers-Kronig dispersion relation Eq. (5.46)

$$\frac{1}{2\pi}\int_{-\infty}^{\infty} \langle X_v(\omega)^2\rangle \mathrm{d}\omega = vkT\frac{1}{\pi}\int_{-\infty}^{\infty} \frac{\alpha''(\omega)}{\omega}\mathrm{d}\omega \tag{6.16}$$

Equating the integrants on both sides yields a second form of the fluctuation-dissipation theorem

$$\langle X_v(\omega)^2\rangle = \frac{2vkT}{\omega}\alpha''(\omega) \tag{6.17}$$

It is this form which is addressed by the name, as the expression relates the spectral density of the fluctuations of X_v to the imaginary part of the associated dynamic susceptibility and as mentioned earlier, the latter describes the energy dissipation.

An analogous relationship holds for the spectral density of the field fluctuations. The general form of the fluctuation dissipation theorem here reads

$$\langle \psi(\omega)^2\rangle = \frac{2kT}{v\omega}\mathrm{a}''(\omega) \tag{6.18}$$

where $a''(\omega)$ represents the imaginary part of the general dynamic modulus, which is defined as

$$\mathrm{a}^* := \frac{1}{\alpha^*} \tag{6.19}$$

The fluctuation-dissipation theorem provides us with an interesting view with regard to the results of the mechanical and dielectric experiments. The theorem teaches us that measurements of response functions equal a spectral analysis of equilibrium fluctuations. Fluctuations generally arise from all dynamical modes in the system, however, different experiments associate them with different weights. Take for example the glass-rubber transition. The transition shows up in measurements of the dynamic compliance, the dynamic modulus and the dielectric function. Although the shapes $J''(\omega)$, $D''(\omega)$, $G''(\omega)$, $E''(\omega)$ and $\epsilon''(\omega)$ differ from each other, the maxima of the absorption curves being located at different frequencies, all these experiments yield spectral densities associated with the dynamics of the same group of motions, namely the α-modes. In the dielectric experiment these are weighted according to the changes of dipole moments, in compliance measurements, the weighting factor relates to the changes in the shape of a polymer and in the dynamic modulus curve, it depends on the internal stresses or moment transfers along a chain. The overall relaxation time spectrum of the α-modes is extremely broad and encompasses more than four orders of magnitude. This is demonstrated in particular by the large difference in the maximum positions between loss compliances and loss moduli, as described by Eq. (5.92). We may conclude that the main contributions to the moduli originate from modes with short relaxation times, whereas the compliances put the major weight on the long time part of the spectrum. The dielectric α-process has its maximum usually in-between the two mechanical processes and thus appears to put main emphasis on the central part.

Most important is the role played by the fluctuation-dissipation theorem in theoretical treatments, as it may be regarded as an interface between the microscopic and the macroscopic properties of a sample. It provides us with a precise prescription of how to proceed when these two are to be related. On the microscopic side, theoretical analysis of dynamical models usually enables us to make a calculation of equilibrium correlation functions for all properties of interest. The fluctuation-dissipation theorem then relates these correlation functions with the results of measurement, as described by the various response functions.

In the following, we will discuss some microscopic dynamical models. We begin with the 'Rouse-model', which describes the dynamics of chains in a non-entangled polymer melt. The effects of entanglements on the motion can be accounted for by the 'reptation model', which we will treat subsequently. Finally, we shall be concerned with the motion of polymer chains in a solvent, when the 'hydrodynamic interaction' between the segments of a chain plays a prominent role.

6.2 The Rouse-Model

If a polymer molecule is stretched out by applying forces to the end groups and then the forces are removed it returns to the initial coiled conformation. The reason for this behavior has already been mentioned: The transition back to an isotropic coil increases the number of available rotational isomeric states and thus the entropy. The recoiling effect can also be expressed in mechanistic terms, by stating that, if the two endgroups of a polymer chain are held fixed at a certain distance, a tensile force arises due to the net moment transfer onto the ends. If, rather than keeping hold of the endgroups, two arbitrary points within a polymer molecule are kept at constant positions, a tensile force arises as well.

We shall derive this force in the next chapter, when dealing with the elasticity of rubbers. If Gaussian properties are assumed, then the result, as given by Eq. (7.13), is as follows: If a sequence within a chain is chosen which has a mean squared end-to-end distance $\langle \Delta \boldsymbol{r}^2 \rangle$, and its endpoints are at a distance $\Delta \boldsymbol{r}$, then the tensile force is

$$\mathbf{f} = b_{\mathrm{R}} \Delta \boldsymbol{r} \tag{6.20}$$

with

$$b_{\mathrm{R}} = \frac{3kT}{\langle \Delta \boldsymbol{r}^2 \rangle} \tag{6.21}$$

The result implies that a sequence behaves like a spring, showing a linear relation between force and extension. The force constant b_{R} is proportional to the absolute temperature T, as is characteristic for forces of entropic origin. Note furthermore that b_{R} decreases on increasing the size of the sequence.

A polymer chain in a melt moves in the surroundings set up by the other chains and at first, this looks like a rather complicated situation. However, as it turns out, one can employ an approximate simple treatment. For particles of at least mesoscopic size, the various interactions with adjacent molecules may be represented in summary by one viscous force. This is well-known from treatments of the dynamics of a colloid in a solvent. There it can be assumed that if a colloid moves with a velocity $\boldsymbol{u}$, the solvent molecules in contact with its surface create a force which is proportional to $\boldsymbol{u}$ and the solvent viscosity η_{s}

$$\mathbf{f} = \zeta \boldsymbol{u} \tag{6.22}$$

with

$$\zeta \sim \eta_{\mathrm{s}} \tag{6.23}$$

ζ is the 'friction coefficient'. We shall utilize this equation also in section 6.4. Then, more explanations will be provided.

Rouse devised a treatment of the dynamics of polymer chains in a melt which makes use of this notion of a viscous force and also takes account of the tensile forces arising in stretched parts of the chain. The procedure used

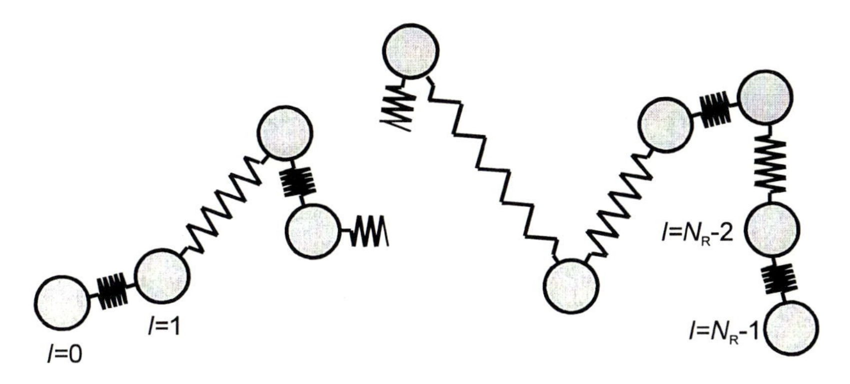

Fig. 6.1. Rouse-chain composed of N_R beads connected by springs

in setting up the 'Rouse-model' is remarkably simple. In a first step the chain is subdivided in N_R 'Rouse-sequences', each sequence being sufficiently long so that Gaussian properties are ensured. Then, in a second step, each Rouse-sequence is substituted by a bead and a spring. The springs are the representatives of the elastic tensile forces, while the beads play the role of centers whereon friction forces apply. The thus emerging 'Rouse-chain' is composed of a series of beads connected by springs, as depicted in Fig. 6.1.

The equations of motion of the 'Rouse-chain' are formulated while neglecting all inertial effects. Then the velocity $\mathrm{d}\boldsymbol{r}_l/\mathrm{d}t$ of the bead l is given by

$$\zeta_\mathrm{R}\frac{\mathrm{d}\boldsymbol{r}_l}{\mathrm{d}t} = b_\mathrm{R}(\boldsymbol{r}_{l+1} - \boldsymbol{r}_l) + b_\mathrm{R}(\boldsymbol{r}_{l-1} - \boldsymbol{r}_l) \tag{6.24}$$

The left-hand side represents the viscous force, the parameter ζ_R designates the friction coefficient per bead. On the right-hand side we have the elastic forces originating from the adjacent beads, which are located at the positions $\boldsymbol{r}_{l-1}$ and $\boldsymbol{r}_{l+1}$. The force constant b_R of the springs depends on the mean squared end-to-end distances of the Rouse-sequences, a_R^2, and follows from Eq. (6.21) as

$$b_\mathrm{R} = \frac{3kT}{a_\mathrm{R}^2} \tag{6.25}$$

It is easy to solve this set of differential equations. Note at the beginning that motions in the three directions of space, x, y, z, decouple and are equivalent. We select for the treatment the z-direction and consider the equations of motion

$$\zeta_\mathrm{R}\frac{\mathrm{d}z_l}{\mathrm{d}t} = b_\mathrm{R}(z_{l+1} - z_l) + b_\mathrm{R}(z_{l-1} - z_l) \tag{6.26}$$

First we discuss chains with infinite length. In this case, we have translational symmetry in terms of l. Then there must be wave-like solutions of the form

$$z_l \sim \exp -\frac{t}{\tau} \cdot \exp \mathrm{i}l\delta \tag{6.27}$$

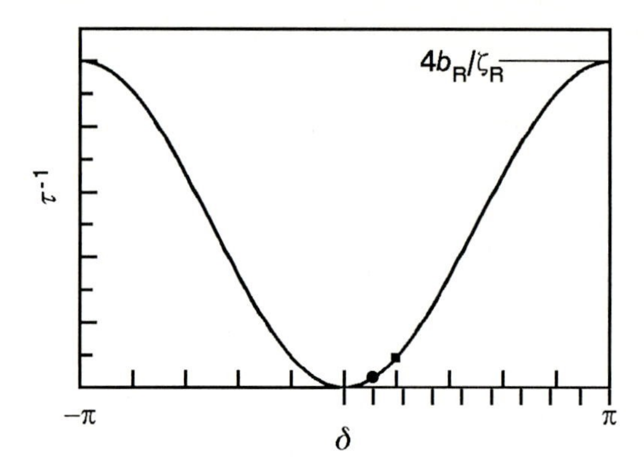

Fig. 6.2. Relaxation rates of Rouse-modes as a function of the phase shift δ. Marks on the inside of the abscissa show the mode positions for a cyclic chain with N_R = 10 beads, the marks on the outside give the modes of a linear chain with the same length. The lowest order Rouse-modes of the two chains with relaxation rates τ_R^{-1} are especially indicated, by a *filled circle* and a *filled square*

They include an exponential time dependence, as can be anticipated for relaxation processes; δ describes the phase shift between adjacent beads. If we take Eq. (6.27) and introduce it in Eq. (6.26) we obtain the dependence of the relaxation rate τ^{-1} on δ:

$$\tau^{-1} = \frac{b_R}{\zeta_R}(2 - 2\cos\delta) = \frac{4b_R}{\zeta_R}\sin^2\frac{\delta}{2} \tag{6.28}$$

Figure 6.2 presents this dependence for δ's between $-\pi$ and $+\pi$. Considerations can be restricted to this range, as other values of δ give nothing new for a discrete chain.

A formal means of accounting for the finite size of a chain while maintaining the wave-like solutions is provided by the introduction of cyclic boundary conditions. For a chain with N_R beads, this implies the equality

$$z_l = z_{l+N_R} \tag{6.29}$$

which is satisfied if

$$N_R\delta = m2\pi \tag{6.30}$$

N_R discrete values of the phase shift are thus selected:

$$\delta_m = \frac{2\pi}{N_R}m \quad , \quad m = -(\frac{N_R}{2} - 1) \quad , \ldots , \quad \frac{N_R}{2} \tag{6.31}$$

Fig. 6.2 shows the locations for $N_R = 10$.

Although for polymers, cyclic boundary conditions can be realized, namely by a synthesis of cyclic macromolecules, common polymer systems are composed of linear chains. These linear chains possess free ends where the tensile forces vanish. The boundary conditions then become

$$z_1 - z_0 = z_{N_\mathrm{R}-1} - z_{N_\mathrm{R}-2} = 0 \tag{6.32}$$

or, in a differential form

$$\frac{\mathrm{d}z}{\mathrm{d}l}(l=0) = \frac{\mathrm{d}z}{\mathrm{d}l}(l = N_\mathrm{R} - 1) = 0 \tag{6.33}$$

The real and the imaginary part of Eq. (6.27)

$$z_l \sim \cos l\delta \cdot \exp -\frac{t}{\tau} \tag{6.34}$$

$$z_l \sim \sin l\delta \cdot \exp -\frac{t}{\tau} \tag{6.35}$$

represent separate solutions of the equation of motion. The boundary condition at $l = 0$ is only fulfilled by the cosine solution. The condition for the upper end, $l = N_\mathrm{R} - 1$, then selects the values of δ, by

$$\frac{\mathrm{d}z_l}{\mathrm{d}l}(l = N_\mathrm{R} - 1) \sim \sin((N_\mathrm{R} - 1)\delta) = 0 \tag{6.36}$$

Since this is solved by

$$(N_\mathrm{R} - 1)\delta = m\pi \tag{6.37}$$

we obtain for a linear chain with free ends the following eigenvalues δ_m

$$\delta_m = \frac{\pi}{N_\mathrm{R} - 1} m \quad , \quad m = 0, 1, 2, \cdots, N_\mathrm{R} - 1 \tag{6.38}$$

Hence, we find for the linear chain N_R independent solutions. They are called 'Rouse-modes' and differentiated by their 'order' m. Figure 6.2 shows the eigenvalues δ_m for a chain with $N_\mathrm{R} = 10$, compared to the cyclic Rouse-chain.

Figure 6.3 presents the displacement pattern associated with the lowest order Rouse-mode with $m = 1$ (the solution for $m = 0$ describes just a free translation). The relaxation rate of this mode, shortly called 'Rouse-rate' τ_R^{-1}, follows from Eq. (6.38) together with Eq. (6.28), as

$$\tau_1^{-1} := \tau_\mathrm{R}^{-1} = \frac{b_\mathrm{R}}{\zeta_\mathrm{R}} \frac{\pi^2}{(N_\mathrm{R} - 1)^2} \tag{6.39}$$

or, with Eq. (6.25), as

$$\tau_\mathrm{R}^{-1} = \frac{3kT\pi^2}{\zeta_\mathrm{R} a_\mathrm{R}^2 (N_\mathrm{R} - 1)^2} \tag{6.40}$$

The result includes the size a_R of the Rouse-sequences and therefore a quantity with a freedom of choice. This arbitrariness can be removed. Since

$$R_0^2 = a_\mathrm{R}^2 (N_\mathrm{R} - 1) \tag{6.41}$$

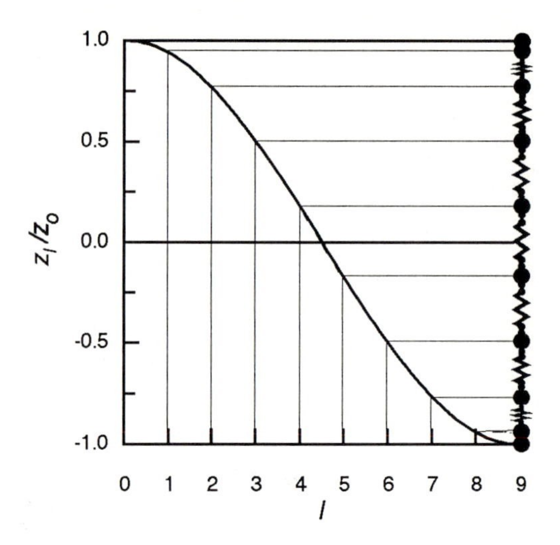

Fig. 6.3. Displacement pattern of the primary Rouse-mode

we obtain for the 'Rouse-time' τ_{R}

$$\tau_{\mathrm{R}} = \frac{1}{3\pi^2}\frac{(\zeta_{\mathrm{R}}/a_{\mathrm{R}}^2)}{kT}R_0^4 \tag{6.42}$$

The result indicates that the ratio $\zeta_{\mathrm{R}}/a_{\mathrm{R}}^2$ should be independent of the choice of the sequence. This is true if the friction coefficient ζ_{R} is proportional to the number of monomer units in the sequence. Strictly speaking, the latter property constitutes a basic requirement for the validity of the Rouse-model: The friction coefficient of a sequence *has* to be proportional to the number of monomer units. In fact, this is not trivial and clear from the very beginning. It seems to be correct in a melt because, as we shall see, here the Rouse-model works quite satisfactorily, if compared with experimental results. On the other hand, the assumption is definitely wrong for isolated polymer chains in a solvent where hydrodynamic interactions strongly affect the motion; we shall be concerned with this point in a subsequent section.

Equation (6.42) gives the dependence of the Rouse-time on the degree of polymerization. Since

$$R_0^2 = a_0^2 N \tag{6.43}$$

we obtain

$$\tau_{\mathrm{R}} \sim N^2 \tag{6.44}$$

The shortest relaxation time in the Rouse-spectrum depends on the choice of the Rouse-sequence and follows, for $m = N_{\mathrm{R}} - 1$, from Eqs. (6.28) and (6.38) as

$$\tau_{N_\mathrm{R}-1} = \frac{\zeta_\mathrm{R} a_\mathrm{R}^2}{12kT} = \frac{(\zeta_\mathrm{R}/a_\mathrm{R}^2)}{12kT} a_\mathrm{R}^4 \tag{6.45}$$

We see that there is a cut-off at the short-time end which depends on a_R. There are, of course, other relaxation modes above this cut-off. They can, however, no longer be described by the Rouse-model as, in this range of high relaxation rates, motions become localized and depend then on the chemical composition of a chain.

As we can see, the motion of a polymer chain in a non-entangled melt, as represented by the Rouse-model, can be described as a superposition of $3N_\mathrm{R}$ linearly independent Rouse-modes, corresponding to N_R modes in x-, y- and z-directions respectively. In a dynamic equilibrium state all these Rouse-modes become thermally excited and it is instructive to calculate their mean-squared amplitudes. The displacement pattern of mode m is given by

$$z_l = Z_m \cos l\delta_m \tag{6.46}$$

Z_m denotes a 'normal coordinate', which determines the mode amplitude. Thermal excitations of modes, oscillatory modes as well as the relaxatory modes discussed here, depend on the associated change in free energy. For our bead-and-spring model, the change in free energy per polymer chain, Δf_p, is given by

$$\Delta f_\mathrm{p} = \frac{b_\mathrm{R}}{2} \sum_{l=0}^{N_\mathrm{R}-2} (z_{l+1} - z_l)^2 \tag{6.47}$$

$$= \frac{b_\mathrm{R}}{2} Z_m^2 \sum_{l=0}^{N_\mathrm{R}-2} (\cos(l+1)\delta_m - \cos l\delta_m)^2 \tag{6.48}$$

$$= \frac{b_\mathrm{R}}{2} Z_m^2 \delta_m^2 \sum_{l=0}^{N_\mathrm{R}-2} \sin^2 \delta_m l = \frac{b_\mathrm{R}}{2} \frac{N_\mathrm{R}-1}{2} Z_m^2 \delta_m^2 \tag{6.49}$$

The function $\Delta f_\mathrm{p}(Z_m)$ determines the probability distribution $p(Z_m)$ for the amplitude Z_m, which follows from Boltzmann statistics as

$$p(Z_m) \sim \exp - \frac{\Delta f_\mathrm{p}(Z_m)}{kT} \tag{6.50}$$

Since

$$\Delta f_\mathrm{p} \sim Z_m^2 \tag{6.51}$$

we find for the normal coordinate Z_m a Gaussian distribution. The variance $\langle Z_m^2 \rangle$ may be derived from the equation

$$\langle \Delta f_\mathrm{p} \rangle = \frac{kT}{2} \tag{6.52}$$

giving

$$\frac{b_{\mathrm{R}}}{2} \cdot \frac{N_{\mathrm{R}}-1}{2} \delta_m^2 \left\langle Z_m^2 \right\rangle = \frac{3kT}{2a_{\mathrm{R}}^2} \cdot \frac{N_{\mathrm{R}}-1}{2} \delta_m^2 \left\langle Z_m^2 \right\rangle = \frac{kT}{2} \tag{6.53}$$

Note that, as expected for an ideal chain, $\langle Z_m^2 \rangle$ is indeed independent of temperature:

$$\left\langle Z_m^2 \right\rangle = \frac{2a_{\mathrm{R}}^2}{3(N_{\mathrm{R}}-1)\delta_m^2} \tag{6.54}$$

or, using Eq. (6.38)

$$\left\langle Z_m^2 \right\rangle = \frac{2}{3\pi^2} \frac{R_0^2}{m^2} \tag{6.55}$$

According to this result, the amplitudes of the Rouse-modes rapidly decrease with increasing mode order m. If we consider the contributions of the different Rouse-modes to the known total mean squared end-to-end distance, $\langle R^2 \rangle = R_0^2$, we find that a large part is already provided by the three lowest order Rouse-modes. Contributions to the end-to-end distance in the z-direction, $\langle R_z^2 \rangle = R_0^2/3$, come from all z-polarized Rouse-modes with odd m's

$$\left\langle (z_{N_{\mathrm{R}}-1} - z_0)^2 \right\rangle = \left\langle (2Z_1)^2 \right\rangle + \left\langle (2Z_3)^2 \right\rangle + \ldots \tag{6.56}$$

which leads to

$$\left\langle (z_{N_{\mathrm{R}}-1} - z_0)^2 \right\rangle = \frac{8}{\pi^2} \cdot \frac{R_0^2}{3} \left(1 + \frac{1}{9} + \ldots \right) = \frac{R_0^2}{3} \tag{6.57}$$

Hence, 90% of the total mean squared end-to-end distance of a chain originates from the lowest order Rouse-modes. In theorical treatments, polymer chains are sometimes substituted by elastic dumbells, set up by two beads connected by a spring. The justification for this simplification follows from the dominant role of the primary Rouse-modes.

We finish this section with the schematic drawing displayed in Fig. 6.4, meant to indicate how the time dependent fluctuations of the amplitude of a Rouse-mode could look-like. The interaction of a chain with its surroundings leads to excitations of this mode at random times. In-between, the mode amplitude decreases exponentially with a characteristic relaxation time as described by the equation of motion. These are the only parts in the time dependent curve which show a well-defined specific behavior; the excitations occur irregularly during much shorter times. We may therefore anticipate that the shape of the time correlation function is solely determined by the repeated periods of exponential decay. Regarding the results of this section, we thus may formulate directly the time correlation function for the normal coordinate Z_m, as

$$\left\langle Z_m(0) Z_m(t) \right\rangle = \left\langle Z_m^2 \right\rangle \cdot \exp -\frac{t}{\tau_m} = \frac{2R_0^2}{3\pi^2 m^2} \exp -\frac{t}{\tau_m} \tag{6.58}$$

the relaxation time τ_m being given by Eqs. (6.38) and (6.28).

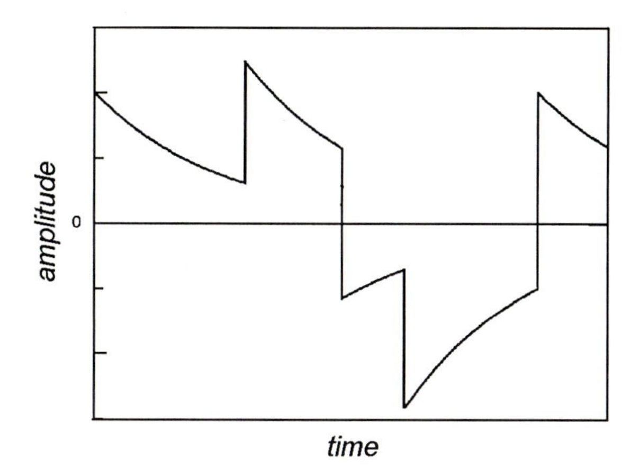

Fig. 6.4. Time dependence of the amplitude Z_m of a Rouse mode (schematic)

6.2.1 Stress Relaxation

Having discussed the microscopic dynamical properties of a system of Rouse-chains, we now inquire about the resulting mechanical behavior and consider as an example the shear stress relaxation modulus, $G(t)$. $G(t)$ can be determined with the aid of the fluctuation-dissipation theorem, utilizing Eq. (6.7)

$$\langle \sigma_{zx}(0)\sigma_{zx}(t)\rangle = kT\frac{G(t)}{v}$$

We have to calculate the fluctuations in the stress field produced by a system of Rouse-chains in thermal equilibrium. As explained above, fluctuations of an intensive variable such as the stress depend upon the size of the chosen subsystems. We select subsystems of volume v, having orthogonal edges with lengths l_x, l_y and l_z, as displayed in Fig. 6.5. Being concerned with the shear stress σ_{zx}, we recognize that contributions arise from all springs which have a non-vanishing component of extension in x-direction. We choose the symbols $\hat{x}_i, \hat{y}_i, \hat{z}_i$ for the three components of the extension of spring i, which may be incorporated in any chain, and designate by $f_{x,i}$ the associated force component along x. The mean value of σ_{zx} in a subsystem is obtained by a summation over the contributions of all included springs i

$$\sigma_{zx} = \sum_i \frac{1}{l_x l_y}\frac{\hat{z}_i}{l_z} \cdot f_{x,i} \tag{6.59}$$

To see the background of this equation, consider a unit area normal to the z-axis, as indicated in Fig. 6.5. Stress on this plane is produced by all springs which cross it. The term

$$\frac{1}{l_x l_y}\frac{\hat{z}_i}{l_z}$$

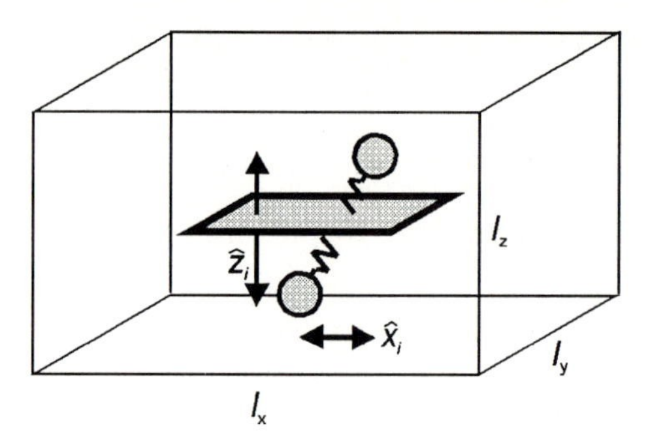

Fig. 6.5. Notions used in the calculation of the tensile stress σ_{zx} associated with a system of Rouse-chains: Reference volume $v = l_x l_y l_z$; unit area, crossed by the spring i with extensions $\hat{x}_i$ and $\hat{z}_i$ along x and z

expresses the probability that spring i, with an extension $\hat{z}_i$ along z, crosses the unit area. When crossing this area, the spring contributes a force $f_{x,i}$ to σ_{zx}. We now adopt Eq. (6.20) and write

$$\sigma_{zx} = \frac{1}{v} \sum_i \hat{z}_i f_{x,i} = \frac{b_\mathrm{R}}{v} \sum_i \hat{z}_i \hat{x}_i \tag{6.60}$$

σ_{zx}, being defined as the spatial average in a volume v, represents a fluctuating quantity which shows different values in different subsystems, or for measurements at different times. The ensemble average vanishes

$$\langle \sigma_{zx} \rangle = 0 \tag{6.61}$$

since

$$\langle \hat{z}_i \hat{x}_i \rangle = \langle \hat{z}_i \rangle \langle \hat{x}_i \rangle = 0 \tag{6.62}$$

regarding that the movements of Rouse-chains along z and x are independent. The time correlation function of the fluctuations of the shear stress, $\langle \sigma_{zx}(0)\sigma_{zx}(t) \rangle$, follows as

$$\langle \sigma_{zx}(0)\sigma_{zx}(t) \rangle = \frac{1}{v^2} b_\mathrm{R}^2 \sum_{k,l,k',l'} \langle \hat{x}_{k,l}(0) \hat{z}_{k,l}(0) \hat{x}_{k',l'}(t) \hat{z}_{k',l'}(t) \rangle \tag{6.63}$$

Here, the extensions along x and z of the spring l on the chain k are denoted $\hat{x}_{k,l}$ and $\hat{z}_{k,l}$ and those of the spring l' on the chain k' correspondingly. The sum includes all chains contained in v. As the extensions of springs in different chains are uncorrelated, we may furthermore write

$$\langle \sigma_{zx}(0)\sigma_{zx}(t) \rangle = \frac{1}{v^2} c_\mathrm{p} v b_\mathrm{R}^2 \sum_{l,l'} \langle \hat{x}_l(0) \hat{x}_{l'}(t) \rangle \langle \hat{z}_l(0) \hat{z}_{l'}(t) \rangle \tag{6.64}$$

where $c_\mathrm{p} v$ gives the number of chains in v.

Chain dynamics may be represented as a superposition of independent Rouse-modes. The displacements of mode m', polarized in z-direction, are given by

$$z_l = Z_{m'} \cos \delta_{m'} l \tag{6.65}$$

The extensions of the springs follow by taking the derivative

$$\hat{z}_l := z_{l+1} - z_l \approx \frac{\mathrm{d}z_l}{\mathrm{d}l} = -Z_{m'} \delta_{m'} \sin \delta_{m'} l \tag{6.66}$$

Equivalently, the extension associated with mode m, polarized along x, is given by

$$\hat{x}_l := x_{l+1} - x_l \approx \frac{\mathrm{d}x_l}{\mathrm{d}l} = -X_m \delta_m \sin \delta_m l \tag{6.67}$$

We thus obtain

$$\langle \hat{x}_l(0) \hat{x}_{l'}(t) \rangle = \sum_m \langle X_m(0) X_m(t) \rangle \sin \delta_m l \cdot \sin \delta_m l' \cdot \delta_m^2 \tag{6.68}$$

and

$$\langle \hat{z}_l(0) \hat{z}_{l'}(t) \rangle = \sum_{m'} \langle Z_{m'}(0) Z_{m'}(t) \rangle \sin \delta_{m'} l \cdot \sin \delta_{m'} l' \cdot \delta_{m'}^2 \tag{6.69}$$

Since

$$\sum_l \sin \delta_m l \cdot \sin \delta_{m'} l \sum_{l'} \sin \delta_m l' \cdot \sin \delta_{m'} l' = \left(\frac{N_\mathrm{R} - 1}{2} \right)^2 \delta_{mm'} \tag{6.70}$$

where $\delta_{mm'}$ denotes the Kronecker-delta, we find

$$\begin{aligned} \langle \sigma_{zx}(0) \sigma_{zx}(t) \rangle &= \frac{1}{v} c_\mathrm{p} b_\mathrm{R}^2 \left(\frac{N_\mathrm{R} - 1}{2} \right)^2 \cdot \\ &\quad \sum_m \delta_m^4 \langle X_m(0) X_m(t) \rangle \cdot \langle Z_m(0) Z_m(t) \rangle \end{aligned} \tag{6.71}$$

If we introduce the mean-squared amplitudes $\langle X_m^2 \rangle$ and $\langle Z_m^2 \rangle$ of the Rouse-modes in thermal equilibrium, as given by Eq. (6.53), we may write

$$\langle \sigma_{zx}(0) \sigma_{zx}(t) \rangle = \frac{1}{v} c_\mathrm{p} (kT)^2 \sum_m \frac{\langle X_m(0) X_m(t) \rangle}{\langle X_m^2 \rangle} \cdot \frac{\langle Z_m(0) Z_m(t) \rangle}{\langle Z_m^2 \rangle} \tag{6.72}$$

Now we utilize the fluctuation-dissipation theorem, i.e. apply Eq. (6.7), to obtain the shear relaxation modulus

$$G(t) = c_\mathrm{p} kT \sum_m \frac{\langle X_m(0) X_m(t) \rangle}{\langle X_m^2 \rangle} \cdot \frac{\langle Z_m(0) Z_m(t) \rangle}{\langle Z_m^2 \rangle} \tag{6.73}$$

Equation (6.73) relates $G(t)$ to the magnitudes and the time dependencies of the fluctuations of the Rouse-modes in thermal equilibrium. The time correlation functions are given by Eq. (6.58)

$$\langle Z_m(0)Z_m(t)\rangle = \langle Z_m^2\rangle \exp -\frac{t}{\tau_m} \tag{6.74}$$

and equivalently

$$\langle X_m(0)X_m(t)\rangle = \langle X_m^2\rangle \exp -\frac{t}{\tau_m} \tag{6.75}$$

Hence, we finally obtain

$$G(t) = c_{\mathrm{p}}kT \sum_m \exp -2\frac{t}{\tau_m} \tag{6.76}$$

Equation(6.76) describes the shear stress relaxation modulus associated with a system of Rouse-chains. The result has a remarkably simple structure, as all Rouse-modes contribute to $G(t)$ with the same weight. Note that the relaxation rates are increased by a factor of two with regard to the Rouse-mode rates τ_m^{-1}.

Equation (6.76) may be further evaluated by carrying out the summation. If we disregard the short-time range, the discussion may be reduced to the contributions of the low order Rouse-modes and we can replace Eqs. (6.28) and (6.38) by the approximate relation

$$\tau_m^{-1} \approx \tau_{\mathrm{R}}^{-1} \cdot m^2 \tag{6.77}$$

Introduction in Eq. (6.76) and a change from the summation to an integral gives

$$G(t) \sim \int_{m=1}^{N_{\mathrm{R}}-1} \mathrm{d}m \exp -2\tau_{\mathrm{R}}^{-1}m^2t \approx \int_{m=0}^{\infty} \mathrm{d}m \exp -2\tau_{\mathrm{R}}^{-1}m^2t \tag{6.78}$$

or, with the substitution

$$u := m\left(\frac{t}{\tau_{\mathrm{R}}}\right)^{1/2} \tag{6.79}$$

the expression

$$G \sim \left(\frac{\tau_{\mathrm{R}}}{t}\right)^{1/2} \int_{u=0}^{\infty} \exp -2u^2 \,\mathrm{d}u \tag{6.80}$$

hence

$$G(t) \sim t^{-1/2} \tag{6.81}$$

The result is a power law which is characteristic for the relaxation of Rouse-chains and it may be compared with experiments on polymer melts. Indeed,

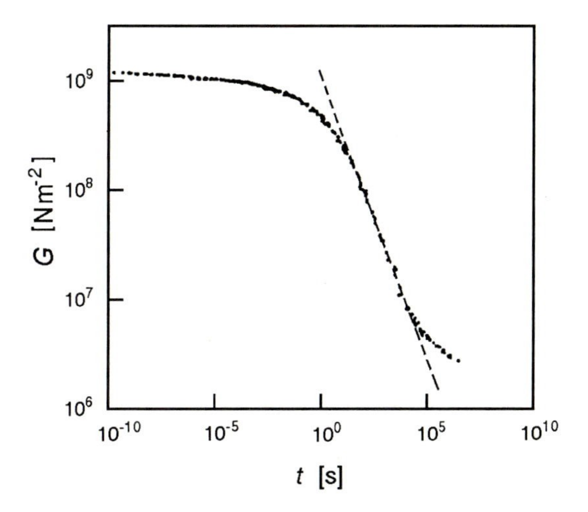

Fig. 6.6. Time dependent shear modulus of PVC. Master curve set up for $T_g = 65\,°C$ as the reference temperature. The *dashed line* indicates the slope predicted by the Rouse-model. Data from Eisele [66]

for several systems one finds a good agreement. The center of the glass-rubber transition, as observed in stress relaxation experiments, is often well-described by Eq. (6.81). Figure 6.6 depicts, as an example, the time dependent shear modulus of poly(vinylchloride) (PVC), presented as a master-curve referred to the glass transition temperature ($T_g = 65\,°C$). The slope of the log-log plot corresponds to an exponent -1/2.

The power law for the time dependent modulus can be transformed into a law valid for the frequency domain, either by applying the general relations of linear response or by using the ω-dependent form of the fluctuation-dissipation theorem. The result is

$$G'(\omega) \sim (\eta_0 \omega)^{1/2} \tag{6.82}$$

The measurement shown in Fig. 5.15 agrees with this prediction.

We can also determine the viscosity at zero shear rate, by application of Eq. (5.107)

$$\begin{aligned} \eta_0 &= \int_0^\infty G(t)dt \\ &= kTc_p \sum_{m=1}^{N_R-1} \frac{\tau_m}{2} \\ &\approx kTc_p \frac{\tau_R}{2} \sum_{m=1}^{\infty} \frac{1}{m^2} = kTc_p \tau_R \frac{\pi^2}{12} \end{aligned} \tag{6.83}$$

Regarding Eq. (6.44), the result indicates a linear dependence of η_0 on the degree of polymerization

$$\eta_0 \sim \frac{c_m}{N} N^2 \sim N \tag{6.84}$$

This is in full agreement with the observations on non-entangled melts displayed in Fig. 5.13.

The Rouse-model is also applicable for entangled melts, however, only in a restricted manner. While a description of the motion of the whole polymer chain is no longer possible, it still can be employed for a treatment of the dynamics of the chain parts between entanglements. We will discuss the resulting overall behavior in the next section.

The Rouse-model has also intrinsic limitations at short times. According to Eq. (6.76), the unrelaxed modulus is determined by the number density of Rouse-sequences, c_{R}, since we find

$$G(0) = c_{\mathrm{p}} kT (N_{\mathrm{R}} - 1) = c_{\mathrm{R}} kT \tag{6.85}$$

At first this dependence on the choice of the sequence associated with an element of the Rouse-chain may look strange, however, the cause of this apparent uncertainty and the solution of the problem is easy to see: One has to realize that the internal degrees of freedom of the sequence give further contributions to the shear modulus; the correct value follows only from both parts together. We therefore have to write in general

$$G(t) = \Delta G_{\mathrm{mic}}(t) + G_{\mathrm{Rouse}}(t) \tag{6.86}$$

where the first part, $\Delta G_{\mathrm{mic}}(t)$, accounts for the short-time properties. In contrast to the Rouse-modes, the internal modes are finally controlled by the microstructure, i.e. the chemical composition. The point of transition from the Rouse representation to the detailed description has, indeed, a freedom of choice, one has only to stay outside the range where specific microscopic effects appear.

6.2.2 Dielectric Normal Mode

Let us return once again to the frequency- and temperature dependent measurements of the dielectric function of polyisoprene (PIP) presented in Sect. 5.3.2. As shown in Figs. 5.19 and 5.20, two relaxation processes exist. The low frequency process, the 'normal mode', is the one of interest here. As already mentioned, it reflects the movements of the end-to-end distance vector $\boldsymbol{R}$ of the chain. The Rouse-model enables these movements to be treated in the case of melts which are not entangled. Earlier, we learned that the motion of the end-to-end distance vector is to a large part due to the superposition of the three lowest order Rouse-modes, polarized in the x, y and z-directions. Therefore, the dielectric normal mode, when measured for samples with molecular

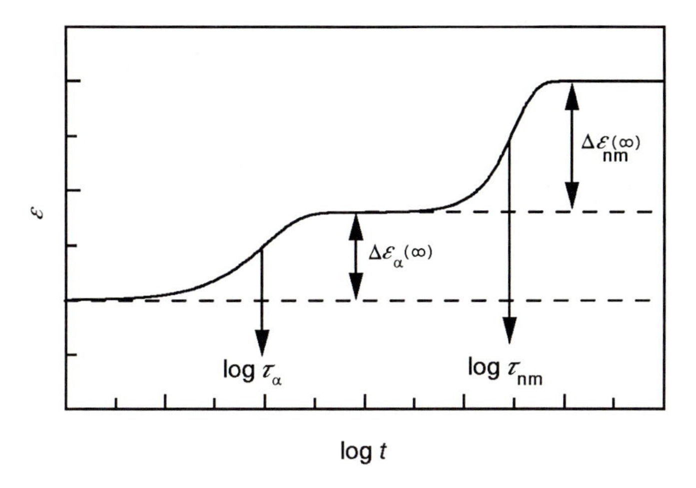

Fig. 6.7. General shape of the time dependent dielectric function $\epsilon(t)$ of PIP showing the α-process and the dielectric normal mode (schematic drawing)

weights below the entanglement limit, may be identified with these primary modes.

For a Rouse-chain built up of N_{R} polar sequences, each one carrying a dipole moment with a longitudinal component $\boldsymbol{p}_{\|}^{l}$, the total dipole moment $\boldsymbol{p}_{\mathrm{p}}$ is given by

$$\boldsymbol{p}_{\mathrm{p}} := \sum_{l=0}^{N_{\mathrm{R}}-1} \boldsymbol{p}_{\|}^{l} \tag{6.87}$$

Let us refer in the discussion to a representation of the dielectric data in the time domain, as expressed by the time dependent dielectric function $\epsilon(t)$. Figure 6.7 depicts its general shape in a schematic drawing. The α-process and the normal mode show up as two subsequent steps located at the times τ_α and τ_{nm}, with heights corresponding to the relaxation strengths $\Delta\epsilon_\alpha$ and $\Delta\epsilon_{\mathrm{nm}}$.

The fluctuation-dissipation theorem provides an exact description of the step $\Delta\epsilon_{\mathrm{nm}}(t)$ associated with the 'normal mode'. Utilizing Eq. (6.6) in combination with Eq. (6.2) and, assuming independent motions of different chains, the relation

$$\langle \boldsymbol{p}_v(0)\boldsymbol{p}_v(t)\rangle = vc_{\mathrm{p}}\,\langle \boldsymbol{p}_{\mathrm{p}}(0)\boldsymbol{p}_{\mathrm{p}}(t)\rangle \tag{6.88}$$

we obtain

$$c_{\mathrm{p}}\,\langle \boldsymbol{p}_{\mathrm{p}}(0)\boldsymbol{p}_{\mathrm{p}}(t)\rangle = 3kT\epsilon_0(\Delta\epsilon_{\mathrm{nm}}(\infty) - \Delta\epsilon_{\mathrm{nm}}(t)) \tag{6.89}$$

The relaxation strength, $\Delta\epsilon_{\mathrm{nm}}(\infty)$, follows from

$$c_{\mathrm{p}}\,\langle \boldsymbol{p}_{\mathrm{p}}^2\rangle = 3kT\epsilon_0\Delta\epsilon_{\mathrm{nm}}(\infty) \tag{6.90}$$

Since

$$\langle \boldsymbol{p}_{\mathrm{p}}^2 \rangle = N_{\mathrm{R}} \langle (\boldsymbol{p}_{\parallel}^l)^2 \rangle \tag{6.91}$$

and

$$\langle \boldsymbol{R}^2 \rangle = N_{\mathrm{R}} a_{\mathrm{R}}^2 \tag{6.92}$$

we have

$$\langle \boldsymbol{p}_{\mathrm{p}}^2 \rangle = \frac{\langle (\boldsymbol{p}_{\parallel}^l)^2 \rangle}{a_{\mathrm{R}}^2} \cdot \langle \boldsymbol{R}^2 \rangle := \beta^2 \cdot \langle \boldsymbol{R}^2 \rangle \tag{6.93}$$

Equation (6.93) relates the variance of the dipole moment of the polymer to the mean-squared end-to-end distance of the chain. We may therefore substitute, in Eq. (6.89), $\boldsymbol{p}_{\mathrm{p}}$ by $\boldsymbol{R}$, thus obtaining

$$c_{\mathrm{p}}\beta^2 \langle \boldsymbol{R}(0)\boldsymbol{R}(t) \rangle = 3kT\epsilon_0(\Delta\epsilon_{\mathrm{nm}}(\infty) - \Delta\epsilon_{\mathrm{nm}}(t)) \tag{6.94}$$

Now we employ the Rouse-model. As the end-to-end distance vector is essentially determined by the lowest order Rouse-modes, we can also represent the time correlation function in good approximation by

$$\langle \boldsymbol{R}(0)\boldsymbol{R}(t) \rangle \approx R_0^2 \exp -\frac{t}{\tau_{\mathrm{R}}} \tag{6.95}$$

Applying Eq. (6.95) in Eq. (6.94) leads us to

$$\epsilon_0 \Delta\epsilon_{\mathrm{nm}}(t) \simeq \frac{c_{\mathrm{p}}\beta^2 R_0^2}{3kT} \left(1 - \exp -\frac{t}{\tau_{\mathrm{R}}}\right) \tag{6.96}$$

Equation (6.96) provides a description of the normal mode, giving the relaxation strength as well as the relaxation time. Notice, in particular, that the observed molecular weight dependence of τ_{nm} for non-entangled melts as shown in Fig. 5.21

$$\tau_{\mathrm{nm}} \sim M^2 \tag{6.97}$$

is in full agreement with the prediction of the Rouse theory

$$\tau_{\mathrm{R}} \sim N^2 \tag{6.98}$$

Further contributions with minor weights originate from the subsequent Rouse-modes with $m > 1$. They may lead to the observed line-broadening on the high frequency side, as described empirically by the Havriliak-Nagami equation.

We finish this discussion with two remarks. Experiments mostly yield the frequency dependent complex dielectric constant, rather than $\epsilon(t)$. The conversion may be carried out straightforwardly by application of the general relationships Eqs. (5.31) and (5.44). We can write

$$\epsilon_0 \Delta\epsilon_{\mathrm{nm}}^*(\omega) = \int_0^\infty \frac{\mathrm{d}}{\mathrm{d}t} \epsilon_0 \Delta\epsilon_{\mathrm{nm}}(t) \cdot \exp -\mathrm{i}\omega t \, \mathrm{d}t \tag{6.99}$$

$$= \epsilon_0 \Delta\epsilon_{\mathrm{nm}}(\infty) \int_0^\infty \frac{1}{\tau_{\mathrm{R}}} \exp -\frac{t}{\tau_{\mathrm{R}}} \cdot \exp -\mathrm{i}\omega t \, \mathrm{d}t \tag{6.100}$$

$$= \epsilon_0 \Delta\epsilon_{\mathrm{nm}}(\infty) \frac{1}{1 + \mathrm{i}\omega\tau_{\mathrm{R}}} \tag{6.101}$$

thus arriving, as expected, at the expression for a Debye process.

The second remark concerns the use of the 'order of magnitude' symbol in Eq. (6.96). In fact, this result is not exact, because we disregarded possible 'inner field'-effects. Complete treatments of dielectric functions of polar liquids must account for the difference between the externally applied electric field and the local field effective at the position of a dipole. These inner fields may well modify the dynamics of polar chains, but to include this effect theoretically is not a simple task and lies outside our scope.

6.3 Entanglement Effects

Entanglements constitute a major feature of the dynamics in polymer melts. Due to their strong interpenetration, which increases with the molecular weight, polymer molecules are highly entangled. Since the chains are linearly connected objects which cannot cross each other, their individual motions become constrained and for the chain as a whole it is therefore impossible to move freely in all directions.

How can one deal with this situation? It is important to recognize that the restrictions are of peculiar nature in that they mainly concern the lateral chain motion, i.e. the motion perpendicular to the chain contour. It is this property which enables us to devise a simple model. Putting the emphasis on the lateral constraints and thus grasping the main point, de Gennes and Edwards suggested we envisage the chain dynamics as a motion in a 'tube'. This tube is set up by those of the adjacent polymers which represent obstacles for the lateral motion and thus models the confinement range. Owing to its simplicity, the tube model enables a theoretical evaluation. As we will see, it can explain the properties of entangled polymers to a large degree in quantitative terms.

We have already met several manifestations of the entanglements, in particular

- the occurrence of the rubber-elastic plateau in the time- and frequency dependent mechanical response functions
- the change in the molecular weight dependence of the viscosity
- the change in the molecular weight dependence of the relaxation time of the dielectric normal mode

Hence, the effect of the entanglements is two-fold since both the elastic and the viscous properties are concerned. The observations all indicate the existence of a critical molecular weight, introduced earlier as the 'critical molecular weight at the entanglement limit', denoted M_{c}. Polymers with low molar mass, $M < M_{\mathrm{c}}$, exhibit no entanglement effects and for $M > M_{\mathrm{c}}$ these show up and become dominant. All properties are affected which are founded on motions on length scales corresponding to molecular weights above M_{c}, in particular those which include the whole polymer chain, such as the viscosity or the dielectric normal mode. As already mentioned, Rouse-dynamics is maintained within the sequences between the entanglement points.

The change in the dynamics from a free Rouse- to a constrained tube-motion, occurring at a certain sequence length, shows up in quasi-elastic neutron scattering experiments. These experiments have the advantage that they combine a spatial resolution in the 1-10 nm range with a frequency resolution in the GHz-range. In addition, experiments may be conducted on mixtures of deuterated and protonated polymers. Since deuterium and hydrogen have different scattering cross-sections, one can study directly the individual properties of the minority species. Studies are usually carried out on a dilute solution of protonated chains in a deuterated matrix. Experiments then yield the intermediate scattering law $S(\boldsymbol{q},t)$. It is related by a Fourier transformation to the time dependent pair correlation function of the monomers in protonated chains, $g(\boldsymbol{r},t)$,

$$S(\boldsymbol{q},t) = \int \exp \mathrm{i}\boldsymbol{q}\boldsymbol{r} \cdot (g(\boldsymbol{r},t) - \langle c_{\mathrm{m}} \rangle)\, \mathrm{d}^3\boldsymbol{r} \tag{6.102}$$

(if necessary for an understanding check the explanations to Eq. (A.34) in the Appendix). For a dilute solution, $g(\boldsymbol{r},t)$ essentially agrees with the pair-correlation function of the monomers of *one* chain. Hence, these experiments inform us about the motion of individual chains.

If the motion of the chains in the melt is not restricted, the limit of the pair correlation function for long periods of time is given by

$$g(\boldsymbol{r}, t \to \infty) = \langle c_{\mathrm{m}} \rangle \tag{6.103}$$

As a consequence, the intermediate scattering law tends asymptotically to zero

$$S(\boldsymbol{q}, t \to \infty) = 0 \tag{6.104}$$

This behavior changes if the motion of the chains remains confined. Then the pair correlation function may differ from $\langle c_{\mathrm{m}} \rangle$ for all times

$$g(\boldsymbol{r}, t \to \infty) \neq \langle c_{\mathrm{m}} \rangle \tag{6.105}$$

Equation (6.105) holds for all distances $\boldsymbol{r}$ which are located within the confinement range. For distances $\boldsymbol{r}$ outside this zone, Eq. (6.104) remains valid, as $g(\boldsymbol{r},t)$ is there determined by the contributions of other chains.

Figure 6.8 presents the results of quasielastic neutron scattering experiments on melts of poly(ethylene-*co*-propylene). The curves belong to different scattering vectors q. We can see that decay times decrease with increasing q, as is generally expected for all kinds of diffusive motions.

A straightforward theoretical analysis proves that, for a system of Rouse-chains, the intermediate scattering law may be expressed as a function of one dimensionless variable only

$$u = q^2 a_{\mathrm{R}}^2 (\tau^{-1}(\delta = \pi) t)^{1/2} = q^2 (12 kT a_{\mathrm{R}}^2 t / \zeta_{\mathrm{R}})^{1/2} \tag{6.106}$$

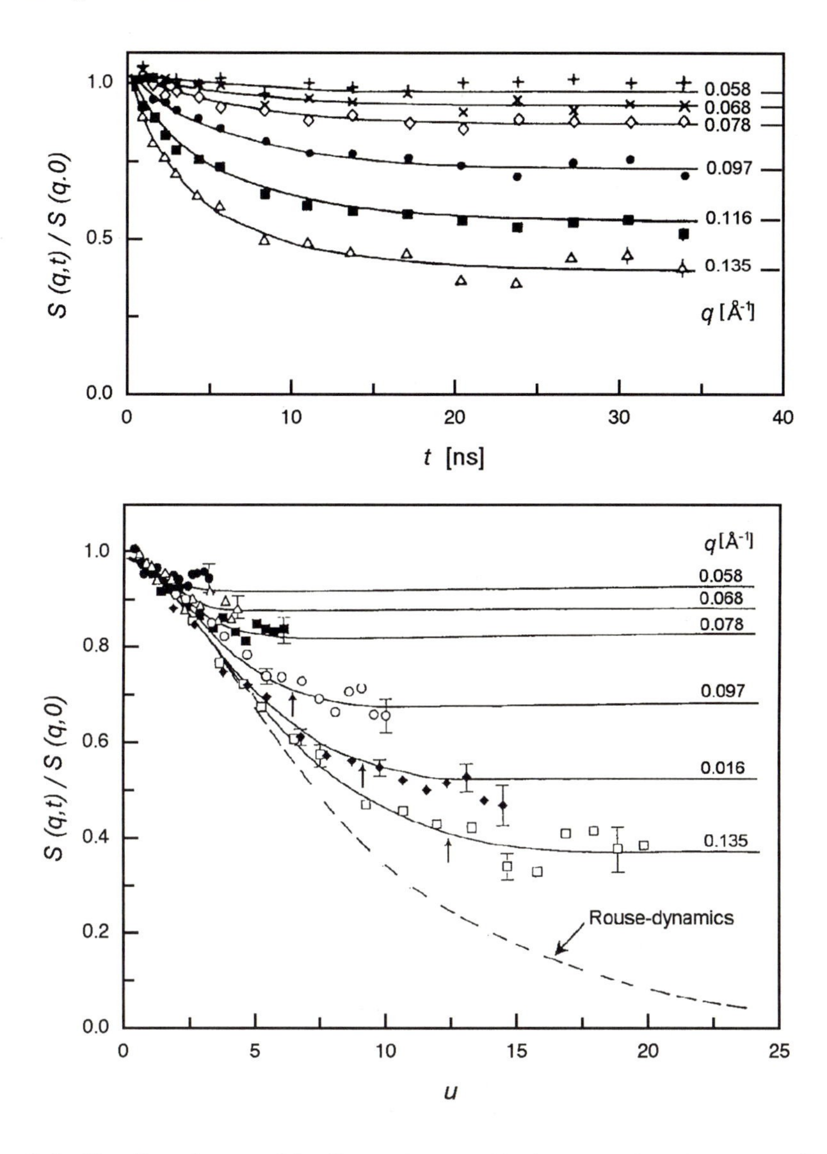

Fig. 6.8. Results of a quasielastic neutron scattering experiment on a melt of poly(ethylene-*co*-propylene) at 199 °C (10% protonated chains dissolved in a deuterated matrix; $M = 8.6 \cdot 10^4$): Intermediate scattering laws measured at the indicated scattering vectors (*top*); data representation using the dimensionless variable $u = q^2(12kTa_R^2 t/\zeta_R)^{1/2}$ (*bottom*). From Richter et al.[67]

$\tau^{-1}(\pi)$ is the maximum relaxation rate of the Rouse-chain, as given by Eq. (6.45). When using the variable u, measurements at different q's must coincide and contribute to one common curve. The lower part of Fig. 6.8 displays this reduced representation. The short term dynamics is indeed Rouse-like. All the curves coincide and agree with the prediction of the Rouse-model. Then however, with longer times, curves deviate from the Rouse scattering law. In particular, in the time range of the experiment the measured curves appear to level-off rather than tending to zero. As explained above, exactly this behavior is indicative of a confinement, at least a temporary one. Hence, we find a result which clearly indicates the change from Rouse-dynamics to a confined motion.

It is possible to derive from the limiting values for long periods the size of the confinement range. $S(q, t \to \infty)$ actually provides a Fourier analysis of its shape. To recognize it, just note that $S(q, t \to \infty)$ is given by

$$S(q, t \to \infty) = \int \exp i\boldsymbol{q}\boldsymbol{r} \cdot (g(\boldsymbol{r}, \infty) - \langle c_{\rm m} \rangle) \, {\rm d}^3\boldsymbol{r} \tag{6.107}$$

and furthermore, that the function

$$g(\boldsymbol{r}, \infty) - \langle c_{\rm m} \rangle$$

has non-vanishing values only for distances $\boldsymbol{r}$ within the confinement range. Consequently, a measurement of the halfwidth Δq of $S(q, \infty)$ enables us to make an estimate of the diameter d of the confinement range, by employing the reciprocity relation af Fourier transforms

$$d \simeq \frac{1}{\Delta q} \tag{6.108}$$

Results, as obtained for different temperatures, are shown in Fig. 6.9.

The neutron scattering experiments indicate that the Rouse-model remains valid for chain sequences below a critical length. This suggests we represent the time dependent shear modulus of an entangled polymer melt as being composed of three parts

$$G(t) = \Delta G_{\rm mic}(t) + c_{\rm p}kT \sum_{m=m^*}^{N_{\rm R}-1} \exp -2\frac{t}{\tau_{\rm m}} + c_{\rm p}kTm^*\Phi\left(\frac{t}{\tau_{\rm d}}\right) \tag{6.109}$$

As above in Eq. (6.86), $\Delta G_{\rm mic}(t)$ describes the short-term contributions determined by the chemical microstructure of the chain. The central part is given by the Rouse-model. It represents the dynamics for chain sequences which are shorter than the chain parts between entanglements and, on the other side, are still long enough to ensure that Gaussian properties hold. The Rouse-mode with the longest relaxation time which is not yet affected by the entanglements is that with the order m^* given by

$$\delta_{m^*}(N_{\rm R,c} - 1) \simeq \pi \tag{6.110}$$

where $N_{\rm R,c}$ is the contour length in Rouse-units corresponding to the critical molecular weight, $M_{\rm c}$. Using Eq. (6.38) we obtain

$$m^* \frac{\pi}{N_{\rm R} - 1} \cdot (N_{\rm R,c} - 1) \simeq \pi \tag{6.111}$$

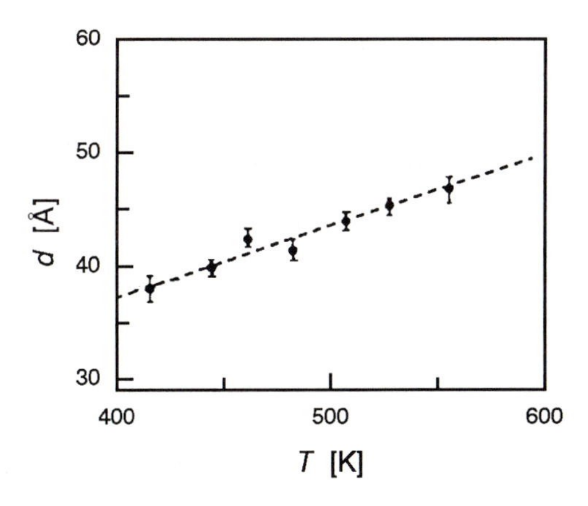

Fig. 6.9. Size d of the confinement range, as derived from the long term limits of the curves shown in Fig. 6.8 [67]

or

$$m^* \simeq \frac{N_{\mathrm{R}} - 1}{N_{\mathrm{R,c}} - 1} \tag{6.112}$$

The associated relaxation time is, according to Eq. (6.77)

$$\tau_{m^*} \approx \tau_{\mathrm{R}} \left(\frac{N_{\mathrm{R,c}} - 1}{N_{\mathrm{R}} - 1} \right)^2 \tag{6.113}$$

We notice that d increases, i.e. the motional constraints get weaker with increasing temperature.

Rouse-modes with $m < m^*$ do not exist. They become replaced by other relaxation processes, and the third term in Eq. (6.109) describes this contribution. The relaxation strength is identical to that of the replaced Rouse-modes, as this part remains unrelaxed after the decay of all modes with $m \geq m^*$. Writing the correlation function for the long-term part in the form $\Phi(t/\tau_{\mathrm{d}})$ implies the assumption that, similar to the Rouse-modes, also this part is controlled by a single characteristic time, the disentangling time τ_{d}, only. As introduced here, $\Phi(t/\tau_{\mathrm{d}})$ is a general normalized function

$$\Phi(0) = 1$$

τ_{d} may be identified, for example, with the integral width of Φ

$$\int\limits_{t=0}^{\infty} \Phi \mathrm{d}t = \tau_{\mathrm{d}}$$

From the experiments, it is known that $\tau_{\rm d}$ exhibits a power law dependence on M

$$\tau_{\rm d} \sim M^{\nu}, \qquad \text{with} \qquad \nu \approx 3.4$$

The main point in behavior, as expressed by Eq. (6.109), is the formation of a gap in the spectrum of relaxation times, arising between the first two contributions and the long-term part. This gap produces the plateau region, and the extension in time is determined by the ratio $\tau_{\rm d}/\tau_{m^*}$. For the dependence of this ratio on the molecular weight we can write

$$\frac{\tau_{\rm d}}{\tau_{m^*}} \simeq \frac{M^{3.4}}{M_{\rm c}^{3.4}} \tag{6.114}$$

considering that $\tau_{\rm d} \simeq \tau_{m^*}$ for $M \simeq M_{\rm c}$.

In order to see the effect of the entanglements on the viscosity, we apply Eq. (5.107)

$$\eta_0 = \int_0^\infty G(t)dt \tag{6.115}$$

to Eq. (6.109). Ignoring the short-term part $\Delta G_{\rm mic}(t)$, we find that the shear viscosity is given by the relaxation strengths and the mean relaxation times of the two contributions, originating from the Rouse-modes and the disentangling processes respectively

$$\eta_0 = c_{\rm p}kT\left[(N_{\rm R} - m^*)\tau_\alpha + m^*\tau_{\rm d}\right] \tag{6.116}$$

or

$$\eta_0 = G(0)\left(\frac{G(0) - G_{\rm pl}}{G(0)} \cdot \tau_\alpha + \frac{G_{\rm pl}}{G(0)}\tau_{\rm d}\right) \tag{6.117}$$

$G_{\rm pl}$ denotes the 'plateau-modulus' and τ_α is the mean relaxation time of the Rouse-mode part, agreeing with the mean relaxation time of a Rouse-system of chains with $N_{\rm R}$ equal to $N_{\rm R,c}$. As the first term on the right-hand side of Eq. (6.117) is constant, we obtain for the molecular weight dependence of the viscosity of entangled melts the expression

$$\eta_0 = \beta_1 + \beta_2\tau_{\rm d}(M) = \beta_1 + \beta_3 M^{\nu} \tag{6.118}$$

In the limit of high molecular weights, $M \gg M_{\rm c}$ we have

$$\eta_0 \sim \tau_{\rm d} \sim M^{\nu} \tag{6.119}$$

6.3.1 The Reptation Model

Starting from the tube concept and considering the motions of the confined chains, Doi and Edwards devised a theory which became well-known as the

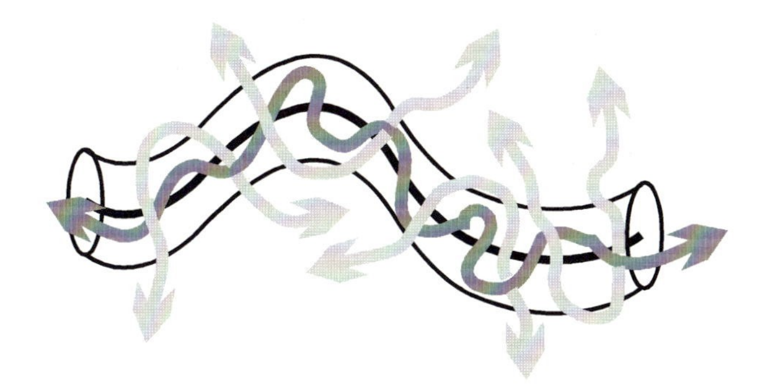

Fig. 6.10. Modelling the lateral constraints on the chain motion imposed by the entanglements by a 'tube'. The average over the rapid wriggling motion within the tube defines the 'primitive path' (*continuous dark line*)

'reptation model'. The model assumptions are indicated in Fig. 6.10. Under the constraints imposed by the tube, the chain motion may be thought as being set up of two different components. First, there is a rapid wriggling motion oriented along the tube cross-sections. It corresponds to the Rouse-part in the spectrum. Averaging over several cycles of this rapid motion gives the mean positions of the monomers along the tube, represented in the figure by the dark line in the tube center. This line, called the 'primitive path', describes the shortest path connecting the endgroups of the chain which is compatible with the topology of the entanglements as modelled by the tube. The second component of the motion is the time dependent evolution of this primitive path, and exactly this process leads to the disentangling of the chain. The Doi-Edwards theory focuses on the latter mechanism and thus reduces the problem of the motion of a chain under the constraints in a melt to the problem of the time dependence of the primitive path.

Both the actual chain as well as the primitive path represent random coils. Since the end-to-end distances are equal, we have

$$R_0^2 = N_{\mathrm{R}} a_{\mathrm{R}}^2 = l_{\mathrm{pr}} a_{\mathrm{pr}} \tag{6.120}$$

Here, we have introduced the contour length of the primitive path, l_{pr}, and an associated sequence length a_{pr}. a_{pr} characterizes the stiffness of the primitive path and is determined by the topology of the entanglement network.

The process of disentangling, as it is envisaged in the reptation model, is sketched in Fig. 6.11. The motion of the 'primitive chain', the name given to the dynamic object associated with the primitive path, is described as a diffusion along its contour, that is to say, a 'reptation'. The associated curvilinear diffusion coefficient can be derived from the Einstein relation, which holds generally, independent of the dimension or the topology. Denoting it $\hat{D}$, we have

$$\hat{D} = \frac{kT}{\zeta_{\mathrm{p}}} \tag{6.121}$$

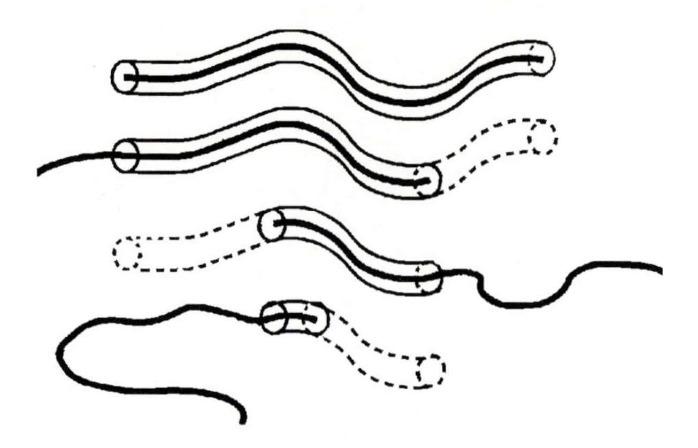

Fig. 6.11. Reptation model: Decomposition of the tube resulting from a reptative motion of the primitive chain. The parts which are left empty disappear

Here, ζ_p is the friction coefficient of the chain. As there are no entanglements within the tube, ζ_p equals the sum of the friction coefficients of all beads

$$\zeta_\mathrm{p} = N_\mathrm{R}\zeta_\mathrm{R} \tag{6.122}$$

Hence, $\hat{D}$ is given by

$$\hat{D} = \frac{kT}{N_\mathrm{R}\zeta_\mathrm{R}} \tag{6.123}$$

The diffusive motion leads to a continuous disentangling of the chain, as indicated in the figure. When parts of the chain have left the original tube, the empty part of the tube is filled with other chains and disappears. The result of the process is a continuous shortening of the initial tube and simultaneously, a continuous increase in the amount of reorientation of the chain. The process of disentangling is finished, when the initial tube has vanished.

The time needed to achieve a complete disentangling can be estimated. In order to get disentangled, chains have to diffuse over a distance l_pr, i.e. the original length of the primitive path, and this requires a time

$$\tau_\mathrm{d} \simeq \frac{l_\mathrm{pr}^2}{\hat{D}} \tag{6.124}$$

If we use Eqs. (6.120) and (6.123) we obtain the molecular weight dependence of the disentangling time

$$\tau_\mathrm{d} \sim \zeta_\mathrm{R} N_\mathrm{R}^3 \tag{6.125}$$

We see that the reptation model predicts a molecular weight dependence which comes near to the experimental result $\tau_\mathrm{d} \sim M^\nu$ with $\nu \approx 3.2 - 3.6$, although the agreement is not perfect. One might conclude, therefore, that the reptation picture provides basically an appropriate approach but should be further developed. There are several suggestions to how an improvement might be accomplished, in particular, by including processes associated with the motion of the neighbouring chains, like a 'constraint-release' or a tube deformation, but it appears that a final conclusion has not yet been reached.

Doi and Edwards analysed the described disentangling process of the primitive chain in more detail. As in the case of the Rouse-motion, the dynamics of the disentangling process can also be represented as a superposition of independent modes. Again, only one time constant, the disentangling time τ_{d}, is included, and it sets the time scale for the complete process. In the Doi-Edwards treatment, τ_{d} is identified with the longest relaxation time. Calculations result in an expression for the time dependent shear modulus in the terminal flow region. It has the form

$$G = G_{\mathrm{pl}} \Phi \left(\frac{t}{\tau_{\mathrm{d}}} \right) \tag{6.126}$$

with

$$\Phi = \frac{8}{\pi^2} \sum_{\text{odd } m} \frac{1}{m^2} \exp - \frac{m^2}{\tau_{\mathrm{d}}} t \tag{6.127}$$

The Einstein relation, already employed in Eq. (6.121) to write down the curvilinear diffusion coefficient in the tube, gives us, when used as normally, also the diffusion coefficient of a polymer in a melt, provided there are no entanglements. We call it D and have

$$D = \frac{kT}{\zeta_{\mathrm{p}}} = \frac{kT}{N_{\mathrm{R}} \zeta_{\mathrm{R}}} \sim \frac{1}{M} \tag{6.128}$$

How does this result change for an entangled melt? The reptation model gives an answer. One has only to realize that the disentangling process is associated with a shift of the center of mass of a polymer molecule over a distance in the order of l_{pr} along the primitive path and therefore leads to a mean-squared displacement

$$\langle \Delta r_{\mathrm{c}}^2 \rangle \simeq R_0^2 = l_{\mathrm{pr}} a_{\mathrm{pr}} \tag{6.129}$$

Since the diffusion coefficient in three dimensions is generally given by

$$D = \frac{\langle \Delta r_{\mathrm{c}}^2 \rangle}{6 \Delta t} \tag{6.130}$$

we obtain

$$D \sim \frac{l_{\mathrm{pr}}}{\tau_{\mathrm{d}}} \sim \frac{N_{\mathrm{R}}}{N_{\mathrm{R}}^3} \sim \frac{1}{M^2} \tag{6.131}$$

Hence, according to the reptation model, the transition from a non-entangled to an entangled polymer melt should be accompanied by a change in the exponent of the power law for the diffusion coefficient, $D \sim M^{\nu}$, from $\nu = -1$ to $\nu = -2$.

Equation (6.131) is in good agreement with experimental results, as is exemplified by the data presented in Fig. 6.12. Here, diffusion coefficients of

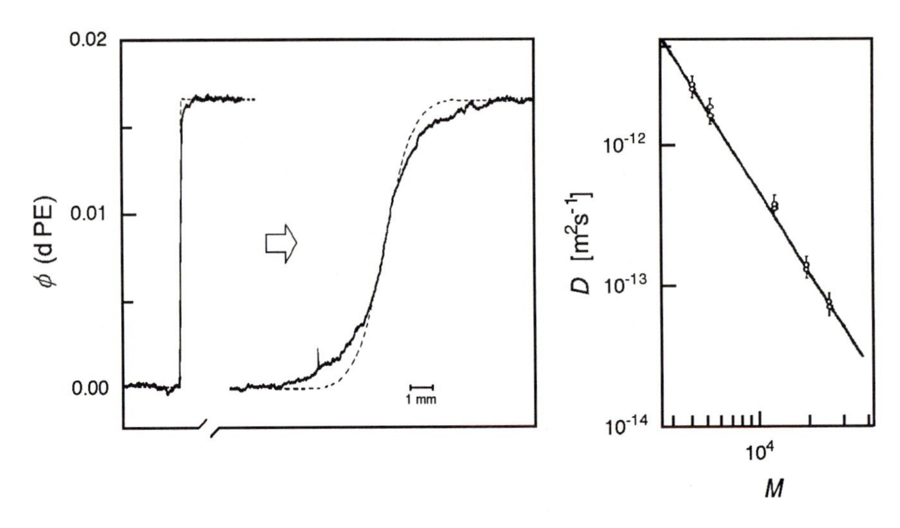

Fig. 6.12. Determination of diffusion coefficients of deuterated PE's in a PE matrix by infrared absorption measurements in a microscope. Concentration profiles $\phi(x)$ obtained in the separated state at the begin of a diffusion run and at a later stage of diffusive mixing (the *dashed lines* were calculated for monodisperse components; the deviations are due to polydispersity) (*left*). Diffusion coefficients at $T = 176\,°\mathrm{C}$, derived from measurements on a series of d-PE's of different molecular weight (*right*). The *continuous line* corresponds to a power law $D \sim M^2$. Work of Klein [68]

deuterated polyethylenes in a matrix of a standard polyethylene were measured in a microscope using infrared radiation. Due to the different vibrational properties of the two species, the infrared radiation discriminates between the two components. Concentration profiles then can be directly determined by an absorption measurement. Following the time dependent evolution of the concentration profile, starting with separated components in two films which are in lateral contact, enables a determination of the diffusion coefficient. The left-hand side of Fig. 6.12 depicts two typical concentration profiles, the curve on the left giving the initial profile at the boundary of the two films, the curve on the right referring to a later stage of development. Experiments were conducted for a series of deuterated polyethylenes with different molecular weights and the right-hand side shows the diffusion coefficients, as obtained for $T = 176\,°\mathrm{C}$. The slope of the broken line exactly agrees with the theoretical prediction.

There is another beautiful experiment which is even more convincing in its support of the reptation model. Fluorescence microscopy enables the motion of fluorescently stained single chains to be directly observed. The technique was applied to a concentrated solution of monodisperse DNA-molecules with ultrahigh molecular weight corresponding to a contour length in the 100 μm-range. On one end of the chain, an especially coated 1 μm-diameter

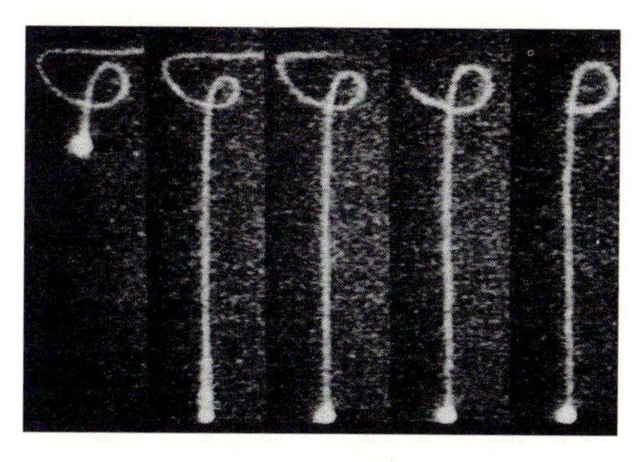

Fig. 6.13. Series of images of a fluorescently stained DNA chain embedded in a concentrated solution of unstained chains: Initial conformation (*left*); partial stretching by a rapid move of the bead at one end (*second from the left*); chain recoil by a reptative motion in the tube (*subsequent pictures to the right*). Reprinted with permission from T.Perkins, D.E.Smith and S.Chu. *Science*, 264:819, 1994. Copyright (1994) American Association for the Advancement of Science

polystyrene sphere was attached, and this could be manipulated and moved with the aid of 'optical tweezers', realized by a laser beam. In one experiment the test chain was rapidly pulled at this end, and Fig. 6.13 shows the subsequent relaxation. We see that the recoil follows exactly the path formed by the chain in its original conformation, i.e. the drawing back occurs within the tube. We thus have direct evidence that tube-like constraints exist in this system and that they are stable for a long time.

6.4 Hydrodynamic Interaction in Solutions

Having discussed the dynamics of polymer chains in the melt, in this section we will examine their motion in solution. As we shall see, here the diffusion coefficient cannot be described by either of the two equations for the melt but is given by

$$D \sim \frac{kT}{\eta_s R} \sim M^{-\nu} \tag{6.132}$$

with $\nu = 3/5$ for good solvents. The reason for the qualitatively different behavior is the dominant effect of 'hydrodynamic interactions'.

Let us first consider an internally rigid colloidal particle suspended in a liquid. Application of an external force will cause the particle to move. The motion necessarily affects the surrounding liquid. The enforced replacement of solvent molecules from the front to the back of the moving colloid results in a disturbance which extends over a larger region. For a constant velocity in a stationary state, a flow field is created which moves together with the particle. Figure 6.14 depicts this flow field. It has a major component in the direction of the velocity of the colloid. Since the flow field is inhomogeneous

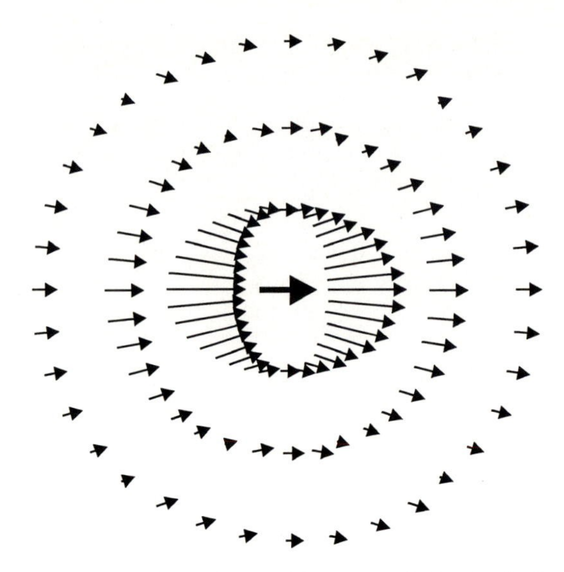

Fig. 6.14. Flow field created in a liquid, if a suspended spherical particle is moved by the action of an external force (Eq. (6.146)). The thicker arrow in the center represents the force

and includes velocity gradients, energy is dissipated. The force which is acting on the particle provides the power.

We can analyse this force in more detail. To begin with, remember the simple case of a shear flow, created if two parallel plates are moved against each other with constant velocity v_x. Here we find a linear velocity profile, like that included in Fig. 6.16. The power which has to be supplied is given by

$$\frac{\mathrm{d}\mathcal{W}}{\mathrm{d}t} = f \cdot v_x = \eta_{\mathrm{s}} \cdot \frac{\partial v_x}{\partial z} L_x L_y \cdot \frac{\partial v_x}{\partial z} L_z = \eta_{\mathrm{s}} V \left(\frac{\partial v_x}{\partial z} \right)^2 \tag{6.133}$$

η_{s} denotes the viscosity of the fluid, L_z is the thickness of the liquid layer, and the product $L_x L_y$ gives the area of the plates. The expression on the right-hand side is quite interesting. It states that the power dissipated per unit volume is proportional to both the square of the velocity gradient and η_{s}. Here, we obtained this result for the particular case of a constant velocity gradient. Indeed, it can be extended to the general case of arbitrary flow fields with varying velocity gradients. Theories derive and utilize the following general expression for the dissipated power

$$\frac{\mathrm{d}\mathcal{W}}{\mathrm{d}t} = \eta_{\mathrm{s}} \int \frac{\partial v_i}{\partial x_j} \cdot \frac{\partial v_i}{\partial x_j} \, \mathrm{d}^3 \boldsymbol{r} \tag{6.134}$$

(adopting the usual convention to perform a summation on repeated indices).

The flow field $\boldsymbol{v}(\boldsymbol{r})$ produced by the particle if it is dragged with a constant velocity $\boldsymbol{u}$ through the liquid may generally be described as

$$\boldsymbol{v}(\Delta\boldsymbol{r} := \boldsymbol{r} - \boldsymbol{r}_{\mathrm{c}}) = H(\Delta\boldsymbol{r}) \cdot \boldsymbol{u} \tag{6.135}$$

in component notation

$$v_i(\Delta\boldsymbol{r}) = H_{ik}(\Delta\boldsymbol{r})u_k \tag{6.136}$$

The velocity gradient tensor included in Eq. (6.134) follows by taking the derivative

$$\frac{\partial v_i}{\partial r_j} = \frac{\partial}{\partial \Delta r_j} H_{ik}(\Delta\boldsymbol{r})u_k := h_{ijk}(\Delta\boldsymbol{r})u_k \tag{6.137}$$

Here we introduced another tensor, of third rank, with components h_{ijk}. By using the distance $\Delta\boldsymbol{r}$ between $\boldsymbol{r}$ and the colloid at $\boldsymbol{r}_{\mathrm{c}}$ as variable, it is expressed that the flow field moves together with the particle, i.e. is stationary in a particle fixed coordinate system. Importantly, the formulated dependence is of the linear type. Hence, it holds for 'Newtonian liquids' which by definition obey linear laws. All low molar mass liquids behave in a Newtonian manner, at least within the limit of low velocities.

If Eq. (6.137) is inserted in Eq. (6.134), and one carries out an integration over the whole flow field, the dissipated power is obtained

$$\frac{\mathrm{d}\mathcal{W}}{\mathrm{d}t} = \eta_{\mathrm{s}} \int h_{ijk}h_{ijl}u_k u_l \, \mathrm{d}^3\boldsymbol{r} := \eta_{\mathrm{s}} u_k \beta_{kl} u_l \tag{6.138}$$

Here, we introduced yet another tensor, with components β_{kl}. It is specific for the particle and depends on its size and shape.

The power required when dragging the particle is supplied by the force which acts on it

$$\frac{\mathrm{d}\mathcal{W}}{\mathrm{d}t} = \mathbf{f} \cdot \boldsymbol{u} = f_l u_l \tag{6.139}$$

Comparing Eqs. (6.138) and (6.139) we obtain

$$f_l = \eta_{\mathrm{s}} u_k \beta_{kl} \tag{6.140}$$

As we can see, $\boldsymbol{u}$ and $\mathbf{f}$ are linearly related. Force and velocity must, in general, not be oriented parallel to each other, but for moving spheres, for symmetry reasons, they do, and we will discuss only this case. Here the tensor β_{ki} reduces to a scalar parameter, β, and we can write

$$\mathbf{f} = \eta_{\mathrm{s}} \beta \cdot \boldsymbol{u} \tag{6.141}$$

The proportionality constant relating the applied force to the resulting velocity is the 'friction coefficient' ζ, and it is given by

$$\zeta = \beta \eta_{\mathrm{s}} \tag{6.142}$$

In order to calculate the friction coefficient of a spherical colloid, the Navier-Stokes equations for liquids have to be solved, with the boundary condition that the liquid layer adjacent to the particle adheres to its surface, thus moving with the same velocity. The problem was solved long ago by Stokes. The result is the famous equation

$$\zeta = 6\pi R\eta_{\mathrm{s}} \tag{6.143}$$

It states that the friction coefficient of a sphere scales with the radius R. When dealing with particles which are isotropic on average but have otherwise an arbitrary structure, it is sometimes convenient to replace them by an 'equivalent sphere'. The replacement implies that we assign to the particle a 'hydrodynamic radius' R_{h}, defined by

$$R_{\mathrm{h}} := \frac{\zeta}{6\pi\eta_{\mathrm{s}}} \tag{6.144}$$

According to the definition, R_{h} is the radius of a spherical particle which possesses the same friction coefficient as the given colloid.

Let us return once again to the flow field depicted in Fig. 6.14. Actually this represents the result of a calculation first carried out by Oseen. Oseen derived the field produced by a point-like particle being dragged through the liquid by a force $\mathbf{f}$. As proved in his treatment, the flow field can be related to $\mathbf{f}$ by the following analytical expression

$$\boldsymbol{v} = \frac{1}{8\pi\eta_{\mathrm{s}}}\left(\frac{\mathbf{f}}{\Delta r} + \frac{(\mathbf{f}\cdot\Delta\boldsymbol{r})\Delta\boldsymbol{r}}{\Delta r^3}\right) \tag{6.145}$$

or, using a tensor notation, by

$$\begin{aligned} v_i &= \frac{1}{8\pi\eta_{\mathrm{s}}\Delta r}\left(\delta_{ij} + \frac{\Delta r_i \Delta r_j}{\Delta r_k \Delta r_k}\right) f_j \\ &:= H_{ij}^{\mathrm{Os}} f_j \end{aligned} \tag{6.146}$$

H_{ij}^{Os} is known as the 'Oseen hydrodynamic interaction tensor'. In a common situation there is a certain flow field, $\boldsymbol{v}_0(\boldsymbol{r})$, and a suspended particle at first moves with the liquid. Application of an external force, $\mathbf{f}$, on the particle produces a disturbance, $\boldsymbol{v}_{\mathrm{d}}$, which becomes superposed on $\boldsymbol{v}_0(\boldsymbol{r})$

$$\boldsymbol{v}(\boldsymbol{r}) = \boldsymbol{v}_0(\boldsymbol{r}) + \boldsymbol{v}_{\mathrm{d}}(\Delta\boldsymbol{r} = \boldsymbol{r} - \boldsymbol{r}_{\mathrm{c}}) \tag{6.147}$$

The Oseen tensor provides an accurate description of $\boldsymbol{v}_{\mathrm{d}}(\Delta\boldsymbol{r})$.

The central point of Oseen's result is the predicted slow decay of the disturbance with the reciprocal of the distance. It implies that the hydrodynamic interaction is of a long-range nature and therefore strong and effective. The consequences for a group of particles are drastic. Since the flow has a major component in the direction of the force, all particles in the group will support each other in the motion. As a result, a group of particles which interact by the flow fields, i.e. are coupled by the 'hydrodynamic interaction', moves faster through the liquid than isolated particles under the same total external force. What happens in the special case of a polymer? The answer is intuitively clear. The monomers of the coil, being concentrated to a volume with a diameter in the order of R_g, strongly interact via the hydrodynamic forces. Indeed, the coupling becomes so strong that all solvent molecules within the coil region are forced to move together with the chain segments, so that the coil can essentially be considered as 'impermeable'. Therefore, polymers in a solvent actually behave like hard spheres, and the hydrodynamic radius R_h attributed to them thus has a real significance.

It is possible to elaborate this intuitive picture in a thorough theory, and this was first done by Kirkwood and Risemann. Their result relates the hydrodynamic radius to the radius of gyration, R_g, of the chain, by

$$R_h \approx \frac{2}{3} R_g \tag{6.148}$$

or the friction coefficient of a polymer molecule, ζ_p, to R_g by

$$\zeta_p = 4\pi\eta_s R_g \tag{6.149}$$

According to the Einstein relation, the friction coefficient of a colloid and its diffusion coefficient are always directly related. If we take Eq. (6.149), we obtain for the diffusion coefficient of a dissolved polymer the expression

$$D = \frac{kT}{4\pi\eta_s R_g} \tag{6.150}$$

Equation (6.150) is quite useful, as it opens another route for a determination of the radius of gyration of chains, and thus, if the effective length per monomer is known, of the molecular weight. For this purpose, a method for measuring D is necessary and there is a standard technique, namely dynamic light scattering experiments. These generally result in a determination of the intermediate scattering law $S(q,t)$, like the quasi-elastic neutron scattering experiments discussed above but with different resolutions in space and time. Figure 6.15 provides an example. Data were obtained for a dilute solution of polystyrene in toluene.

The result may be explained as follows. First note that for the given spatial resolution, in the order of the wavelength of light, polymers appear point-like. A dilute solution of polymers therefore resembles an ensemble of independently moving point-like colloids. The intermediate scattering law then relates to the diffusive motion of the center of mass of a single polymer, as described by the self-correlation part of the time-dependent pair correlation function, $\hat{g}(\boldsymbol{r},t)$ (compare Eq. (A.32) in the Appendix). For a diffusing colloid, $\hat{g}(\boldsymbol{r},t)$ is given by

$$\hat{g} = \frac{1}{(4\pi Dt)^{3/2}} \exp -\frac{r^2}{4Dt} \tag{6.151}$$

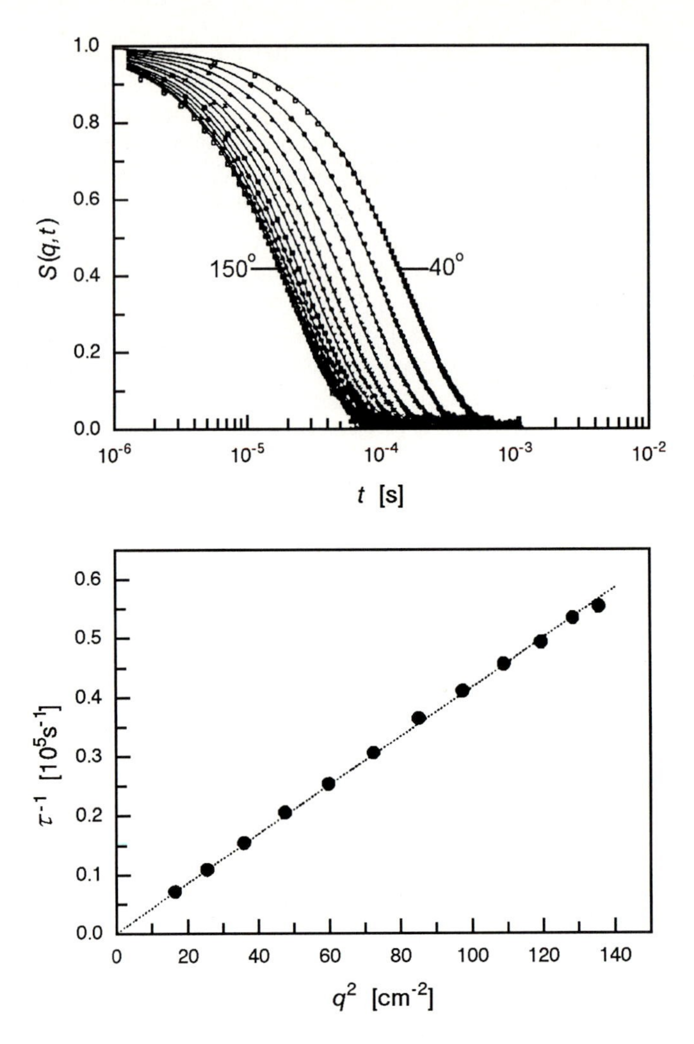

Fig. 6.15. Results of a dynamic light scattering experiment on a solution of PS ($M = 4.9 \cdot 10^4, c_{\mathrm{w}} = 0.053$ g cm^{-3}) in dioxane: Intermediate scattering law $S(q,t)$ for different values of the scattering angle in the range $2\vartheta_{\mathrm{B}} = 40°$ to $150°$ (in steps of $10°$)(*top*); derived q-dependence of the rate of decay τ_{-1}(*bottom*) [69]

as this expression is the solution of the basic differential equation for diffusion

$$\frac{\partial}{\partial t}\hat{g}(\boldsymbol{r},t) = D\Delta\hat{g}(\boldsymbol{r},t) \tag{6.152}$$

under the condition

$$\hat{g}(\boldsymbol{r},t=0) = \delta(\boldsymbol{r}) \tag{6.153}$$

The intermediate scattering law therefore has the form

$$\begin{aligned} S(q,t) &= \int \exp i\boldsymbol{q}\boldsymbol{r} \cdot \hat{g}(\boldsymbol{r},t)\, \mathrm{d}^3\boldsymbol{r} \\ &= \exp -2Dq^2 t \end{aligned} \tag{6.154}$$

Hence, a simple exponential decay is expected and indeed verified by the experiments, like those presented in Fig. 6.15. The signature of diffusive motions is the predicted increase of the relaxation rate with the square of the scattering vector, the factor $2D$ representing the proportionality constant. This dependence is also shown in the figure. The diffusion coefficient can be derived from the slope of the line in a plot of τ^{-1} versus q^2.

6.4.1 Intrinsic Viscosity

A second, even more simple and also less expensive technique for determining the hydrodynamic radius is viscosimetry. Generally, the suspension of colloids or the dissolution of polymers leads to an increase in the measured macroscopic viscosity. It is easy to see the principal reason for this effect, when considering the simple shear flow situation depicted in Fig. 6.16. Let us choose a spherical volume element of the liquid and follow the changes imposed by the flow field. As we see, it becomes translated, rotated and deformed. If this volume element is replaced now by a rigid spherical colloid, the translation and rotation are still possible, the deformation, however, is inhibited. The inability of the rigid sphere to realize the deformation which would be necessary to keep the linear profile of the flow field unperturbed, results in an additional local flow around the sphere. The local flow is associated with high velocity gradients and therefore causes an extra energy dissipation. This additional dissipation of energy becomes apparent in an increase in the force required to move the upper plate, when compared to the pure solvent at the same plate

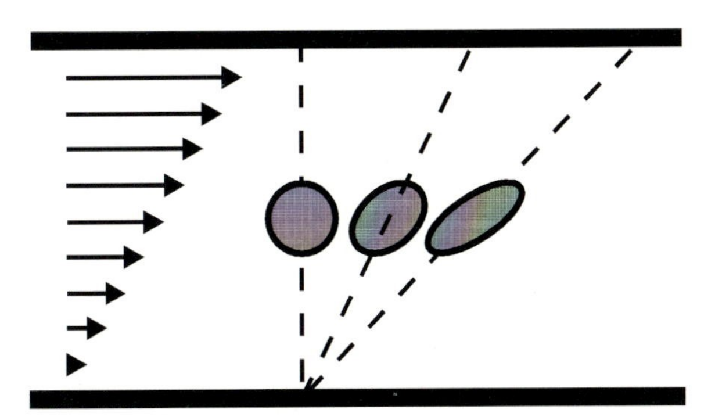

Fig. 6.16. Translation, rotation and deformation of a spherical volume element in a liquid under the conditions of simple shear flow

velocity. If this force is measured, one can derive an effective 'macroscopic viscosity' η, when utilizing the same equation as for a pure fluid

$$\mathbf{f} = \eta L_x L_y \frac{\partial v_x}{\partial z} \tag{6.155}$$

The quantity of interest is the excess of η over the viscosity of the pure solvent, η_s. It was Einstein, who demonstrated that this excess can be directly interpreted when dealing with a suspension of spheres. He derived the following power series

$$\eta = \eta_s(1 + \gamma\phi + \ldots) \tag{6.156}$$

Here, ϕ denotes the volume fraction occupied by spheres and γ is a numerical factor ($\gamma = 2.5$). The result is remarkably simple, as it implies that the extra viscosity is only dependent on the volume fraction of the spheres, irrespective of whether there are many small spheres or larger spheres in smaller numbers.

Equation (6.156) can be applied to a solution of polymers since the macromolecules, being hydrodynamically impermeable, behave like hard spheres with volumes as given by the hydrodynamic radius. Detailed theoretical treatments suggest a minor correction because it is found that the hydrodynamic radius to be used in viscosity measurements differs slightly from that applied in the representation of the diffusion coefficient. While the latter is given by Eq. (6.148), viscosity measurements have to be based on the relation

$$R_h \approx \frac{7}{8} R_g \tag{6.157}$$

ϕ in Eq. (6.156) then is given by

$$\phi = c_p \frac{4\pi}{3} R_h^3 \approx c_p \frac{7\pi}{6} R_g^3 \tag{6.158}$$

In works on polymer solutions a particular quantity has been introduced in order to specify the extra viscosity. It is called 'intrinsic viscosity', denoted $[\eta]$, and defined as

$$[\eta] := \lim_{c_w \to 0} \frac{\eta - \eta_s}{\eta_s} \cdot \frac{1}{c_w} \tag{6.159}$$

which includes a passage to the limit of vanishing polymer concentrations. Applying Eq. (6.156), the intrinsic viscosity follows as

$$[\eta] = \frac{\gamma\phi}{c_w} \tag{6.160}$$

Since

$$c_p \sim \frac{c_w}{M} \tag{6.161}$$

and generally

$$R_g \sim a N^\nu \sim a \left(\frac{M}{M_m} \right)^\nu \tag{6.162}$$

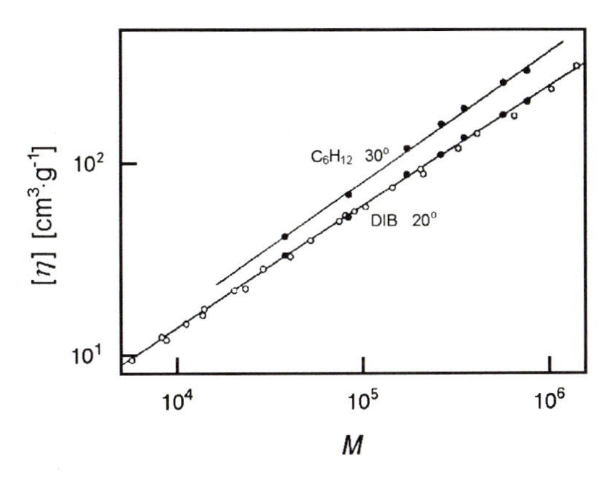

Fig. 6.17. Intrinsic viscosity-molecular weight relationship for PIB in diisobutylene (DIB) at 20 °C and in cyclohexane at 30 °C. Data collected in Flory's book [70]

with

$a = a_{\mathrm{F}} \quad , \quad \nu = 3/5$ for very long chains in good solvents

$a = a_0 \quad , \quad \nu = 1/2$ for chains in theta solvents

the intrinsic viscosity results as

$$[\eta] = \mathrm{const}\frac{a^3}{M}\left(\frac{M}{M_{\mathrm{m}}}\right)^{3\nu} \tag{6.163}$$

This is usually expressed by the formula

$$[\eta] = K \cdot M^{\mu} \tag{6.164}$$

whereby, as comparison shows, the prefactor is

$$K = \mathrm{const}\left(\frac{a}{M_{\mathrm{m}}^{\nu}}\right)^3 \tag{6.165}$$

and the exponent is given by

$$\mu = 3\nu - 1 \tag{6.166}$$

Equation (6.164) is known as the 'Mark-Houwink-Sakurada relation'. It generally holds very well, as is also exemplified by the data obtained for two different solutions of poly(isobutylene) presented in Fig. 6.17.

It is important to notice that with the aid of viscosity measurements one can discriminate between ideal and expanded chains. One finds an exponent $\mu = 0.5$ corresponding to $\nu = 0.5$ for ideal chains, and larger values for expanded chains, up to the limit of $\mu = 0.8$ corresponding to $\nu = 0.6$ expected for a perfectly expanded chain with ultrahigh molecular weight. In addition, the prefactor K can be used for determining the chain stiffness, as expressed by the effective length per monomer, a. Since the constant in Eq. (6.165) is known, a can be derived. In particular, when carrying out viscosity measurements in theta-solvents, the characteristic ratio $C_\infty = a^2/a_\mathrm{b}^2$ (Eq. (2.32)) may be determined.

Once K and μ have been established for a given polymer-solvent system, and often μ is found to lie somewhere in-between the two limiting cases of an ideal and a perfectly expanded chain, molecular weights of samples can be derived from measured intrinsic viscosities. Owing to these relationships viscosity measurements became a standard analytical tool which is useful and simple, and therefore frequently applied.

6.5 Further Reading

R.B. Bird, R.C. Armstrong, O. Hassager: *Dynamics of Polymeric Liquids, Vol.2 Kinetic Theory*, John Wiley & Sons, 1977

M. Doi, S.F. Edwards: *The Theory of Polymer Dynamics*, Clarendon Press, 1986

P.-G. de Gennes: *Scaling Concepts in Polymer Physics*, Cornell University Press, 1979

Chapter 7

Non-linear Mechanical Behavior

The observations, notions and model calculations discussed in the previous two chapters concerned the range of linear responses only. In fact, when working with polymers under realistic conditions one is frequently reaching the limits of these treatments since non-linear effects appear and have to be properly accounted for. We shall deal in this chapter with two cases of special importance:

- When using rubbers, non-linear mechanical properties are encountered in all situations of practical interest. Rubbers exhibit large deformations even under comparatively weak external forces and thus are mostly found outside the range of small strains.
- During processing of polymer melts, strain rates are usually so high that a characterization of the flow properties by a constant viscosity coefficient, as for a low molar mass Newtonian liquid, is no longer adequate.

Let us begin with a look at three typical examples. The first one, presented in Figure 7.1, depicts the load-extension curve observed for a piece of natural rubber. The extension is described by the ratio between the lengths in the stressed and the natural state, denoted λ, and the load is given in terms of the force per cross-section area in the undeformed state, denoted $\hat{\sigma}_{zz}$. The measurement goes up to an extension ratio of about 7, thus demonstrating the unique deformability of rubbers. Non-linearity is evident in the curve. It exhibits a sigmoidal shape; a linear law holds only for a negligible small range around the origin.

The other two examples deal with the flow properties of polymer melts as they are encountered under ordinary processing conditions. Figure 7.2 presents results of measurements of the viscosity of a melt of polyethylene, obtained at steady state for simple shear flows under variation of the shear rate. Data were collected for a series of different temperatures. At low strain rates one finds a constant value for the viscosity coefficient, i.e. a strict proportionality between shear stress and shear rate, but then a decrease sets in. This deviation from linearity is commonly found in polymers and begins even at moderate strain rates. As one observes a decrease in the viscosity, i.e. the ratio between shear stress and shear rate, the effect is usually called 'shear thinning'.

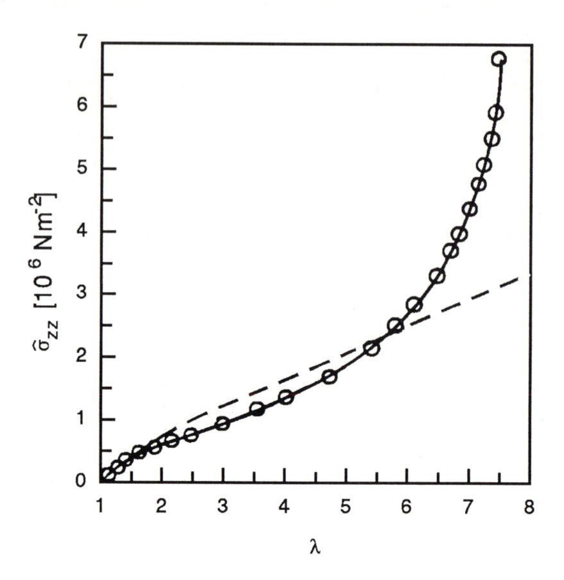

Fig. 7.1. Load-extension curve registered for a sample of natural rubber (λ: extension ratio; $\hat{\sigma}_{zz}$: tensile stress). Comparison with the function Eq. (7.31) derived for an ideal rubber (*broken line*). Results from Treloar [71]

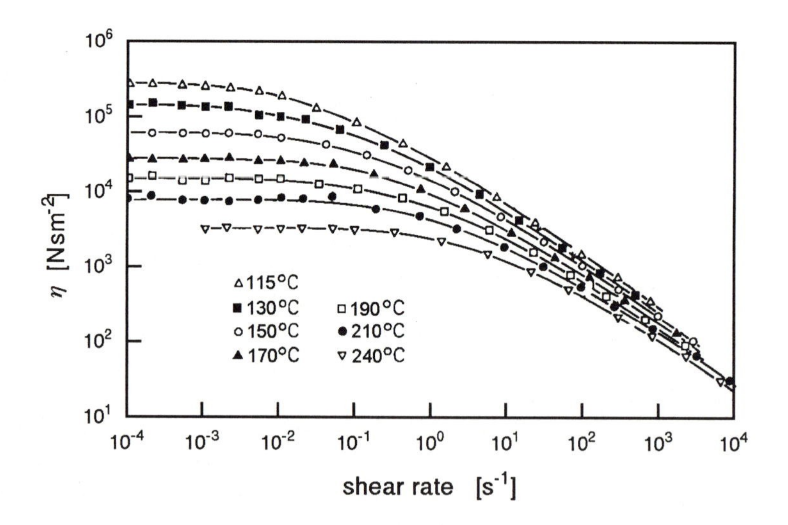

Fig. 7.2. Shear rate dependence of the viscosity η, observed for a melt of PE at various temperatures. Measurements by Meissner [72]

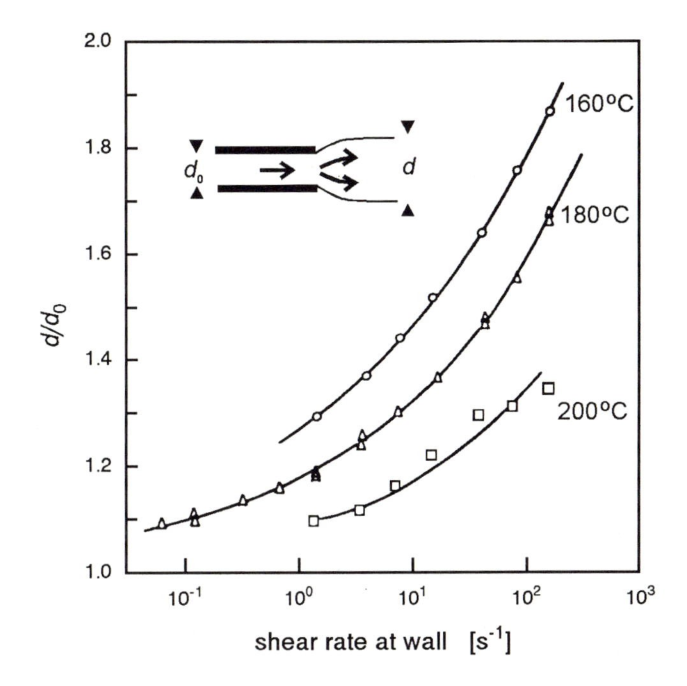

Fig. 7.3. Extrudate swell observed for a melt of PS for various shear rates and temperatures. Data from Burke and Weiss [73]

The third and last example concerns another characteristic property of flowing polymer melts. If during an extrusion process a polymer melt is forced to flow through a capillary, the extrudate swells at the exit. Figure 7.3 shows the swelling ratio of a polystyrene melt, i.e. the ratio d/d_0 between the diameter of the extrudate and the diameter of the capillary, as a function of the extrusion rate, which is specified here by the shear rate measured at the wall. Again the magnitude of the effect increases with the shear rate. The behavior indicates that shear flow in polymer melts is accompanied by the development of a 'normal stress' acting perpendicular to the shear stress and the direction of the flow. In the interior of the capillary, this pressure is provided by the wall. Swelling occurs at the moment when the end of the wall is reached. The building-up of normal stresses in simple shear flows is a non-linear phenomenon, being absent in Newtonian liquids. Indeed, if normal stresses are found, they are proportional to the square of the shear rate in lowest order.

Qualitative understanding of the origin of the normal stress phenomenon is not difficult. We learned in the previous chapters that a polymer melt resembles a transient network of entangled polymers. High shear rates, as they are encountered in a capillary, result in a deformation of the chains and, therefore, also of the formed network. As a consequence, a tension builds up along the lines of flow and draws the extrudate back when the confinement provided by the walls of the tube terminates. As the tension increases with the strain rate, the chain recoil at the exit increases as well and the extrudate rearranges to a larger diameter.

In this qualitative consideration, we have already addressed a principal point. If strains in a rubber or strain rates in flowing polymer melts are not really small, the chains are displaced significantly from their equilibrium conformations corresponding to isotropic random coils. If structures are altered by external stresses or during flow, then, as was already emphasized in the previous chapters, we are leaving the range of linear responses and non-linear effects appear. Hence, we have a clear-cut criterion and may judge from the presence of non-linearities whether or not structure changes have been induced.

The two other experiments also exemplify this general relationship. The non-linear sigmoidal stress-strain curve of a rubber under a tensile force shown in Fig. 7.1 is indicative of and caused by the significant changes in the conformational distribution of the chains building up the rubber network. Also the 'shear thinning' demonstrated by the data in Fig. 7.2 originates from induced conformational changes. We learned in the last chapter that, according to Eq. (6.119), the viscosity coefficient η is directly related to the time required by the chains for a disentangling. Apparently this time is reduced in flowing melts, as a consequence of a preorientation of chains which removes some of the entanglements and thus facilitates the disentangling.

That we are considering rubber behavior and the non-Newtonian flow properties of melts here in one common chapter is for good reasons. The elastic part in the response of a polymer melt, which is a main cause for its peculiar flow properties, is due to the existing network of entanglements. Understanding the origin of rubber elasticity thus also provides us with a basis for understanding the elastic forces in polymer melts. Some of the microscopic models dealing with the non-Newtonian flow properties of polymer melts actually describe the melt as a network of entanglements which are continuously created and destroyed.

Setting up microscopic models helps in the basic understanding, however, perfect agreement with experimental data is not usually reached. In this situation, a different approach becomes important. One can aim for constructing empirical 'constitutive equations' with the objective of having expressions in hand which can describe the associated stresses for any kind of strain or flow fields. Here, rubbers and melts can again be treated by the same formalism which enables finite deformations to be described without the limitations imposed in the linear treatments. This is achieved by the introduction of the 'Cauchy strain tensor', as will be explained in a forthcoming section. Using this tool we will first formulate 'Finger's constitutive equation', which is valid for all isotropic elastic bodies, and apply it to rubbers, and then present Lodge's rheological equation of state of rubber-like liquids, to be used for a description of the flow properties of polymer melts.

We begin with a discussion of the physical basis of rubber elasticity.

7.1 Rubber Elasticity

Rubbers, as they find use in a large variety of products such as elastic foams, films, bands, or tires, possess unique mechanical properties which sets them apart from all other materials. With regard to the microscopic state of order and the local molecular dynamics, rubbers are in the liquid state, the flow properties, however, differ qualitatively from those of fluids. Rubbers are built up of cross-linked polymers, and the cross-links completely suppress any irreversible flow. A piece of rubber can actually be envisaged as one huge polymer molecule of macroscopic size which possesses a high internal flexibility. The cross-links stabilize the shape of a sample, but this shape can be greatly changed by the application of stress. There are opposing forces which balance the external stress, however, compared to the internal forces in crystalline or glassy solids, these are very weak, resulting in elastic moduli which are four orders of magnitude smaller. As we shall see and discuss in more detail in this chapter, these restoring forces are mainly of an entropic nature.

Let us begin by considering a prismatic piece of rubber with orthogonal edges $L_z, L_x = L_y$, and inquire about the force produced by an extension ΔL_z in z-direction. We will use as independent variable the extension ratio λ, defined as

$$\lambda := \frac{L_z + \Delta L_z}{L_z} \tag{7.1}$$

Thermodynamics provides the general tool to be applied in order to obtain the force and we first have to select the appropriate thermodynamic potential. In dealing with a rubber, we choose the Helmholtz free energy, considering that one is usually interested in the force under isothermal conditions and, moreover, that rubbers are essentially incompressible. For a piece of rubber which is to be extended in one direction, the Helmholtz free energy $\mathcal{F}$ is a function of λ. Knowing $\mathcal{F}$, the force f required to obtain an extension λ follows by taking the derivative

$$f = \left(\frac{\partial \mathcal{F}}{\partial \Delta L_z}\right)_{V,T} = \frac{1}{L_z}\left(\frac{\partial \mathcal{F}}{\partial \lambda}\right)_{V,T} \tag{7.2}$$

In general this force is set up of two different contributions

$$f = \frac{1}{L_z}\left(\frac{\partial \mathcal{E}}{\partial \lambda}\right)_{V,T} - \frac{T}{L_z}\left(\frac{\partial \mathcal{S}}{\partial \lambda}\right)_{V,T} \tag{7.3}$$

$$= f_{\mathcal{E}} + f_{\mathcal{S}} \tag{7.4}$$

$f_{\mathcal{E}}$ denotes the energetic contribution, $f_{\mathcal{S}}$ gives the entropic part of the force. Under the condition that the volume V is not only constant on varying λ, but also on changing T, a temperature dependent measurement of the force f enables us to make a separate determination of the two parts. The entropic part follows as

$$f_{\mathcal{S}} = -\frac{T}{L_z}\left(\frac{\partial \mathcal{S}}{\partial \lambda}\right)_{V,T} = \frac{T}{L_z}\left(\frac{\partial^2 \mathcal{F}}{\partial T \partial \lambda}\right)_V = T\left(\frac{\partial f}{\partial T}\right)_{V,\lambda} \tag{7.5}$$

and the energetic part as

$$f_{\mathcal{E}} = f - T\left(\frac{\partial f}{\partial T}\right)_{V,\lambda} = -Tf\frac{\partial}{\partial T}\ln\left(\frac{f}{T}\right)\bigg|_{V,\lambda} \tag{7.6}$$

Many kinds of loading experiments have been performed on a large variety of different rubbers and they all agree on the general conclusion that the retractive forces are mainly of entropic origin. This is at least true for the interesting range of moderate to large deformations; at the very begin of the deformation process, i.e. at low extensions, the energetic part may give more significant contributions. The observations suggest to introduce the concept of an 'ideal rubber', as a body with the property

$$f_{\mathcal{E}} = 0$$

hence

$$f = f_{\mathcal{S}}$$

As we shall see, 'ideal rubbers' possess properties which reproduce at least qualitatively the main features in the behavior of real rubbers and thus can provide an approximate first description. Equation (7.6) implies that for an ideal rubber, force and temperature are linearly related

$$f \sim T$$

Indeed, a strict proportionality to the absolute temperature is the characteristic signature for all forces of entropic origin. Just remember the ideal gas which is probably the better known system, where we find $p \sim T$.

Temperature measurements can be used in order to detect and specify deviations from the ideal case for a given sample. If a piece of rubber is thermally isolated and stretched, its temperature increases. The relation between an elongational step $\mathrm{d}\lambda$ and the induced change $\mathrm{d}T$ in temperature follows from

$$\mathrm{d}\mathcal{E} = c_v \mathrm{d}T = fL_z \mathrm{d}\lambda \tag{7.7}$$

as

$$\mathrm{d}T = \frac{fL_z}{c_v}\mathrm{d}\lambda \tag{7.8}$$

where c_v denotes the heat capacity. For a non-vanishing energetic contribution to the force, the measured temperature increase becomes smaller. Observed reductions are typically in the order of $10 - 20\%$.

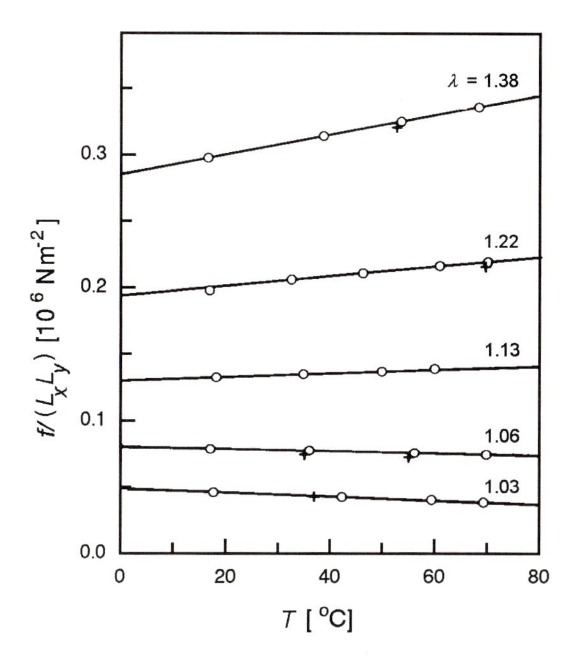

Fig. 7.4. Observation of a thermoelastic inversion point for natural rubber: The temperature dependence of the force at constant extension exhibits a reversal in slope. Measurements by Anthony et al.[74]

One might think at first that the energetic part of the force, $f_{\mathcal{E}}$, could be derived also from a temperature dependent measurement of the force on the basis of Eq. (7.6). In fact, direct application of this equation is experimentally difficult since the volume does not remain constant under the normally given constant pressure conditions. Indeed, thermal expansion is observed and this is also the reason for the occurrence of a 'thermoelastic inversion point'. It shows up in temperature dependent measurements on rubbers which are kept at a fixed length. Figure 7.4 shows a series of measurements which were performed at different values of λ. For high extensions, we find the signature of ideal rubbers, i.e. an increase $f \sim T$. For low extensions, on the other hand, thermal expansion overcompensates this effect, and then even leads to a decrease of the force.

Evidently, the entropic forces must originate from the polymer chains which set up the network. It is easy to see the physical basis of the retraction mechanism: When chains are extended on stretching the network, the number of available rotational isomeric states and thus the entropy decreases, and this produces a retractive force. Statistical thermodynamics can describe this effect in more detail, employing model considerations.

7.1.1 The Fixed Junction Model of Ideal Rubbers

Let us first consider a single polymer chain with a mean-squared end-to-end distance R_0^2 and inquire about the force which arises if the two ends become separated. As sketched in Fig. 7.5, we assume that one end-group of the chain is located at the origin of a cartesian coordinate system and the second end-group can be moved along the y-axis. In order to keep this second end fixed at a distance y, a non-vanishing force has to be applied. It can be calculated using the same equation as for the macroscopic piece of rubber and follows as

$$f = \frac{\partial f_\mathrm{p}}{\partial y} = -T\frac{\partial s_\mathrm{p}}{\partial y} \tag{7.9}$$

Here, f_p and s_p denote the free energy and the entropy of the chain which, being set up of freely jointed segments, has an invariant internal energy ε_p, being associated with the kinetic energy of motion only. Polymer chains in rubbers possess Gaussian properties as in a melt and this can be used for a calculation of the entropy s_p. Applying a basic law in statistical thermodynamics, the entropy can be derived from the partition function, $Z_\mathrm{p}(y)$, by

$$s_\mathrm{p} = k \ln Z_\mathrm{p}(y) \tag{7.10}$$

In our case, the partition function is determined by the number of conformations available for the chain if the second end is at a distance y. $Z_\mathrm{p}(y)$ can be directly written down. Remember that for a mobile free end the probability distribution in space is given by the Gaussian function, Eq. (2.10)

$$p(x, y, z) = \left(\frac{3}{2\pi R_0^2}\right)^{3/2} \exp -\frac{3(x^2 + y^2 + z^2)}{2R_0^2}$$

with

$$R_0^2 = \langle x^2 + y^2 + z^2 \rangle$$

This probability distribution just reflects the number of conformations available for the chain if the end-to-end distance vector is kept fixed at (x, y, z).

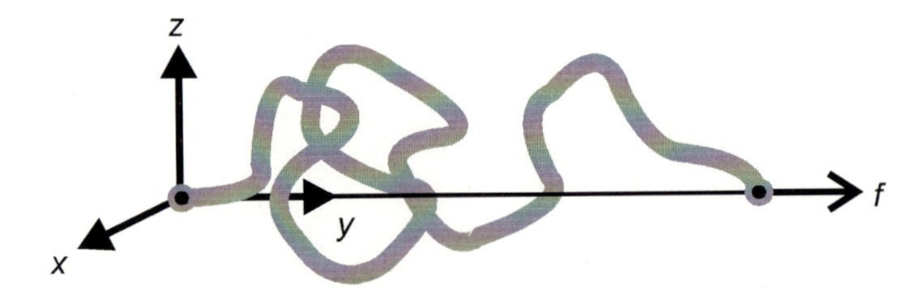

Fig. 7.5. External force f, required to keep the ends of a polymer chain at a fixed distance

Therefore, the quantity of interest, $Z_{\mathrm{p}}(y)$, is given by

$$Z_{\mathrm{p}}(y) \sim p(0,y,0) \sim \exp -\frac{3y^2}{2R_0^2} \tag{7.11}$$

The entropy follows from Eq. (7.10) as

$$s_{\mathrm{p}}(y) = s_{\mathrm{p}}(0) + k \cdot \left(-\frac{3y^2}{2R_0^2}\right) \tag{7.12}$$

which yields, for the force, the simple expression

$$f = \frac{3kT}{R_0^2} \cdot y = b \cdot y \tag{7.13}$$

This in an interesting result. It tells us that this entropic force increases linearly with the distance between the two end-groups, just as if they were connected by a mechanical spring. The stiffness constant, denoted b, increases with temperature and decreases with an increasing size of the chain.

The force disappears only for $y = 0$. This, however, does not imply that in a thermal equilibrium the end-groups are coupled together . A harmonic oscillator with a stiffness constant b which is in contact with a heat bath at a temperature T shows a non-vanishing mean-squared displacement $\langle y^2 \rangle$. Straightforward application of Boltzmann-statistics yields

$$\langle y^2 \rangle = \frac{kT}{b} \tag{7.14}$$

Here we obtain correspondingly

$$\langle y^2 \rangle = \frac{R_0^2}{3} \tag{7.15}$$

which is the result expected for a Gaussian chain.

A rubber represents an ensemble of polymer chains, each one running between two cross-links. The Helmholtz free energy of a sample is related to the distribution function of the conformational states. Since the chains are not free, the ends being fixed at cross-links, a change in the external shape of a sample necessarily modifies the distribution function. We have to consider this change and calculate the resulting change in free energy.

To keep the treatment simple we include a number of assumptions:

- All the chains which compose the network have the same degree of polymerization N, and thus identical values for the stiffness constant $b \sim R_0^{-2} \sim N^{-1}$.
- The conformational distribution in the undeformed state agrees with that of an uncross-linked melt, i.e. is given by an isotropic Gaussian function.

- The cross-link points are fixed within the sample. Any sample deformation becomes directly transferred to the cross-links and changes their positions in an affine manner.

The model so set up is the simplest one to treat and known as the 'fixed junction model' with reference to the last point. Let us refer again to a prismatic piece of rubber, with edges L_x, L_y and L_z, and consider a homogeneous orthogonal deformation, which changes the lengths of the edges as follows

$$L_x \rightarrow \lambda_1 L_x \quad (7.16)$$

$$L_y \rightarrow \lambda_2 L_y \quad (7.17)$$

$$L_z \rightarrow \lambda_3 L_z \quad (7.18)$$

The extension ratios λ_1, λ_2 and λ_3 also determine the shifts in the locations of all junction points and therefore the changes of the end-to-end distance vectors of all chains. More specifically, the end-to-end distance vector of chain i in the unstrained state

$$\boldsymbol{r}'_i = \begin{pmatrix} x'_i \\ y'_i \\ z'_i \end{pmatrix}$$

transforms affinely into a vector $\boldsymbol{r}_i$ in the deformed sample, given by

$$\boldsymbol{r}_i = \begin{pmatrix} x_i = \lambda_1 x'_i \\ y_i = \lambda_2 y'_i \\ z_i = \lambda_3 z'_i \end{pmatrix}$$

We wish to calculate the change in entropy resulting from the deformation and first write down the entropy in the stress free state. For a single chain with ends fixed at a distance $\boldsymbol{r}'_i$, the entropy is given by Eq. (7.12), replacing y^2 by r'^2_i

$$s'_i = s_{\mathrm{p}}(0,0,0) - \frac{3}{2R_0^2}(x'^2_i + y'^2_i + z'^2_i) \quad (7.19)$$

To obtain the total entropy of the sample, we include in the summation all chains of a Gaussian ensemble. For a volume V and a chain density c_{p} we write

$$\begin{aligned} \mathcal{S}' = \sum_i s'_i &= Vc_{\mathrm{p}} \int s_{\mathrm{p}}(x',y',z')p(x',y',z')\mathrm{d}x'\mathrm{d}y'\mathrm{d}z' \\ &= Vc_{\mathrm{p}} \int \left(s_{\mathrm{p}}(0,0,0) - k\frac{3}{2R_0^2}(x'^2 + y'^2 + z'^2) \right) \\ &\quad \cdot \left(\frac{3}{2R_0^2\pi} \right) \exp -\frac{3}{2R_0^2}(x'^2 + y'^2 + z'^2)\mathrm{d}x'\mathrm{d}y'\mathrm{d}z' \\ &= Vc_{\mathrm{p}} \left(s_{\mathrm{p}}(0,0,0) - \frac{3k}{2R_0^2}(\langle x'^2 \rangle + \langle y'^2 \rangle + \langle z'^2 \rangle) \right) \\ &= Vc_{\mathrm{p}} \left(s_{\mathrm{p}}(0,0,0) - \frac{3k}{2} \right) \end{aligned} \quad (7.20)$$

Next, we derive the entropy in the deformed state, $\mathcal{S}$. It follows as

$$\begin{aligned}\mathcal{S} &= Vc_{\rm p}\int\left(s_{\rm p}(0,0,0) - k\frac{3}{2R_0^2}(\lambda_1^2x'^2+\lambda_2^2y'^2+\lambda_3^2z'^2)\right)\\ &\quad\cdot\left(\frac{3}{2R_0^2\pi}\right)^{3/2}\exp-\frac{3}{2R_0^2}(x'^2+y'^2+z'^2)\mathrm{d}x'\mathrm{d}y'\mathrm{d}z'\\ &= Vc_{\rm p}\left(s_{\rm p}(0,0,0)-\frac{3k}{2R_0^2}(\lambda_1^2\langle x'^2\rangle+\lambda_2^2\langle y'^2\rangle+\lambda_3^2\langle z'^2\rangle)\right)\\ &= Vc_{\rm p}\left(s_{\rm p}(0,0,0)-\frac{1}{2}k(\lambda_1^2+\lambda_2^2+\lambda_3^2)\right)\end{aligned}\tag{7.21}$$

In the calculation of the integral we refer to the distribution function in the natural state, and the deformation is accounted for by introducing into the expression for $\mathcal{S}$ the modified single chain entropies. The quantity of interest is the change in entropy, following as

$$\Delta\mathcal{S} = \mathcal{S}-\mathcal{S}' = -Vc_{\rm p}\frac{k}{2}(\lambda_1^2+\lambda_2^2+\lambda_3^2-3)\tag{7.22}$$

Consider now the case of an uniaxial extension in z-direction, with an extension ratio λ

$$\lambda_3 = \lambda\tag{7.23}$$

The induced changes in the two lateral directions are equal to each other

$$\lambda_1 = \lambda_2\tag{7.24}$$

Assuming incompressibility, as expressed by

$$\lambda_1\lambda_2\lambda_3 = 1\tag{7.25}$$

we have

$$\lambda_1^2\lambda = 1\tag{7.26}$$

Using these relations, we obtain for the change in entropy resulting from an uniaxial extension λ the expression

$$\Delta\mathcal{S} = -\frac{Vc_{\rm p}k}{2}\left(\frac{2}{\lambda}+\lambda^2-3\right)\tag{7.27}$$

The force follows as above, by taking the derivative of the associated free energy

$$f = -\frac{T}{L_z}\frac{\partial\mathcal{S}}{\partial\lambda} = -\frac{T}{L_z}\frac{\partial\Delta\mathcal{S}}{\partial\lambda} = \frac{Vc_{\rm p}kT}{L_z}\left(-\frac{1}{\lambda^2}+\lambda\right)\tag{7.28}$$

Replacement of the force by the stress yields a form which is independent of the sample dimensions. The stress in the deformed state is obtained, when referring to the actual cross-section

$$\sigma_{zz} := \frac{f}{(L_xL_y/\lambda)}\tag{7.29}$$

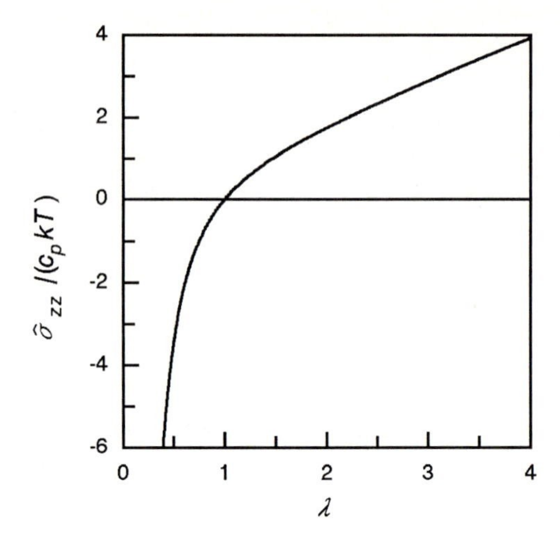

Fig. 7.6. (Nominal) stress-strain curve of an uniaxially deformed ideal rubber according to Eq. (7.31)

and this leads to

$$\sigma_{zz} = c_{\mathrm{p}} kT \left(\lambda^2 - \frac{1}{\lambda} \right) \tag{7.30}$$

Sometimes, for direct comparisons with measured load-extension curves, the 'nominal stress', sometimes called also 'engineering stress', is used. Here the force is referred to the cross-section $L_x L_y$ in the natural state, giving

$$\hat{\sigma}_{zz} := \frac{f}{L_x L_y} = c_{\mathrm{p}} kT \left(\lambda - \frac{1}{\lambda^2} \right) \tag{7.31}$$

As we can see, the model considerations result in a short analytical expression for the stress-strain relation of an uniaxially deformed ideal rubber. Equations (7.30) and (7.31) are valid for all λ's, including both extensions and compressions for $\lambda > 1$ and $\lambda < 1$ respectively. Important to note, apart from T the stress depends on the chain density c_{p} only. The fact that there is no explicit effect of the degree of polymerization teaches us that the initial assumption of a uniform value of N for the chains between cross-links is not a necessary prerequisite for the result and can therefore be omitted. As long as c_{p} is constant, the same elastic force emerges for any distribution of N's; each chain gives in the isotropic average an equal contribution to the stress which is independent of the length. Figure 7.6 shows the dependence as given by Eq. (7.31). In the limit of large extensions the retractive force increases linearly with λ, while in the compression range the behavior is mostly non-linear. Clearly the pressure must diverge for $\lambda \to 0$. The elastic modulus is related to small strains only, being determined by the slope at the origin, $\lambda = 1$, as

$$E := \frac{d\sigma_{zz}}{de_{zz}}(e_{zz} = 0) = \frac{d\sigma_{zz}}{d\lambda}(\lambda = 1) = \frac{d\hat{\sigma}_{zz}}{d\lambda}(\lambda = 1) = 3c_p kT \qquad (7.32)$$

We will compare these model predictions with load-deformation curves of real rubbers, but beforehand a remark is necessary. It is clear that there is no basis for a comparison of absolute values. These depend in the model on the density c_p of chains between chemical cross-links, however, in a rubber not only the chemical junction points, but in addition the entanglements act as cross-links. In a polymer melt, in the region of the rubbery plateau, the latter are indeed the only ones present. Chemical cross-linking stabilizes the topological cross-links, as these become trapped if a permanent network is created. The relative weight of the contribution of the entanglements to the force can be estimated, in the simplest way by comparing the retractive forces before and after the chemical cross-linking of a melt. As it turns out, for the usual low cross-link densities, chain entangling is even dominant and accounts for the larger part of the total force. Entanglements differ from the chemical cross-links in that they are not fixed and may slip along chains, which diminishes their efficiency. Actually, a certain reduction in efficiency on similar grounds is also to be anticipated for the chemical cross-links. We have assumed a perfect fixing in space, however, in reality, the junction points fluctuate performing a restricted Brownian motion around their mean positions (which is, in fact, accounted for in an improved model, known as the 'phantom network'). Hence, in conclusion, one normally has to relinquish a prediction of absolute stress values and reduce comparisons with models to the shape of measured curves. Values for c_p, as derived from a data fitting, have to be addressed as an 'effective density' of active elements. c_p may be formally translated into a 'mean molecular weight of chains between chemical or physical junction points', denoted M_{eff}, through

$$c_p = \frac{\rho}{M_{\text{eff}}} \cdot \mathrm{N}_L \qquad (7.33)$$

but M_{eff} is just another empirical parameter. It depends in an unpredictable manner on the cross-link density, the functionality of the cross-linking groups (they can couple together the ends of three, four or even more chains), the chain stiffness etc.

We should now look at two sets of data, depicted previously in Fig. 7.1 and now in Fig. 7.7. Both figures include adjusted model curves. As we can see, the data in Fig. 7.7, which cover the range of moderate extensions together with compression states, are remarkably well represented. On the other hand, pronounced differences show up in Fig. 7.1 which goes to large extensions. The reason for the discrepancy becomes clear when recalling a basic property of the Gaussian model chain. As pointed out previously, in Sect. 2.3.1, the Gaussian description implies an infinite contour length of chains, however, in reality their lengths are finite. Each chain possesses a maximum extension as given

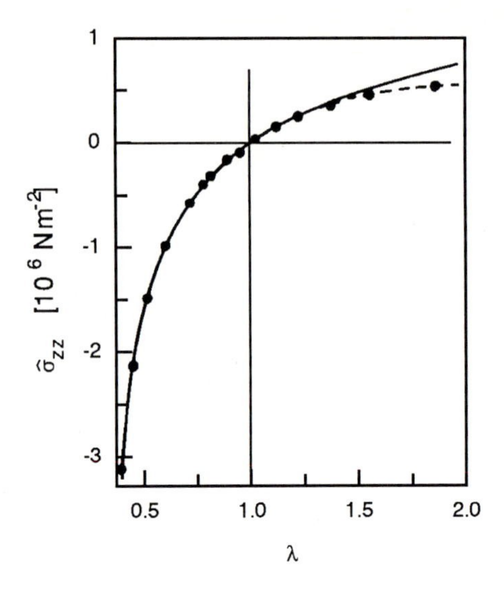

Fig. 7.7. Extension ($\lambda > 1$)-compression ($\lambda < 1$)-curve of natural rubber, compared to the theoretical function Eq. (7.31) derived for an ideal rubber (*continuous line*). Data from Treloar [71]

by its length in the straight helical conformation. Obviously this limitation must produce at larger extensions deviations from the Gaussian behavior, and this is the cause of the observed increase in stress. In fact, it is possible to account for this effect and so to improve the agreement. One has to replace the Gaussian distribution in Eq. (7.11) by real distribution functions, either by the exact distribution function given in Sect. 2.3.1 or a suitable approximation and then modify the entropy calculations correspondingly. This corrected curves usually provide a satisfactory representation of measured curves even in the range of larger extensions.

Last but not least, models which assume ideal behavior do not account for the always present energetic effects. Therefore, with these models alone, a comprehensive representation of thermoelastic data encompassing a larger temperature range, like the ones displayed in Fig. 7.4, cannot be achieved.

Regarding these complications, one might feel that attempts to account for all the remaining deviations by proper modifications of statistical thermodynamic models of ideal rubbers do not look very promising. In this situation a certain drawback might be considered as appropriate. Rather than searching for a detailed microscopic understanding one can look for phenomenological treatments enabling us to make an empirical representation of the mechanical properties of a given rubbery material. One might think at first that this is a simpler goal to achieve. However, even this is not a trivial task, in particular if one's aim is a comprehensive description of all modes of deformation employing only a small number of material parameters. This is exactly the desire of engineers who need reliable and directly applicable material functions for working with rubbers. Continuum mechanics provides us with tools for generally dealing with large deformations, not only those found in rubbers, but also those encountered in flowing polymer melts. In the next section, we will discuss briefly the basis of these concepts.

7.1.2 The Cauchy Strain Tensor

Elasticity theory for solids is formulated with the assumption that strains remain small. One then finds the linear relations between stress and strain as they are described by Hook's law. For rubbers, deformations are generally large and the linear theory then becomes invalid. We have to ask how strain can be characterized in this general case and how it can be related to the applied stress.

First, let us recall the definition of stress. In deformed rubbers the state of stress can be described in the same manner as in the case of small deformations of solids, by giving the stress tensor $\boldsymbol{\sigma} = (\sigma_{ij})$. The meaning of the components σ_{ij} is indicated in Fig. 7.8. Imagine we pick out a cubic volume element in the deformed body at the position $\boldsymbol{r}$, with edges parallel to the laboratory fixed cartesian coordinate system. Then $\boldsymbol{\sigma}$ specifies the forces which the material outside the cube exerts through the surfaces on the particles in the interior. More specifically, component σ_{ij} denotes the force per unit area acting along the j-axis on the face oriented perpendicular to the i-axis. Knowledge of $\boldsymbol{\sigma}$ also enables us to calculate the force $\mathbf{f}$ acting on any plane with normal vector $\boldsymbol{n}$. $\mathbf{f}$ follows as the product

$$\mathbf{f} = \boldsymbol{\sigma} \cdot \boldsymbol{n} \tag{7.34}$$

The stress tensor is symmetric

$$\sigma_{ij} = \sigma_{ji} \tag{7.35}$$

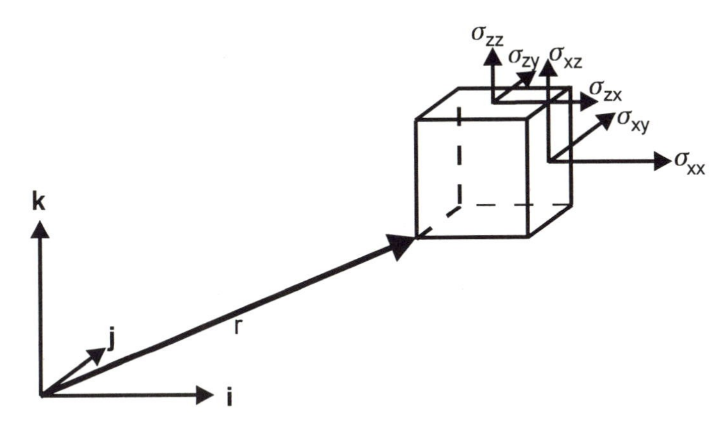

Fig. 7.8. Description of the stress around a material point at r in a deformed body: Stress tensor components σ_{ij} specifying the forces acting on the faces of a cubic volume element

since static equilibrium is only attained for a vanishing torque on the cube. Sometimes, in special cases, the stress tensor is calculated for the forces being referred to the cross-section in the undeformed state. For rubbers under a large strain, this leads to altered values and these, as mentioned earlier, are called 'nominal' or 'engineering' stresses.

A deformation of a body of rubber displaces all material points in the sample. There is a one-to-one correspondence between the locations of a material point in the deformed and the unstrained body. We again refer to the laboratory-fixed coordinate system and describe the relation between the locations in the deformed state

$$\boldsymbol{r} = \begin{pmatrix} x \\ y \\ z \end{pmatrix}$$

and the natural state

$$\boldsymbol{r}' = \begin{pmatrix} x' \\ y' \\ z' \end{pmatrix}$$

by the displacement function

$$\boldsymbol{r}'(\boldsymbol{r})$$

Choosing this function rather than the reverse relation $\boldsymbol{r}(\boldsymbol{r}')$ implies that one refers to the deformed body in the description of the strain, in agreement with the description of the stress.

Not all functions $\boldsymbol{r}'(\boldsymbol{r})$ result in stress: If a sample is translated or rotated as a rigid body, no stress arises. Clearly, the general prerequisite for stress is a change in internal distances. For a check one can pick out a material point at $\boldsymbol{r}$, select a neighboring material point at $\boldsymbol{r} + \mathrm{d}\boldsymbol{r}$ and inquire about the change in their distance on removing the load. If $\mathrm{d}\boldsymbol{r}$ transforms into $\mathrm{d}\boldsymbol{r}'$, the difference of the squares of the lengths is given by

$$\mathrm{d}\boldsymbol{r}' \cdot \mathrm{d}\boldsymbol{r}' - \mathrm{d}\boldsymbol{r} \cdot \mathrm{d}\boldsymbol{r} \tag{7.36}$$

As the relation between $\mathrm{d}\boldsymbol{r}'$ and $\mathrm{d}\boldsymbol{r}$ is determined by the vector gradient of the mapping function $\boldsymbol{r}'(\boldsymbol{r})$

$$\mathrm{d}\boldsymbol{r}' = \frac{\partial \boldsymbol{r}'}{\partial \boldsymbol{r}} \cdot \mathrm{d}\boldsymbol{r} \tag{7.37}$$

with

$$\left(\frac{\partial \boldsymbol{r}'}{\partial \boldsymbol{r}}\right)_{ij} := \frac{\partial r'_i}{\partial r_j} \tag{7.38}$$

we can write for the squared lengths difference

$$\begin{aligned} \mathrm{d}r'_i \mathrm{d}r'_i - \mathrm{d}r_i \mathrm{d}r_i &= \frac{\partial r'_i}{\partial r_j} \mathrm{d}r_j \frac{\partial r'_i}{\partial r_k} \mathrm{d}r_k - \mathrm{d}r_i \mathrm{d}r_i \\ &= \mathrm{d}r_j (C_{jk} - \delta_{jk}) \mathrm{d}r_k \end{aligned} \tag{7.39}$$

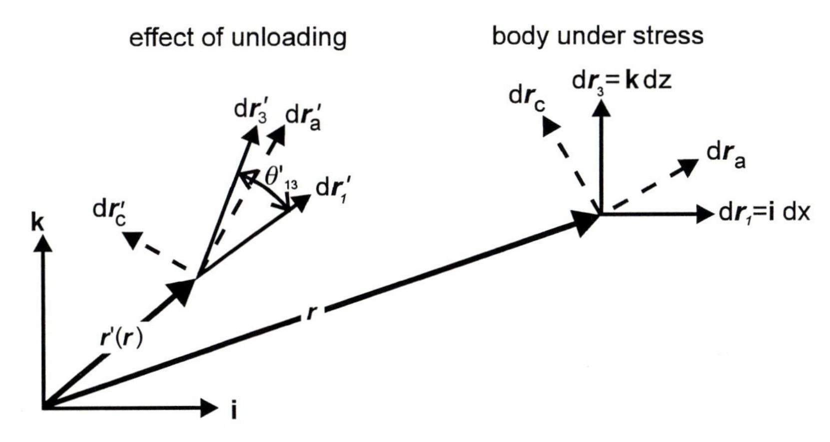

Fig. 7.9. Notions used in the definition of the Cauchy strain tensor: The material point at r in the deformed body with its neighborhood shifts on unloading to the position r'. The orthogonal infinitesimal distance vectors $\mathrm{d}r_1$ and $\mathrm{d}r_3$ in the deformed state transform into the oblique pair of distance vectors $\mathrm{d}r'_1$ and $\mathrm{d}r'_3$. Orthogonality is preserved for the distance vectors $\mathrm{d}r_a$, $\mathrm{d}r_c$ oriented along the principal axes

Here, we have introduced a tensor $\mathbf{C}$, defined as

$$C_{jk} := \frac{\partial r'_i}{\partial r_j}\frac{\partial r'_i}{\partial r_k} \tag{7.40}$$

$\mathbf{C}$ is called the 'Cauchy strain tensor'. Any deformation, it may be large or small, homogeneous or non-uniform, can be characterized by a space dependent function $\mathbf{C}(\boldsymbol{r})$. Knowing $\mathbf{C}(\boldsymbol{r})$, the local extensions in all directions can be obtained for each material point of a deformed body just by a calculation of the two-fold product Eq. (7.39). Recall that we have introduced the Cauchy strain tensor by a two step procedure starting from the complete description of the deformation by the displacement function $\boldsymbol{r}'(\boldsymbol{r})$, going at first to the deformation gradient function $\partial\boldsymbol{r}'/\partial\boldsymbol{r}$ and then proceeding to $\mathbf{C}(\boldsymbol{r})$. The first step eliminated rigid translations and the second step the rotations, overall as well as local ones. $\mathbf{C}(\boldsymbol{r})$ therefore does indeed include only those parts of the body motion which are true deformations giving rise to the development of stress. For homogeneous states of strain, $\mathbf{C}(\boldsymbol{r})$ reduces to a unique tensor and in the following, we will deal with this simpler case only.

It is easy to see the significance of the various components of the Cauchy strain tensor, and we refer here to Fig. 7.9. First consider the infinitesimal differential vector $\mathrm{d}\boldsymbol{r}_1$ parallel to the x-axis. On unloading it transforms into $\mathrm{d}\boldsymbol{r}'_1$, which has the squared length

$$\mathrm{d}\boldsymbol{r}'_1 \cdot \mathrm{d}\boldsymbol{r}'_1 = \left(\left(\frac{\partial x'}{\partial x}\right)^2 + \left(\frac{\partial y'}{\partial x}\right)^2 + \left(\frac{\partial z'}{\partial x}\right)^2\right)\mathrm{d}x^2 \tag{7.41}$$

The right-hand side equals the product $C_{11}\mathrm{d}x^2$. Hence, generalizing this result, the diagonal elements C_{ii} describe the length changes of the edges of a cubic volume element, oriented parallel to the axes of the laboratory fixed cartesian coordinate system, the changes being expressed in terms of the squared length ratios.

As the second, consider the angle θ'_{13} enclosed by $\mathrm{d}\boldsymbol{r}'_1$ and $\mathrm{d}\boldsymbol{r}'_3$, the transforms of $\mathrm{d}\boldsymbol{r}_1 = i\mathrm{d}x$, $\mathrm{d}\boldsymbol{r}_3 = j\mathrm{d}y$. Its cosine follows from

$$|\,\mathrm{d}\boldsymbol{r}'_1\,|\,|\,\mathrm{d}\boldsymbol{r}'_3\,|\cos\theta'_{13} = \mathrm{d}\boldsymbol{r}'_1 \cdot \mathrm{d}\boldsymbol{r}'_3 \tag{7.42}$$

The right-hand side agrees with the product $C_{13}\mathrm{d}x\mathrm{d}y$. Hence, we obtain

$$\cos\theta'_{13} = \frac{C_{13}}{(C_{11}\cdot C_{33})^{1/2}} \tag{7.43}$$

and learn from this example that the non-diagonal elements of the Cauchy strain tensor describe the changes of the three rectangular angles of the cube.

The Cauchy strain tensor is symmetric because of its definition. Therefore, it can be converted into a diagonal form by an appropriate rotation of the coordinate system. We deal with these conditions as indicated in Fig. 7.9, by attaching, to each selected material point, a triple of orthogonal infinitesimal distance vectors $\mathrm{d}\boldsymbol{r}_a, \mathrm{d}\boldsymbol{r}_b$ and $\mathrm{d}\boldsymbol{r}_c$, oriented parallel to the 'principal axes', which diagonalize the Cauchy tensor. The zero non-diagonal elements of the Cauchy tensor in the principal axes system imply that all angles θ'_{ij}, enclosed by the transforms of the infinitesimal distance vectors, designated as $\mathrm{d}\boldsymbol{r}'_a, \mathrm{d}\boldsymbol{r}'_b$ and $\mathrm{d}\boldsymbol{r}'_c$, are 90°. Hence, on unloading the orthogonal triple $\mathrm{d}\boldsymbol{r}_a, \mathrm{d}\boldsymbol{r}_b, \mathrm{d}\boldsymbol{r}_c$ may become rotated and the three vectors may change their lengths, orthogonality, however, is preserved. Previously we have introduced factors λ_i as denoting the extension ratios in orthogonal deformations, with reference to the natural unstrained state (Eqs. (7.16) - (7.18)). To agree with this first definition, we now have to represent the Cauchy strain tensor in diagonal form as

$$C_{ij} = \delta_{ij}\lambda_i^{-2} \tag{7.44}$$

using the reciprocal values λ_i^{-1}.

Instead of employing the Cauchy strain tensor one can also utilize the 'Eulerian strain tensor' defined as

$$2E_{ij} := \delta_{ij} - C_{ij} \tag{7.45}$$

In the limit of infinitesimal deformations the Eulerian strain tensor becomes identical with the strain tensor used in the linear elasticity theory, as can be easily shown. The components e_{ij} of the linear strain tensor are defined by

$$e_{ij} := \frac{1}{2}\left(\frac{\partial s_i}{\partial r'_j} + \frac{\partial s_j}{\partial r'_i}\right) \tag{7.46}$$

whereby $\boldsymbol{s}$ denotes the shift of the material point $\boldsymbol{r}'$ in the undeformed state induced by the load:

$$\boldsymbol{s} = \boldsymbol{r} - \boldsymbol{r}' \tag{7.47}$$

For small deformations it makes no difference, whether the undeformed or the deformed body is chosen as the reference frame. We can therefore also write

$$e_{ij} = \frac{1}{2}\left(\frac{\partial s_i}{\partial r_j} + \frac{\partial s_j}{\partial r_i}\right) \tag{7.48}$$

E_{ij} may be reformulated as

$$\begin{aligned} E_{ij} &= \frac{1}{2}\left(\delta_{ij} - \frac{\partial(r_k - s_k)}{\partial r_i}\frac{\partial(r_k - s_k)}{\partial r_j}\right) \\ &= \frac{1}{2}\left(\delta_{ij} - \left(\delta_{ki} - \frac{\partial s_k}{\partial r_i}\right)\left(\delta_{kj} - \frac{\partial s_k}{\partial r_j}\right)\right) \\ &= \frac{1}{2}\left(\frac{\partial s_i}{\partial r_j} + \frac{\partial s_j}{\partial r_i} - \frac{\partial s_k}{\partial r_i}\cdot\frac{\partial s_k}{\partial r_j}\right) \end{aligned} \tag{7.49}$$

Hence, in linear approximation, neglecting in the limit of infinitesimal deformations the second order term, we do indeed obtain

$$E_{ij} = e_{ij} \tag{7.50}$$

7.1.3 Finger's Constitutive Equation

We are now in the position to formulate empirical stress-strain relations in the form of 'constitutive equations'. The Cauchy strain tensor focuses on those parts in the deformation of a body which produce stress. We can therefore assume that the stress, $\boldsymbol{\sigma}$, is a function of the Cauchy strain tensor, $\mathbf{C}$, only. As this basic assumption alone is certainly too general for a promising approach, one may wonder if there are any restrictions with regard to the possible form of the functional dependence. In fact, theoretical analysis led to the conclusion that such restrictions do exist. The problem aroused much interest as early as at the end of the last century and a main result was due to Finger. He succeeded in deriving a constitutive equation which is generally valid for all elastic isotropic bodies. The term 'elastic' here is used in the general sense, implying that the body reacts in a well-defined way to an externally applied force and returns completely to its natural state upon unloading. Finger arrived at the conclusion that the stress-strain relation for this kind of general-elastic or, as they are sometimes called, 'hyperelastic' bodies depends on one scalar function only, namely the relation between the free energy density and the state of strain as characterized by the 'strain invariants'.

As any second rank tensor, the Cauchy strain tensor possesses three invariants. These are expressions in terms of the tensor components C_{ij} which remain invariant under all rotations of the coordinate system. The three invariants of the Cauchy strain tensor are given by the following expressions

$$I_C = C_{11} + C_{22} + C_{33} \tag{7.51}$$

$$II_C = C_{11}C_{22} + C_{22}C_{33} + C_{33}C_{11} - C_{12}C_{21} - C_{13}C_{31} - C_{23}C_{32} \tag{7.52}$$

$$III_C = \text{Det } \mathbf{C} \tag{7.53}$$

For the diagonal form Eq. (7.44), they reduce to

$$I_C = \lambda_1^{-2} + \lambda_2^{-2} + \lambda_3^{-2} \tag{7.54}$$

$$II_C = \lambda_1^{-2}\lambda_2^{-2} + \lambda_2^{-2}\lambda_3^{-2} + \lambda_3^{-2}\lambda_1^{-2} \tag{7.55}$$

$$III_C = \lambda_1^{-2} \cdot \lambda_2^{-2} \cdot \lambda_3^{-2} \tag{7.56}$$

There are different choices for the invariants since any combination gives new invariant expressions but the most common is the one cited here. As the free energy density depends on the local strain only and, being a scalar quantity, must be invariant under all rotations of the coordinate system, one can readily assume for the free energy density a functional dependence

$$f(I_C, II_C, III_C) \tag{7.57}$$

We now have the ingredients to formulate Finger's constitutive equation. It relates the Cauchy strain tensor to the stress tensor in the form

$$\boldsymbol{\sigma} = c_{-1}\mathbf{C}^{-1} + c_0\mathbf{1} + c_1\mathbf{C} \tag{7.58}$$

$\mathbf{C}^{-1}$ denotes the inverse of the Cauchy tensor, $\mathbf{1}$ is the unit tensor. c_{-1}, c_0, c_1 describe functions of the three invariants, and these are directly related to the free energy density. The relations are

$$c_{-1} = 2 \cdot III_C^{3/2} \frac{\partial f}{\partial II_C} \tag{7.59}$$

$$c_0 = -2 \cdot III_C^{1/2} \left(II_C \frac{\partial f}{\partial II_C} + III_C \frac{\partial f}{\partial III_C} \right) \tag{7.60}$$

$$c_1 = -2 \cdot III_C^{1/2} \frac{\partial f}{\partial I_C} \tag{7.61}$$

Finger derived this equation on the basis of general arguments. As we see, it provides us with a powerful tool: Once one succeeds in determining the strain dependence of the free energy of a body, the stresses produced in all kinds of deformations can be predicted.

There is an alternative form of Finger's equation which gives us a choice and is, indeed, to be preferred when dealing with rubbers. We introduce the 'Finger strain tensor' $\mathbf{B}$, being defined as the reciprocal of the Cauchy strain tensor

$$\mathbf{B} := \mathbf{C}^{-1} \tag{7.62}$$

The substitution of $\mathbf{C}$ by $\mathbf{B}$ implies, as the main point, that one is now choosing the invariants of $\mathbf{B}$ as independent variables, rather than those associated with $\mathbf{C}$. The second form of Finger's constitutive equation is

$$\boldsymbol{\sigma} = b_1\mathbf{B} + b_0\mathbf{1} + b_{-1}\mathbf{B}^{-1} \tag{7.63}$$

with

$$b_1 = \frac{2}{III_B^{1/2}}\frac{\partial f}{\partial I_B} \tag{7.64}$$

$$b_0 = \frac{2}{III_B^{1/2}}\left(II_B\frac{\partial f}{\partial II_B} + III_B\frac{\partial f}{\partial III_B}\right) \tag{7.65}$$

$$b_{-1} = -2\cdot III_B^{1/2}\frac{\partial f}{\partial II_B} \tag{7.66}$$

As the directions of the principal axes of $\mathbf{B}$ and $\mathbf{C}$ must coincide, the diagonal form of the Finger strain tensor is simply

$$B_{ij} = \lambda_i^2\delta_{ij} \tag{7.67}$$

The invariants then follow as

$$I_B = \lambda_1^2 + \lambda_2^2 + \lambda_3^2 \tag{7.68}$$

$$II_B = \lambda_1^2\lambda_2^2 + \lambda_2^2\lambda_3^2 + \lambda_3^2\lambda_1^2 \tag{7.69}$$

$$III_B = \text{Det}\,\mathbf{B} \tag{7.70}$$

Whether to use the first or the second form of Finger's constitutive equation is just a matter of convenience, depending on the expression obtained for the free energy density in terms of the one or the other set of invariants. For the system under discussion, a body of rubbery material, the choice is clear: The free energy density of an ideal rubber is most simply expressed when using the invariants of the Finger strain tensor. Equation (7.22), giving the result of the statistical mechanical treatment of the fixed junction model, exactly corresponds to

$$f = \frac{G}{2}(I_B - 3) \tag{7.71}$$

As we can see, the second and third invariants, II_B and III_B, are not included. The third invariant relates generally to the relative volume change. For incompressible bodies, this is equal to unity

$$III_B = \lambda_1^2\lambda_2^2\lambda_3^2 = 1 \tag{7.72}$$

and, therefore, it can be omitted in our further treatments. When writing down Eq. (7.71) we introduced a parameter G. As it turns out, G corresponds to the shear modulus.

Knowing f, we can formulate the constitutive equation of an ideal rubber. Since only b_1 gives a contribution, we simply obtain

$$\boldsymbol{\sigma} = G \cdot \mathbf{B} \tag{7.73}$$

This result, however, is not yet complete. For an incompressible solid like a rubber, superposition of a hydrostatic pressure onto the other applied external forces leaves the shape of the sample and thus the state of strain unchanged. We can account for this arbitrariness by including the undetermined hydrostatic pressure, denoted p, as a further component in the equation and rewrite it as

$$\boldsymbol{\sigma} = G \cdot \mathbf{B} - p\mathbf{1} \tag{7.74}$$

The constitutive equation thus yields stresses which are indeterminate to the extent of an arbitrary hydrostatic pressure.

The same modification is necessary when dealing in general with incompressible hyperelastic bodies. Introducing the additional term in Finger's constitutive equation, Eq. (7.63), and regarding the absence of III_B we obtain

$$\boldsymbol{\sigma} = 2\frac{\partial f}{\partial I_B} \cdot \mathbf{B} - 2\frac{\partial f}{\partial II_B} \cdot \mathbf{B}^{-1} - p\mathbf{1} \tag{7.75}$$

This is now a constitutive equation which looks most appropriate for our purposes. Being generally valid for all incompressible elastic bodies, it can be applied in particular for the treatment of real rubbers.

So far, we have been concerned with uniaxial deformations of rubbers only. Utilizing Eq. (7.75) we can analyze in a straightforward manner any kind of deformation and, in particular, the important case of a simple shear. To begin with, we check once again for the results obtained for the known case of uniaxial strain or compression of an ideal rubber. Here, the transformation equations are

$$z' = \lambda^{-1} z \tag{7.76}$$

$$x' = \lambda^{1/2} x \tag{7.77}$$

$$y' = \lambda^{1/2} y \tag{7.78}$$

The Cauchy strain tensor follows from Eq. (7.40) as

$$\mathbf{C} : \begin{pmatrix} \lambda & 0 & 0 \\ 0 & \lambda & 0 \\ 0 & 0 & \lambda^{-2} \end{pmatrix} \tag{7.79}$$

and the Finger tensor therefore as

$$\mathbf{B} : \begin{pmatrix} \lambda^{-1} & 0 & 0 \\ 0 & \lambda^{-1} & 0 \\ 0 & 0 & \lambda^{2} \end{pmatrix} \tag{7.80}$$

In this case inversion of the Cauchy tensor, which is necessary when deriving the Finger tensor, is trivial. For non-orthogonal deformations this is more complicated, and here one can make use of a direct expression for the components of **B** which reads

$$B_{ij} = \frac{\partial r_i}{\partial r'_k} \frac{\partial r_j}{\partial r'_k} \tag{7.81}$$

The proof of Eq. (7.81) is straightforward. We have

$$B_{ij} C_{jk} = \frac{\partial r_i}{\partial r'_l} \cdot \frac{\partial r_j}{\partial r'_l} \cdot \frac{\partial r'_m}{\partial r_j} \cdot \frac{\partial r'_m}{\partial r_k} \tag{7.82}$$

Since

$$\frac{\partial r'_m}{\partial r_j} \cdot \frac{\partial r_j}{\partial r'_l} = \frac{\partial r'_m}{\partial r'_l} = \delta_{ml} \tag{7.83}$$

we indeed obtain

$$B_{ij} C_{jk} = \frac{\partial r_i}{\partial r'_l} \cdot \frac{\partial r'_l}{\partial r_k} = \delta_{ik} \tag{7.84}$$

Due to the unknown hydrostatic pressure, p, the individual normal stresses σ_{ii} are indeterminate. However, the normal stress differences are well-defined. Let us consider the difference between σ_{zz} and σ_{xx}. Insertion of the Finger strain tensor associated with uniaxial deformations, Eq. (7.80), in the constitutive equation (7.74) yields

$$\sigma_{zz} - \sigma_{xx} = G(\lambda^2 - \lambda^{-1}) \tag{7.85}$$

As σ_{xx} vanishes, the result is in agreement with the previous Eq. (7.30). Comparison reveals the equality

$$G = c_{\mathrm{p}} kT \tag{7.86}$$

In the uniaxial deformation mode both lateral dimensions change. Other types of strain are obtained if one of the lateral edges is kept fixed. ‘Shear’ is the general name used for this group of deformations. We first consider an orthogonal deformation known as ‘pure shear’, which corresponds to the following transformation relations:

$$z' = \lambda^{-1} z \tag{7.87}$$

$$x' = \lambda x \tag{7.88}$$

$$y' = y \tag{7.89}$$

Figure 7.10 depicts schematically this deformation, together with the two other modes considered here, uniaxial loading and simple shear. The drawings indicate for all modes the changes in shape of a cubic volume element ’cut out’

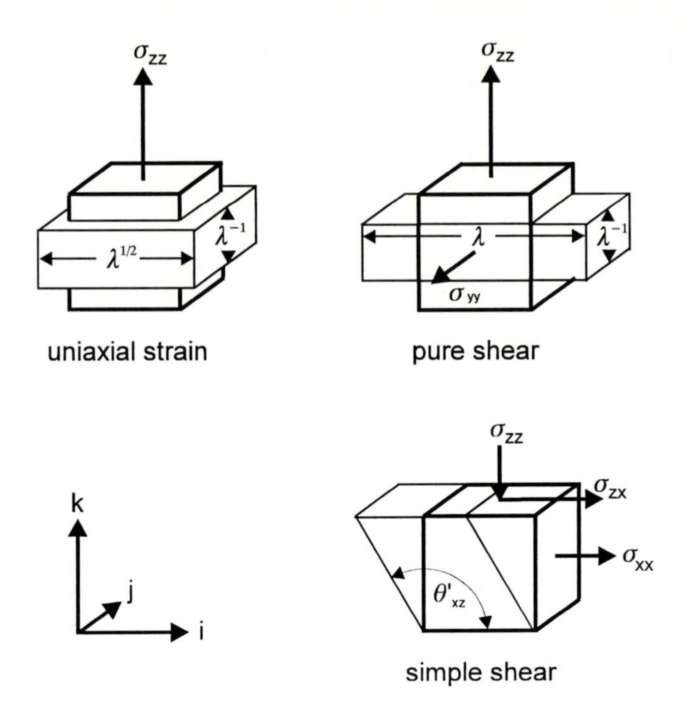

Fig. 7.10. Several modes of deformation of a rubber. Changes in shape of a cubic volume element picked out in the stressed state, as it results from the unloading. Acting components of stress are indicated

in the stressed state, resulting from an unloading. The Finger strain tensor associated with pure shear has the form

$$\mathbf{B} : \begin{pmatrix} \lambda^{-2} & 0 & 0 \\ 0 & 1 & 0 \\ 0 & 0 & \lambda^2 \end{pmatrix} \tag{7.90}$$

Insertion of **B** in Eq. (7.74) results in the following expressions for the normal stress differences

$$\sigma_{zz} - \sigma_{xx} = G(\lambda^2 - \lambda^{-2}) \tag{7.91}$$

$$\sigma_{zz} - \sigma_{yy} = G(\lambda^2 - 1) \tag{7.92}$$

$$\sigma_{yy} - \sigma_{xx} = G(1 - \lambda^{-2}) \tag{7.93}$$

Hence, for this mode we find non-vanishing values for all three normal stress differences.

As our third example we consider the important case of 'simple shear', in the form indicated in Fig. 7.10. The transformation relations are

$$x' = -\gamma \cdot z + x \quad (7.94)$$
$$y' = y \quad (7.95)$$
$$z' = z \quad (7.96)$$

Here, the symbol γ stands for the shear strain, defined as

$$\gamma := \tan(\theta'_{xz} - 90^0) \quad (7.97)$$

(γ is identical with the component e_{zx} of the linear strain tensor). Application of Eq. (7.81) yields the associated Finger strain tensor

$$\mathbf{B} : \begin{pmatrix} 1+\gamma^2 & 0 & \gamma \\ 0 & 1 & 0 \\ \gamma & 0 & 1 \end{pmatrix} \quad (7.98)$$

Use in Eq. (7.74) gives

$$\sigma_{zx} = G\gamma \quad (7.99)$$
$$\sigma_{xx} - \sigma_{zz} = G\gamma^2 \quad (7.100)$$
$$\sigma_{yy} - \sigma_{zz} = 0 \quad (7.101)$$

These are noteworthy results. First, we find a linear relation between the shear stress σ_{zx} and the shear strain γ which is unlimited. Thus, for this special kind of load, linearity is retained up to arbitrarily large deformations. We also can identify G, introduced at first as an empirical constant, as indeed representing the shear modulus. The predicted linearity is largely corroborated by the experimental findings. Figure 7.11 depicts the dependence $\sigma_{zx}(\gamma)$ as measured for a sample of natural rubber. A steady increase in shear strain is observed which is essentially linear, apart from a slight curvature at moderate deformations.

Non-linearity shows up in the second finding. We observe that simple shear is accompanied by the development of a non-vanishing normal stress difference $\sigma_{xx} - \sigma_{zz}$. The effect is non-linear since it is proportional to the square of γ. The result tells us that in order to establish a simple shear deformation in a rubber, application of shear stress alone is insufficient. One has to apply in addition either pressure onto the shear plane ($\sigma_{zz} < 0$) or a tensile force onto the plane normal to the x-axis ($\sigma_{xx} > 0$), or an appropriate combination of both. The difference $\sigma_{xx} - \sigma_{zz}$ is called the primary normal stress difference; likewise $\sigma_{yy} - \sigma_{zz}$ is commonly designated as the secondary normal stress difference. As we can see, the latter vanishes for an ideal rubber.

Having found such a simple description for ideal rubbers, enabling us to make predictions of the stress for all kinds of deformations in a straightforward way, one might presume that the modifications in behavior, as they are observed for real rubbers, can be accounted for by suitable alterations performed in the framework of the general equation for incompressible solids Eq. (7.75). In a first step one may consider the effects of an inclusion of the

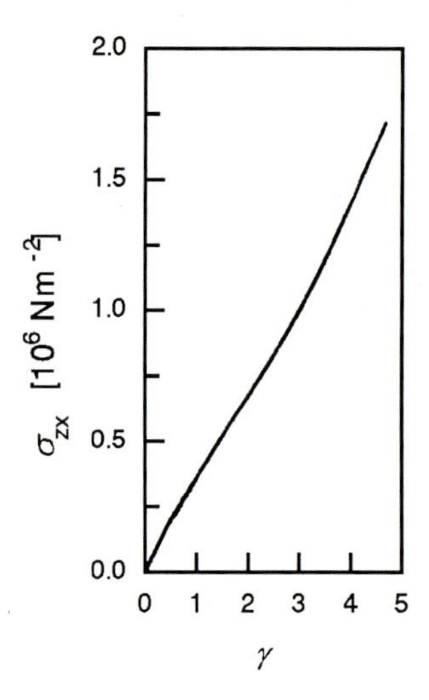

Fig. 7.11. Shear stress-shear strain curve observed for a sample of natural rubber during simple shear. Data of Treloar [71]

second term proportional to $\mathbf{B}^{-1}$ in the simplest possible form, by assuming constant values for both, the derivative $\partial f/\partial I_B$ and the derivative $\partial f/\partial II_B$. Such a choice is equivalent to a free energy function

$$f = \beta_1(I_B - 3) + \beta_2(II_B - 3) \tag{7.102}$$

where β_1 and β_2 denote the two constants. The resulting constitutive equation is

$$\boldsymbol{\sigma} = -p\mathbf{1} + 2\beta_1\mathbf{B} - 2\beta_2\mathbf{B}^{-1} \tag{7.103}$$

In the consideration of uniaxial extensions we can proceed in analogous manner as above, and we are led to the result

$$\sigma_{zz} - \sigma_{xx} = 2\beta_1(\lambda^2 - \lambda^{-1}) - 2\beta_2(\lambda^{-2} \quad \lambda) \tag{7.104}$$

$$= (2\beta_1\lambda + 2\beta_2)(\lambda - \lambda^{-2}) \tag{7.105}$$

For a derivation of the state of stress under simple shear, we need the form of $\mathbf{B}^{-1} = \mathbf{C}$. It follows by applying Eq. (7.40) to Eqs. (7.94) to (7.96). The result is

$$\mathbf{B}^{-1} : \begin{pmatrix} 1 & 0 & -\gamma \\ 0 & 1 & 0 \\ -\gamma & 0 & 1+\gamma^2 \end{pmatrix} \tag{7.106}$$

Equation (7.103) then yields

$$\sigma_{zx} = (2\beta_1 + 2\beta_2)\gamma \tag{7.107}$$

$$\sigma_{xx} - \sigma_{zz} = (2\beta_1 + 2\beta_2)\gamma^2 \tag{7.108}$$

$$\sigma_{yy} - \sigma_{zz} = 2\beta_2 \cdot \gamma^2 \tag{7.109}$$

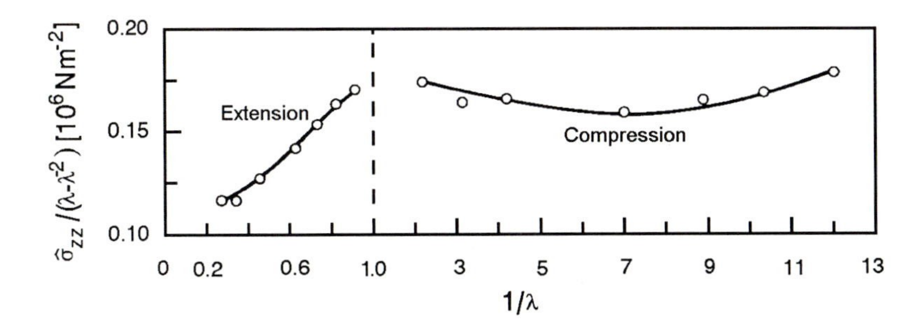

Fig. 7.12. Mooney-plot of compression and extension data obtained for natural rubber. Results from Rivlin and Saunders [75]

For simple shear, we have thus obtained again linearity for σ_{zx}, as for ideal rubbers, but non-vanishing values now for both, the primary and the secondary normal stress difference.

The suggestion to expand the treatment in this way is due to Mooney. It was readily accepted and applied with success for the description of tensile stress-strain curves. Figure 7.12 presents an example which also reveals, however, a major deficiency. Data are shown in the form of a 'Mooney-plot', which is based on Eq. (7.105), written as

$$\frac{\hat{\sigma}_{zz}}{\lambda - \lambda^{-2}} = 2\beta_1 + \frac{2\beta_2}{\lambda} \tag{7.110}$$

employing the nominal tensile stress and setting $\hat{\sigma}_{xx} = 0$. Results encompass both the extension and the compression range. We notice that the Mooney equation yields a satisfactory data representation for the extension range, with a value $\beta_2 > 0$ as determined by the slope, however, there is no continuation into the range of compressions, where we find $\beta_2 \approx 0$. Hence, the simple modification of the free energy function, as expressed by Eq. (7.102) is actually inadequate for the desired general representation of real rubber behavior.

There are various suggestions for a better choice, however, they are all rather complicated and their discussion is outside our scope. It appears today that a short analytical expression for the free energy density of a real rubber in the form of a simple extension of the free energy density of an ideal rubber does not exist. Even so, the general constitutive equation, Eq. (7.75), certainly provides us with a sound basis for treatments. Once the functional dependence of the free energy density, $f(I_B, II_B)$, has been mapped by a suitable set of experiments, and one succeeds in representing the data by an empirical expression, one can predict the stresses for any kind of deformation.

7.2 Non-Newtonian Melt Flow

Polymers, to a large extent, owe their attractivity as materials for a wide range of applications to their ease of processing. Manufacturing processes such as injection molding, fiber spinning or film formation are mainly conducted in the melt and it is a great advantage of polymers that melting temperatures are comparatively low. On the other hand, the flow properties of polymer melts are complicated and process control requires a broad knowledge. To deal with simple Newtonian liquids, one needs just one parameter, namely the viscosity. In principle, knowing it permits us to calculate stresses for any given flow pattern. Polymeric liquids are more complex in behavior. From the very beginning, i.e. even at low frequencies or low shear rates, one has to employ two coefficients for a characterization. As we have already seen in the discussion of linear responses, in addition to the shear viscosity, one needs to know the recoverable shear compliance which relates to the always present elastic forces. At higher strain rates, more complications arise. There, both parameters are no longer constants but change with the shear rate. In this section, we will briefly discuss these phenomena.

The problem which has to be solved in order to set the processing of polymer melts on a solid basis is the formulation of rheological equations of state. The rheologists' approach in treating the problems of melt flow is generally a purely empirical one. The objective is the finding of analytical expressions which describe correctly the relationships between the time dependent strains or velocities in a given fluid, and the emerging stresses. Ideally, equations should be generally applicable, i.e. hold for all possible strain histories, and at the same time analytically simple, to facilitate their use. As we will see, for polymer melts, this aim amounts to a difficult task. We can enter into this field only in a very restricted manner, and will present solely the simplest model, known as the 'Lodge fluid'. The main purpose of this section is to introduce and discuss some standard experiments which are commonly employed for a characterization of polymer flow behavior. They refer to situations frequently encountered under practical conditions.

7.2.1 Rheological Material Functions

Two of the various flow patterns realized in polymer melts during processing are of particular importance. First, there is the flow through a tube. Here, all volume elements move along straight lines parallel to the axis. The velocities are non-uniform, the maximum velocity being found at the center and the minimum at the wall where the fluid particles are fixed. The consequence of the varying velocities are frictional forces, acting between the volume elements incorporated in adjacent stream lines. Evidently, flow through tubes is locally equivalent to shear flows. Similarly to simple shear, layers of liquid slide over each other without changing their form. A different kind of deformation is found in 'extensional flows'. These are encountered, for example,

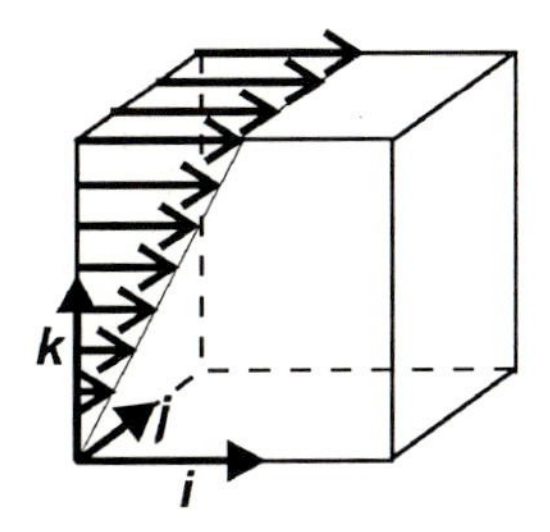

Fig. 7.13. Pattern of simple shear flow. Arrows indicate the velocity field

during melt spinning in the fiber production. The dominant feature now is the stretching of all volume elements, which means a great change in form; the sliding of adjacent layers against each other, giving the main effect in shear flow, is less important. Hence, we have, at least, two different classes of deformations, which both have to be analysed. As a minimum, one needs for a first rheological characterization of a given polymer two kinds of experimental arrangements, and we discuss them now one after the other.

'Simple shear flow' is identical with the pattern depicted in Fig. 7.13. The velocity field is given by

$$v_x = \dot{\gamma} z \tag{7.111}$$

$$v_y = v_z = 0 \tag{7.112}$$

The only parameter included is the 'shear rate' $\dot{\gamma}$. There is an instrument which is widely used because it realizes simple shear flow conditions in a manner convenient for experiments. This is the 'cone-and-plate rheometer' sketched on the left-hand side of Fig. 7.14. The polymer melt is placed in the gap and, if the gap angle is kept small, then the shear rate is the same everywhere. With the aid of this apparatus, various experiments may be carried out, as for example

- determinations of the steady state properties at constant shear rates
- measurements under oscillatory shear
- observations of the stress growth for linearly increasing shear
- detection of the shear stress relaxation after a sudden deformation of the melt
- studies of the elastic recovery after unloading

The two material parameters which characterize polymeric fluids at low strain rates, the viscosity, η, and the recoverable shear compliance, J_e, can be directly determined. η follows from the measurement of the torque under steady state conditions, J_e shows up in the reverse angular displacement subsequent to an unloading, caused by the retraction of the melt. From the discussion of the properties of rubbers we know already that simple shear is associated with the building-up of normal stresses. More specifically, one finds a non-vanishing

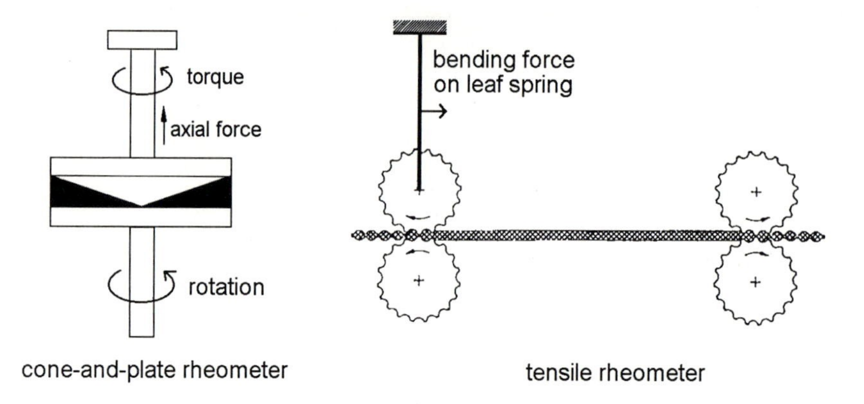

Fig. 7.14. Standard arrangements used for the characterization of shear flows (*left*) and extensional flows (*right*). Shear viscosities can be derived from the torque measured in a cone-and-plate rheometer; the primary normal stress difference is deduced from the axial force. Elongational viscosities follow from the tensile force required for the drawing of a molten fibre in the tensile rheometer, as monitored by a leaf spring

value of the first normal stress difference. We may anticipate, therefore, that the deformation of the entanglement network in a shear-deformed polymer melt likewise results in the emergence of normal stresses. The cone-and-plate rheometer permits us to make a direct determination, as this causes a thrusting axial force. Being proportional to the square of the shear in rubbers and equivalent, as we shall learn, to the square of the shear rate in polymer melts, normal stresses constitute a non-linear phenomenon. Although they can be small for low shear rates, they are present from the very beginning.

The properties of melts under extensional flows can be studied with the aid of a 'tensile rheometer'. The basic experimental arrangement is sketched on the right-hand side of Fig. 7.14. A cylindrical rod of polymer melt, usually floating on a liquid substrate, is drawn by two pairs of ribbed rollers. One of the rollers is mounted on a leaf spring, so that the force required for the drawing process can be monitored by the deflection. It is important to recognize the conditions imposed if the rollers circulate with a constant frequency. Evidently, the procedure realizes a constant rate of elongation relative to the momentary length, i.e. a constant value for the relative length change per second. Expressed in terms of the stretching ratio, λ, we have

$$\frac{1}{\lambda(t)}\frac{\mathrm{d}\lambda}{\mathrm{d}t} = \text{const} \tag{7.113}$$

The process thus differs from a drawing where clamps attached to the two ends of a sample move with a constant relative speed, as this leads to

$$\frac{d\lambda}{dt} = \text{const} \tag{7.114}$$

The constant in Eq. (7.113) is known as the 'Hencky rate of extension' and we choose for it the symbol $\dot{e}_H$

$$\dot{e}_H := \frac{d \ln \lambda}{dt} \tag{7.115}$$

Whereas experiments conducted with a constant rate of elongation yield a linear extension with time, those carried out with a constant Hencky rate result in an exponential time dependence of λ

$$\lambda(t) = \exp(\dot{e}_H t) \tag{7.116}$$

For solids one may choose either of the two drawing processes. For liquids a drawing with a constant Hencky rate of extension is to be preferred, because the original length included in λ then becomes irrelevant, as is desired.

We now introduce the major rheological material functions, with illustrations provided by typical experimental results. Figure 7.15 depicts data obtained for low density polyethylene under steady shear flow conditions, employing a cone-and-plate rheometer. Curves display both the shear rate dependence of the viscosity, with similar results as in Fig. 7.1, and the shear rate dependence of the first normal stress difference. The stresses arising for simple shear flows may be generally expressed by the following set of equations

$$\sigma_{zx} = \eta(\dot{\gamma})\dot{\gamma} \tag{7.117}$$

$$\sigma_{xx} - \sigma_{zz} = \Psi_1(\dot{\gamma})\dot{\gamma}^2 \tag{7.118}$$

$$\sigma_{yy} - \sigma_{zz} = \Psi_2(\dot{\gamma})\dot{\gamma}^2 \tag{7.119}$$

Equations include and thus define three rheological material functions

- the shear viscosity η
- the primary normal stress coefficient Ψ_1
- the secondary normal stress coefficient Ψ_2

All three functions are dependent on the shear rate $\dot{\gamma}$. Ψ_2 does not appear in the figure, because it is difficult to deduce it from measurements in a cone-and-plate rheometer. It can be obtained using other devices and results indicate that Ψ_2 is usually much smaller than Ψ_1, thus having only a minor effect on the flow behavior. The important property showing up in the measurements is the 'shear thinning', observed for both η and Ψ_1, and setting in for strain rates above $\dot{\gamma} \simeq 10^{-2}\text{s}^{-1}$.

The third curve in the figure, denoted γ_e, gives the amount of shear strain recovery subsequent to a removal of the external torque. There is an interesting observation: The linear increase of γ_e with $\dot{\gamma}$ at low shear rates is exactly

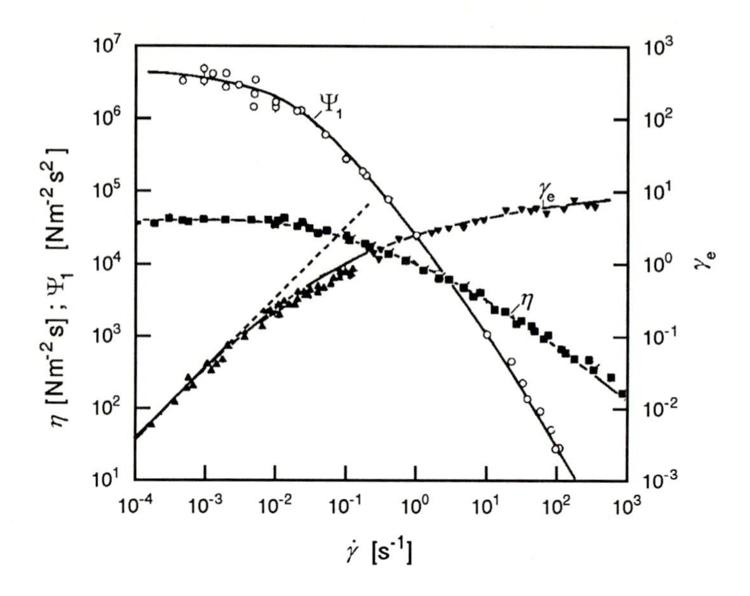

Fig. 7.15. PE under steady state shear flow at 150 °C: Strain rate dependencies of the viscosity η, the primary normal stress coefficient Ψ_1 and the recoverable shear strain γ_e. The *dotted line* represents Eq. (7.122). Results obtained by Laun [76]

determined by the zero shear rate values of the viscosity and the primary normal stress coefficient, η_0 and $\Psi_{1,0}$. This is revealed by the coincidence of the limiting curve $\gamma_c(\dot{\gamma} \to 0)$ with the dotted line which represents the linear function

$$\frac{\Psi_{1,0}}{2\eta_0}\dot{\gamma} = \frac{\sigma_{xx} - \sigma_{zz}}{2\sigma_{zx}} \tag{7.120}$$

The agreement implies a relationship between the zero shear rate value of the recoverable shear compliance, following from γ_e as

$$J_e^0 := \lim_{\dot{\gamma} \to 0} \frac{\gamma_e}{\sigma_{zx}} \tag{7.121}$$

and the zero shear values of the viscosity and the primary normal stress coefficient. It has the following form

$$J_e^0 = \frac{\Psi_{1,0}}{2\eta_0^2} \tag{7.122}$$

The relationship tells us that, in the limit of low shear rates, we still have only two parameters which are independent. The primary normal stress coefficient does not represent an independent property, but is deducable from the two parameters controlling the linear response, J_e^0 and η_0.

Figure 7.16 refers to two other standard experiments. It depicts the results of 'stress growth experiments', conducted again on a polyethylene melt. The

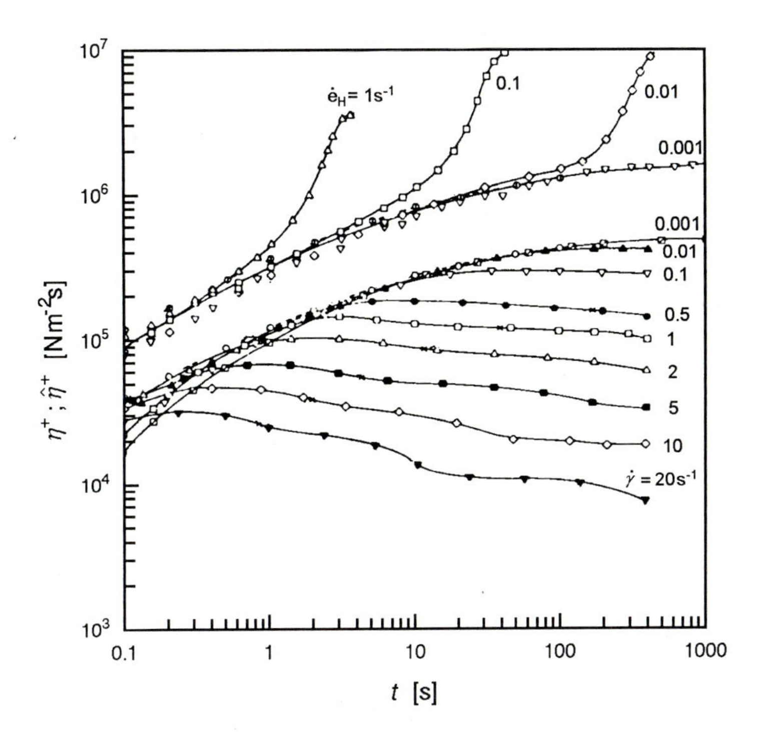

Fig. 7.16. Results of stress growth experiments in shear, $\eta^+(t)$, and extension, $\hat{\eta}^+(t)$, at different deformation rates $\dot{\gamma}$ and $\dot{e}_{\mathrm{H}}$ respectively, carried out for LDPE at 150 °C. Measurements by Meissner [77]

figure includes both, measurements probing shear and tensile properties thus facilitating a direct comparison. Curves show the building-up of shear stress upon inception of a steady state shear flow at zero time and the development of tensile stress upon inception of a steady state extensional flow. Measurements were carried out for various values of the shear rate $\dot{\gamma}$ or the Hencky rate of extension $\dot{e}_{\mathrm{H}}$.

The results of the shear experiments are described with the aid of the 'time dependent viscosity' defined as

$$\eta^+(t, \dot{\gamma}) := \frac{\sigma_{zx}(t)}{\dot{\gamma}} \tag{7.123}$$

In an analoguous manner the 'time dependent extensional viscosity', defined as

$$\hat{\eta}^+(t) := \frac{\sigma_{zz}(t)}{\dot{e}_{\mathrm{H}}} \tag{7.124}$$

is employed for a description of the properties of extensional flow. The steady

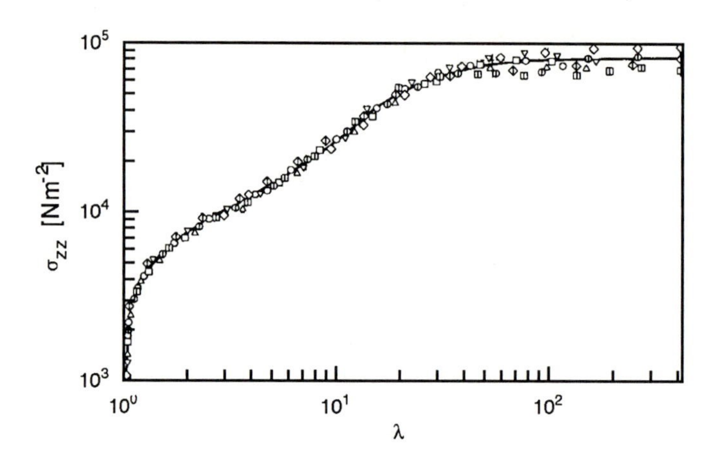

Fig. 7.17. Tensile stress growth curves observed for PE at 150 °C, obtained for a constant Hencky strain rate $\dot{e}_{\rm H} = 0.1\ {\rm s}^{-1}$. Data from Laun and Münstedt [78]

state value reached after the period of growth

$$\hat{\eta} := \hat{\eta}^{+}(t \to \infty) = \frac{\sigma_{zz}}{\dot{e}_{\rm H}} \tag{7.125}$$

is known as the 'extensional viscosity' or 'Trouton-viscosity'. A look at the results is quite instructive and shows us characteristic differences between shear and tensile deformations. First consider the limit of low strain rates. We find here a simple result, as both curves agree in shape, just differing by a factor of three. As we shall see below, the two curves can be deduced from the flow properties in the linear regime, by application of Lodge's equation of state for rubber-like liquids. The reason for the difference and the fact that the larger forces arise in the extensional flow is obvious: For equal values of $\dot{\gamma}$ and $\dot{e}_{\rm H}$, the tensile strains produce much larger deformations. Most interesting is the comparison of the behavior at larger strain rates. In both cases, one observes characteristic deviations from the low strain rate curves. The shear stressing leads to deviations to lower values, i.e. to a 'shear thinning'. The asymptotic values reached at long duration give the steady state shear viscosities shown in Fig. 7.15. A quite different behavior appears in the tensile deformation mode, where we find deviations to higher values, indicating a 'strain hardening'. In the latter experiments it is often difficult to reach the steady state, as this is frequently preceded by a fracture of the melt fibre. Sometimes one is successful and then curves like the one shown in Fig. 7.17 are observed. The shape in the range before the steady state plateau is reached reminds one of stress-strain curves of rubbers under uniaxial load, like that shown in Fig. 7.1. The similarity might suggest that, like the rubber, here in the melt again a strain hardening arises from the limited extensibility of the mobile chain sequences. As it turns out, however, this is not the primary cause. As

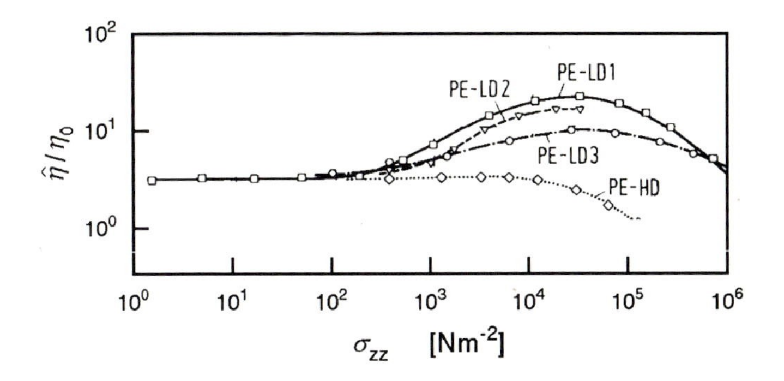

Fig. 7.18. Steady state extensional viscosities $\hat{\eta}$ as a function of the applied tensile stress, as observed for various samples of PE. Data are given in reduced form, with reference to the respective zero shear rate viscosities η_0. From Münstedt and Laun [79]

will be shown in the next section, an increase in stress is also predicted for the 'Lodge liquid' if strain rates become large compared to intrachain relaxation rates.

A further typical result is shown in Fig. 7.18, giving the plateau values of the extensional viscosity at steady state as a function of the associated drawing stress σ_{zz}. Measurements were carried out for various low density polyethylenes ('PE-LD') and one linear polyethylene ('PE-HD'). For the majority of samples, one observes in the non-linear range an increase at first and then for higher strain rates and stresses, after passing over a maximum, a decrease. We may understand this behavior as being the result of competition between two effects. After the first period of strain hardening, there follows a viscosity reduction on further increasing the strain rates owing to a decrease in the entanglement density in the oriented fiber occurring for reasons which are similar to the shear thinning case.

7.2.2 The Lodge Liquid

Treatments of flows of polymer melts require a rheological equation of state which enables us to calculate the stress tensor for all points $\boldsymbol{r}$ of the flow field for a given pattern. We have a complete description of the flow kinetics if we specify for each material point, located at $\boldsymbol{r}$ at time t, the full trajectory in the past. The latter may be expressed by the function

$$\boldsymbol{r}'(\boldsymbol{r}, t, t') \tag{7.126}$$

which describes the positions $\boldsymbol{r}'$ at the previous times t'. Clearly we may start from the assumption that the stress field at time t, $\boldsymbol{\sigma}(\boldsymbol{r}, t)$, is a functional of the trajectories of all material points. This statement, however, is certainly too general to be useful for applications. In the discussion of finite deformations of rubbers, we have learned how the problem can be reduced. There an essential reduction was achieved by the introduction of the Cauchy tensor $\mathbf{C}$, or alternatively the Finger tensor $\mathbf{B}$. As we have seen, the state of stress of a hyperelastic body may be described by functions $\boldsymbol{\sigma}(\mathbf{C})$ or $\boldsymbol{\sigma}(\mathbf{B})$. This notion can be extended and generalized in an obvious manner to treat the time dependent deformations occurring in polymer melts. Referring to the Finger tensor, we may introduce a function

$$\mathbf{B}(\boldsymbol{r}, t, t')$$

depending on the present time t and a previous time t', with components B_{ij} being given by

$$B_{ij}(t, t') = \frac{\partial r_i}{\partial r'_k} \frac{\partial r_j}{\partial r'_k} \tag{7.127}$$

This is analogous to the previous definition, Eq. (7.81), which is now applied to the state of the melt at time t', given by the displacement function $\boldsymbol{r}'(\boldsymbol{r}, t, t')$, rather than to the natural state of a rubber. According to the definition, the 'time dependent Finger strain tensor', $\mathbf{B}(\boldsymbol{r}, t, t')$, characterizes the deformation of the neighborhood of the material point at $\boldsymbol{r}$ which has taken place between the past time t' and the present time t. With the aid of the Finger tensor, we have formulated a general equation of state valid for all hyperelastic bodies. Likewise, we may assume now that the stress in a flowing melt can be expressed as a functional of the tensors $\mathbf{B}(\boldsymbol{r}, t, t')$:

$$\boldsymbol{\sigma}(\boldsymbol{r}, t) = \boldsymbol{\sigma}(\mathbf{B}(\boldsymbol{r}, t, t')) \qquad \text{with} \qquad t' < t \tag{7.128}$$

This general expression first accounts for the principal of causality by stating that the state of stress at a time t is dependent on the strains in the past only. Secondly, by using the time dependent Finger tensor $\mathbf{B}$, one extracts from the flow fields only those properties which produce stress and eliminates motions like translations or rotations of the whole body which leave the stress invariant. Equation (7.128) thus provides us with a suitable and sound basis for further considerations.

The problem is to find expressions which can describe the experimental observations. We have a perfect solution if we succeed in constructing an equation of state which correctly formulates the stresses for an arbitrary strain history. To reach this final goal is certainly very difficult. It might appear, however, that we have a suitable starting point. Polymer fluids have much in common with rubbery materials. In a simplified view, we find, as the only difference, that cross-links are permanent in rubbers whereas, in fluids, they are only temporary with lifetimes which are still large compared to all the internal equilibration processes. Hence, it makes sense to search for a formula which includes, from the very beginning, the properties of rubbers as expressed by the equation of state of ideal rubbers, Eq. (7.74).

Lodge was the first to formulate such an equation by a combination of Eq. (7.74) with the Boltzmann superposition principle as expressed by Eq. (5.111). Explicitly, the 'Lodge equation of state of rubber-like liquids', when written for homogeneous deformations, has the following form

$$\boldsymbol{\sigma}(t) = -\int\limits_{t'=-\infty}^{t} G(t-t')\frac{\mathrm{d}\mathbf{B}(t,t')}{\mathrm{d}t'}\mathrm{d}t' \tag{7.129}$$

The Boltzmann superposition principle represents the stress as a result of changes in the state of strain at previous times. In the linear theory valid for small strains, these can be represented by the linear strain tensor. In Lodge's equation the changes in the latter are substituted by changes in the time dependent Finger tensor

$$-\mathrm{d}\mathbf{B}(t,t')$$

thus enabling us to describe finite strains. The meaning of the function G remains unchanged. It still represents the stress relaxation modulus.

To see the consequences implied by this equation of state, it is instructive to consider first simple shear flow conditions. We may write down the time dependent Finger tensor immediately, just by replacing in Eq. (7.98), derived for a deformed rubber, γ by the increment $\gamma(t)-\gamma(t')$. This results in

$$\mathbf{B}(t,t') := \begin{pmatrix} 1+(\gamma(t)-\gamma(t'))^2 & 0 & \gamma(t)-\gamma(t') \\ 0 & 1 & 0 \\ \gamma(t)-\gamma(t') & 0 & 1 \end{pmatrix} \tag{7.130}$$

Next, we introduce $\mathbf{B}$ into Lodge's equation of state, calculate the derivatives and thus obtain all stress components. The result for the shear stress is

$$\sigma_{zx}(t) = \int\limits_{t'=-\infty}^{t} G(t-t')\frac{\mathrm{d}\gamma}{\mathrm{d}t'}\mathrm{d}t' \tag{7.131}$$

and for the primary normal stress difference we obtain

$$(\sigma_{xx}-\sigma_{zz})(t) = 2\int\limits_{t'=-\infty}^{t} G(t-t')(\gamma(t)-\gamma(t'))\frac{\mathrm{d}\gamma}{\mathrm{d}t'}\mathrm{d}t' \tag{7.132}$$

The secondary normal stress difference vanishes, as was the case for an ideal rubber

$$\sigma_{yy}-\sigma_{zz} = 0 \tag{7.133}$$

Representing a combination of the equation of state of ideal rubbers and Boltzmann's superposition principle, Lodge's equation provides an interpolation between the properties of rubbers and viscous liquids. The limiting cases of an elastic rubber and the Newtonian liquid are represented by

$$G = \mathrm{const} \tag{7.134}$$

and

$$G(t-t') = \eta\delta(t-t') \tag{7.135}$$

respectively. Insertion of a constant G in Lodge's equation reproduces the elastic properties of a rubber under simple shear deformation, as we find

$$\sigma_{zx}(t) = G \cdot \gamma(t) \tag{7.136}$$

$$\begin{aligned}(\sigma_{xx} - \sigma_{zz})(t) &= 2G \int_{t'=0}^{t} (\gamma(t) - \gamma(t')) \frac{\mathrm{d}\gamma}{\mathrm{d}t'} \mathrm{d}t' \\ &= G\gamma^2(t) \end{aligned} \tag{7.137}$$

If we take Eq. (7.135), we obtain

$$\sigma_{zx}(t) = \eta \frac{\mathrm{d}\gamma}{\mathrm{d}t} \tag{7.138}$$

$$(\sigma_{xx} - \sigma_{zz})(t) = 0 \tag{7.139}$$

i.e. the flow properties of a Newtonian liquid.

Next, let us consider the predictions for a shear stress growth experiment. In a stress growth experiment, a linearly increasing shear is imposed, i.e.

$$\gamma(t) = \dot{\gamma} \cdot t \qquad \text{for} \qquad t > 0 \tag{7.140}$$

Applying Eq. (7.131), we obtain, for the time dependent shear stress,

$$\sigma_{zx}(t) = \dot{\gamma} \int_{t'=0}^{t} G(t-t')\mathrm{d}t' \tag{7.141}$$

and for the primary normal stress difference

$$\begin{aligned}(\sigma_{xx} - \sigma_{zz})(t) &= 2\dot{\gamma}^2 \int_{t'=0}^{t} G(t-t')(t-t')\mathrm{d}t' \\ &= 2\dot{\gamma}^2 \int_{0}^{t} G(t'')t''\mathrm{d}t'' \end{aligned} \tag{7.142}$$

The results give us the following expressions for the material functions $\eta^+(t)$ and $\Psi_1^+(t)$:

$$\eta^+(t) = \int_{0}^{t} G(t'')\mathrm{d}t'' \tag{7.143}$$

and

$$\Psi_1^+(t) = 2\int_0^t G(t'')t''\mathrm{d}t'' \tag{7.144}$$

The steady state values reached after passing through the period of stress growth equal the asymptotic limits for $t \to \infty$

$$\eta = \eta^+(t \to \infty) \tag{7.145}$$

$$\Psi_1 = \Psi_1^+(t \to \infty) \tag{7.146}$$

We thus obtain, for the steady state viscosity and the steady state primary normal stress coefficient, the expressions

$$\eta = \int_0^\infty G(t'')\mathrm{d}t'' \tag{7.147}$$

$$\Psi_1 = 2\int_0^\infty G(t'')t''\mathrm{d}t'' \tag{7.148}$$

As we can see, both are independent of the strain rate $\dot{\gamma}$. Hence, as a first conclusion, Lodge's equation of state cannot describe the shear thinning phenomenon. Equation (7.147) is in fact identical with Eq. (5.107) derived in the framework of linear response theory. The new result contributed by Lodge's formula is the expression Eq. (7.148) for the primary normal stress difference. It is interesting to note that the right-hand side of this equation has already appeared in Eq. (5.108) of the linear theory, formulating the relationship between $G(t)$ and the recoverable shear compliance J_e^0. If we take the latter equation, we realize that the three basic parameters of the Lodge's rubber-like liquid, η_0, J_e^0 and $\Psi_{1,0}$ are indeed related, by

$$\Psi_{1,0} = 2J_\mathrm{e}^0\eta_0 \tag{7.149}$$

The result exactly reproduces the experimental observations in the limit of low shear rates as presented in Fig. 7.15; the straight dotted line is based on this relation. We may conclude, therefore, that Lodge's equation describes correctly the development of normal stresses in the initial period of shear-stressed polymeric liquids.

Now, we consider the second class of experiments and check for the predictions of Lodge's model with regard to extensional flows. Using again an equation from the ideal rubbers we can directly write down the time dependent Finger tensor $\mathbf{B}(t,t')$. It has the form

$$\mathbf{B}(t,t') := \begin{pmatrix} \left(\frac{\lambda(t)}{\lambda(t')}\right)^{-1} & 0 & 0 \\ 0 & \left(\frac{\lambda(t)}{\lambda(t')}\right)^{-1} & 0 \\ 0 & 0 & \left(\frac{\lambda(t)}{\lambda(t')}\right)^{2} \end{pmatrix} \tag{7.150}$$

and follows from Eq. (7.80) by a substitution of the stretching ratio λ referring to the natural state, by the relative length change between times t' and t as given by the ratio $\lambda(t)/\lambda(t')$. Tensile stress growth experiments are conducted with a constant Hencky rate of extension, starting the drawing process at zero time. This leads to an exponential time dependence of λ as expressed by Eq. (7.116) and therefore to the following expression for the time dependent Finger tensor:

$$\mathbf{B}(t,t') := \begin{pmatrix} \exp -\dot{e}_{\mathrm{H}}(t-t') & 0 & 0 \\ 0 & \exp -\dot{e}_{\mathrm{H}}(t-t') & 0 \\ 0 & 0 & \exp -2\dot{e}_{\mathrm{H}}(t-t') \end{pmatrix} \tag{7.151}$$

Insertion of $\mathbf{B}(t,t')$ in the equation of state for rubber-like liquids yields the normal stress difference

$$\begin{aligned} (\sigma_{zz}-\sigma_{xx})(t) &= -\int\limits_{t'=0}^{t} G(t-t')\frac{\mathrm{d}}{\mathrm{d}t'}(\exp 2\dot{e}_{\mathrm{H}}(t-t') \\ &\qquad - \exp -\dot{e}_{\mathrm{H}}(t-t'))\mathrm{d}t' \\ &= \dot{e}_{\mathrm{H}}\int\limits_{t'=0}^{t} G(t-t')(\exp 2\dot{e}_{\mathrm{H}}(t-t') \\ &\qquad + \exp -\dot{e}_{\mathrm{H}}(t-t'))\mathrm{d}t' \end{aligned} \tag{7.152}$$

In the absence of lateral pressures, i.e.

$$\sigma_{xx} = 0$$

we can formulate the result as

$$\sigma_{zz}(t) = \dot{e}_{\mathrm{H}}\hat{\eta}^{+}(t,\dot{\epsilon}) \tag{7.153}$$

with

$$\hat{\eta}^{+}(t,\dot{e}_{\mathrm{H}}) = \int\limits_{t''=0}^{t} G(t'')(2\exp 2\dot{e}_{\mathrm{H}}t'' + \exp -\dot{e}_{\mathrm{H}}t'')\mathrm{d}t'' \tag{7.154}$$

Figure 7.19 shows the time dependent viscosities derived from Eqs. (7.143) and (7.154) for both simple shear and extensional flow. For simplicity a single exponential relaxation with a relaxation time τ is assumed for $G(t'')$. The dotted line represents the time dependent viscosity for simple shear, $\eta^{+}(t)$, which is independent of $\dot{\gamma}$. A qualitatively different result is found for the extensional flow. As we see, the time dependent extensional viscosity $\hat{\eta}^{+}(t)$ increases with $\dot{e}_{\mathrm{H}}$ and for $\dot{e}_{\mathrm{H}} > 0.5\tau^{-1}$ a 'strain hardening' arises.

It is interesting to compare these calculations with the experimental results presented in Fig. 7.16. Indeed, there is a perfect agreement in the limit of low

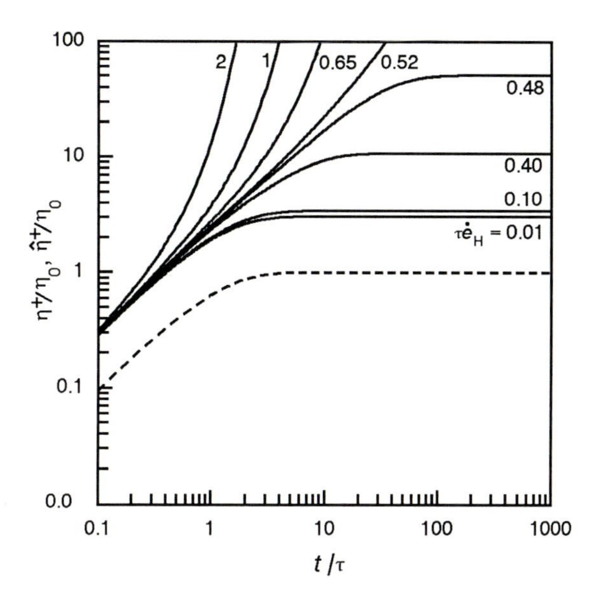

Fig. 7.19. Time dependent viscosities for shear and extensional flow, η^+ and $\hat{\eta}^+$, as predicted by Lodge's equation of state. Calculations are performed for different Hencky strain rates $\dot{e}_{\mathrm{H}}$, assuming a single exponential relaxation modulus $G(t) \sim \exp -t/\tau$

deformation rates, $\dot{\gamma} \to 0$ and $\dot{e}_{\mathrm{H}} \to 0$. In both the model and the experiment, we find two parallel curves for $\eta(t)$ and $\hat{\eta}(t)$ which are separated by a factor 3. Notice that the factor 3 also emerges for a rubber under static conditions, where we found $E = 3G$ (Eqs. (7.32), (7.86)). There is no agreement at all for the shear flow at finite strain rates since shear thinning phenomena are not accounted for. Similarly, the agreement is lacking for the tensile stress growth experiments at higher strain rates. Although the Lodge model predicts an increase with the strain rate, the experimental curves differ from the calculated data. In particular, the viscosity reduction at high strain rates indicated by the final values is not accounted for by the model. Hence, we have to conclude that the capacity of Lodge's model is limited when compared to the behavior of real polymer melts. What it provides is a correct prediction of the normal stress phenomena at low strain rates, and it might also describe part of the stress enhancement observed in extensional flows.

As it is obviously necessary to alter the assumptions, one might check at first if moderate modifications of Lodge's model could lead to an improvement of the agreement with the experiment. Still, Lodge's model can be regarded as a good starting point, as it furnishes a perfect representation of data for low strain rates, independent of the total deformation. There are various proposals in the literature on how an improvement could be achieved, however, a completely satisfactory equation of state in the sense of providing all material functions in quantitative terms for a given polymer, using a small number of adjustable parameters, has, so far, not been found. Rather than giving an overview, which lies definitely outside our scopes, we will just cite, for illustration, one of the better known approaches. One generalization of Eq. (7.129) reads as follows

$$\boldsymbol{\sigma} = - \int\limits_{t'=-\infty}^{t} G(t-t') \frac{\mathrm{d}\Phi(\mathbf{B}(t,t'))}{\mathrm{d}t'} dt' \tag{7.155}$$

Here, while retaining the relaxation function $G(t-t')$, the Finger tensor is replaced by a functional Φ depending on $\mathbf{B}(t,t')$ and its invariants $I_B(t,t')$ and $II_B(t,t')$. The modification enables us to account for the viscosity reduction effects. As an example, one can factorize Φ and write

$$\Phi(\mathbf{B}(t,t')) = \exp -\beta(I_B(t,t') - 3) \cdot \mathbf{B}(\mathbf{t},\mathbf{t}') \tag{7.156}$$

The exponential acts like a damping function and causes stress reductions if the deformations become large, as is desired.

7.2.3 Stress-Optical Rule and Network Model

Figures 7.20 and 7.21 present experimental results which may look at first view quite astonishing. Flowing polymer melts are birefringent, for obvious reasons, and the experiment compares for tensile stress growth experiments on polystyrene the evolution with time of the birefringence

$$\Delta \mathrm{n} = \mathrm{n}_c - \mathrm{n}_a \tag{7.157}$$

(n_c and n_a are the indices of refraction along and normal to the stretching direction) with that of the tensile stress. The time dependencies shown in Fig. 7.20 have a similar appearance and indeed, as proved by the plot of $\Delta\mathrm{n}$ versus σ_{zz} in Fig. 7.21, obtained by elimination of t from all curves, there is a strict relationship, even common to all temperatures. Moreover, over a wide range of stresses this relation agrees with a linear equation

$$\Delta \mathrm{n} = \mathrm{const} \cdot \sigma_{zz} \tag{7.158}$$

As only differences between normal stresses are well-defined, we write more accurately

$$\Delta \mathrm{n} = C_{\mathrm{opt}}(\sigma_{zz} - \sigma_{xx}) \tag{7.159}$$

Equation (7.159) is known as the 'linear stress-optical rule' and is generally valid for all polymer melts. The proportionality constant C_{opt} is called the 'stress-optical coefficient', and its value describes a characteristic property of each polymer.

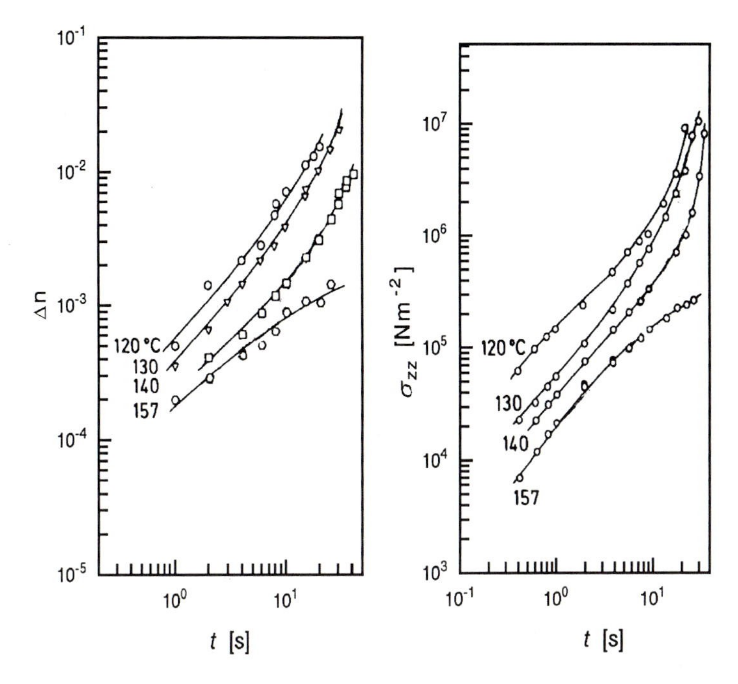

Fig. 7.20. Time dependence of tensile stress observed in stressing experiments ($\dot{e}_{\mathrm{H}} = 0,075\ \mathrm{s}^{-1}$) carried out on polystyrene at the indicated temperature (*right*). Simultaneous measurement of the birefringence as a function time (*left*). Data of Matsumoto and Bogue [80]

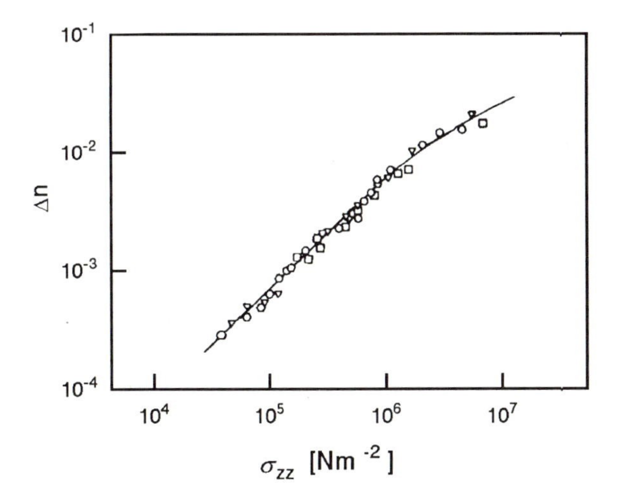

Fig. 7.21. Relation between birefringence and tensile stress deduced from the data of Fig. 7.20 after elimination of time [80]

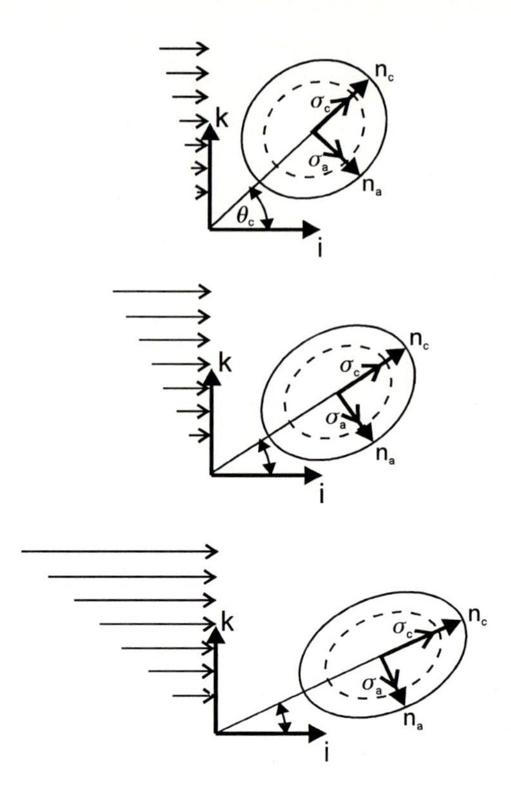

Fig. 7.22. Stress-optical relation as observed in polymer melts under simple shear flow: Optical indicatrix *(ellipsoids drawn with continuous lines)* and stress tensor *(ellipsoids depicted with broken lines)* show equal orientations of the principal axes and proportionality between the birefringence $\mathrm{n}_c - \mathrm{n}_a$ and the principal stress difference $\sigma_c - \sigma_a$. The inclination angle θ_c decreases with increasing shear rate

The linear stress-optical rule also holds under the conditions of simple shear flow. Observed data comply with the scenario depicted schematically in Fig. 7.22. The drawings show the principal axes of the stress tensor and of the optical indicatrix, for different shear rates. Data evaluation proves that the orientations of the two tripels of principal axes always coincide, as is indicated in the sketches. The inclination angle of the primary axis, θ_c, is 45° for infinitesimally small shear rates, and then decreases towards zero on increasing $\dot{\gamma}$. The stress optical rule here reads

$$\Delta\mathrm{n} := \mathrm{n}_c - \mathrm{n}_a = C_{\mathrm{opt}}(\sigma_c - \sigma_a) \tag{7.160}$$

It can be verified by simultaneous measurements of the birefringence $\mathrm{n}_c - \mathrm{n}_a$, the inclination angle θ_c and the shear stress σ_{zx}. Straightforward calculations yield the following relation between the principal stresses σ_a, σ_c and the shear

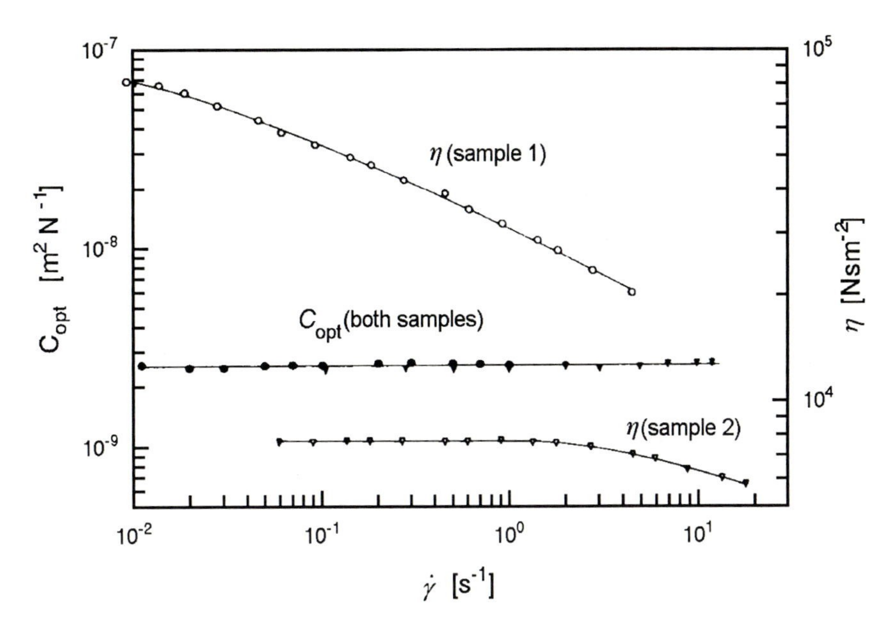

Fig. 7.23. Steady state shear viscosities and stress optical coefficient observed for two samples of PE with different molecular weight distributions at 150 °C. Measurements by Wales [81]

stress σ_{zx} in the laboratory-fixed coordinate system:

$$2\sigma_{zx} = (\sigma_c - \sigma_a) \sin 2\theta_c \tag{7.161}$$

Combination of Eqs. (7.160) and (7.161)gives

$$\frac{\Delta \text{n} \cdot \sin 2\theta_c}{2\sigma_{zx}} = C_{\text{opt}} \tag{7.162}$$

Figure 7.23 shows the results of measurements of C_{opt} based on this equation, carried out for two different samples of polyethylene over a wide range of shear rates $\dot{\gamma}$. As we note, C_{opt} is strictly constant, in striking contrast to the shear viscosities η which are included in the figure for comparison. We find here a linear relation which is valid within the range of non-linear mechanical behavior. Surely we are in the non-linear range: Birefringence is indicative of changes in the conformational distribution of the chains and this, in turn, leads to the non-linearity.

Indeed, validity of the linear stress-optical rule is a key observation with regard to the physical nature of the stresses created in flowing polymer melts. Generally speaking, stress in a polymer fluid arises from all forces acting between monomers on alternate sides of a reference plane. In polymers, we may divide them into two parts. We have first the strong valence bond forces

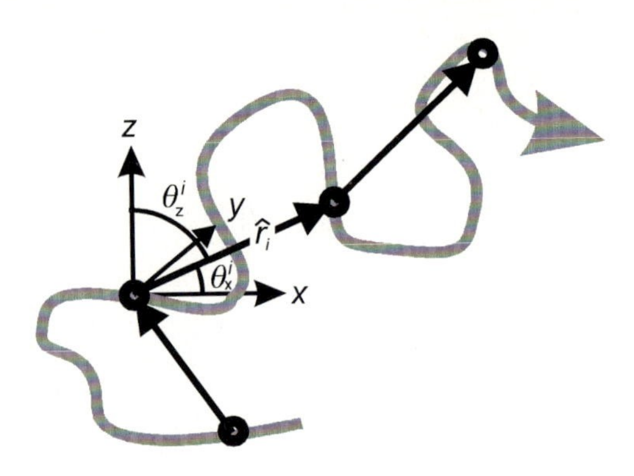

Fig. 7.24. Representation of a polymer chain as a series of freely jointed springs with extensions $\hat{r}_i$ and orientation angles $\theta_x^i, \theta_y^i, \theta_z^i$

which are effective along a polymer chain and secondly the non-bonded interactions active between all monomers on adjacent positions, they may belong to the same or to different chains. Having two contributions to the stress, the question arises as to how this can be accomodated with the validity of the stress-optical rule. As we can only anticipate optical effects for intrachain deformations, we must conclude that the intramolecular forces dominate the situation in flowing polymer melts, the effects of the non-bonded interactions being negligible.

Indeed, starting from this picture, one can verify the stress-optical rule and set it on a microscopic basis. We describe a chain in the spirit of the Rouse-model, as sketched in Fig. 7.24. Each polymer is subdivided in sequences of equal size, long enough to ensure that they behave like elastic springs, with a force constant b_{R} given by Eq. (6.18)

$$b_{\mathrm{R}} = \frac{3kT}{a_{\mathrm{R}}^2} \tag{7.163}$$

Using this representation of the chains in a melt, they may be entangled or not, we can deduce the linear stress-optical rule. Previously, in Eq. (6.60), we formulated an expression for the contribution of stretched springs to the shear stress σ_{zx}

$$\sigma_{zx} = \frac{b_{\mathrm{R}}}{v} \sum_i \hat{x}_i \hat{z}_i$$

The summation included all springs i in the volume v, with extensions $\hat{x}_i$ and $\hat{z}_i$ along the $x-$ and $z-$directions. As all springs behave equivalently we can also write

$$\sigma_{zx} = b_{\mathrm{R}} c_{\mathrm{spr}} \langle \hat{x}_i \hat{z}_i \rangle \tag{7.164}$$

where c_{spr} is the number density of springs. We now employ the same arguments to find an expression for the tensile stress. It reads, in full analogy to

Eq. (7.164)

$$\sigma_{zz} = b_{\mathrm{R}} c_{\mathrm{spr}} \langle \hat{z}_i \hat{z}_i \rangle \quad (7.165)$$

For the normal stress difference we obtain correspondingly

$$\sigma_{zz} - \sigma_{xx} = c_{\mathrm{spr}} b_{\mathrm{R}} (\langle \hat{z}_i^2 \rangle - \langle \hat{x}_i^2 \rangle) \quad (7.166)$$

The averaging carried out on the right-hand sides involves two steps, a time average over the fluctuations of each spring and subsequently the average over all springs. $\langle \hat{z}_i^2 \rangle$ and $\langle \hat{x}_i^2 \rangle$ denote the resulting mean-squared extensions along z (the unique axis) and x.

Introducing the total extension $\hat{r}_i$ and the orientation angles θ_z^i and θ_x^i as described in Fig. 7.24, we may write for each spring i

$$\hat{z}_i = \hat{r}_i \cos \theta_z^i \quad (7.167)$$

$$\hat{x}_i = \hat{r}_i \cos \theta_x^i \quad (7.168)$$

We insert also the above expression for b_{R} and obtain

$$\sigma_{zz} - \sigma_{xx} = c_{\mathrm{spr}} \frac{3kT}{a_{\mathrm{R}}^2} \left(\langle \cos^2 \theta_z^i \rangle - \langle \cos^2 \theta_x^i \rangle \right) \langle \hat{r}_i^2 \rangle \quad (7.169)$$

with the assumption of independent fluctuations in $\hat{r}_i$ and the orientation angles. Since we have for each spring i

$$\cos^2 \theta_x^i + \cos^2 \theta_y^i + \cos^2 \theta_z^i = 1 \quad (7.170)$$

for the average over all springs correspondingly

$$\langle \cos^2 \theta_x^i \rangle + \langle \cos^2 \theta_y^i \rangle + \langle \cos^2 \theta_z^i \rangle = 1 \quad (7.171)$$

furthermore, due to the overall uniaxial symmetry

$$\langle \cos^2 \theta_x^i \rangle = \langle \cos^2 \theta_y^i \rangle \quad (7.172)$$

and therefore

$$\langle \cos^2 \theta_x^i \rangle = \frac{1}{2} (1 - \langle \cos^2 \theta_z^i \rangle) \quad (7.173)$$

we obtain for the normal stress difference

$$\sigma_{zz} - \sigma_{xx} = c_{\mathrm{spr}} \frac{3kT}{a_{\mathrm{R}}^2} \left(\frac{3 \langle \cos^2 \theta_z^i \rangle - 1}{2} \right) \langle \hat{r}_i^2 \rangle \quad (7.174)$$

The expression in parentheses on the right-hand side characterizes the degree of orientation of the springs in the sample and is known as 'orientational order parameter' (more about this parameter will be said in the next chapter). We denote it $\mathrm{S}_{\mathrm{or}}^{\mathrm{spr}}$ and write shortly

$$\sigma_{zz} - \sigma_{xx} = c_{\mathrm{spr}} \frac{3kT}{a_{\mathrm{R}}^2} \langle \hat{r}_i^2 \rangle \mathrm{S}_{\mathrm{or}}^{\mathrm{spr}} \quad (7.175)$$

Next, we consider the optical properties of a spring. A stretched spring is an optically anisotropic object, and we may write down a general expression for the associated polarizability tensor, denoted $\boldsymbol{\beta}_{\mathrm{spr}}$. It must have the form

$$\boldsymbol{\beta}_{\mathrm{spr}} = n_{\mathrm{m}}\boldsymbol{\beta} \cdot \Phi(f) \tag{7.176}$$

Here, $\boldsymbol{\beta}$ stands for the polarizability tensor of one monomer unit in a perfect orientation parallel to the stretching direction of the spring, and n_{m} is the number of monomers per spring. $\Phi(f)$ is a certain function of the tensile force in the spring, of even character

$$\Phi(f) = \Phi(-f) \tag{7.177}$$

and with the limiting values

$$\Phi(f \rightarrow 0) \longrightarrow 0 \tag{7.178}$$

$$\Phi(f \rightarrow \infty) \longrightarrow 1 \tag{7.179}$$

There are good arguments in support of Eq. (7.176). Application of a force f causes simultaneously with the stretching of the spring also an orientation of the incorporated segments. The degree of orientation must be an unambiguous function of f only, not dependent on n_{m}. The latter point becomes clear when we consider the change in $\boldsymbol{\beta}_{\mathrm{spr}}$ resulting from a coupling in series of n_{spr} springs with equal forces. One expects, of course,

$$\boldsymbol{\beta}_{\mathrm{spr}} \sim n_{\mathrm{spr}} \tag{7.180}$$

and this is only fulfilled by the form Eq. (7.176). The even symmetry of $\Phi(f)$ and the limits $\Phi(f \rightarrow 0, \infty)$ have obvious reasons.

We may also formulate an expansion in powers of f valid for small forces. In the absence of linear terms we have

$$\boldsymbol{\beta}_{\mathrm{spr}} = n_{\mathrm{m}}\boldsymbol{\beta}\Phi_2 f^2 \tag{7.181}$$

with

$$\Phi_2 := \frac{1}{2}\frac{\mathrm{d}^2\Phi}{\mathrm{d}f^2} \tag{7.182}$$

In the next chapter we will discuss again the birefringence of uniaxially deformed samples. As will be shown, one can relate $\Delta\mathrm{n}$ to the degree of orientation of the monomers in a sample, as expressed by the associated orientational order parameter, $\mathrm{S}_{\mathrm{or}}^{\mathrm{m}}$, and the anisotropy of the polarizability per monomer,$\Delta\beta$, using

$$\Delta\mathrm{n} = \frac{\bar{\mathrm{n}}^2 + 2}{\bar{\mathrm{n}}}\frac{1}{6\epsilon_0}c_{\mathrm{m}}\Delta\beta\mathrm{S}_{\mathrm{or}}^{\mathrm{m}} \tag{7.183}$$

(see Eqs. (8.37) and (8.17)). We may employ the same equation to formulate $\Delta\mathrm{n}$ for an ensemble of springs instead of the monomers, thus obtaining

$$\Delta\mathrm{n} = \frac{\bar{\mathrm{n}}^2 + 2}{\bar{\mathrm{n}}}\frac{1}{6\epsilon_0}c_{\mathrm{spr}}\Delta\beta_{\mathrm{spr}}\mathrm{S}_{\mathrm{or}}^{\mathrm{spr}} \tag{7.184}$$

As the spring extensions and thus the forces fluctuate, we write for the anisotropy, $\Delta\beta_{\mathrm{spr}}$, of the polarizability of a spring

$$\Delta\beta_{\mathrm{spr}} = n_{\mathrm{m}}\Delta\beta\,\langle\Phi(f)\rangle \tag{7.185}$$

Now we can express the stress-optical ratio. Adopting the series expansion Eq. (7.181) and using Eqs. (7.175), (7.184), and (7.185) we obtain

$$\frac{\Delta \mathrm{n}}{\sigma_{zz}-\sigma_{xx}} = \frac{\bar{\mathrm{n}}^2+2}{\bar{\mathrm{n}}}\frac{1}{6\epsilon_0}\frac{n_{\mathrm{m}}\Delta\beta a_{\mathrm{R}}^2}{3kT}\frac{\langle f^2\rangle}{\langle \hat{r}_i^2\rangle}\cdot\Phi_2 \tag{7.186}$$

Since

$$\mathbf{f} = \frac{3kT}{a_{\mathrm{R}}^2}\hat{\boldsymbol{r}}_i \tag{7.187}$$

and we have for an ideal chain

$$a_{\mathrm{R}}^2 = n_{\mathrm{m}}a_0^2$$

there finally results

$$\frac{\Delta \mathrm{n}}{\sigma_{zz}-\sigma_{xx}} := C_{\mathrm{opt}} = \frac{1}{2\epsilon_0}\frac{\bar{\mathrm{n}}^2+2}{\bar{\mathrm{n}}}\frac{kT}{a_0^2}\Phi_2\Delta\beta \tag{7.188}$$

Thus, we have indeed found an expression for C_{opt} which is constant for a given polymer. We learn that the stress-optical coefficient includes three microscopic parameters, the size a_0 of a monomer, expressing the chain stiffness, the optical anisotropy per monomer, $\Delta\beta$, and the coefficient Φ_2, which relates to the elastic restoring forces.

We have presented the linear stress-optical rule here as a basic property of polymer melts but, of course, it also holds for rubbers, with unchanged stress-optical coefficients. This must be the case, since stresses arise from a network of chains in both melts and rubbers, so that the arguments presented above apply for both systems equally. Figure 7.25 shows as an example the relation between birefringence and tensile stress as observed for a sample of natural rubber.

With this unambiguous evidence for the dominant rule of the network forces in mind, we return once again to Lodge's equation of state. We introduced it as an empirical expression constructed with the objective of combining the properties of rubbers with those of viscous liquids. It is possible to associate the equation with a microscopic model. As the entanglement network, although temporary in its microscopic structure, yields under steady state conditions stationary viscoelastic properties, we have to assume a continuous destruction and creation of the stress-bearing chain sequences. This implies that at any time the network will consist of sequences with different ages. As long as a sequence exists, it can follow all imposed deformations.

To describe this situation, we may proceed as follows. Let $n(t,t')\mathrm{d}t'$ be the number of stress-bearing sequences per unit volume, created during an

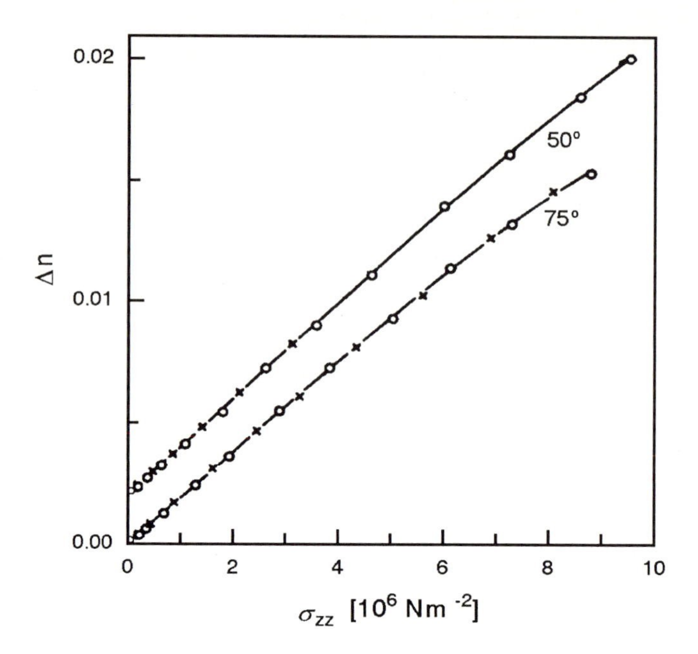

Fig. 7.25. Relation between birefringence and tensile stress for natural rubber. The upper curve is shifted in vertical direction. Work of Treloar [71]

intervall $\mathrm{d}t'$ at a past time t' and persisting up to the present time t. We furthermore assume that at the moment of creation these sequences do not bear any stress. Their contribution to the stress at t then may be expressed as

$$n(t,t')kT\mathbf{B}(t,t')\mathrm{d}t' \tag{7.189}$$

if we adopt ideal rubber behavior (Eqs. (7.74) and (7.86)). An equation in agreement with Lodge's equation of state is obtained, when writing

$$(\boldsymbol{\sigma} + p\mathbf{1})(t) = \int\limits_{t'=-\infty}^{t} n(t,t')kT\mathbf{B}(t,t')\mathrm{d}t' \tag{7.190}$$

which implies that the contributions of all active chain sequences with different ages are superimposed. To show the agreement, we rewrite the Lodge-equation

$$(\boldsymbol{\sigma} + p\mathbf{1})(t) = -\int\limits_{t'=-\infty}^{t} G(t-t')\frac{\mathrm{d}\mathbf{B}(t,t')}{\mathrm{d}t'}\mathrm{d}t' \tag{7.191}$$

by carrying out an integration by parts

$$(\boldsymbol{\sigma} + p\mathbf{1})(t) = -\mathbf{B}(t, t' = t)\frac{\mathrm{d}G}{\mathrm{d}t'}(t' = t) + \int_{t'=-\infty}^{t} \frac{\mathrm{d}G(t-t')}{\mathrm{d}t'} \cdot \mathbf{B}(t, t')\mathrm{d}t' \quad (7.192)$$

Since

$$\mathbf{B}(t, t) = \mathbf{1} \quad (7.193)$$

we may incorporate the first term on the right hand side in $p\mathbf{1}$ and write

$$(\boldsymbol{\sigma} + p\mathbf{1})(t) = \int_{t'=-\infty}^{t} m(t-t')\mathbf{B}(t, t')\mathrm{d}t' \quad (7.194)$$

with

$$m(t-t') = \frac{\mathrm{d}G(t-t')}{\mathrm{d}t'} \quad (7.195)$$

This is a second form for the equation of state of rubber-like liquids. $m(t-t')$ is called a 'memory function', since it characterizes in a general view the fading of the memory on the past. As Eqs. (7.190) and (7.194) have equal forms we find

$$m(t-t') = kTn(t, t') \quad (7.196)$$

The network model thus provides us with a microscopic interpretation of the memory function.

7.3 Further Reading

R.B. Bird, R.C. Armstrong, O. Hassager: *Dynamics of Polymeric Liquids,* Vol.1 *Fluid Mechanics*, John Wiley & Sons, 1977

R.B. Bird, R.C. Armstrong, O. Hassager: *Dynamics of Polymeric Liquids,* Vol.2 *Kinetic Theory*, John Wiley & Sons, 1977

P.J. Flory: *Principles of Polymer Chemistry*, Cornell University Press, 1953

W.W. Graessley: *Viscoelasticity and Flow in Polymer Melts and Concentrated Solutions* in J.E. Mark, A. Eisenberg W.W. Graessley, L. Mandelkern, J.L. Koenig: *Physical Properties of Polymers*, Am.Chem.Soc., 1984

H. Janeschitz-Kriegl: *Polymer Melt Rheology and Flow Birefringence*, Springer, 1983

R.G. Larson: *Constitutive Equations for Polymer Melts and Solutions*, Butterworths, 1988

W. Retting, H.M. Laun: *Kunststoff-Physik*, Hanser, 1991

L.R.G. Treloar: *The Physics of Rubber Elasticity*, Clarendon Press, 1975

Chapter 8

Yield Processes and Fracture

A prerequisite for the use of polymeric materials in daily life is their mechanical stability. Goods made of plastics must keep their form under the permitted loads and one has to be sure that under ordinary conditions they will not break. Therefore, a good knowledge of the limits of mechanical stability is of utmost importance. This chapter deals with these 'ultimate properties'. We will discuss the mechanisms of plastic flow in polymeric solids and also present some of the generally used basic concepts of fracture mechanics. Large parts of this chapter are only descriptive since the present microscopic understanding is not very advanced. From the long experience in polymer manufacturing there exists a wealth of empirical knowledge. It is far too broad to treat it here in a survey. So we will just consider some of the main observations.

Deformations of a polymeric solid may include, in addition to the reversible part, an irreversible flow. This plastic flow or 'yielding' sets in, when the stress becomes large enough and surmounts the 'yield point'. For polymers, this threshold is low compared to metals or ceramics. Moderate forces are often sufficient to induce yield processes. Temperature plays a big role and there are also effects from the environment when fluids or gases are present which permeate into the polymer.

From all the observations on various polymeric solids, it has become clear that there are two mechanisms of yielding. They have a quite different appearance and are easily discriminated. The first mechanism is known as 'shear yielding'. Figure 8.1 gives an example and depicts the behavior of polyethylene. Stretching a sample with a constant rate results in the load-extension curve shown. The force increases at first but then, reaching the 'yield-point', it passes over a maximum and somewhere a neck appears. Continuing the drawing, the neck extends up to the full length of the sample. This 'cold-drawing' takes place under an essentially constant tensile force and finally elongates the sample by several times its original length. If the stretching is continued after completion of the cold-drawing, the force increases up to the point of break. The sketches included in the figure illustrate all these changes. 'Shear yielding' is typical for partially crystalline polymers and is also found for several amorphous polymers, as for example polycarbonate (PC). Figure 8.2 is a photograph of the begin of necking in polycarbonate and explains the

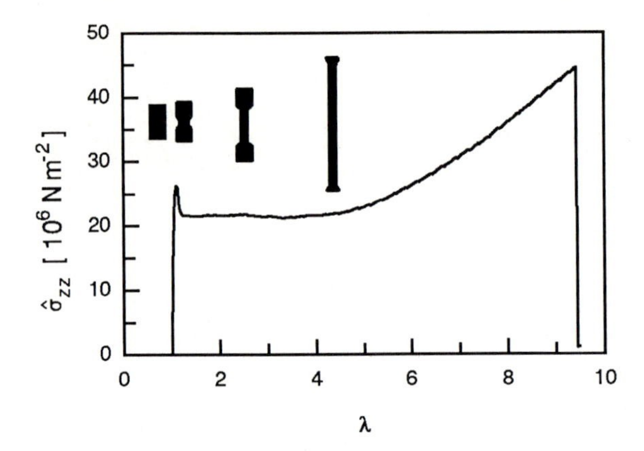

Fig. 8.1. Load ($\hat{\sigma}_{zz}$: nominal tensile stress)-extension (λ: extension ratio) curve of a sample of PE ($M = 3.6 \cdot 10^5$, drawing velocity $d\lambda/dt = 2.4 \cdot 10^{-2} s^{-1}$). The changes in the shape of the sample are schematically indicated

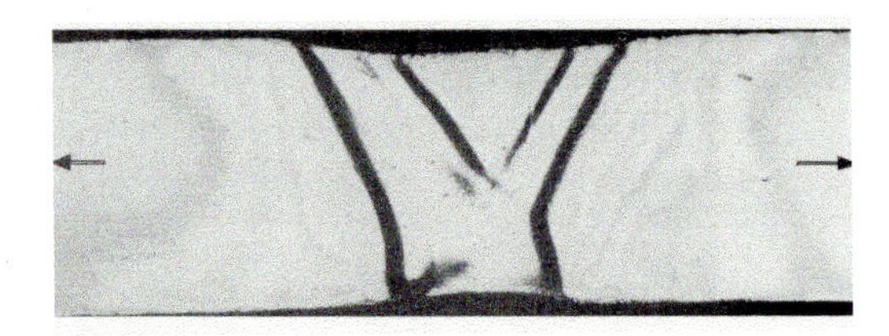

Fig. 8.2. Formation of shear bands at the begin of necking, observed for a sample of PC. The *arrows* indicate the direction of the applied tensile stress. Micrograph obtained by Morbitzer [82]

origin of the name 'shear yielding': We observe here shear bands, and they are preferentially oriented along the directions of maximum shear stress.

A quite different behavior is observed for polystyrene. Figure 8.3 depicts again a load-extension curve. The force increases at first linearly but then, after a slight bending, the sample breaks before a maximum is reached. In the bending range just before fracture, a 'whitening' is observed, as shown by the photograph. Closer inspection reveals the formation of many void containing microdeformation zones. These localized zones of plastic flow are called 'crazes', and 'crazing' is the term used to address this second mechanism of yield. 'Shear yielding' and 'crazing' are not alternative processes which exclude each other. Often both show up together, or one follows the other. Which one is dominant depends on the stress conditions and the temperature.

The amount of flow before fracture determines the 'ductility' of a polymer sample. In 'tough' materials, a considerable amount of energy is dissipated by yield processes prior to fracture. In contrast, 'brittle' samples break without showing much preceding flow. The difference becomes apparent in the fracture

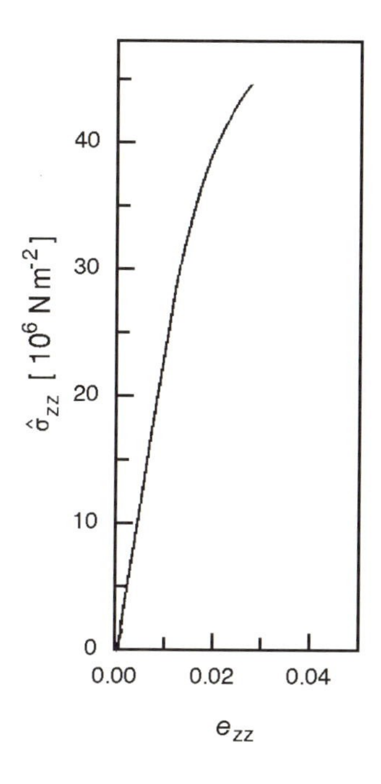

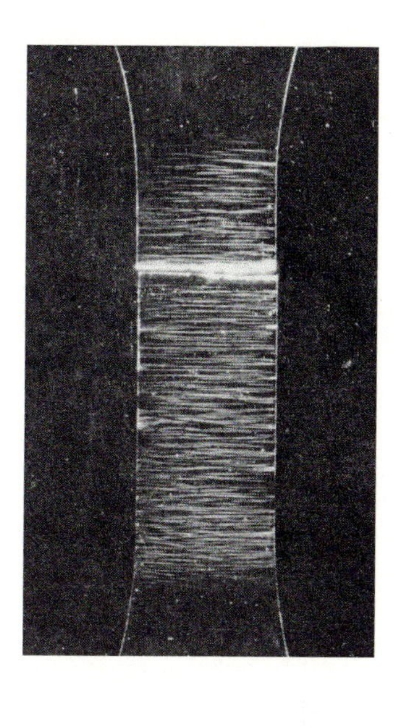

Fig. 8.3. Tensile stress-strain curve of a sample of PS and photograph showing crazes

energy which is much lower for brittle samples than for tough compounds. It is clear that the fracture energy, as determined by the area under the load-extension curve, is the appropriate measure for the ductility of a given sample. Therefore, regarding our examples, polyethylene is to be considered as tough and polystyrene as brittle.

The ductility may change with temperature. This is demonstrated by Fig. 8.4, showing measurements on poly(vinylchloride) at different temperatures and, in addition, a series of load-extension curves obtained at one temperature for various strain rates. As we can see, the brittleness of the sample increases continuously upon lowering the temperature, and also on increasing the strain rate. These tendencies, which are generally found, have obvious reasons. Yielding is based on specific relaxation processes and thus depends on the ratio of strain rates to the respective relaxation rates. Large values of this ratio imply that plastic flow cannot take place and this results in brittle behavior. Sometimes, an early break at high strain rates is also caused by adiabatic heating effects. If the heat is not conducted away rapidly enough, crystallites may melt and the sample fails by a kind of melt fracture. Toughness is a property which is appreciated in many applications. The desire for high-performance polymer materials is to have, simultaneously, a high yield point, high values of the elastic modulus and a high ductility. To achieve this synergism is a difficult task and a main goal of industrial research.

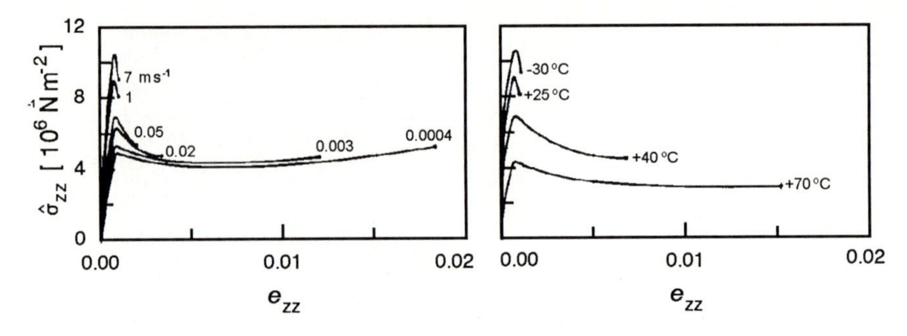

Fig. 8.4. Load-extension curves of PVC measured at room temperature for the indicated strain rates (*left*) and at different temperatures for a constant strain rate ($\dot{e}_{zz} = 1\ \mathrm{ms}^{-1}$) (*right*). From Retting [83]

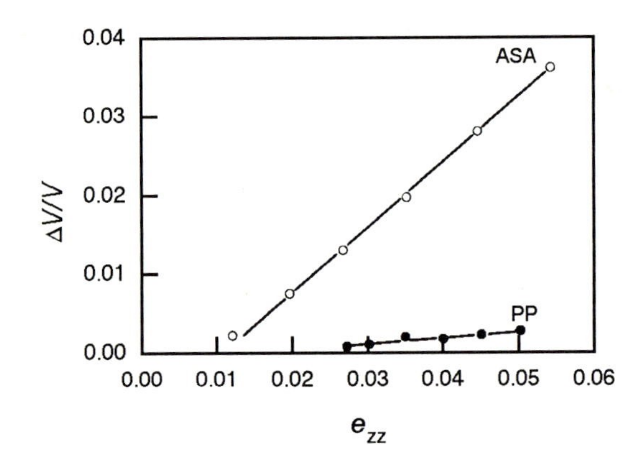

Fig. 8.5. Relationship between volume changes and strains, observed for creep experiments on PP and ASA. Work of Bucknall [84]

At the end of this introductory section we present in Fig. 8.5 results of a measurement which can be used to discriminate between the two yield processes and to determine the respective weights. The method involves a simultaneous monitoring of the extension and the volume change during a tensile creep experiment. Since shear yielding takes place at constant volume, any volume changes are indicative of crazing and enable us to detect it. The two chosen polymers, polypropylene (PP) and 'ASA', a polymer composite with complex multiphase microstructure, represent extreme cases. A slope of unity in the plot of $\Delta V/V$ versus strain, as observed for ASA, indicates a deformation entirely by crazing, whereas the slope close to zero shown by polypropylene indicates nearly pure shear yielding.

8.1 Shear Yielding

8.1.1 Mechanics of Neck Formation

Let us come back to Fig. 8.1 with the observations on a sample of polyethylene drawn at room temperature with a constant extension rate. Having reached a critical value for the extension, flow sets in at some point of the sample and is accompanied by a drop in the load. The lateral sample dimensions narrow and finally a neck is formed, with a smaller, but again stable cross-section area. Between the original parts and the neck, we find a 'shoulder' with a continuous profile. Further extension of the sample is achieved by an increase in the neck, i.e. a move of the two shoulders over the sample. The external force which has to be applied for this 'cold-drawing' remains constant as long as the length of the neck increases.

It is interesting to analyse in more detail the deformation process accomplished by the passage of the shoulders across the sample. We select a sample with cylindrical shape, consider a volume element at the centerline and follow the changes in its shape from the original state into the extended state in the neck. The states of deformation imposed on the volume element if the shoulder is moved over its position are well-defined. Since shear flow occurs without a volume change, the sequence of deformation states can be directly derived from the profile of the shoulder. We choose a moving coordinate system fixed at the center of the shoulder and describe the profile of the shoulder by the function $b(z)$, giving the radius b of the sample as a function of the coordinate z, as indicated in Fig. 8.6. For a volume element at the centerline, with radius l_ρ and length l_z, we have due to its incompressibility the relation

$$l_z(z) \cdot \pi l_\rho^2(z) = \text{const} \tag{8.1}$$

Its extension, λ, depends on z only, being given by

$$\lambda(z) = \frac{l_z(z)}{l_z(-\infty)} = \frac{l_\rho^2(-\infty)}{l_\rho^2(z)} = \frac{b^2(-\infty)}{b^2(z)} \tag{8.2}$$

The profile also determines the strain rate, by

$$\frac{\mathrm{d}\lambda}{\mathrm{d}t} = \frac{\mathrm{d}\lambda}{\mathrm{d}z} \cdot \frac{\mathrm{d}z}{\mathrm{d}t} = -\frac{2b^2(-\infty)}{b^3(z)} \cdot \frac{\mathrm{d}b}{\mathrm{d}z} \cdot \frac{\mathrm{d}z}{\mathrm{d}t} \tag{8.3}$$

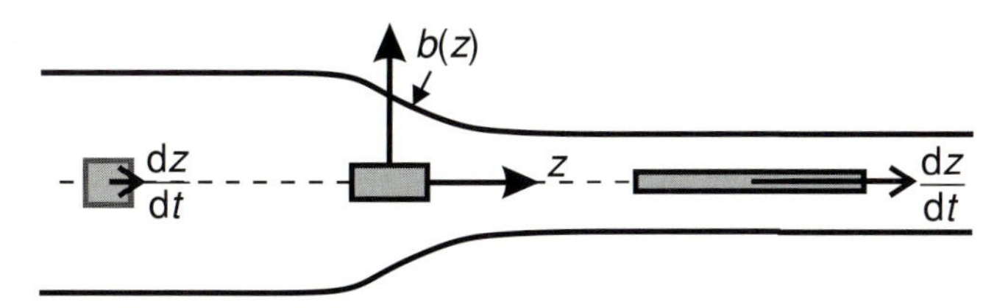

Fig. 8.6. States of deformation of a volume element passed over by a shoulder with the profile $b(z)$, and the velocity $\mathrm{d}z/\mathrm{d}t$ relative to the shoulder

Here, $\mathrm{d}z/\mathrm{d}t$ denotes the velocity of the volume element relative to the shoulder as measured in the moving coordinate system. Since the material flow through the shoulder is a constant, we may write

$$\pi b^2(z)\frac{\mathrm{d}z}{\mathrm{d}t} = \text{const}$$

and therefore obtain

$$\frac{\mathrm{d}\lambda}{\mathrm{d}t} \sim -\frac{1}{b^5(z)}\frac{\mathrm{d}b}{\mathrm{d}z} \tag{8.4}$$

or, in terms of the Hencky strain rate employed in rheological treatments

$$\dot{e}_{\mathrm{H}} := \frac{1}{\lambda}\frac{\mathrm{d}\lambda}{\mathrm{d}t} \sim -\frac{1}{b^3}\frac{\mathrm{d}b}{\mathrm{d}z} \tag{8.5}$$

Equation (8.5) tells us, that the extension rate of the volume element is not a constant, but strongly time-dependent. It starts from zero, then increases, passes over a maximum near to the position of the largest slope in the profile and finally comes back to zero.

It is important to note that this peculiar time-dependence of the deformation rate is associated with a simple law for the evolution of stress. Since the external force during cold-drawing is constant, the tensile stress acting on the volume element follows from

$$\sigma_{zz}(z)\pi b^2(z) = \text{const} \tag{8.6}$$

if we assume a uniform distribution over the cross-section. As a consequence stress and extension become linearly related

$$\sigma_{zz}(z) \sim \lambda(z) \tag{8.7}$$

Equation (8.7) is imposed by the conditions of the drawing-experiment. Indeed, it is exactly this requirement by which the time-dependent deformation $\lambda(t)$ is selected and thus the profile determined.

In order to understand better how this selection mechanism works we need a knowledge of elementary stress-extension curves, for example those which are measured for constant Hencky strain rates. If flow sets in locally and there the radius b of the sample begins to decrease, the extension λ of the volume element at the centerline is given by

$$\lambda(t) = \frac{b^2(0)}{b^2(t)} \tag{8.8}$$

The extension rate follows as

$$\frac{\mathrm{d}\lambda}{\mathrm{d}t} = -\frac{2b^2(0)}{b^3(t)}\cdot\frac{\mathrm{d}b}{\mathrm{d}t} \tag{8.9}$$

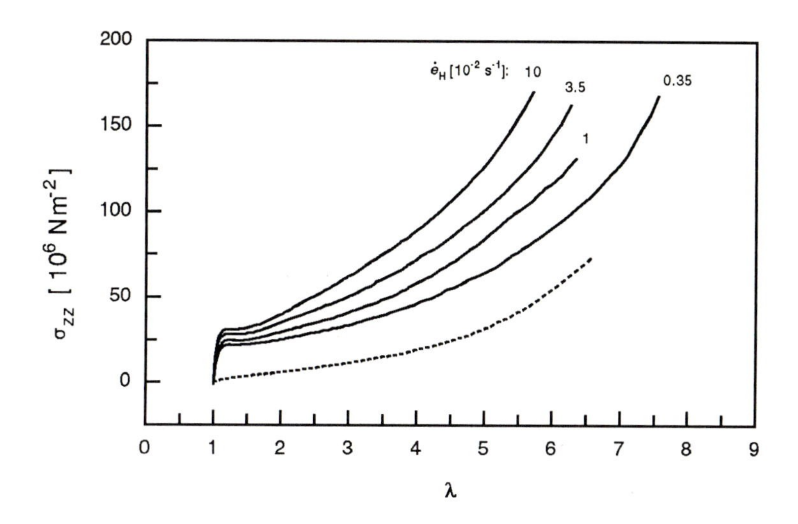

Fig. 8.7. Stress-extension curves measured for a sample of PE ($M = 3.6 \cdot 10^5$) at the indicated Hencky strain rates. Constant strain rates were realized by a registration of the strain at the location of a developing neck and a continuous readjustment of the applied tensile force, using an electronically controlled feedback circle. The *broken line* gives the $\sigma_{zz}(\lambda)$-curve measured for a poly(ethylene-*co*-vinylacetate)(27% vac-units, $\phi_c = 0.30$). No strain rate dependence is observed for this rubbery material [85]

and the Hencky strain rate as

$$\dot{e}_H := \frac{1}{\lambda}\frac{d\lambda}{dt} = -\frac{2}{b(t)} \cdot \frac{db}{dt} \tag{8.10}$$

A deformation of the central volume element with a constant Hencky strain rate, $\dot{e}_H$, thus corresponds to an exponential time-dependence of the minimum radius in the flow zone

$$b(t) = b(0) \exp -\frac{\dot{e}_H}{2} t \tag{8.11}$$

It is possible to realize this time-dependence by introducing a feedback-loop into a standard stress-elongation measuring device. Figure 8.7 presents a series of stress-strain curves $\sigma_{zz}(\lambda)$ thus obtained for polyethylene, for different values of the Hencky-strain rate. We see that all the curves exhibit a plateau at intermediate deformations. With increasing Hencky strain rate the stress increases throughout the whole range.

Figure 8.8 shows the relation between the plateau stress, σ_y, and the Hencky-strain rate. It is well represented by the expression

$$\sigma_y = \sigma_0 \log(\text{const} \cdot \dot{e}_H) \tag{8.12}$$

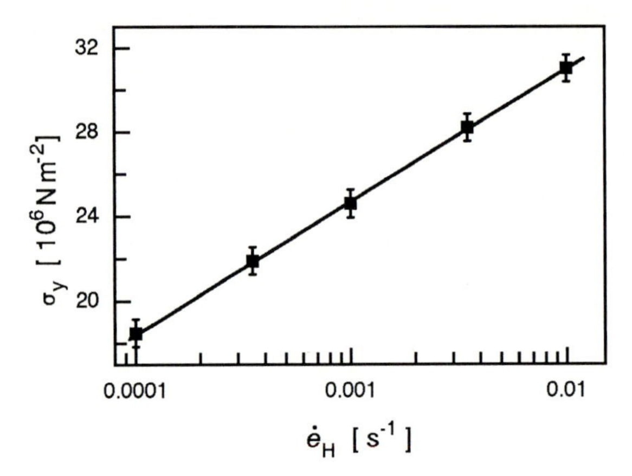

Fig. 8.8. Dependence of the stress at the plateau, σ_y, on the Hencky strain rates, extracted from the data in Fig. 8.7 [85]

or reversely by

$$\dot{e}_{\mathrm{H}} \sim \exp \frac{\sigma_y}{\sigma_0} \tag{8.13}$$

Equation (8.13) suggests an activated process, whereby the stress controls the activation energy.

Let us for the moment forget about the strain rate effects and assume a rate independent stress-strain curve of the same type as the experimental curves. We may ask for the expected result of an experiment where the externally applied tensile stress increases linearly with the extension, $\sigma_{zz} \sim \lambda$, which is the condition experienced by each volume element during cold-drawing. Figure 8.9 depicts the situation. If the load is high enough to induce shear flow at a critical extension, λ_{A}, the stress will follow the straight line included in the figure. Obviously, equilibrium cannot be established over a large range of deformations up to the point where the line and the stress-extension curve cross each other again. This point, λ_{B}, then constitutes the new equilibrium. Between λ_{A} and λ_{B}, there is a gap in the sequence of accessible states of deformation. As we can see, this gap is a consequence of the non-linearity of the stress-extension curve.

Absence of deformation states between λ_{A} and λ_{B} is the ideal case. The observation of a shoulder with finite height in cold-drawn polymer samples tells us that reality is different, as this implies a continuous transition from λ_{A} to λ_{B}. The establishment of force-balanced states also in the transition region is evidently accomplished by the strain rate effects. Quite generally, stresses grow and decay with increasing and decreasing strain rate respectively, thus allowing for a continuous readjustment of the internal forces to the linear relation, $\sigma_{zz} \sim \lambda$, imposed by the drawing-experiment. According to Eq. (8.5),

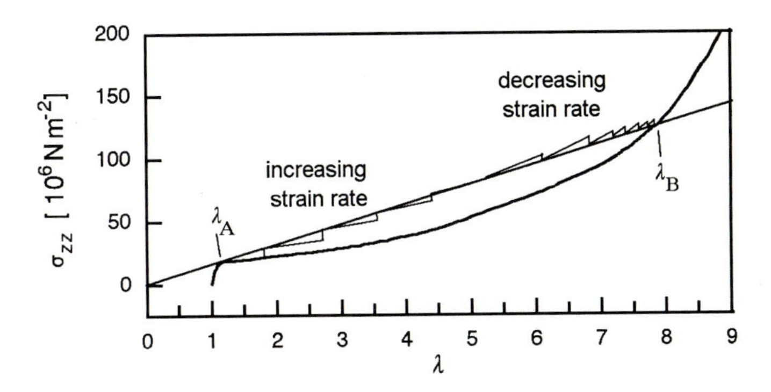

Fig. 8.9. Mechanical response of a volume element in a fiber with the shown $\sigma(\lambda)$ dependence, if subjected to a linearly increasing stress as represented by the *straight line.* For a hypothetical sample without internal friction there is no equilibrium between λ_A and λ_B. In the real sample, a balance is achieved by the strain rate dependent viscous forces (*thin line with steps*)

the shoulder profile determines the strain rate at each point, and it possesses exactly that shape which is necessary for this balance.

So far we have considered the states of deformation of a volume element which is passed over by the moving shoulder but have not directly touched the primary question: Why does necking occur at all and what are the prerequisites? Actually, our discussion of the factors determining the shoulder profile has focused already on the main point which gives the answer: Necking is a direct consequence of the occurrence of a gap in the sequence of accessible states. Figure 8.10 may help to further explain the situation. Being concerned about the equilibrium properties of an incompressible fiber with a sigmoidal $\sigma_{zz}(\lambda)$ dependence, we also should have a look at the shape of the associated force-extension curve, given by $f(\lambda) \sim \sigma_{zz}/\lambda$. As we can see, in contrast to $\sigma_{zz}(\lambda)$ which possesses a plateau, $f(\lambda)$ exhibits a maximum and a minimum. The extrema are located at the positions, where

$$\frac{\mathrm{d}}{\mathrm{d}\lambda}\frac{\sigma_{zz}}{\lambda} = 0 \tag{8.14}$$

or

$$\frac{\mathrm{d}\sigma_{zz}}{\mathrm{d}\lambda} = \frac{\sigma_{zz}}{\lambda} \tag{8.15}$$

and these can be determined by the tangent-construction shown in the figure. The procedure is known as the 'Considère-construction'. In a next step, we may calculate the free energy of the fiber in dependence on λ. It follows as

$$\mathcal{F}(\lambda) = \int_1^\lambda f \mathrm{d}\lambda' \tag{8.16}$$

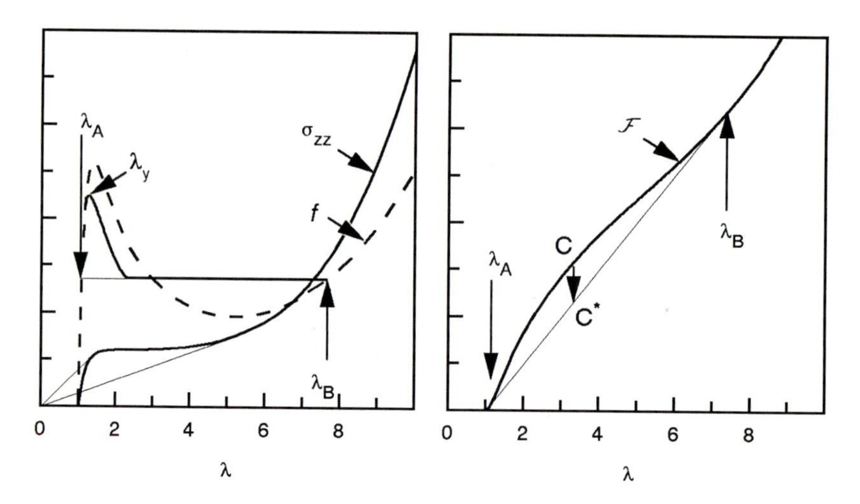

Fig. 8.10. λ-dependence of the retracting force f (*left, broken line*) and the free energy $\mathcal{F}$ (*right*) of an elastic fiber with a sigmoidal stress-extension relation $\sigma_{zz}(\lambda)$ (*left, continuous line*). The locations of the extrema in $f(\lambda)$ follow from the Considère tangent-construction, as indicated on the *left*

and the result is also included in the figure. Importantly, the shape of $\mathcal{F}(\lambda)$ indicates that we have three ranges which differ in the mechanical stability, namely

$$\begin{aligned} \lambda < \lambda_{\rm A} &\quad : \quad \text{a (small) stable range} \\ \lambda_{\rm A} < \lambda < \lambda_{\rm B} &\quad : \quad \text{an instable intermediate zone} \\ \lambda > \lambda_{\rm B} &\quad : \quad \text{a final stable range} \end{aligned}$$

$\lambda_{\rm A}$ and $\lambda_{\rm B}$ are determined by the common tangent on two points of $\mathcal{F}$ (λ), as is shown in the drawing. The cause of the instability and the consequences are evident: A transition from the homogeneous state C to a mixture of states A and B, as represented by the point C* on the connecting line, results in a decrease of $\mathcal{F}$ and will therefore take place spontaneously. The situation is perfectly analogous to the transition to a two-phase state found for a binary mixture within the miscibility gap which was discussed earlier (compare Fig. 3.16 and the related explanations). Now we have a gap in the sequence of accessible states of deformation. Hence, a hyperelastic fiber with a sigmoidal $\sigma_{zz}(\lambda)$-curve which leads to a maximum and minimum in the load-extension curve must disintegrate within a certain range of λ's into a two-phase structure, with parts extended to $\lambda_{\rm A}$ and other parts extended to $\lambda_{\rm B}$. The 'equilibrium phases' either follow from the common tangent-construction applied on $\mathcal{F}(\lambda)$, or equivalently by searching for the Maxwell-line solution for $f(\lambda)$ (equal values for the integral between $\lambda_{\rm A}$ and $\lambda_{\rm B}$, independent of whether it is calculated for the horizontal 'Maxwell-line' or along the curve $f(\lambda)$). If these conditions are given, an experimental load-extension curve might look like the one indicated in the figure: After the first increase, the load drops at some point $\lambda_y > \lambda_A$ and decays to the horizontal line. λ_y may be identified with the yield point. The load drop is accompanied by the first formation of the second phase, with elongation λ_B, i.e. the appearance of the neck. A move along the horizontal line, i.e. the drawing at constant load, is accomplished by the neck extension. After having reached $\lambda = \lambda_B$, the second phase extends furthermore.

Of course, cold-drawing of a polymeric solid with an amorphous or partially crystalline structure does not exactly agree with the drawing of this hypothetical hyperelastic fiber, as there are large contributions of friction by internal forces and also irreversible structural changes. As mentioned above in the discussion of the shoulder profile, this produces strain rate effects, which lead in the dynamic drawing process to the occurrence of transition states with deformations intermediate between the two coexisting states. However, a qualitative change in the stability criteria is not to be expected and we may therefore hold onto the criterion as it is also stated in the literature: Necking can be anticipated if the Considère-construction applied to the elementary stress-extension curves, $\sigma_{zz}(\lambda)$, finds two points with $d\sigma_{zz}/d\lambda = \sigma_{zz}/\lambda$.

8.1.2 Structure Changes on Cold-Drawing

The discussion of the mechanics of the cold-drawing process also leads us to the next question: What are the structural changes in a volume element when converted from the original state into the extended state in the neck, or, what is the microscopic background of the peculiar shape of the measured $\sigma_{zz}(\lambda)$ curves? Evidently the extension introduces anisotropy in an originally isotropic sample due to the chains becoming preferentially aligned in drawing direction and this occurs for both partially crystalline and amorphous polymers. Interestingly enough, this orientation process is associated with a stress-extension curve which shows similarities to the $\sigma_{zz}(\lambda)$-curve of a rubber. Figure 8.7 includes, for a comparison, the stress-extension curve of an ethylene-vinylacetate statistical copolymer, a rubber-like material where crystallites act as physical cross-links. One observes an offset to higher stresses for the solid samples but the increases at higher λ's look quite comparable. Hence, it appears that the structural changes during cold-drawing include the stretching of a network as a major ingredient. This network is built up by the entanglements, as in the melt. At low stresses the elements of the solid structure, the crystallites in the partially crystalline samples, or the solid amorphous regions in a glass prevent any network deformation. If the applied stress overcomes the yield point these solid elements are no longer stable. The lamellar crystallites rotate and disintegrate into mosaic blocks and the solid glassy structure also is continuously reorganized. The reorganisation of the solid structure continues throughout the plateau region, being accompanied by an increasing chain alignment. In this range, the main part of the stress is

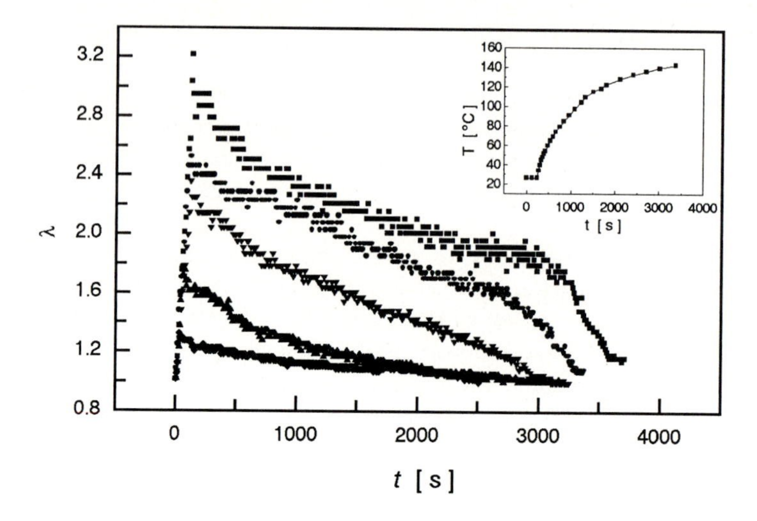

Fig. 8.11. Shrinkage of cold-drawn PE-samples (same as Fig. 8.7) upon heating. The heating-curve is given in the insert [85]

attributed to the work suspended in the structure reorganisation. The situation changes at high λ's, where the network becomes dominant. The observed strain hardening, with an appearance similar to a rubber, is most probably also due to finite extensibilities, here of chain parts between entanglements. In partially crystalline samples like polyethylene, the effect might be further enhanced by the formation of a stable crystallite texture able to sustain high stresses.

As we can see, even in polymeric solids and not just in melts, entanglements play an important role. There is more supporting evidence. The behavior of a cold-drawn polymer sample is quite spectacular when it is heated up to temperatures above the melting point or the glass transition temperature: A shrinkage is observed which finally results in a recovery of the shape before cold-drawing. Figure 8.11 presents observations on the polyethylene sample of Fig. 8.7, drawn at ambient temperature to various values of λ ($\lambda = 1.6 - 3.2$). Unloading at the drawing temperature results in a small contraction only, but on heating the shrinking continues and finally brings the sample back to its original length. Obviously the memory is preserved by the entanglement network. If crystallites melt or become too small to sustain the entropic chain forces, the network extension decreases and at the end the original isotropic state is recovered.

There is a second observation which is also directly related to the existing entanglement network: A reduction in the entanglement density greatly increases the drawability of samples. One possible route for a reduction is the 'gel-spinning' of polyethylene. High molecular weight polyethylene is dis-

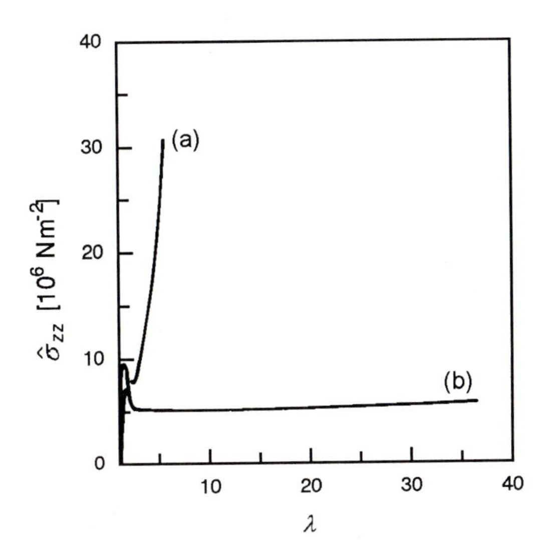

Fig. 8.12. Load-extension curves observed on drawing at room temperature a melt crystallized sample of PE (*a*) and a gel-spun PE-fiber (*b*). From Lemstra et al.[86]

solved in a solvent, and the solution is spun through a die into a cooling bath, thus producing a gel. Then the solvent is removed by drying or extraction and the fiber is drawn again. Figure 8.12 shows a typical load-extension curve obtained in this way, in comparison with the usual cold-drawing behavior: We can see that the drawability has increased tremendously. The reason for this drastic change is the reduction of the number of entanglements by the dissolution process and the transfer of this lowered number to the gel and the dried fiber. For this mechanism, one expects an inverse relation between the polymer concentration in the solution and the drawability of the fiber and this is indeed observed: Figure 8.13 depicts the dependence of the maximum draw ratio, $\lambda_{\max}$, on the volume fraction ϕ of polyethylene in the solution, and it corresponds to a function $\lambda_{\max} \sim \phi^{-1/2}$.

The chain orientation produced by drawing generally increases the tensile modulus. Gel-spinning has gained technical importance since the 'ultra-drawing' attained in this special process yields particularly high stiffness values. Elastic moduli up to $2 \cdot 10^{11}$ Nm^{-2} have been obtained for draw ratios of about 100, which comes rather near to the theoretical limit given by the stiffness of an *all-trans* polyethylene chain ($E = 3 \cdot 10^{11}$ Nm^{-2}).

As the tensile properties of drawn samples are dependent on the achieved degree of orientation, a measure is required for its description. Most appropriate is the orientational order parameter $\mathrm{S}^{\mathrm{m}}_{\mathrm{or}}$, defined by the average

$$\mathrm{S}^{\mathrm{m}}_{\mathrm{or}} = \left\langle \frac{3\cos^2\theta - 1}{2} \right\rangle \tag{8.17}$$

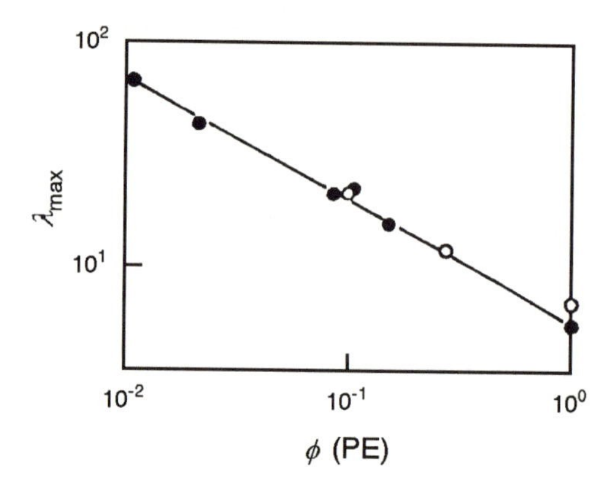

Fig. 8.13. Relation between the maximum draw ratio of a gel-spun PE fiber ($M > 10^6$) and the concentration in the solution. Data from Iguchi and Kyotani [87]

Here θ denotes the angle between the long axis of a monomer and the direction of drawing, i.e. the unique axis of the sample. S_{or}^{m} has the advantage that it can be directly deduced from the birefringence Δn, using the relation

$$S_{or}^{m} = \frac{\Delta n}{\Delta n_{max}} \tag{8.18}$$

Δn_{max} denotes the maximum birefringence realized in the ideal case of a perfectly oriented sample. A derivation of Eq. (8.18) is given at the end of this section.

Alternatively, for characterizing the state of orientation one may also employ NMR spectroscopy, with the best conditions being found for deuterated samples. The general procedure was sketched previously at the end of Sect. 5.3.4: As ^{2}H-spectra depend on the orientational distribution of the C-H bonds relative to the external magnetic field, they may be evaluated by model calculations under variation of the orientational distribution function of the monomers. NMR can discriminate between deuterons in crystallites and amorphous regions on the basis of the different mobilities which result in different 'spin-lattice relaxation times' and thus enables the degrees of orientation in both phases of a partially crystalline samples to be determined separately. Figure 8.14 gives an example and presents NMR spectra obtained for the crystalline and amorphous regions of a cold-drawn polyethylene registered for two different sample orientations relative to the magnetic field. Evaluation of the data indicates a substantially lower degree of orientational order in the amorphous regions than for the crystallites. Spectra can be reproduced, when assuming Gaussian orientational distribution functions with half widths of about 3 degrees for the crystallites and 24 degrees for the monomers in the amorphous regions.

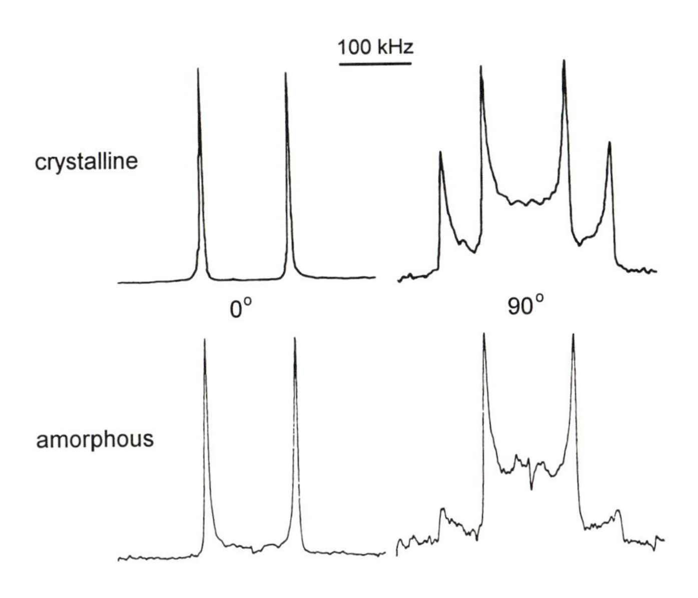

Fig. 8.14. ^{2}H NMR spectra registered for the crystalline and the amorphous phase of cold-drawn PE. Spectra were obtained for orientations of the unique axis of the samples parallel (0°) and perpendicular (90°) to the magnetic field. From Spiess et al. [88]

Fig. 8.15. Morphological changes on cold-drawing, observed for a film of PE in a polarizing optical microscope. Micrographs obtained by Hay and Keller [89]

More detailed insight into the morphology changes in partially crystalline polymers which accompany the transformation from the original isotropic structure into the extended oriented state comes from microscopy and X-ray scattering. The general conclusion drawn from numerous studies is that the original spherulitic structure, being set up of an isotropic distribution of stacks of layer-like crystallites, is converted into a fibrillar morphology. The latter is composed of microfibrils with transverse diameters in the order of 10 nm. Figure 8.15 presents for illustration micrographs obtained in a polarizing

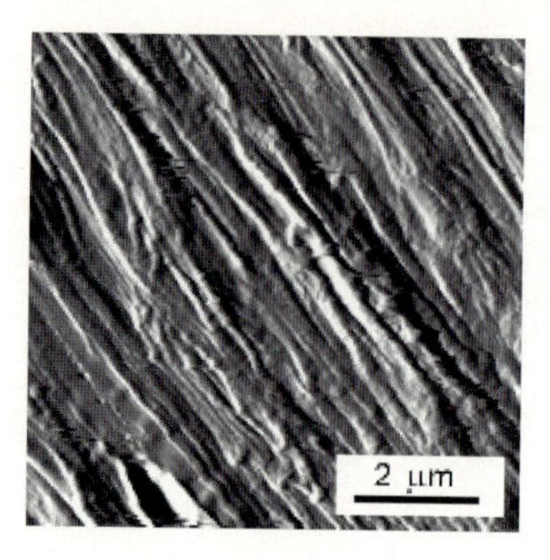

Fig. 8.16. Surface structure of a cold-drawn PE film registered with an atomic force microscope [85]

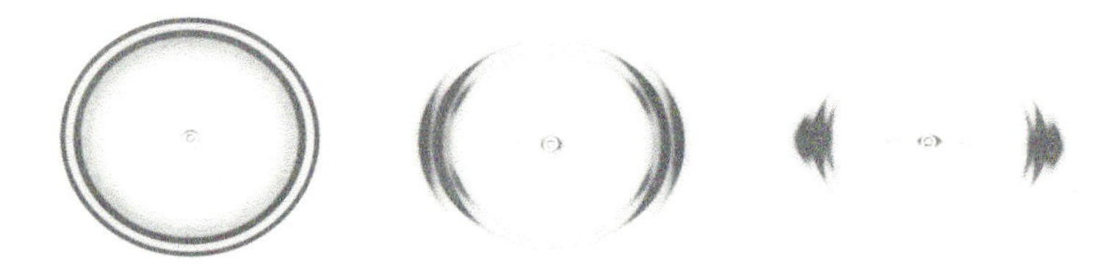

Fig. 8.17. X-ray scattering patterns of PE, as registered for the isotropic part (*left*), the center of the shoulder (*center*) and the neck (*right*) of a cold-drawn sample. The two reflections in the isotropic pattern are to be assigned to the 110- and the 200-lattice planes of orthorhombic polyethylene. The third reflection with a larger lattice plane spacing emerging in the textured patterns is due to another metastable crystalline phase which forms during drawing [85]

microscope before and after drawing a film of polyethylene. The morphology in the extended state is set up of fibrous elements. Optical microscopy cannot resolve the microfibrils. These show up, for example, in the picture of Fig. 8.16 obtained with an atomic force microscope. It displays the surface structure of a cold-drawn polyethylene film, and the fibrils are clearly observable.

The complete change in morphology necessitates a disintegration of the original crystallites into small blocks, slippage and rotations of the blocks, and finally a new assembly as parts of the oriented microfibrils. The texture changes may be followed by X-ray scattering, by registration of scattering patterns for various positions in the shoulder. This is exemplified by Fig. 8.17 which includes in addition to the isotropic pattern of the original structure of a polyethylene sample and the final pattern in the neck also a pattern obtained within the shoulder region. The two latter patterns are typical fiber-diagrams, and may serve as a basis for considerations about the microscopic steps in the structure change. In particular, they may help to identify crystal-

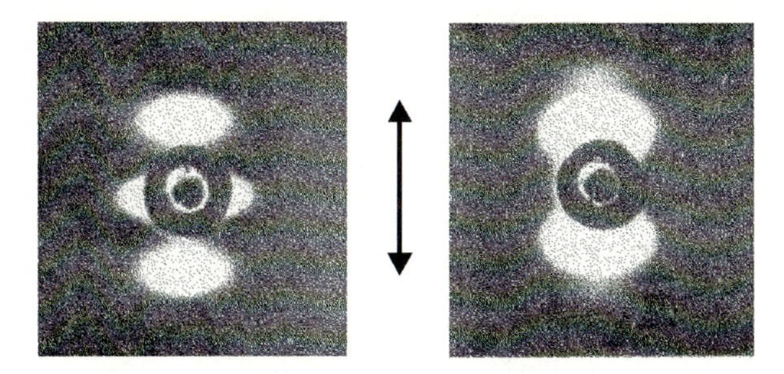

Fig. 8.18. SAXS diagrams registered for PE drawn in the solid state at 70 °C directly after the drawing (*left*) and after a subsequent heat treatment at 120 °C (*right*). The *arrow* displays the drawing direction. Experiment by Fischer et al. [90]

lographic planes which are preferred in slippage processes. As it appears, the reorganisation process is essentially a two-step procedure. The first step leads to the special texture indicated by the second X-ray scattering pattern, with different orientational distributions for different lattice planes, and the second step then to the final chain alignment in drawing direction, as indicated by the 'equatorial' positions of the Bragg-reflections. So far the causes of this peculiar mechanism of structure change are not understood.

As was discussed extensively in Sect. 4.1, X-ray scattering experiments in the small-angle range can be used for an investigation of the crystalline-amorphous superstructure. The conversion of the spherulitic structure composed of isotropically distributed stacks of crystallites with quasi-periodic order into an oriented state which is still partially crystalline (the degree of crystallinity remains largely unchanged), leads to a scattering pattern like that shown on the left-hand side of Fig. 8.18. This 'two-point diagram' obtained for a drawn polyethylene, with two intensity maxima on the 'meridian' parallel to the direction of drawing, was registered on an X-ray film using a high-resolution camera. The maxima indicate a quasi-periodic density fluctuation along the oriented microfibrils. An interesting effect is observed when samples drawn in the solid state are annealed at temperatures which are elevated but still below the temperature of final melting. As shown by the scattering pattern on the right-hand side of Fig. 8.18, this leads to an increase in the reflection intensities, which is indicative of an improvement in the crystalline-amorphous ordering. In the discussion of mechanisms of crystallization in Chapter 4 we dealt in Sect. 4.2.2 with a 'spinodal mode'. This special mode of crystallization is observed, for example, for cold-drawn amorphous poly(ethylene terephtalate) if it becomes annealed in the glass-transition range. The partially crystalline structure which forms under these conditions is similar to that observed here for the annealed polyethylene and the two processes do indeed appear to have a common physical basis. In both cases, the structural reorganisation corresponds to a process of defect cluster-

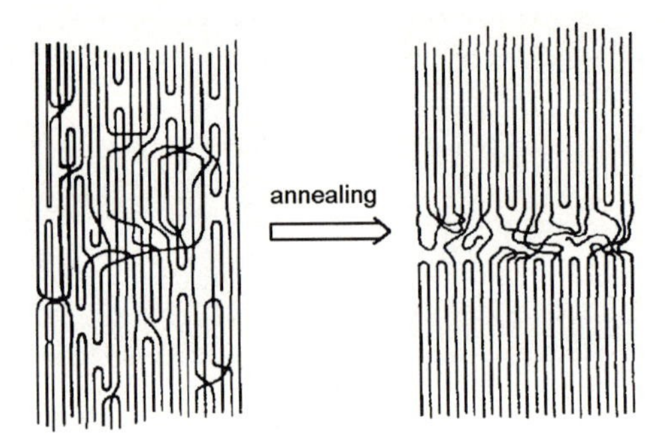

Fig. 8.19. Improvement of the ordering between crystalline and amorphous regions resulting from a heat treatment of a cold-drawn sample. From Fischer et al.[90]

ing, which assembles all entanglements which cannot be incorporated into a crystal in laterally extended layer-like disordered regions. Figure 8.19 due to Fischer illustrates this view.

Relation Between Birefringence and Orientational Order

We give here a derivation of Eqs. (8.17) and (8.18) which relate the birefringence, $\Delta n = n_c - n_a$, of an uniaxially drawn sample to the orientational order parameter of the monomers, S^{or}_m. Two coordinate systems are introduced. The first one, with coordinates x, y, z, is fixed on the sample, with the z-axis oriented in drawing direction. The second, with coordinates x', y', z', is anchored on a monomer unit and varies in orientation between different monomers. We choose the latter local coordinate system so that the polarizibility tensor per monomer, $\boldsymbol{\beta}'$, has a diagonal form

$$\boldsymbol{\beta}' = \begin{pmatrix} \beta_\perp & 0 & 0 \\ 0 & \beta_\perp & 0 \\ 0 & 0 & \beta_\perp + \Delta\beta \end{pmatrix} \tag{8.19}$$

Hereby, we assume an uniaxial local symmetry with the symmetry axis in chain direction which for many polymers is a good approximation. $\boldsymbol{\beta}'$ can be transformed into the sample-fixed coordinate system by

$$\boldsymbol{\beta} = \boldsymbol{\Omega}^{-1} \cdot \boldsymbol{\beta}' \cdot \boldsymbol{\Omega} \tag{8.20}$$

$\boldsymbol{\Omega}$ is the rotation matrix which accomplishes the transformation. We calculate the diagonal elements of $\boldsymbol{\beta}$. β_{xx} follows as

$$\beta_{xx} = \sum_l \Omega^{-1}_{xl} \beta'_{ll} \Omega_{lx}$$

$$= \beta_\perp \cos^2\theta_{x',x} + \beta_\perp \cos^2\theta_{y',x} + (\beta_\perp + \Delta\beta)\cos^2\theta_{z',x}$$
$$= \beta_\perp + \Delta\beta\cos^2\theta_{z',x} \tag{8.21}$$

where $\theta_{i',j}$ denotes the angle between the axes i' and j. The results for β_{yy} and β_{zz} are correspondingly

$$\beta_{yy} = \beta_\perp + \Delta\beta\cos^2\theta_{z',y} \tag{8.22}$$

$$\beta_{zz} = \beta_\perp + \Delta\beta\cos^2\theta_{z',z} \tag{8.23}$$

As the monomer orientations vary, we take the averages

$$\langle\beta_{xx}\rangle = \langle\beta_{yy}\rangle = \beta_\perp + \Delta\beta\langle\cos^2\theta_{z',x}\rangle \tag{8.24}$$

$$\langle\beta_{zz}\rangle = \beta_\perp + \Delta\beta\langle\cos^2\theta_{z',z}\rangle \tag{8.25}$$

All non-diagonal elements, $\langle\beta_{i\neq j}\rangle$, vanish due to the uniaxial symmetry of the sample. Since

$$\cos^2\theta_{z',x} + \cos^2\theta_{z',y} + \cos^2\theta_{z',z} = 1 \tag{8.26}$$

hence

$$2\langle\cos^2\theta_{z',x}\rangle = 1 - \langle\cos^2\theta_{z',z}\rangle \tag{8.27}$$

we have

$$\langle\beta_{zz}\rangle - \langle\beta_{xx}\rangle = \Delta\beta\cdot\frac{3\langle\cos^2\theta_{z',z}\rangle - 1}{2} \tag{8.28}$$

or shortly, introducing the orientational order parameter defined by Eq. (8.17),

$$\langle\beta_{zz}\rangle - \langle\beta_{xx}\rangle = \Delta\beta\cdot \mathrm{S}_{\mathrm{or}}^{\mathrm{m}} \tag{8.29}$$

Next we search for an expression for the dielectric tensor

$$\epsilon = \begin{pmatrix} \epsilon_\perp & 0 & 0 \\ 0 & \epsilon_\perp & 0 \\ 0 & 0 & \epsilon_{||} \end{pmatrix} \tag{8.30}$$

Application of the Clausius-Mosotti equation yields

$$\epsilon_{||} - 1 = (\epsilon_{||} + 2)\frac{1}{3\epsilon_0}c_{\mathrm{m}}\langle\beta_{zz}\rangle \approx (\bar{\epsilon} + 2)\frac{1}{3\epsilon_0}c_{\mathrm{m}}\langle\beta_{zz}\rangle \tag{8.31}$$

and

$$\epsilon_\perp - 1 \approx (\bar{\epsilon} + 2)\frac{1}{3\epsilon_0}c_{\mathrm{m}}\langle\beta_{xx}\rangle, \tag{8.32}$$

with

$$\bar{\epsilon} := (2\epsilon_\perp + \epsilon_{||})/3 \tag{8.33}$$

The anisotropy of the dielectric constant

$$\Delta\epsilon := \epsilon_{||} - \epsilon_\perp \tag{8.34}$$

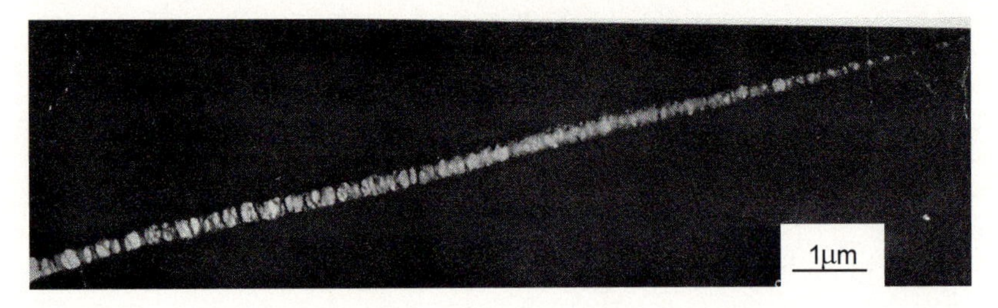

Fig. 8.20. Electron micrograph of a craze in PS, obtained by Kambour [91]

follows as

$$\begin{aligned}\Delta\epsilon &= (\bar{\epsilon}+2)\frac{1}{3\epsilon_0}c_{\rm m}(\langle\beta_{zz}\rangle - \langle\beta_{xx}\rangle) \\ &= (\bar{\epsilon}+2)\frac{1}{3\epsilon_0}c_{\rm m}\Delta\beta\cdot {\rm S}_{\rm or}^{\rm m}\end{aligned} \tag{8.35}$$

Finally, the birefringence is obtained as

$$\Delta\epsilon = \Delta({\rm n}^2) \approx 2\bar{\rm n}\Delta{\rm n} \tag{8.36}$$

which results in

$$\Delta{\rm n} = \frac{\bar{n}^2+2}{\bar{n}}\frac{1}{6\epsilon_0}c_{\rm m}\Delta\beta\cdot {\rm S}_{\rm or}^{\rm m} \tag{8.37}$$

$$:= \Delta{\rm n}_{\rm max}\cdot {\rm S}_{\rm or}^{\rm m} \tag{8.38}$$

We thus have proved, that measurements of the birefringence yield the order parameter of the monomers $\mathrm{S}_{\mathrm{or}}^{\mathrm{m}}$.

8.2 Crazing

Electron microscopy can be used to have a direct look on the structure of a craze. Figure 8.20 presents a micrograph obtained from a microtomed ultrathin slice which cuts through a craze in polystyrene. In addition, for further clarification, Fig. 8.21 gives a sketch of a craze, drawn for a location on an edge. Crazes are found both at the surface and in the interior of samples, and represent, as shown by the figures, localized zones of deformation. As indicated by the sketch, they have the form of a lens which contains fibrils in a void matrix. The micrograph shows these fibrils, which connect the two surfaces of the craze. The fibrils have diameters in the order of 10 nm and fill the craze volume to about 50%. The resulting large density difference to the matrix is the cause of the strong light scattering by the crazes and leads to the stress whitening. Diameters of crazes when they become visible in an optical

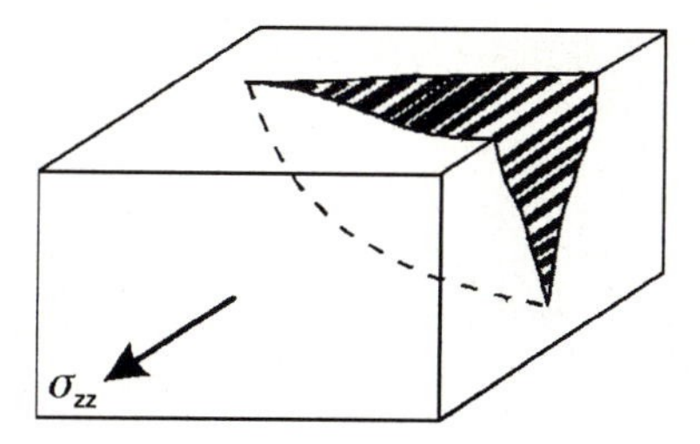

Fig. 8.21. Schematic of an edge-located craze. The surfaces of the deformation zone are connected by fibrils which fill the craze only partially

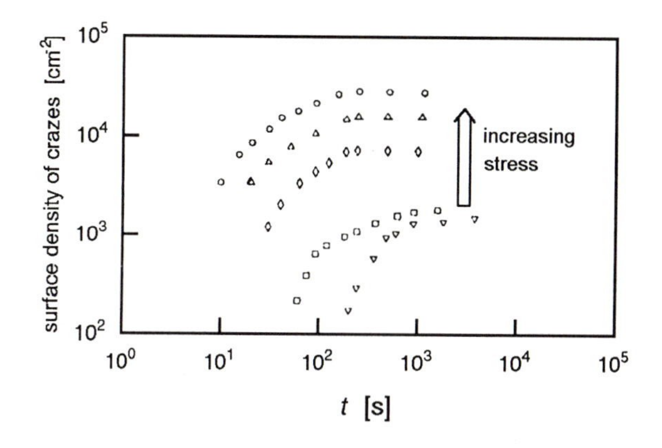

Fig. 8.22. Increase of the number of crazes with time, observed for a sample of PS, subject to various states of tensile stress. Work of Argon and Hannoosh [92]

microscope are in the order of μm, but then they can grow up to macroscopic dimensions, up to the cross-section of the whole sample.

For which polymers and under which conditions do crazes occur? Crazes form primarily in amorphous polymers, for molecular weights above the entanglement limit. There is no craze formation under compression or under pure shear. The typical situation leading to craze initiation is the imposition of an uniaxial or biaxial tensile stress. If such stresses are applied and fulfill certain threshold conditions, crazes form statistically, preferentially at first at the sample surface. The initiation rate depends on the applied stress, as is shown in Fig. 8.22. The higher the stress imposed, the shorter is the time for the observation of the first crazes. After the initial increase with time, the craze density saturates. Removing the stress, the crazes close their openings somewhat, but survive. They disappear only if the sample is annealed at temperatures above the glass transition.

Regarding the statistical nature of their appearance craze formation looks like an activated process. Although the details of the initiation step are not yet

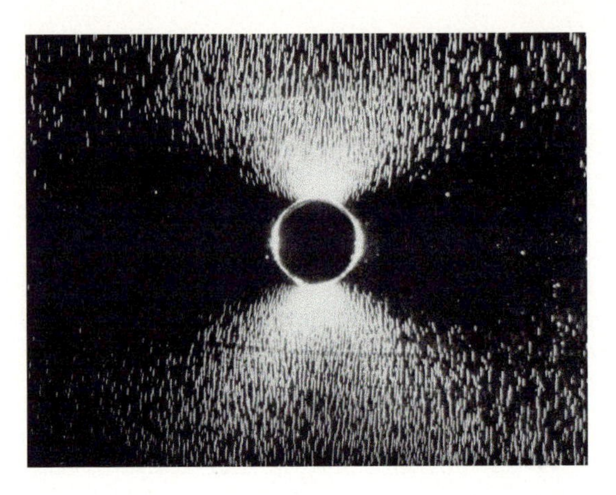

Fig. 8.23. Craze pattern in the vicinity of a hole in a plate of PMMA subjected to a tensile stress in horizontal direction. Micrograph from Sternstein et al. [93]

clarified, it is reasonable to assume that cavitation, i.e. the development of microvoids, must play a primary role. As the creation of pores in a homogeneous material is exceedingly difficult, it seems also clear that local heterogeneities, resulting in a stress intensification, are a necessary prerequisite. Hence, the activation energy of the process could be related to the formation of pores, at locations where tensile stresses are sufficiently magnified.

Figure 8.23 presents a nice experiment, where a planar stress field ($\sigma_3 = 0$) with varying values of the two in-plane principal stresses, σ_1 and σ_2, was produced by a tensile force on a plate of poly(methylmethacrylate), into which a circular hole had been cut. One observes regions without any crazes and other regions where crazes occur, showing a systematically varying density. Observations of this kind suggest we formulate a criterion for crazing, giving the boundary line between planar stress states which lead to crazing and those which do not. For the experiment presented here, which referred to surface crazes only, the condition for crazing may be written as

$$|\sigma_1 - \sigma_2| \geq A + \frac{B}{\sigma_1 + \sigma_2} \tag{8.39}$$

with $A < 0$ and $B > 0$. A and B are material parameters which depend on temperature. The important point expressed by the criterion is that both shear, as given by the difference $\sigma_1 - \sigma_2$, and a dilatational component, as described by the sum $\sigma_1 + \sigma_2$, are necessary prerequisites for the formation of crazes.

Crazes under a constant stress are not stable and grow in length and thickness with a constant rate. The direction of growth is well-defined with the craze tip moving perpendicular to the maximum principal stress. If the direction of the maximum stress changes, the direction of the craze changes accordingly. This is clearly demonstrated by the photograph presented in Fig. 8.24.

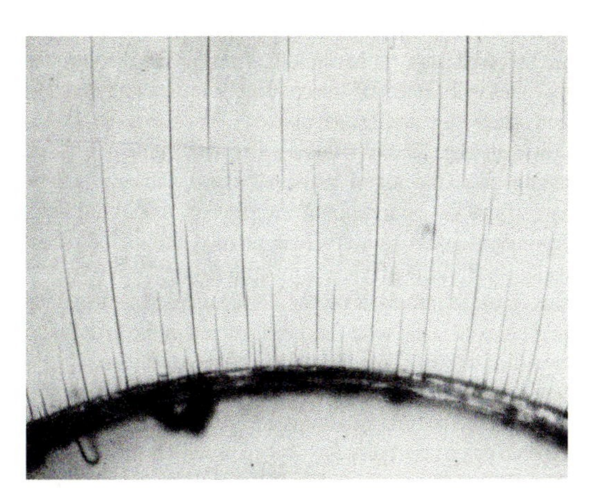

Fig. 8.24. Curvilinear crazes developed in the high stress region near a hole in a plate of PMMA, following lines perpendicular to the maximum principal stress. From Sternstein et al. [93]

The crazes start at the points of largest stress enhancement at the edge of the hole and then proceed perpendicular to the maximum principal stress.

Simultaneously with the lateral dimensions, the thickness increases. This is accompanied by an equivalent increase in the length of the fibrils connecting the craze surfaces. Since observations indicate a constant thickness of the fibrils subsequent to a short initial period, their growth obviously occurs by a drawing-in of fresh material from the surfaces. It thus appears that the fibril drawing process is equivalent to the macroscopic cold-drawing by neck extension, now taking place in mesoscopic dimensions, with the 'shoulders' being fixed onto and moving together with the craze surfaces. Indeed, the extension ratios found for the microfibrils, which can be directly derived from the volume fraction of material within the craze, essentially agree with the draw ratios associated with neck formation.

It is possible to provide an estimate for the draw ratio reached. It may be identified with the point where strain hardening sets in and, as for an entangled polymer melt, the hardening is caused by the approach of the chain sequences between entanglements to their limits of extensibility. First we need knowledge of the molecular weight of these sequences. This molecular weight, denoted M_{e}, can be obtained with the aid of Eq. (7.86)

$$G = c_{\mathrm{p}} kT$$

which gives the shear modulus of an ideal rubber. M_{e} may be calculated if we identify G with the plateau value of the dynamic shear modulus of the melt, G_{pl}, and write for the density of active chains

$$c_{\mathrm{p}} = \frac{\rho}{M_{\mathrm{e}}} \mathrm{N_L} \tag{8.40}$$

M_e then follows as

$$M_e = \frac{N_L kT\rho}{G_{pl}} \tag{8.41}$$

The degree of polymerization, N_e, and the contour length $l_{ct,e}$ of chains with molecular weight M_e are

$$N_e = \frac{M_e}{M_m} \tag{8.42}$$

and

$$l_{ct,e} = a_m N_e \tag{8.43}$$

where a_m and M_m stand for the length and the molecular weight per monomer respectively. Chains with an end-to-end distance vector oriented along the drawing direction experience the highest draw ratios. Their average end-to-end distance in the isotropic state, R_e, is according to Eq. (2.35)

$$R_e = a_0 N_e^{1/2} \tag{8.44}$$

With the knowledge of $l_{ct,e}$ and R_e, we can now estimate the draw ratio at the onset of strain hardening. Calling it λ_{max}, we may write

$$\lambda_{max} \simeq \frac{l_{ct,e}}{R_e} = \frac{a_m}{a_0} N_e^{1/2} \tag{8.45}$$

Figure 8.25 shows a comparison between experimentally determined draw ratios of craze fibrils and values of λ_{max}. The agreement is not perfect, but the general tendency of the data is satisfactory. We may conclude from this analysis that the crazes are stabilized by the strain hardened fibrils which bear the imposed load. Without the strain hardening effect, stable crazes would not exist.

An appealing model explaining the mechanism of lateral craze growth was suggested by Argon. It is indicated in Fig. 8.26, with a schematic drawing of the edge of a craze and its development with time. The main point in the model is an assumed instability of the air-polymer interface, which may arise under the conditions of the sharply decaying dilatational stress effective at the craze tip. The phenomenon is known from fluids, when menisci advance under the action of a suction gradient. Then the surface may become instable to wave-like perturbations above a critical wavelength. As indicated in the drawing, fibrils may be produced at the advancing craze tip by the break-up of the interface subsequent to the repeated establishment of a transverse corrugation. The wavelength of the corrugation determines the distance and thus also the final thickness of the fibrils. A prerequisite for the process is plastic flow in a limited zone in front of the moving edge, as it may readily occur for the given high stress values. Material enters this plastic zone at the outer end, then advances through the zone under increasing strain and finally becomes included in the interface convolution process forming the fibrillar craze matter. Argon's model explains many of the observations and thus appears quite reasonable.

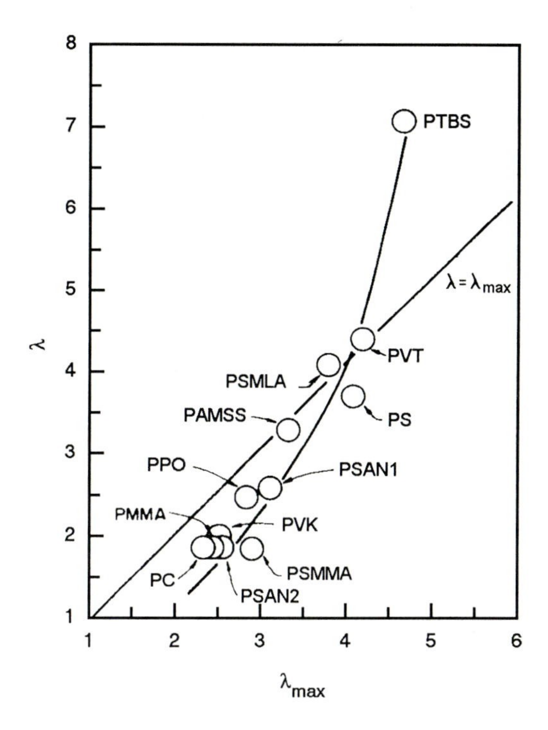

Fig. 8.25. Extension ratios λ of craze fibrils observed for various compounds, plotted against the draw ratio at the onset of strain hardening, λ_{max}, as derived from the plateau modulus. From Kramer [94]

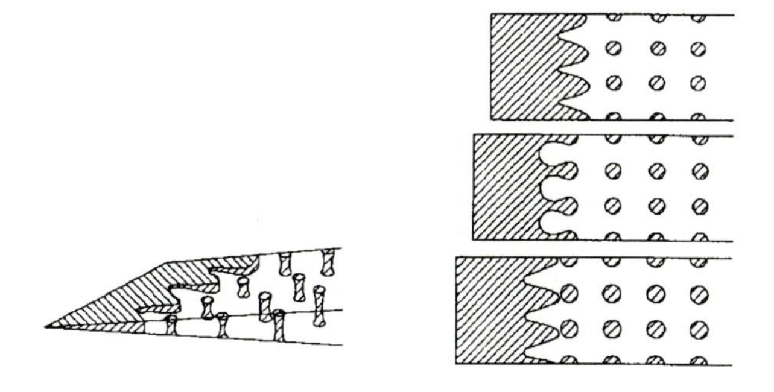

Fig. 8.26. Argon's model of lateral craze growth based on the phenomenon known as 'meniscus instability'. Side view on the corrugated polymer-air interface (*left*). Advances of the craze front by a repeated break-up of the interface (view in fibril direction, *right*) [95]

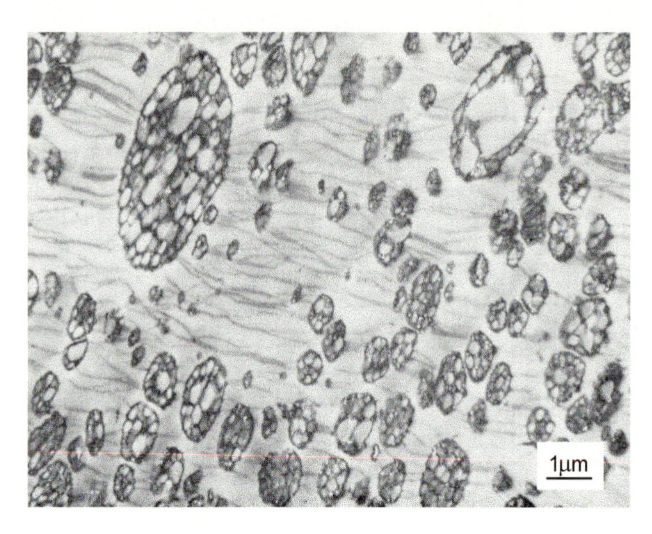

Fig. 8.27. Electron micrograph of a stained microtomed section of HIPS, showing crazes running between the elastomeric precipitates. From Kambour[91]

We finish this chapter with a look at the structure of 'high impact polystyrene', abbreviated 'HIPS', which is a widely used product, belonging to the larger class of 'rubber toughened' thermoplastics. Mixing polystyrene with an elastomer like polybutadiene results in a two-phase structure. Figure 8.27 displays the structural details as they appear in an electron micrograph. Embedded in the polystyrene matrix are spherical inclusions of polybutadiene, and the picture shows that the structure is even more complex, since there are again polystyrene inclusions within the inclusion. High impact polystyrene is a tough material and owes this toughness to the initiation of crazes in large numbers. These are observable in the picture. If stress is applied, crazes are generated at the surfaces of the inclusions, which produce a stress intensification. One might think at first that for the given high density of crazes the material would also fracture more easily. However, since each craze runs only to an adjacent rubber particle and there becomes arrested, this is not the case.

8.3 Brittle Fracture

If the stress applied to a polymeric solid is sufficiently high, so that yield processes set in, and then is further enhanced, it will ultimately come to the point of break. In tough samples there is extensive plastic flow before final fracture resulting in a large amount of energy being dissipated, whereas, in brittle compounds, failure occurs much earlier, immediately after the first yielding. Samples of polystyrene or of poly(methylmethacrylate) break after

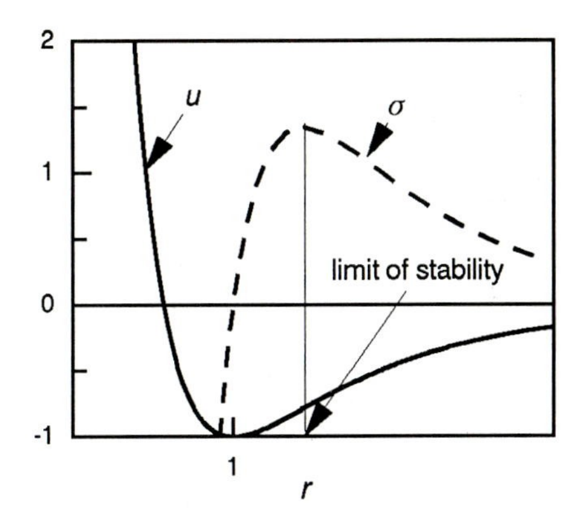

Fig. 8.28. Theoretical limit for mechanical stability under an isotropic tensile stress σ. $u(r)$ is the pair potential in a low molar mass crystal

the first observation of crazes and thus exemplify brittleness. In the last section of this chapter we will consider this process of 'brittle fracture' in more detail.

It is a general observation for all kinds of solid materials, metals, ceramics or polymers, that they never reach their theoretical limits of strength. How is this 'theoretical limit' determined? For a low molar mass crystal this is easy to see when considering, for example, the change of the free energy associated with a homogeneous dilatation. Basically, the free energy is determined by the pair interactions between the atoms or molecules in the crystal, and this function, $u(r)$, has a general shape as sketched in Fig. 8.28. Enlarging the volume by applying an external isotropic dilatational stress, $\sigma = -p$, first requires an increasing stress, but then σ passes over a maximum and one comes into a region of mechanical instability, where crystals would disintegrate spontaneously. The location of the maximum thus determines the theoretical limit of strength. If we neglect the minor phononic contributions, it follows from $u(r)$ only, the condition being

$$\frac{\mathrm{d}\sigma}{\mathrm{d}V} = \frac{\mathrm{d}^2\mathcal{F}}{\mathrm{d}V^2} \sim \frac{\mathrm{d}^2 u}{\mathrm{d}r^2} = 0 \tag{8.46}$$

We derived this result for an applied isotropic dilatational stress. Using similar arguments, one could deduce also the theoretical limits for other kinds of stress.

In a polymeric solid we find more complex conditions. Here, we have both intramolecular covalent forces and intermolecular van der Waals forces and, furthermore, a heterogeneous microscopic structure with a low degree of order. Molecular fracture mechanisms are therefore more involved and include chain slippage as well as chain scission. The latter process can only occur for fully extended sequences which are anchored with their ends in crystallites or immobile entanglements, by a combined effect of mechanical stress and thermal fluctuations. Even if the situation is complicated, on the basis of a knowledge of the relevant energies and forces one can still obtain a rough estimate of the theoretical limits of strength. As already mentioned, these theoretical limits are never reached, for both low molecular mass crystals and polymeric solids. What is the cause of the large discrepancies? This is easily revealed: The lack of agreement is due to the presence of flaws which are always present in solid bulk matter. Near to flaws, which may be microvoids, microcracks, inclusions of foreign particles, or other structural heterogeneities, stresses become greatly intensified. As a consequence, when applying stress, the limits of stability may be locally exceeded, and fracture is then initiated.

Being controlled by the flaws in a sample, fracture strength is not a well-defined bulk property like the parameters and functions describing the viscoelastic behavior. This has consequences for the characterization of strength by measurements. As different samples never possess an equal distribution of flaws, an exact reproducibility cannot be expected and large variations are the rule. Predictions with regard to the strength of bulk samples can therefore be only of statistical nature, based on series of measurements and the deduction of probabilities that fracture occurs under given conditions of stress. A peculiar feature is the volume dependence. Since fracture can start at any flaw, the increase in their total number with increasing volume reduces the mechanical stability of a sample correspondingly. Hence in conclusion, dealing with the property 'strength' is not a conventional task and requires special measures.

An obvious way to overcome some of the difficulties and to have a frame for reproducible measurements is the preparation of special samples which have a macroscopic flaw where the fracture then starts, followed by a propagation of the crack in a controlled manner. Although this procedure cannot remove the principal difficulty encountered in the assessment of the fracture behavior of a given sample, it can provide true material parameters for use in comparisons of different materials. Figure 8.29 depicts a standard configuration, with a sharp crack introduced in a plate. Applying stress, the fracture starts at the two edges of the crack. 'Linear fracture mechanics' deals with these conditions and describes the behavior of sharp cracks cut into a linear elastic material.

8.3.1 Linear Fracture Mechanics

The analysis was first carried out by Griffith in a treatment of the brittle fracture of metals. Actually, the considerations are of general nature and can also be applied to polymers, after introducing some physically important but formally simple modifications. Griffith's approach is based on linear elasticity theory and its utility for polymers may look questionable at first, as those are neither elastic nor linear under the conditions near to failure. However, as we will see, the theory is indeed applicable and provides also here a satisfactory description of crack growth.

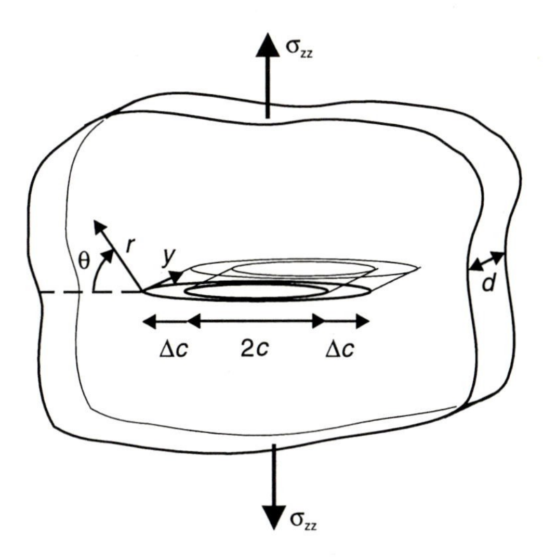

Fig. 8.29. Standard configuration ('opening mode I') considered in linear fracture mechanics: Infinite plate containing a crack of length $2c$, subject to a tensile stress σ_{zz}

We consider the plate containing a crack of length $2c$ shown in Fig. 8.29. The plate has a thickness d and its area is assumed as infinite. If a tensile force is applied perpendicular to the crack direction, we find far away from the crack a uniform uniaxial stress, denoted σ_{zz}. Following Griffith's approach, we first inquire about the decrease in the elastic free energy of the plate which results if the length of the crack is increased by $\Delta 2c$. There is an exact solution of the problem, being given by the expression

$$\Delta \mathcal{F} = \frac{\pi c \sigma_{zz}^2}{E} d \Delta 2c \tag{8.47}$$

(E is the tensile elastic modulus). The increase of the crack length produces additional surfaces on both sides, and this requires a work, $\Delta \mathcal{W}$, which is proportional to their area

$$\Delta \mathcal{W} = 2wd\Delta 2c \tag{8.48}$$

The equation includes w as proportionality constant. Griffith, treating perfectly brittle solids, identified w with the surface free energy. For polymers, the meaning of w has to be modified and, as will be discussed below, more contributions are included and even take over control. Independent of the physical background of w, the condition for fracture follows from a comparison of the two quantities, $\Delta \mathcal{F}$ and $\Delta \mathcal{W}$: Crack growth occurs, if the 'strain energy release rate' $\mathrm{G_I}$, introduced as

$$\mathrm{G_I} := \frac{1}{d}\frac{\Delta \mathcal{F}}{\Delta 2c} = \frac{\pi c \sigma_{zz}^2}{E} \tag{8.49}$$

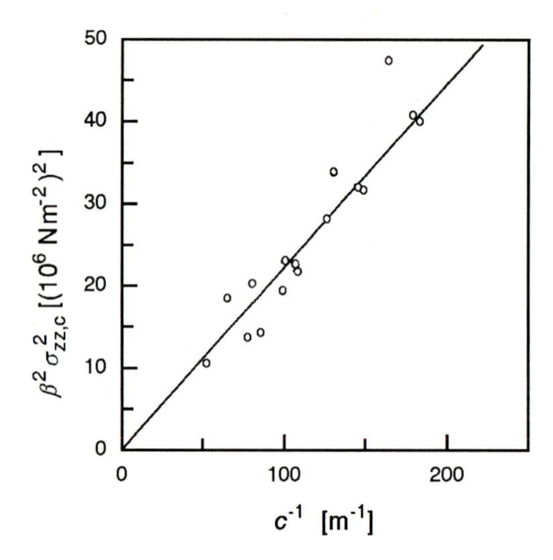

Fig. 8.30. Values of the tensile stress $\sigma_{zz,c}$ at the beginning of crack growth in samples of PMMA containing a crack of length $2c$. β^2 is a geometrical constant, equal to π for infinite plates and with slightly different values for finite bodies. Data from Williams [96]

can provide the work which is necessary to create the new surface, and this happens for

$$\mathrm{G_I} \geq 2w := \mathrm{G_{Ic}} \tag{8.50}$$

Equation (8.50) is known as Griffith's fracture criterion. It includes a critical value of the strain energy release rate, denoted $\mathrm{G_{Ic}}$, and the latter is determined by the surface parameter w. The subscript 'I' is used in the literature to indicate reference to the 'opening mode I' of Fig. 8.30 and to discriminate it from other possible modes of loading. From Eqs. (8.49) and (8.50) follows the critical value of the tensile stress, as

$$\sigma_{zz,\mathrm{c}} = \left(\frac{2Ew}{\pi c}\right)^{1/2} \tag{8.51}$$

$\sigma_{zz,\mathrm{c}}$ sets the stability limit for the crack. For tensile stresses below the limit, the crack just opens maintaining a constant length $2c$. If the tensile load exceeds $\sigma_{zz,\mathrm{c}}$, the crack starts growing, up to the point of total break of the sample. Note that only two material coefficients enter into the fracture criterion Eq. (8.51), namely the tensile elastic modulus E and the surface parameter w.

One result of the treatment is the predicted dependence of $\sigma_{zz,\mathrm{c}}$ on the crack length c. One expects $\sigma_{zz,\mathrm{c}} \sim 1/\sqrt{c}$, and this is indeed verified by experimental observations. Figure 8.30 depicts, as an example, fracture data obtained for poly(methylmethacrylate) at room temperature, represented in a plot of $\sigma^2_{zz,c}$ versus c^{-1}.

There is a second, equivalent form of the fracture criterion. It is possible to calculate the stress field near to the crack tip for the cracked plate subject to a tensile load. The result has the form

$$\sigma_{ij}(r,\theta) = \frac{\mathrm{K_I}}{(2\pi r)^{1/2}}\Phi_{ij}(\theta) \qquad (8.52)$$

when employing the cylindrical coordinates r, θ and y introduced in Fig. 8.28. According to this expression, the spatially varying stress tensor depends on the distance r from the tip and the angle θ only, and not on y. We have here a state of plane stress, with $\sigma_{yy} = 0$ and non-vanishing values of the stress only parallel to the plate surfaces. In Eq. (8.52), the stress distribution is given by a certain angular dependent tensor function $\Phi_{ij}(\theta)$, together with the distance factor $1/(2\pi r)^{1/2}$. The expression implies that the relative stress distribution around the crack tip does not vary with the external load, which just affects the magnitude. The effect of the loading is completely accounted for by the parameter $\mathrm{K_I}$, called the 'stress intensity factor'. It is possible to calculate the stress intensity factor for the standard system with the central crack, and the result is

$$\mathrm{K_I} = \sigma_{zz}(\pi c)^{1/2} \qquad (8.53)$$

Equation (8.52) predicts a singularity for the stress at the crack tip, as $\sigma_{ij} \to \infty$ for $r \to 0$. Therefore, the local stresses, as described by this expression, cannot be utilized in the formulation of a fracture criterion. Instead, one may try to employ the stress intensity factor and consider the validity of a stability condition with the general form

$$\mathrm{K_I} \leq \mathrm{K_{Ic}} \qquad (8.54)$$

where $\mathrm{K_{Ic}}$ denotes again a critical value. The prospects for such a criterion are favorable, since theoretical analysis proves that, in contrast to the diverging stress at the tip, the total elastically stored energy remains finite. It is therefore not too surprising that $\mathrm{K_I}$ may be directly related to the strain energy release rate, $\mathrm{G_I}$, introduced above. As the comparison of Eq. (8.53) with Eq. (8.49) shows, we have indeed

$$\mathrm{K_I}^2 = E\mathrm{G_I} \qquad (8.55)$$

As it turns out, this relation holds not only for the considered configuration with a crack in the center of an infinite plate, but also for finite plates and even for other locations of the crack, for example, for a semi-infinite plate with a notch at the edge. The critical value of the stress intensity factor $\mathrm{K_{Ic}}$ follows from a combination of Eqs. (8.55) and (8.50), as

$$\mathrm{K_{Ic}}^2 = 2wE \qquad (8.56)$$

It is determined by two material coefficients only, as is the case for $\mathrm{G_{Ic}}$.

In conclusion, Griffith's analysis of the crack stability provides two true material parameters, the critical values at failure of the strain energy release rate, G_{Ic}, and of the stress intensity factor, K_{Ic}. Either of them can be used for a description of the fracture properties of common, i.e. not pre-cracked samples, in the sense that higher values of G_{Ic} or K_{Ic} indicate a higher resistance to failure in general.

8.3.2 Slow Mode of Crack Growth

Griffith's analysis predicts a stable crack for stresses below the critical limit of failure and an uncontrolled accelerated growth above the limit, due to the permanent excess of the released strain energy over the required surface free energy. Polymer behavior is different, as is exemplified by the results of fracture experiments conducted on poly(methylmethacrylate) depicted in Fig. 8.31. Studies were carried out under variation of the external load, i.e. K_I, with crack velocities being registered with the aid of a high speed camera. Cracks have a constant length up to a certain value of the stress intensity factor or the strain energy release rate, and we denote the latter G_{Ii}. At this point crack growth sets in, however, not in an accelerated manner but with a slow stationary growth rate (in order to establish a constant stress intensity factor for growing cracks, the tensile load has to be permanently readjusted to fulfill Eq. (8.53)). The growth rate increases with K_I until another point is reached where a second transition takes place, now to the truly unstable mode of accelerated growth. Usually this second transition is called the 'crit-

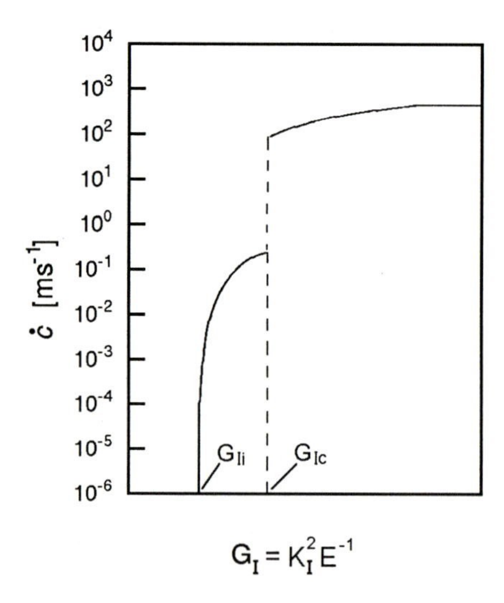

Fig. 8.31. Crack growth as a function of G_I, as observed for PMMA ($M = 2 \cdot 10^6$) at room temperature. From Döll [97]

ical' transition and its location in terms of the stress intensity factor and the strain energy release rate is denoted K_{Ic} and G_{Ic} respectively. Also in the second range constant crack velocities $\dot{c}$ may be reached after a first period of acceleration, and they are determined by inertial forces and the speed of elastic waves.

The polymer specific feature is the existence of a range of stationary slow growth preceding the uncontrollable rapid mode of failure, i.e. the occurrence of the 'subcritical crack growth' in the range $G_{Ii} < G_I < G_{Ic}$. Its origin becomes basically clear, if we look at Fig. 8.31 from the other perspective, as giving the strain energy release rate G_I as a function of an imposed crack velocity $\dot{c}$. Under stationary conditions, G_I must equal the work suspended to form new surfaces, as expressed by the parameter $2w$. According to the curves, for subcritical growths, w increases with $\dot{c}$. In Griffith's original treatment of brittle fracture, w corresponded to the surface free energy only, which just relates to the abscence of binding forces in one direction. In the case under discussion, the fracture of polymers, the situation is qualitatively different. The work required for the formation of a surface during fracture here is obviously dominated by a preceding plastic flow. The observed rate dependence would not arise for a surface free energy determined w, but is indicative of a dominant role of flow processes. Indeed, the values of w deduced from Griffith's fracture criterion for polymers are much larger than the surface free energy alone. For example, for poly(methylmethacrylate) one finds $G_{Ic} \simeq 100 - 1000\ \mathrm{Jm}^{-2}$, compared to a surface free energy $2w_s \simeq 0.1\ \mathrm{Jm}^{-2}$. Hence, although the fracture of materials like polystyrene and poly(methylmethacrylate) is still called 'brittle', it does not comply with this name in the strict sense.

The growth rate dependence of the work suspended in plastic flows is conceivable if we recall, for example, the strain rate dependence of the force applied for the shear yielding during cold-drawing. Experimentally one finds a power law for the relation between K_I and $\dot{c}$ in the range of subcritical growth

$$\dot{c} \sim {K_I}^{\beta} \tag{8.57}$$

with a large exponent β (≈ 25).

In fact, there is a detailed picture about the flow processes controlling the slow growth of cracks starting from a macroscopic notch or being initiated somewhere in a sample near to a flaw, and it directly relates to crazing. Insight into the structure comes again from electron microscopy. Figure 8.32 presents micrographs of the region around a crack tip and we find a plastically deformed zone in front of the tip. The picture with higher resolution on the right side shows that this zone is filled with fibrils and thus equals the interior structure of a craze. As we also find fibrils attached to the surfaces of the crack, a picture of crack growth emerges as indicated schematically in the upper part of Fig. 8.33. If a tensile load is applied, the high stresses in the immediate neighborhood of the crack tip result in a crossing of the yield point and the formation of a zone with plastic deformation. In polymers, the pertinent mode of local plastic flow is 'crazing'. Hence, fibrils form in the plastic zone, being

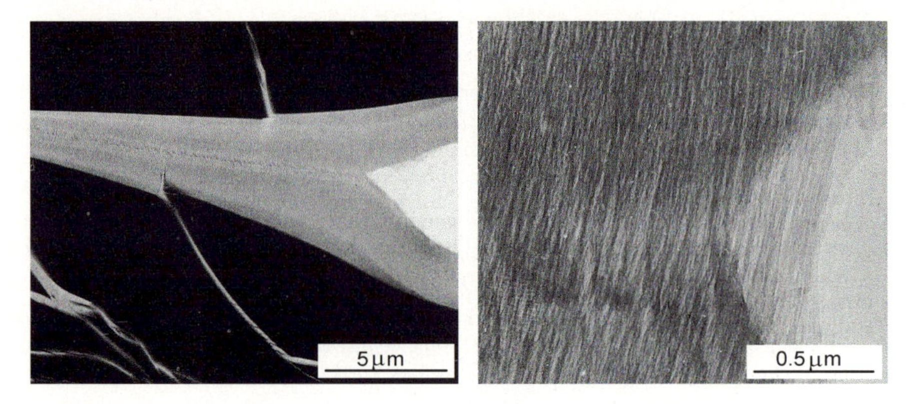

Fig. 8.32. Crack tip (bright sections on the right in both pictures) within a craze in PS. Electron micrographs obtained by Michler [98]

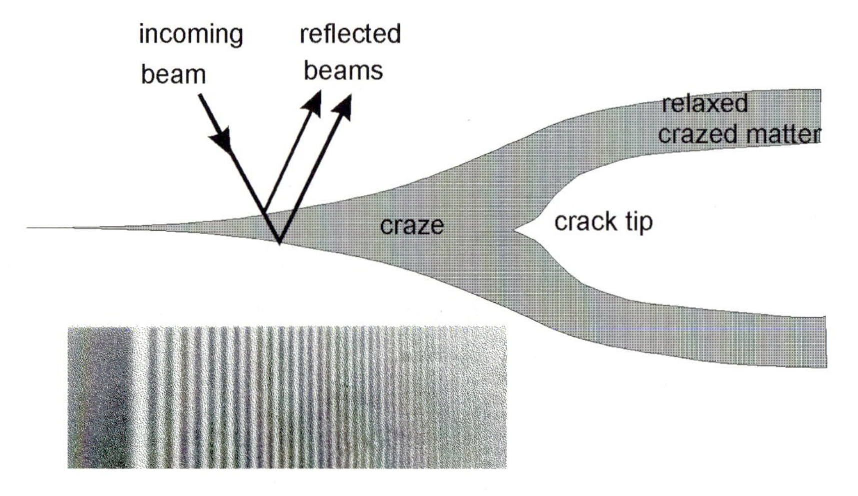

Fig. 8.33. Crack propagation behind a simultaneously moving zone with crazed matter. The profile of the preceding craze tip can be derived from the interference pattern produced by the two reflected beams. The fringe pattern presented here was obtained for PMMA by Doyle [99]

contained in a void matrix. If the stress applied is enhanced further, the fibrils elongate and may reach at a certain point a limit of internal stability, at which they break in two parts. At this moment, a crack has formed within the craze and subcritical crack growth starts. Under the applied stress, the craze growth is in a radial direction followed by the crack tip within the craze. For fixed K_I, we have stationary conditions, with a constant 'cusp-like' shape of the zone with crazed matter in front of the tip and with a constant crack velocity. Failure is completed when the crack has expanded over the sample cross-section.

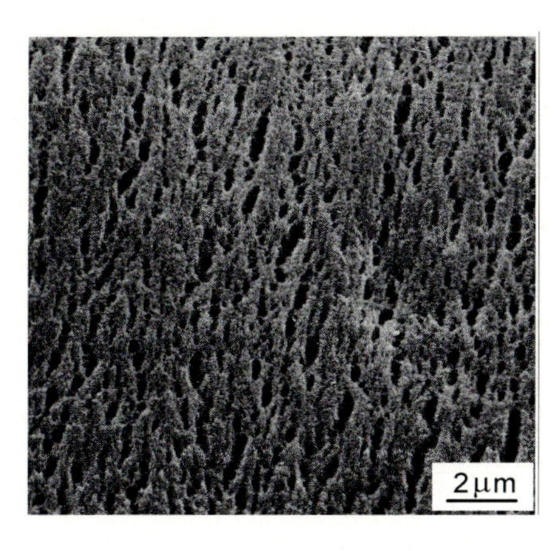

Fig. 8.34. Fracture surface of PMMA as observed in a scanning electron microscope. The structure originates from collapsed crazed matter. Micrograph obtained by Döll and Könczöl [100]

Additional proof for the crack growth subsequent to a craze follows from the surface structure of the broken sample. A thin film remains on the surface formed by the collapsed crazed matter, as can be observed by scanning electron microscopy. Figure 8.34 presents as an example a picture of the fracture surface of poly(methylmethacrylate), obtained for a resolution in the μm-range. The appearance of fracture surfaces may show considerable variations between different polymers, what is never observed, however, is a microscopically smooth surface, as it would result for a cleavage in a truly brittle fracture.

The cusp-like profile of the craze zone in front of the crack may be deduced from another experiment on poly(methylmethacrylate), presented in the lower part of Fig. 8.33. The displayed fringe pattern is observed in a microscope, when the region around the crack tip is illuminated by a light beam from above. As a result of the interference of two light beams, one reflected at the upper and the other at the lower surface of the craze with crack, an interference pattern emerges. Analysis enables the profile of the craze in front of the crack to be determined. Typical profiles so obtained are cusp-like, similar to the schematic drawing.

With the knowledge about the structure of the plastic zone in front of the crack tip, one may write down a simple equation for the surface parameter, known also as 'material resistance' $2w$. The total expended work equals that consumed on drawing the fibrils in the craze up to their maximum length at the point of break. As was discussed in the previous section, fibrils elongate by an incorporation of fresh material at the craze surface, which occurs under a constant tensile stress. Hence, we may formulate

$$2w \approx \sigma_{\mathrm{y}} \cdot \Delta z_{\mathrm{b}} \tag{8.58}$$

where σ_{y} stands for the tensile stress effective at the craze surface and Δz_{b}

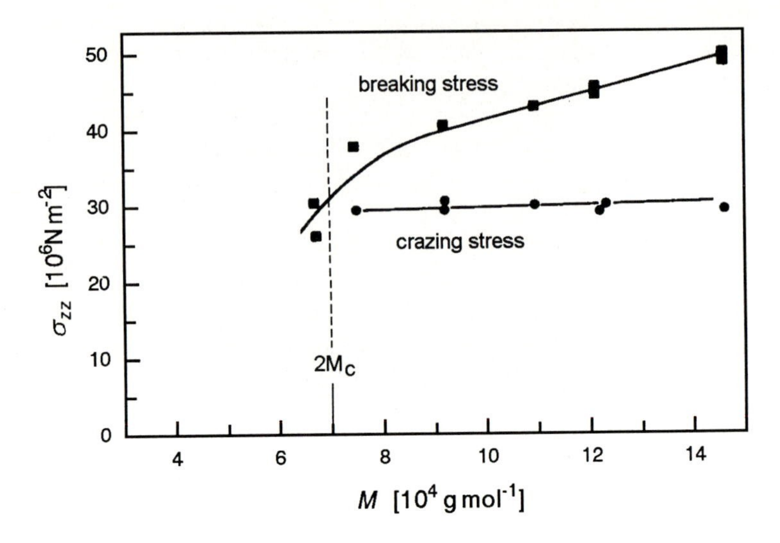

Fig. 8.35. Crazing and fracture of fractions of PS, subject to uniaxial stress. Molecular weight dependencies of the minimum stress for the observation of crazes and the stress at failure. Data from Fellers and Kee [101]

denotes the maximum fibril length, reached at the point of break. This leads us to a specific expression for the stress intensity factor to be employed during slow crack growth. Stationary growth implies equality of G_I with $2w$ and consequently, with Eq. (8.55), the relation

$$K_I = EG_I = E\sigma_y \cdot \Delta z_b \quad (8.59)$$

One may wonder about the cause of the fibril rupture. Ideally, for a fibril with constant width, elongation by the incorporation of additional material at both ends could continue infinitely. Reality is different, since the fibril diameter is not truly stable but narrows due to an ongoing creep. Away from the crack tip this is a slow process, but adjacent to the tip it becomes accelerated by the intensification of stress. Rather that breaking somewhere in the central part, fibrils often also fail at the craze surface, for example, if an obstacle prevents further fibrillation. Generally, the stability of the fibrils against rupture increases with the molecular weight, as the necessary chain disentangling requires more and more time, and both σ_y and Δz_b go to higher values. As a consequence, the strength of samples also increases with molecular weight, and this is shown by the data in Fig. 8.35. They were obtained from measurements on fractions of polystyrene. In addition to the critical stress at break, the figure includes also the minimum stresses for the observation of crazes. In contrast to the stresses at failure, the latter ones are M-independent, telling us that craze initiation is a local event, depending primarily on the interaction between neighboring chain sequences. Furthermore, as crazes are only observed for $M > 2M_c$ ($M_c = 3.6 \cdot 10^4$ is the critical molecular weight at the entanglement limit), we see that as a prerequisite for crazing the entanglement network has to be well established, as already stated previously.

On the basis of Eq. (8.59), we can understand how the balance between the released strain energy and the energy consumption in work of plastic deformation is realized over a finite range of crack velocities. This becomes accomplished by the rate dependence of the drawing stress σ_y, which goes up with $\dot{c}$, and that of Δz_b. By the increase of $2w = \sigma_y \Delta z_b$ with $\dot{c}$, a new equilibrium with G_I can be established following any upward change in G_I or K_I. The limit of stationary slow growth is reached, when $2w$ is no longer rate sensitive enough to compensate for the increase in G_I. Then growth is no longer stationary and the crack expands in an uncontrolled fashion.

We should not finish this last chapter without a short remark on environmental effects. Indeed, influences of environmental gases or liquids on crazing and failure of polymeric materials can be rather strong and have to be properly accounted for in experiments. The primary action of active agents involves absorption and thus swelling of the material, with the result of a plastification, i.e. a decrease of T_g, and a reduction of surface energies. Both effects obviously facilitate craze initiation, as cavitation processes are involved and the chain mobility is increased. Craze growth is promoted as well, for the same reasons, and so is the tendency for failure, as the load carrying capacity of the fibrils may be diminished. Presence of agents which diffuse into the polymer may therefore greatly reduce the barriers for crazing and fracture. In extreme cases, bodies may even break spontaneously on coming into contact with an absorbing agent, since then frozen strains are released and produce internal stresses. It is obvious that these are important influences and they have to be carefully considered to be sure that materials are safe under all possible environmental conditions.

8.4 Further Reading

E.H. Andrews, P.E. Reed, J.G. Williams, C.B. Bucknall: Advances in Polymer Science Vol.27 *Failure in Polymers*, Springer, 1978

H.H. Kausch: *Polymer Fracture*, Springer, 1978

H.H. Kausch (Ed.): Advances in Polymer Science Vol.91/92 *Crazing in Polymers*, Springer, 1990

G.H. Michler: *Kunststoff-Mikromechanik*, Hanser, 1992

I.M. Ward: *Mechanical Properties of Solid Polymers*, John Wiley & Sons, 1971

Appendix

Scattering Experiments

A.1 Fundamentals

As for bulk condensed matter in general, analysis of the microscopic structure of polymer systems is mostly carried out by scattering experiments. This chapter in the Appendix is meant to provide the reader with a summary of results of scattering theory, including both general and specific equations, in a selection suggested by the needs of the considerations in this book.

Depending on the system under study and the desired resolution, photons in the X-ray and light scattering range, or neutrons are used. The general set up of a scattering experiment is indicated schematically in Fig. A.1. We have an incident beam of monochromatic radiation with wavelength λ and an intensity I_0. It becomes scattered by a sample and the intensity I of the scattered waves is registered by a detector (D) at a distance A, under variation of the direction of observation. Employing the 'scattering vector' $\boldsymbol{q}$, defined as

$$\boldsymbol{q} := \boldsymbol{k}_{\mathrm{f}} - \boldsymbol{k}_{\mathrm{i}} \tag{A.1}$$

where $\boldsymbol{k}_{\mathrm{f}}$ and $\boldsymbol{k}_{\mathrm{i}}$ denote the wave vectors of the incident and the scattered plane waves, the result of a scattering experiment is usually expressed by giving the 'intensity distribution in $\boldsymbol{q}$-space', $I(\boldsymbol{q})$. In the majority of scattering experiments on polymers the radiation frequency remains practically unchanged. Then we have

$$|\,\boldsymbol{k}_{\mathrm{f}}\,| \approx |\,\boldsymbol{k}_{\mathrm{i}}\,| = \frac{2\pi}{\lambda} \tag{A.2}$$

and $|\,\boldsymbol{q}\,|$ is related to the 'Bragg scattering angle' ϑ_{B} by

$$|\,\boldsymbol{q}\,| = \frac{4\pi}{\lambda}\sin\vartheta_{\mathrm{B}} \tag{A.3}$$

(ϑ_{B} is identical to half of the angle enclosed by $\boldsymbol{k}_{\mathrm{i}}$ and $\boldsymbol{k}_{\mathrm{f}}$).

A.1.1 Basic Equations

Two different functions can be used for representing scattering data in reduced forms. The first one, denoted $\Sigma(\boldsymbol{q})$, is the differential scattering cross-section

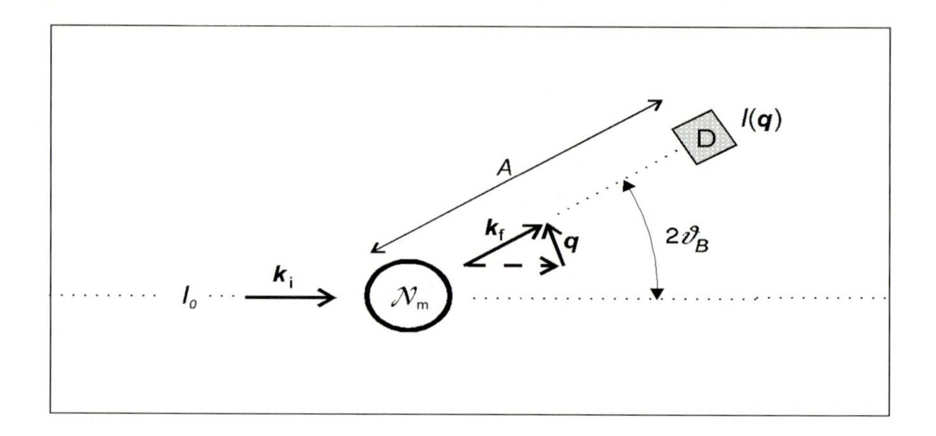

Fig. A.1. General set-up of a scattering experiment

per unit volume of the sample

$$\Sigma(\boldsymbol{q}) := \frac{1}{V}\frac{\mathrm{d}\sigma}{\mathrm{d}\Omega} = \frac{1}{V}\frac{I(\boldsymbol{q})A^2}{I_0} \tag{A.4}$$

In light scattering experiments this function is called the 'Rayleigh-ratio'. While the effect of the volume is removed, $\Sigma(\boldsymbol{q})$ remains dependent on the scattering power of the particles in the sample, which varies with the applied radiation. For X-rays, the scattering power is related to the electron densities, for light scattering to the associated refractive indices and for neutron scattering to the 'scattering length' densities.

This dependence on the applied radiation is eliminated in the second function which, however, can only be employed if the scattering can be treated as being due to just one class of particles. In polymer systems these can be identified with the monomeric units. For equal particles the scattering properties can be described by the 'interference function' $S(\boldsymbol{q})$, also called 'scattering function' or 'scattering law', which is defined as

$$S(\boldsymbol{q}) := \frac{I(\boldsymbol{q})}{I_{\mathrm{m}}\mathcal{N}_{\mathrm{m}}} \tag{A.5}$$

Here $\mathcal{N}_{\mathrm{m}}$ represents the total number of particles/monomers in the sample, and I_{m} is the scattering intensity produced by one particle, if placed in the same incident beam. The interference function expresses the ratio between the actual intensity and the intensity which would be measured, if all particles in the sample were to scatter incoherently. It thus indeed describes the interference effect.

As may be directly verified, $\Sigma(\boldsymbol{q})$ and $S(\boldsymbol{q})$ are related by the equation

$$\Sigma(\boldsymbol{q}) = \langle c_{\mathrm{m}} \rangle \left(\frac{\mathrm{d}\sigma}{\mathrm{d}\Omega}\right)_{\mathrm{m}} S(\boldsymbol{q}) \tag{A.6}$$

Here $(\mathrm{d}\sigma/\mathrm{d}\Omega)_\mathrm{m}$ denotes the scattering cross section per particle or monomer and $\langle c_\mathrm{m}\rangle$ stands for their mean density

$$\langle c_\mathrm{m}\rangle = \frac{\mathcal{N}_\mathrm{m}}{V} \tag{A.7}$$

Scattering diagrams generally emerge from the superposition and interference of the scattered waves emanating from all the particles in the sample. If we describe the amplitudes of single scattered waves at the point of registration by the detector in normalized form, by complex numbers of modulus unity and phases φ_i, the total scattering amplitude is obtained as

$$C = \sum_{i=1}^{\mathcal{N}_\mathrm{m}} \exp \mathrm{i}\varphi_i \tag{A.8}$$

Simple geometrical considerations which are presented in all textbooks dealing with scattering theory show that the phases φ_i are determined by the particle positions $\boldsymbol{r}_i$ and the scattering vector $\boldsymbol{q}$ only, being given by

$$\varphi_i = \boldsymbol{q}\boldsymbol{r}_i \tag{A.9}$$

Hence, the scattering amplitude produced by a set of particles at locations $\boldsymbol{r}_i$ may be formulated as a $\boldsymbol{q}$-dependent function

$$C(\boldsymbol{q}) = \sum_{i=1}^{\mathcal{N}_\mathrm{m}} \exp \mathrm{i}\boldsymbol{q}\boldsymbol{r}_i \tag{A.10}$$

The scattering intensity is proportional to the squared modulus of C. Since measurements require a certain time, average values are generally obtained and we write

$$I(\boldsymbol{q}) \sim \langle |\, C(\boldsymbol{q})\,|^2 \rangle \tag{A.11}$$

The brackets indicate an ensemble average which involves, as always in statistical treatments of physical systems, all microscopic states of the sample. For ergodic systems the time average carried out by the detector equals the theoretical ensemble average.

As the normalization of the amplitudes of the single scattered waves is already implied in the definition of the interference function Eq. (A.5), Eq. (A.11) may be completed to

$$S(\boldsymbol{q}) = \frac{1}{\mathcal{N}_\mathrm{m}} \langle |\, C(\boldsymbol{q})\,|^2 \rangle \tag{A.12}$$

This is a basic equation of general validity and it may serve as starting point for the derivation of other forms of scattering equations.

Our task is the calculation of the above average of the squared scattering amplitude. A first formula follows directly by insertion of Eq. (A.10), leading to

$$S(\boldsymbol{q}) = \frac{1}{\mathcal{N}_\mathrm{m}} \sum_{i,j=1}^{\mathcal{N}_\mathrm{m}} \langle \exp \mathrm{i}\boldsymbol{q}(\boldsymbol{r}_i - \boldsymbol{r}_j) \rangle \tag{A.13}$$

Instead of specifying the discrete positions $\boldsymbol{r}_i$ of all particles, one can also use a continuum description and introduce the particle density distribution $c_\mathrm{m}(\boldsymbol{r})$. First we write down the scattering amplitude for a single microstate, as represented by the associated density distribution

$$C(\boldsymbol{q}) = \int_V \exp \mathrm{i}\boldsymbol{q}\boldsymbol{r} \cdot (c_\mathrm{m}(\boldsymbol{r}) - \langle c_\mathrm{m} \rangle) \mathrm{d}^3\boldsymbol{r} \tag{A.14}$$

As scattering occurs only, if c_m varies within the sample, we subtract here the mean value $\langle c_\mathrm{m} \rangle$, thus relating the scattering directly to the fluctuations. As we can see, $C(\boldsymbol{q})$ equals the Fourier-transform of the fluctuations in the particle density. Insertion of Eq. (A.14) into Eq. (A.12) and carrying out the ensemble average yields

$$S = \frac{1}{\mathcal{N}_\mathrm{m}} \int_V \int_V \exp \mathrm{i}\boldsymbol{q}(\boldsymbol{r}' - \boldsymbol{r}'') \langle (c_\mathrm{m}(\boldsymbol{r}') - \langle c_\mathrm{m} \rangle) \cdot (c_\mathrm{m}(\boldsymbol{r}'') - \langle c_\mathrm{m} \rangle) \rangle \mathrm{d}^3\boldsymbol{r}' \mathrm{d}^3\boldsymbol{r}'' \tag{A.15}$$

For all macroscopically homogeneous systems, where

$$\langle c_\mathrm{m}(\boldsymbol{r}') c_\mathrm{m}(\boldsymbol{r}'') \rangle = \langle c_\mathrm{m}(\boldsymbol{r}' - \boldsymbol{r}'') c_\mathrm{m}(0) \rangle \tag{A.16}$$

this equation may be reduced to a single integral. Substitution of $\boldsymbol{r}' - \boldsymbol{r}''$ by $\boldsymbol{r}$ yields

$$S(\boldsymbol{q}) = \frac{1}{\langle c_\mathrm{m} \rangle} \int_V \exp \mathrm{i}\boldsymbol{q}\boldsymbol{r} \cdot (\langle c_\mathrm{m}(\boldsymbol{r}) c_\mathrm{m}(0) \rangle - \langle c_\mathrm{m} \rangle^2) \mathrm{d}^3\boldsymbol{r} \tag{A.17}$$

Equation (A.17) expresses $S(\boldsymbol{q})$ as the Fourier-transform of the space dependent correlation function of the particle density.

A third form of the basic scattering equation is obtained if structures are characterized with the aid of the 'pair distribution function' $g(\boldsymbol{r})$. Per definition, the product

$$g(\boldsymbol{r}) \mathrm{d}^3\boldsymbol{r}$$

gives the probability that starting from a given particle, the particle itself or some other particle is found in the volume element $\mathrm{d}^3\boldsymbol{r}$ at a distance $\boldsymbol{r}$. The pair distribution function $g(\boldsymbol{r})$ is composed of two parts

$$g(\boldsymbol{r}) = \delta(\boldsymbol{r}) + g'(\boldsymbol{r}) \tag{A.18}$$

the first one giving the self-contribution and the second, g', the contributions of the other particles. For fluid systems with short-range order, the limiting value of the pair distribution function at large distances equals the mean density:

$$g(|\boldsymbol{r}| \to \infty) \to \langle c_\mathrm{m} \rangle \tag{A.19}$$

As follows directly from the definitions, density distribution and pair distribution function are related by

$$\langle c_\mathrm{m}(\boldsymbol{r}) c_\mathrm{m}(0) \rangle = \langle c_\mathrm{m} \rangle \cdot g(\boldsymbol{r}) \tag{A.20}$$

Insertion of Eq. (A.20) in Eq. (A.17) gives

$$S(\boldsymbol{q}) = \int_V \exp \mathrm{i}\boldsymbol{q}\boldsymbol{r} \cdot (g(\boldsymbol{r}) - \langle c_\mathrm{m} \rangle) \mathrm{d}^3 \boldsymbol{r} \tag{A.21}$$

As we notice, the scattering function once again equals a Fourier-transform, now of the pair distribution function.

From Eq. (A.21) there follows the asymptotic value of S for large values of $\boldsymbol{q}$. In the limit $\boldsymbol{q} \to \infty$ only the contribution of the self-correlation part, $\delta(\boldsymbol{r})$, is left, and we find

$$S(\boldsymbol{q} \to \infty) \to 1 \tag{A.22}$$

We may conclude that, for large $\boldsymbol{q}$, there are neither constructive nor destructive interferences between the particles so that they behave like incoherent scatterers.

For isotropic systems with

$$g(\boldsymbol{r}) = g(r := |\boldsymbol{r}|) \tag{A.23}$$

the scattering function is also isotropic

$$S(\boldsymbol{q}) = S(q := |\boldsymbol{q}|) \tag{A.24}$$

The Fourier-relation between $g(r)$ and $S(q)$ then has the form

$$S(q) = \int_{r=0}^{\infty} \frac{\sin qr}{qr} 4\pi r^2 (g(r) - \langle c_\mathrm{m} \rangle) \mathrm{d}r \tag{A.25}$$

We have formulated three equivalent relations, Eqs. (A.13), (A.17) and (A.21), which may all be employed in the evaluation of scattering data. All three equations express a Fourier-relation between $S(\boldsymbol{q})$ and functions which describe properties of the microscopic structure in statistical terms. To put special emphasis on this well defined structural background, $S(\boldsymbol{q})$, firstly introduced as the 'scattering function', is often also addressed as the 'structure function' or 'structure factor'. We will use all these different names, chosen freely.

A.1.2 Time-Resolved Scattering Experiments

Since the particles in a sample are moving, the interference pattern fluctuates in time. The fluctuations can be included in the treatment, by representing the scattering amplitude as a function with a statistical time dependence, $C(\boldsymbol{q},t)$. So far we have dealt with the case of 'static scattering experiments', where the fluctuations are not registered and the detector only furnishes the mean value of the scattering intensity. It is also possible to conduct scattering experiments with neutrons or light in such a way that the time dependent fluctuations are monitored. These time-resolved measurements provide information on the internal dynamics in a sample and thus may expand considerably the information content of a scattering experiment.

We give here a summary of the main results as they follow from a purely classical treatment. These are applicable for 'quasi-elastic' scattering experiments, i.e. experiments where the amount of energy exchange between the neutrons or photons and the sample remains small compared to both the initial energy of the scattered particles and the thermal energy kT of the particles in the sample. In studies of relaxation processes in polymer systems these requirements are generally fulfilled.

If the particles in the sample are mobile, their positions or, in the continuum description, the density distribution become functions of time, $\boldsymbol{r}_i(t)$ or $c_{\mathrm{m}}(\boldsymbol{r},t)$. As a consequence the scattering amplitude C fluctuates, and we may write either

$$C(\boldsymbol{q},t) = \sum_{i=1}^{\mathcal{N}_{\mathrm{m}}} \exp \mathrm{i}\boldsymbol{q}\boldsymbol{r}_i(t) \tag{A.26}$$

or

$$C(\boldsymbol{q},t) = \int_V \exp \mathrm{i}\boldsymbol{q}\boldsymbol{r} \cdot (c_{\mathrm{m}}(\boldsymbol{r},t) - \langle c_{\mathrm{m}}\rangle)\mathrm{d}^3\boldsymbol{r} \tag{A.27}$$

There are techniques which provide a statistical analysis of these time dependent fluctuations. The main result of the analysis is a determination of the time correlation function of the scattering amplitude, being defined as

$$S(\boldsymbol{q},t) := \frac{1}{\mathcal{N}_{\mathrm{m}}} \langle C(\boldsymbol{q},t'+t)C^*(\boldsymbol{q},t')\rangle \tag{A.28}$$

The chosen form relates to the previous Eq. (A.12) and expands it by an inclusion of the time dependence. $S(\boldsymbol{q},t)$ expresses the correlation in time of the scattering amplitudes measured at two times separated by t. The stationary character of thermal equilibrium states implies that the right-hand side is independent of t'. In 'dynamic light scattering' experiments the determination of $S(\boldsymbol{q},t)$ is accomplished by the photon correlation technique. In 'dynamic neutron scattering' experiments, spin-echo measurement can be used in order to obtain the correlation function.

$S(\boldsymbol{q},t)$ is called 'intermediate scattering function' or 'intermediate scattering law', and plays a central role in the analysis of dynamic scattering experiments. The importance is due to $S(\boldsymbol{q},t)$ being related in a well-defined way to the equilibrium dynamics in the system. The relation may be directly established, by insertion of the time dependent scattering amplitudes, as described by Eq. (A.26) or Eq. (A.27), in Eq. (A.28). This results in

$$S(\boldsymbol{q},t) = \frac{1}{\mathcal{N}_{\mathrm{m}}} \sum_{i,j=1}^{\mathcal{N}_{\mathrm{m}}} \langle \exp \mathrm{i}\boldsymbol{q}(\boldsymbol{r}_i(t) - \boldsymbol{r}_j(0)) \rangle \tag{A.29}$$

and

$$S(\boldsymbol{q},t) = \frac{1}{\mathcal{N}_{\mathrm{m}}} \int\limits_V \int\limits_V \exp \mathrm{i}\boldsymbol{q}(\boldsymbol{r}' - \boldsymbol{r}'') \cdot \left(\langle c_{\mathrm{m}}(\boldsymbol{r}',t) c_{\mathrm{m}}(\boldsymbol{r}'',0) \rangle - \langle c_{\mathrm{m}} \rangle^2 \right) \mathrm{d}^3\boldsymbol{r}' \mathrm{d}^3\boldsymbol{r}'' \tag{A.30}$$

For a macroscopically homogeneous system the latter equation reduces to

$$S(\boldsymbol{q},t) = \frac{1}{\langle c_{\mathrm{m}} \rangle} \int\limits_V \exp \mathrm{i}\boldsymbol{q}\boldsymbol{r} \cdot \left(\langle c_{\mathrm{m}}(\boldsymbol{r},t) c_{\mathrm{m}}(0,0) \rangle - \langle c_{\mathrm{m}} \rangle^2 \right) \mathrm{d}^3\boldsymbol{r} \tag{A.31}$$

In the discussion of the structure of samples we employed the pair distribution function $g(\boldsymbol{r})$. This notion can be generalized to include also the dynamics, by introducing the 'time dependent pair distribution function' $g(\boldsymbol{r},t)$. Per definition, the product

$$g(\boldsymbol{r},t)\mathrm{d}^3\boldsymbol{r}$$

describes the probability that, starting from a given particle at zero time, the same or another particle is found in the volume element $\mathrm{d}^3\boldsymbol{r}$ at a distance $\boldsymbol{r}$, if the check is performed after a time delay t. As in the case of the static equation (A.18), one finds two contributions

$$g(\boldsymbol{r},t) = \hat{g}(\boldsymbol{r},t) + g'(\boldsymbol{r},t) \tag{A.32}$$

The first one, $\hat{g}$, gives the probability that a particle will become displaced by $\boldsymbol{r}$ during the time t; the second contribution, g', is furnished by the other particles.

As follows directly from the definitions, $g(\boldsymbol{r},t)$ is proportional to the space- and time-dependent density correlation function

$$\langle c_{\mathrm{m}}(\boldsymbol{r},t) c_{\mathrm{m}}(0,0) \rangle = \langle c_{\mathrm{m}} \rangle g(\boldsymbol{r},t) \tag{A.33}$$

Consequently, Eq. (A.31) is equivalent to

$$S(\boldsymbol{q},t) = \int\limits_V \exp \mathrm{i}\boldsymbol{q}\boldsymbol{r} \cdot (g(\boldsymbol{r},t) - \langle c_{\mathrm{m}} \rangle) \mathrm{d}^3\boldsymbol{r} \tag{A.34}$$

Equation (A.34) is the generalization Eq. (A.21).

Rather than measuring the time correlation function of the scattering amplitude, one can also use a monochromator and determine the spectral density. As stated by the Wiener-Chinchin theorem, the spectral density of a fluctuating quantity and its time correlation function are related by Fourier-transformations. Applied to our case we may write

$$S(\boldsymbol{q},\omega) = \int\limits_{t=-\infty}^{\infty} \exp -\mathrm{i}\omega t \cdot S(\boldsymbol{q},t)\mathrm{d}t \tag{A.35}$$

and reversely

$$S(\boldsymbol{q},t) = \frac{1}{2\pi} \int\limits_{\omega=-\infty}^{\infty} \exp \mathrm{i}\omega t \cdot S(\boldsymbol{q},\omega)\mathrm{d}\omega \tag{A.36}$$

$S(\boldsymbol{q},\omega)$ is known as the 'dynamic scattering law' or, emphasizing the structural background, as the 'dynamic structure factor'. If we use Eq. (A.34) we obtain

$$S(\boldsymbol{q},\omega) = \int\limits_V \int\limits_{t=-\infty}^{\infty} \exp \mathrm{i}(\boldsymbol{q}\boldsymbol{r} - \omega t) \cdot (g(\boldsymbol{r},t) - \langle c_{\mathrm{m}}\rangle)\mathrm{d}^3\boldsymbol{r}\mathrm{d}t \tag{A.37}$$

As we can see, dynamic scattering experiments carried out under variation of the scattering vector and the frequency provide a Fourier-analysis of the time dependent pair distribution function.

Usually, experiments yield the partial differential cross-section per unit volume

$$\Sigma(\boldsymbol{q},\omega) := \frac{1}{V}\frac{\mathrm{d}^2\sigma}{\mathrm{d}\omega\mathrm{d}\Omega} \tag{A.38}$$

It is related to the dynamic scattering law by

$$\Sigma(\boldsymbol{q},\omega) = \langle c_{\mathrm{m}}\rangle \left(\frac{\mathrm{d}\sigma}{\mathrm{d}\Omega}\right)_{\mathrm{m}} S(\boldsymbol{q},\omega) \tag{A.39}$$

in full analogy to Eq. (A.6).

Finally, we formulate relations between static and dynamic scattering functions. According to the definitions, the static scattering law is identical to the dynamic scattering law at zero time

$$S(\boldsymbol{q}) = S(\boldsymbol{q},t=0) \tag{A.40}$$

Applying Eq. (A.36) we can introduce $S(\boldsymbol{q},\omega)$ and obtain

$$S(\boldsymbol{q}) = \frac{1}{2\pi} \int\limits_{\omega=-\infty}^{\infty} S(\boldsymbol{q},\omega)\mathrm{d}\omega \tag{A.41}$$

For the differential cross-section we find correspondingly

$$\Sigma(\boldsymbol{q}) = \frac{1}{2\pi} \int_{\omega=-\infty}^{\infty} \Sigma(\boldsymbol{q},\omega)\mathrm{d}\omega \tag{A.42}$$

A.2 Absolute Intensities in Light-, X-ray- and Neutron-Scattering Experiments

In standard light scattering experiments on polymer solutions, usually absolute scattering intensities are determined since this is necessary for the determination of molecular weights. Measurements of absolute intensities also provide additional information in X-ray or neutron diffraction studies on multicomponent- and multiphase-polymer systems. In the following we will give the equations to be used.

For a complete evaluation of absolute scattering intensities one requires a knowledge of the differential cross-section per monomer, as this is included in Eq. (A.6). Let us first deal with light scattering experiments on dilute polymer solutions. The quantity of interest here is the effective differential cross-section of the dissolved monomers. Clearly it must be related to the difference in the polarizabilities of the solute and the solvent, as no scattering at all would arise for equal polarizabilities. Since a dilute solution scatters light like a gas, we can use Rayleigh's scattering equation, thereby substituting the polarizability of the gas atoms by the difference in the polarizabilities between solute and solvent molecules, called $\delta\beta$:

$$\left(\frac{\mathrm{d}\sigma}{\mathrm{d}\Omega}\right)_{\mathrm{m}} = \frac{\pi^2 \delta\beta^2}{\varepsilon_0^2 \lambda_0^4} \tag{A.43}$$

More accurately, $\delta\beta$ expresses the difference between the polarizability of a monomer and the total polarizability of all displaced solvent molecules which together occupy an equal volume. λ_0 denotes the wavelength of the light in vacuum. Equation (A.43) is valid for the scattering of isotropic particles if both the incoming and the scattered beam are polarized perpendicular to the scattering plane.

For dilute systems, $\delta\beta$ can be related to the difference between the indices of refraction of the solution, n, and the pure solvent, $\mathrm{n_s}$, by

$$c_{\mathrm{m}}\delta\beta = \varepsilon_0(\mathrm{n}^2 - \mathrm{n_s^2}) \tag{A.44}$$

We apply this relation in Eq. (A.43) and obtain

$$\left(\frac{\mathrm{d}\sigma}{\mathrm{d}\Omega}\right)_{\mathrm{m}} = \frac{\pi^2}{\lambda_0^4}\frac{(\mathrm{n}^2 - \mathrm{n_s^2})^2}{c_{\mathrm{m}}^2} \tag{A.45}$$

Use of the approximation

$$\mathrm{n}^2 - \mathrm{n}_\mathrm{s}^2 \approx \frac{\mathrm{dn}^2}{\mathrm{d}c_\mathrm{m}} c_\mathrm{m} = 2\mathrm{n}_\mathrm{s} \frac{\mathrm{dn}}{\mathrm{d}c_\mathrm{m}} c_\mathrm{m} \tag{A.46}$$

yields

$$\left(\frac{\mathrm{d}\sigma}{\mathrm{d}\Omega}\right)_\mathrm{m} = \frac{4\pi^2\mathrm{n}_\mathrm{s}^2}{\lambda_0^4} \left(\frac{\mathrm{dn}}{\mathrm{d}c_\mathrm{m}}\right)^2 \tag{A.47}$$

The Rayleigh-ratio follows from Eq. (A.6) as

$$\Sigma(\boldsymbol{q}) = \frac{4\pi^2\mathrm{n}_\mathrm{s}^2}{\lambda_0^4} c_\mathrm{m} \left(\frac{\mathrm{dn}}{\mathrm{d}c_\mathrm{m}}\right)^2 S(\boldsymbol{q}) \tag{A.48}$$

Replacement of the number density of monomers, c_m, by their concentration by weight, c_w, using

$$c_\mathrm{m} = c_\mathrm{w}\mathrm{N}_\mathrm{L}/M_\mathrm{m} \tag{A.49}$$

finally yields

$$\Sigma(\boldsymbol{q}) = K_\mathrm{l} M_\mathrm{m} c_\mathrm{w} S(\boldsymbol{q}) \tag{A.50}$$

with the 'contrast factor for light', K_l, being given by

$$K_\mathrm{l} = \frac{4\pi^2\mathrm{n}_\mathrm{s}^2}{\lambda_0^4\mathrm{N}_\mathrm{L}} \left(\frac{\mathrm{dn}}{\mathrm{d}c_\mathrm{w}}\right)^2 \tag{A.51}$$

Light scattering experiments are usually evaluated on the basis of this equation.

Next, the respective relations for X-ray scattering experiments are given. In the small angle range, which is of special interest for polymer studies, the effective differential cross-section per monomer is

$$\left(\frac{\mathrm{d}\sigma}{\mathrm{d}\Omega}\right)_\mathrm{m} = r_\mathrm{e}^2(\Delta Z)^2 \tag{A.52}$$

Here r_e denotes the classical electron radius

$$r_\mathrm{e} = 2.81 \cdot 10^{-15}\ \mathrm{m} \tag{A.53}$$

ΔZ describes again a difference between monomer and displaced solvent molecules, now the difference in the total number of electrons. ΔZ may be deduced from the electron densities of the single components, the electron density of the monomeric unit, $\rho_\mathrm{e,m}$, and the electron density of the solvent, $\rho_\mathrm{e,s}$. We may write

$$(\Delta Z)^2 = (\rho_\mathrm{e,m} - \rho_\mathrm{e,s})^2 v_\mathrm{m}^2 \tag{A.54}$$

(v_m stands for the monomer volume). If we now take Eqs. (A.52) and (A.54), introduce them into Eq. (A.6) and also exchange c_m against c_w with the aid of Eq. (A.49) we obtain a result analogous to the previous Eq. (A.50)

$$\Sigma(\boldsymbol{q}) = K_\mathrm{x} M_\mathrm{m} c_\mathrm{w} S(\boldsymbol{q}) \tag{A.55}$$

The contrast factor K_x now relates to X-ray scattering and is given by

$$K_x = r_e^2(\rho_{e,m} - \rho_{e,s})^2 v_m^2 \frac{N_L}{M_m^2} \tag{A.56}$$

For the evaluation of X-ray scattering experiments on multicomponent- and multiphase systems such as polymer blends or partially crystalline polymer systems, a fourth form of scattering equations is often appropriate. We first refer to a one-component system composed of particles with Z_m electrons, corresponding to a differential cross-section

$$\left(\frac{d\sigma}{d\Omega}\right)_m = Z_m^2 r_e^2 \tag{A.57}$$

We now introduce the electron density $\rho_e(\boldsymbol{r})$. It is given by

$$Z_m c_m(\boldsymbol{r}) = \rho_e(\boldsymbol{r}) \tag{A.58}$$

Replacing $c_m(\boldsymbol{r})$ by $\rho_e(\boldsymbol{r})$ in Eq. (A.17) and applying Eq. (A.6) results in

$$\Sigma(\boldsymbol{q}) = r_e^2 \int\limits_V \exp i\boldsymbol{q}\boldsymbol{r} \cdot (\langle \rho_e(\boldsymbol{r})\rho_e(0)\rangle - \langle \rho_e \rangle^2) d^3\boldsymbol{r} \tag{A.59}$$

Considerations on a more general basis show that Eq. (A.59) not only holds for one-component systems but is generally valid in X-ray scattering experiments. It describes Σ as the Fourier-transform of the space dependent electron density correlation function. Equation (A.59) is fully equivalent to the Eqs. (A.13), (A.17) and (A.21), when used together with Eq. (A.6).

Finally, we turn to neutron scattering experiments and write down the corresponding equations directly. The cross-section per monomer or per solvent molecule follows from the sum over the 'scattering-lengths' b_i of the respective constituent atoms, when calculating the square

$$\left(\frac{d\sigma}{d\Omega}\right)_m = \left(\sum_i b_i\right)^2 \tag{A.60}$$

It is obvious that neutron scattering experiments on dilute solutions can also be described by an equation with the form of Eqs. (A.50), (A.55)

$$\Sigma(\boldsymbol{q}) = K_n M_m c_w S(\boldsymbol{q}) \tag{A.61}$$

Now we have to use the contrast factor for neutron scattering, denoted K_n, and it is given by

$$K_n = (\sum_i b_i - \sum_j b_j)^2 \frac{N_L}{M_m^2} \tag{A.62}$$

There are two sums on the right-hand side. The first one with running index i refers to the monomer, the second one, with index j, is meant to represent the total scattering length of the displaced solvent molecules. Clearly, for neutron scattering, similarly to the scattering of photons, it is the difference in scattering power between monomers and solvent molecules which determines the absolute scattering intensity.

We can also formulate a general equation for neutron scattering in analogy to Eq. (A.59) valid for X-rays. This is achieved by substituting the product $r_e\rho_e$ by the 'scattering length density' ρ_n, resulting in

$$\Sigma(\boldsymbol{q}) = \int_V \exp i\boldsymbol{q}\boldsymbol{r} \cdot \left(\langle \rho_n(\boldsymbol{r})\rho_n(0) \rangle - \langle \rho_n \rangle^2 \right) \mathrm{d}^3\boldsymbol{r} \tag{A.63}$$

$\rho_n(\boldsymbol{r})$ denotes the 'scattering-length density'. All these equations concern a coherent process of neutron scattering, since they all express interferences and these can only arise for well-defined phase relations between the scattered waves emanating from the different particles. In fact, in neutron scattering experiments, one encounters a complication since there always exists an additional contribution of incoherent scattering processes. The latter take place without a regular phase relation between different particles. Although this second part cannot contain any information on the structure, it is nevertheless important since it can be used in quasi-elastic scattering experiments for the study of single particle motions. Complete treatments and experiments show that one can measure an intermediate scattering law associated with the incoherent part only and that this directly relates to the Fourier-transform of the self-correlation function of a moving particle, denoted $\hat{g}(\boldsymbol{r}, t)$ in Eq. (A.32). Explicitly, the relation has the form

$$S^{\mathrm{inc}}(\boldsymbol{q}, t) = \int_V \exp i\boldsymbol{q}\boldsymbol{r} \cdot \left(\hat{g}(\boldsymbol{r}, t) - \hat{g}(\boldsymbol{r}, t \to \infty) \right) \mathrm{d}^3\boldsymbol{r} \tag{A.64}$$

The relative weights of the coherent and the incoherent process in a neutron scattering experiment vary greatly between different atoms. For hydrogen atoms, for example, the incoherent scattering is dominant, and dynamic neutron scattering experiments on organic materials therefore frequently focus on the dynamics of the single particles. For deuterium we find the reverse situation, with a dominance of the coherent process, which opens the way for investigations of the structure and of collective dynamical processes.

A.3 Low Angle Scattering Properties

A.3.1 Guinier's Law

Scattering experiments at low angles on dilute colloidal systems, polymer solutions included, can be applied for a determination of the molecular weight and the size of colloids or polymers. The relation may be explained as follows.

We start from Eq. (A.13) and consider a dilute system of colloids, each one composed of N monomers. For $\mathcal{N}_\mathrm{p}$ colloids in the sample we have

$$\mathcal{N}_\mathrm{m} = N\mathcal{N}_\mathrm{p} \tag{A.65}$$

For a dilute system we can neglect all interferences between different colloids. Equation (A.13) can then be rewritten

$$\begin{aligned} S(\boldsymbol{q}) &= \frac{1}{\mathcal{N}_\mathrm{m}} \sum_{i,j=1}^{\mathcal{N}_\mathrm{m}} \langle \exp \mathrm{i}\boldsymbol{q}(\boldsymbol{r}_i - \boldsymbol{r}_j) \rangle \\ &= \frac{1}{\mathcal{N}_\mathrm{p} N} \mathcal{N}_\mathrm{p} \sum_{i,j=1}^{N} \langle \exp \mathrm{i}\boldsymbol{q}(\boldsymbol{r}_i - \boldsymbol{r}_j) \rangle \end{aligned} \tag{A.66}$$

In the low angle range close to the origin we may use a series expansion up to the second order, giving

$$S(\boldsymbol{q}) \approx \frac{1}{N} \sum_{i,j=1}^{N} \langle 1 - \mathrm{i}\boldsymbol{q}(\boldsymbol{r}_i - \boldsymbol{r}_j) + \frac{1}{2}[\boldsymbol{q}(\boldsymbol{r}_i - \boldsymbol{r}_j)]^2 \rangle \tag{A.67}$$

For isotropic systems the linear term vanishes, and the quadratic term may be transformed to

$$\langle [\boldsymbol{q} \cdot (\boldsymbol{r}_i - \boldsymbol{r}_j)]^2 \rangle = \frac{1}{3} q^2 \langle | \boldsymbol{r}_i - \boldsymbol{r}_j |^2 \rangle \tag{A.68}$$

This leads us to

$$S(\boldsymbol{q}) \approx \frac{1}{N} \left(N^2 - \frac{q^2}{6} \sum_{i,j=1}^{N} | \boldsymbol{r}_i - \boldsymbol{r}_j |^2 \right) \tag{A.69}$$

If we introduce the 'radius of gyration', defined by

$$R_\mathrm{g}^2 := \frac{1}{2N^2} \sum_{i,j=1}^{N} | \boldsymbol{r}_i - \boldsymbol{r}_j |^2 \tag{A.70}$$

we obtain an expression for the structure factor in the limit of low $\boldsymbol{q}$'s

$$S(\boldsymbol{q}) \approx N \left(1 - \frac{q^2 R_\mathrm{g}^2}{3} + \cdots \right) \tag{A.71}$$

The equation, often adressed in the literature as 'Guinier's-law', tells us that measurements in the low angle range can be used for a determination of the size of a colloid, as characterized by R_g and the mass, as given by N.

An equivalent, probably better known definition of R_g is

$$R_\mathrm{g} := \frac{1}{N} \sum_{i=1}^{N} \langle | \boldsymbol{r}_i - \boldsymbol{r}_\mathrm{c} |^2 \rangle \tag{A.72}$$

whereby $\boldsymbol{r}_{\mathrm{c}}$ denotes the center of mass of the colloid, given by

$$\boldsymbol{r}_{\mathrm{c}} := \frac{1}{N}\sum_{i=1}^{N} \boldsymbol{r}_i \tag{A.73}$$

The equivalence follows by noting that

$$\begin{aligned}
\sum_{i,j=1}^{N} (\boldsymbol{r}_i - \boldsymbol{r}_j)^2 &= \sum_{i,j=1}^{N} [(\boldsymbol{r}_i - \boldsymbol{r}_{\mathrm{c}}) - (\boldsymbol{r}_j - \boldsymbol{r}_{\mathrm{c}})]^2 \\
&= N\Big(\sum_{i}^{N}(\boldsymbol{r}_i - \boldsymbol{r}_{\mathrm{c}})^2 + \sum_{j}^{N}(\boldsymbol{r}_j - \boldsymbol{r}_{\mathrm{c}})^2\Big) \\
&\quad -2\sum_{i}^{N}(\boldsymbol{r}_i - \boldsymbol{r}_{\mathrm{c}})\sum_{j}^{N}(\boldsymbol{r}_j - \boldsymbol{r}_{\mathrm{c}}) \\
&= 2N\sum_{i}^{N}(\boldsymbol{r}_i - \boldsymbol{r}_{\mathrm{c}})^2
\end{aligned} \tag{A.74}$$

thereby taking into account that

$$\sum_{i}^{N}(\boldsymbol{r}_i - \boldsymbol{r}_{\mathrm{c}}) = 0 \tag{A.75}$$

A.3.2 Forward Scattering

A general relation associates the limiting value of the structure factor in the forward direction, $S(\boldsymbol{q} \to 0)$, with the fluctuation of the number of particles in a given volume V and, furthermore, in a second step, with the isothermal compressibility of the sample.

The relation follows directly from an application of Eq. (A.15):

$$\begin{aligned}
S(\boldsymbol{q} \to 0) &= \frac{1}{\langle \mathcal{N}_{\mathrm{m}} \rangle}\left\langle \left(\int_V (c_{\mathrm{m}}(\boldsymbol{r}) - \langle c_{\mathrm{m}} \rangle)\mathrm{d}^3\boldsymbol{r} \right)^2 \right\rangle \\
&= \frac{1}{\langle \mathcal{N}_{\mathrm{m}} \rangle}\langle (\mathcal{N}_{\mathrm{m}} - \langle \mathcal{N}_{\mathrm{m}} \rangle)^2 \rangle
\end{aligned} \tag{A.76}$$

The ratio

$$\frac{\langle (\mathcal{N}_{\mathrm{m}} - \langle \mathcal{N}_{\mathrm{m}} \rangle)^2 \rangle}{\langle \mathcal{N}_{\mathrm{m}} \rangle} = \frac{\langle \mathcal{N}_{\mathrm{m}}^2 \rangle - \langle \mathcal{N}_{\mathrm{m}} \rangle^2}{\langle \mathcal{N}_{\mathrm{m}} \rangle} \tag{A.77}$$

is independent of the chosen volume. This must be the case, since the left-hand side of the equation, i.e. $S(\boldsymbol{q})$, does not depend on V, and it follows also from the treatment of fluctuations in statistical thermodynamics. Fluctuation theory relates the particle number fluctuation to the isothermal compressibility

$$\kappa_T := \left(\frac{\partial \langle c_{\mathrm{m}} \rangle}{\partial p}\right)_T \tag{A.78}$$

by

$$\frac{\langle \mathcal{N}_{\mathrm{m}}^2 \rangle - \langle \mathcal{N}_{\mathrm{m}} \rangle^2}{\langle \mathcal{N}_{\mathrm{m}} \rangle} = kT\kappa_T \tag{A.79}$$

Combination of Eqs. (A.76) and (A.79) yields

$$S(\boldsymbol{q} \to 0) = kT\kappa_T \tag{A.80}$$

Equation (A.80) is generally valid for all one-component systems, independent of the state of order, it may be gaseous, liquid-like or crystalline.

An equation fully equivalent to Eq. (A.80) can be applied in studies of polymer solutions. The limiting value for $\boldsymbol{q} \to 0$ of the structure factor of a solution is given by

$$S(\boldsymbol{q} \to 0) = kT\kappa_{\mathrm{osm}} \tag{A.81}$$

as it follows by a replacement of κ_T in Eq. (A.80) by the osmotic compressibility

$$\kappa_{\mathrm{osm}} = \left(\frac{\partial \langle c_{\mathrm{m}} \rangle}{\partial \Pi}\right)_T \tag{A.82}$$

Π denotes the osmotic pressure and $\langle c_{\mathrm{m}} \rangle$ gives the mean monomer density in the solution. The replacement is justified by fluctuation theory which derives for the fluctuation of the number of colloids in a fixed volume of a solution an equation equivalent to Eq. (A.79), substituting κ_T by κ_{osm}.

A.4 Special Polymer Systems

A.4.1 Binary Mixtures and Block Copolymers

Scattering experiments play a prominent role in the analysis of structures of polymer mixtures and block copolymers. Of special importance are studies of the homogeneous states since here, scattering curves may be evaluated quantitatively by a comparison with theoretical scattering functions. They are derived in this section using general relations between fluctuations and response functions.

In the discussion of concentration fluctuations in polymer blends in Sect. 3.2.3 we derived Eq. (3.148), which expresses the increase in the Gibbs free energy associated with the formation of a concentration wave of the A-chains. Denoting the amplitude of a wave with wave vector $\boldsymbol{k}$ by $\phi_{\boldsymbol{k}}$, the energy increase $\delta\mathcal{G}$ may be expressed by the quadratic form

$$\delta\mathcal{G} = \frac{1}{2} a_{\boldsymbol{k}} \phi_{\boldsymbol{k}}^2 \tag{A.83}$$

a_k represents a general modulus. One can also write

$$\mathrm{d}\mathcal{G} = \psi_k \mathrm{d}\phi_k \tag{A.84}$$

with

$$\psi_k = a_k \phi_k \tag{A.85}$$

ψ_k represents a potential which is the energetic conjugate to ϕ_k. The linear relation between ψ_k and ϕ_k can also be expressed reversely, by

$$\phi_k := \alpha_k \psi_k \tag{A.86}$$

thereby introducing the response coefficient α_k. α_k is the reciprocal of a_k

$$\alpha_k = \frac{1}{a_k} \tag{A.87}$$

The mean-squared fluctuations in thermal equilibrium can be calculated as for any pair of energy conjugated variables, by applying Boltzmann statistics. The probability distribution for ϕ_k is

$$p(\phi_k) \sim \exp -\frac{a_k \phi_k^2}{2kT} \tag{A.88}$$

and we therefore obtain

$$\langle \phi_k^2 \rangle = \frac{kT}{a_k} = kT\alpha_k \tag{A.89}$$

The fluctuation $\langle \phi_k^2 \rangle$ shows up in a scattering experiment, and determines the scattering intensity at $\boldsymbol{q} = \boldsymbol{k}$. In Sect. 3.2.3, we used a lattice model and introduced the scattering function per cell, $S_c(\boldsymbol{q})$ (Eq. (3.153)). As proved by a short calculation at the end of this section, $S_c(\boldsymbol{q})$ and $\langle \phi_k^2 \rangle$ are related by

$$v_c S_c(\boldsymbol{q} = \boldsymbol{k}) = \langle \phi_k^2 \rangle \tag{A.90}$$

(v_c denotes the cell volume). Hence, we find a direct relationship between the scattering function $S_c(\boldsymbol{q})$ and the response coefficients α_k

$$S_c(\boldsymbol{q}) v_c = kT\alpha_{k=q} \tag{A.91}$$

This is a most important result. It teaches us that the problem of calculating the structure factors of mixtures of polymer chains is equivalent to the problem of calculating response functions. De Gennes was the first to apply this relation in a theoretical analysis. The procedure is usually addressed as the 'random phase approximation', shortly 'RPA', adopting the name given by Bohm, Pines and Nozieres in a work on electron properties in metals.

Equation (A.91) can be applied in two directions. If response coefficients are known, scattering functions can be calculated, and reversely, if we know about the structure factors, this knowledge may be used for the derivation of response coefficients. In the following, both procedures will be combined.

We first inquire about the scattering function of an *athermal* mixture of A- and B-chains. We assume that the mixture is densely packed and incompressible. Under this condition we may choose either the A- or the B-monomers as the representative particles solely responsible for the scattering, and we select the A's.

It is now our task to calculate the response coefficient associated with an excitation of a concentration wave with wave vector $\boldsymbol{k}$. Imagine that this excitation is due to the action of a sinusoidally varying potential which interacts exclusively with the A's. If this potential has the amplitude $\psi_{\boldsymbol{k}}$, the response is described by

$$\phi_{\boldsymbol{k}} = \alpha^{0}_{\boldsymbol{k}} \psi_{\boldsymbol{k}} \tag{A.92}$$

Although the potential only interacts directly with the A's, there are, as we shall see, more effects which contribute to $\alpha^{0}_{\boldsymbol{k}}$. $\alpha^{0}_{\boldsymbol{k}}$ represents a 'collective response coefficient'. The upper index '0' is meant to indicate, that we are dealing with an athermal system where $\chi = 0$.

Owing to the incompressibility, an excitation of a concentration wave of the A's with amplitude $\phi_{\boldsymbol{k}}$ is necessarily associated with a simultaneous excitation of a concentration wave of the B's. The latter has the reverse amplitude

$$\phi^{\mathrm{B}}_{\boldsymbol{k}} = -\phi^{\mathrm{A}}_{\boldsymbol{k}} := -\phi_{\boldsymbol{k}} \tag{A.93}$$

Formally, the induced displacement of the B's may be regarded as a result of the action of a second potential, an 'internal field' with amplitude $\hat{\psi}_{\boldsymbol{k}}$. Since this internal field arises from the requirement of a constant total monomer density, it cannot discriminate between A- and B-monomers, but must act on both species equally. The external potential $\psi_{\boldsymbol{k}}$ and the induced internal field $\hat{\psi}_{\boldsymbol{k}}$ produce together the following responses

$$\phi_{\boldsymbol{k}} = \alpha^{\mathrm{AA}}_{\boldsymbol{k}} (\psi_{\boldsymbol{k}} + \hat{\psi}_{\boldsymbol{k}}) \tag{A.94}$$

$$\phi^{\mathrm{B}}_{\boldsymbol{k}} = -\phi_{\boldsymbol{k}} = \alpha^{\mathrm{BB}}_{\boldsymbol{k}} \hat{\psi}_{\boldsymbol{k}} \tag{A.95}$$

The A-chains interact with both the external potential and the induced internal field, whereas the B-chains interact with the internal field only.

We have introduced the two response coefficients $\alpha^{\mathrm{AA}}_{\boldsymbol{k}}$ and $\alpha^{\mathrm{BB}}_{\boldsymbol{k}}$ into these equations. According to their appearance in the equations they represent response coefficients of the A- and B-chains, expressing how chains of both types respond to the acting fields. As all many-chain effects which ensure the constant total density are included in $\hat{\psi}_{\boldsymbol{k}}$, $\alpha^{\mathrm{AA}}_{\boldsymbol{k}}$ and $\alpha^{\mathrm{BB}}_{\boldsymbol{k}}$ represent true single-chain response coefficients. This conclusion is very important, as we are now only concerned with the responses of non-interacting single chains and this problem is easy to solve. Recall that chains are ideal in a melt, so that we have to inquire about the response coefficients of Gaussian chains. These are indeed known since response coefficients are proportional to structure factors, and the structure factors of ideal chains are given by the Debye equation,

Eq. (2.61). It can be applied separately for the A- and B-chains, thus yielding $\alpha_{\boldsymbol{k}}^{\mathrm{AA}}$ and $\alpha_{\boldsymbol{k}}^{\mathrm{BB}}$.

With this step we have essentially solved our problem. The arguments apply strictly to athermal mixtures only but we shall retain the representation of the single chain responses by ideal chain response coefficients also in the subsequent treatment of other mixtures, assuming that this is still a good approximation.

For a calculation of $S_{\mathrm{c}}(\boldsymbol{q})$ we have to compute the collective response coefficient $\alpha_{\boldsymbol{k}}^{0}$. First we eliminate the induced field $\hat{\psi}_{\boldsymbol{k}}$. Addition of Eqs. (A.94) and (A.95) leads to

$$\hat{\psi}_{\boldsymbol{k}} = -\frac{\alpha_{\boldsymbol{k}}^{\mathrm{AA}}}{\alpha_{\boldsymbol{k}}^{\mathrm{AA}} + \alpha_{\boldsymbol{k}}^{\mathrm{BB}}} \psi_{\boldsymbol{k}} \tag{A.96}$$

Insertion of Eq. (A.96) in Eq. (A.94) gives

$$\phi_{\boldsymbol{k}} = \frac{\alpha_{\boldsymbol{k}}^{\mathrm{AA}} \alpha_{\boldsymbol{k}}^{\mathrm{BB}}}{\alpha_{\boldsymbol{k}}^{\mathrm{AA}} + \alpha_{\boldsymbol{k}}^{\mathrm{BB}}} \psi_{\boldsymbol{k}} \tag{A.97}$$

Comparison of this equation with Eq. (A.92) yields

$$\frac{1}{\alpha_{\boldsymbol{k}}^{0}} = \frac{1}{\alpha_{\boldsymbol{k}}^{\mathrm{AA}}} + \frac{1}{\alpha_{\boldsymbol{k}}^{\mathrm{BB}}} \tag{A.98}$$

Hence, we obtain an explicit expression for the collective response coefficient $\alpha_{\boldsymbol{k}}^{0}$, in terms of the known single chain response coefficients $\alpha_{\boldsymbol{k}}^{\mathrm{AA}}$ and $\alpha_{\boldsymbol{k}}^{\mathrm{BB}}$.

Next, we turn to a non-athermal mixture. The difference in the interaction between like and unlike chains may be approximately accounted for in the spirit of the Flory-Huggins treatment, by introduction of the χ-parameter. This is achieved by changing Eq. (A.92) into

$$\phi_{\boldsymbol{k}} = \alpha_{\boldsymbol{k}}^{0} (\psi_{\boldsymbol{k}} + \chi' \phi_{\boldsymbol{k}}) \tag{A.99}$$

with

$$\chi' = \frac{2\chi kT}{v_{\mathrm{c}}} \tag{A.100}$$

The idea behind Eq. (A.99) is easy to see, as it corresponds to a sequence of two processes. If an external potential $\psi_{\boldsymbol{k}}$ initiates a concentration wave, this wave in turn produces a molecular field $\chi' \phi_{\boldsymbol{k}}$, which for $\chi' > 0$ further reinforces and for $\chi' < 0$ weakens the external potential. Equation (A.99) is identical in form with the well-known mean-field equation of ferromagnetism, where the 'Weiss-field' λM produced by the magnetization M reinforces the primary magnetic field H:

$$M = \alpha_H (H + \lambda M) \tag{A.101}$$

(α_H is the magnetic susceptibility). Equation (A.99) gives

$$\phi_{\boldsymbol{k}} = \frac{\alpha_{\boldsymbol{k}}^{0}}{1 - \chi' \alpha_{\boldsymbol{k}}^{0}} \psi_{\boldsymbol{k}} \tag{A.102}$$

hence

$$\alpha_k := \frac{\phi_k}{\psi_k} = \frac{\alpha_k^0}{1 - \chi' \alpha_k^0} \tag{A.103}$$

Taking the reciprocals on both sides yields

$$\frac{1}{\alpha_k} = \frac{1}{\alpha_k^0} - \chi' \tag{A.104}$$

Equation (A.104) relates the collective response coefficient α_k^0 of an athermal polymer mixture to that of a mixture with non-vanishing χ-parameter.

We can now combine Eqs. (A.98),(A.100) and (A.104) to obtain the final result of the RPA-treatment, which has the form

$$\frac{1}{\alpha_k} = \frac{1}{\alpha_k^{\mathrm{AA}}} + \frac{1}{\alpha_k^{\mathrm{BB}}} - \frac{2\chi kT}{v_{\mathrm{c}}} \tag{A.105}$$

Block copolymers constitute a second system which was successfully treated in the RPA-framework. We notice at the beginning that we have only to discuss the athermal situation since the effect of a non-vanishing χ-parameter can then be treated in the same manner as for the polymer mixtures. The response equations for a block copolymer melt in the homogeneous phase can be directly formulated, writing

$$\phi_k = \alpha_k^{\mathrm{AA}}(\psi_k + \hat{\psi}_k) + \alpha_k^{\mathrm{AB}} \hat{\psi}_k \tag{A.106}$$

$$\phi_k^B = -\phi_k = \alpha_k^{\mathrm{BA}}(\psi_k + \hat{\psi}_k) + \alpha_k^{\mathrm{BB}} \hat{\psi}_k \tag{A.107}$$

Incompressibility is accounted for in the same way as for polymer mixtures, by introduction of the internal field $\hat{\psi}_k$. The new feature in the block copolymer system is the occurrence of the coefficients α_k^{AB} and α_k^{BA}. They describe cross-responses, given by the reaction of the A-chains to a force which acts on the B-chains and vice-versa. As is obvious, cross-responses are brought into the system by the chemical coupling of the A- and B-chains. It is exactly this coupling which sets block copolymers apart from the usual binary mixtures, and it also changes, of course, the response conditions.

The two coupled equations may be evaluated as follows. First note that the ratio between ψ and $\hat{\psi}_k$ is fixed by the assumed incompressibility

$$0 = \psi_k(\alpha_k^{\mathrm{AA}} + \alpha_k^{\mathrm{BA}}) + \hat{\psi}_k(\alpha_k^{\mathrm{AB}} + \alpha_k^{\mathrm{BA}} + \alpha_k^{\mathrm{AA}} + \alpha_k^{\mathrm{BB}}) \tag{A.108}$$

$\hat{\psi}_k$ can be eliminated from the Eqs. (A.106) and (A.108) and we obtain the collective response coefficient of the A-blocks

$$\alpha_k^0 := \frac{\phi_k^A}{\psi_k} = \frac{\alpha_k^{\mathrm{AA}}\alpha_k^{\mathrm{BB}} - \alpha_k^{\mathrm{AB}}\alpha_k^{\mathrm{BA}}}{\alpha_k^{\mathrm{AB}} + \alpha_k^{\mathrm{BA}} + \alpha_k^{\mathrm{AA}} + \alpha_k^{\mathrm{BB}}} \tag{A.109}$$

Equation (A.109) represents the RPA-result for athermal block copolymers. The effect of a non-vanishing χ-parameter can be accounted for as above in Eq. (A.104), by writing again

$$\frac{1}{\alpha_{\boldsymbol{k}}} = \frac{1}{\alpha^0_{\boldsymbol{k}}} - \frac{2\chi kT}{v_c} \tag{A.110}$$

$\alpha^0_{\boldsymbol{k}}$ is now given by Eq. (A.109).

In order to obtain the full expressions for the scattering functions of polymer mixtures and block copolymers, we have to substitute the collective and single chain response coefficients by the corresponding scattering laws. Response coefficients and scattering functions are related by Eq. (A.91). The single chain AA- and BB-response coefficients therefore are given by the corresponding Debye scattering functions:

$$\frac{kT}{v_c}\alpha^{AA}_{\boldsymbol{k}=\boldsymbol{q}} = \phi N_A S_D(R_A q) \tag{A.111}$$

$$\frac{kT}{v_c}\alpha^{BB}_{\boldsymbol{k}=\boldsymbol{q}} = (1-\phi) N_B S_D(R_B q) \tag{A.112}$$

N_A, N_B and R_A, R_B are the degrees of polymerization and the mean squared end-to-end distances of the two polymers. The factors ϕ and $1-\phi$ account for the dilution of the respective chains in the mixture or the block copolymer.

We still need an expression for the cross-response coefficients. The Debye-structure functions $S_D(R_A q)$ and $S_D(R_B q)$ are the Fourier-transforms of the pair distribution functions $g_{AA}(\boldsymbol{r})$ and $g_{BB}(\boldsymbol{r})$ for the A- and B-monomers within their blocks. Considering this definition, it is clear how the coefficients for the cross responses $\alpha^{AB}_{\boldsymbol{k}}$ and $\alpha^{BA}_{\boldsymbol{k}}$ should be calculated. Obviously, they must correspond to the Fourier-transforms of the pair distribution functions $g_{AB}(\boldsymbol{r})$ and $g_{BA}(\boldsymbol{r})$ which describe the probability of finding a B- or A-monomer at a distance $\boldsymbol{r}$ from a A- or B-monomer respectively. Actually, both are identical

$$g_{AB}(\boldsymbol{r}) = g_{BA}(\boldsymbol{r}) \tag{A.113}$$

We will not present the calculation here. It is straightforward and leads to

$$\frac{1}{2}[N_{AB} S_D(R_0 q) - \phi N_A S_D(R_A q) - (1-\phi) N_B S_D(R_B q)] = \frac{kT}{v_c}\alpha^{AB}_{\boldsymbol{k}=\boldsymbol{q}} \tag{A.114}$$

with

$$N_{AB} = N_A + N_B \quad , \quad R_0^2 = R_A^2 + R_B^2 \tag{A.115}$$

Introduction of the Eqs. (A.111) - (A.114) into Eq. (A.98) and Eq. (A.109) respectively, together with a use of Eq. (A.104), yields the RPA scattering functions given in Sects. 3.2.3 and 3.3.3.

As a supplement, at the end of this section we derive Eq. (A.90)

$$S_c(\boldsymbol{q}) = \frac{\langle \phi^2_{\boldsymbol{q}=\boldsymbol{k}} \rangle}{v_c}$$

We start from Eq. (A.14)

$$C(\boldsymbol{q}) = \int\limits_V \exp \mathrm{i}\boldsymbol{q}\boldsymbol{r} \cdot (c_{\mathrm{m}}(\boldsymbol{r}) - \langle c_{\mathrm{m}} \rangle) \mathrm{d}^3\boldsymbol{r}$$

For polymer mixtures, the fluctuation of the concentration of A-monomers can be described as a superposition of independent waves with amplitude $C_{\boldsymbol{k}}$

$$c_{\mathrm{m}}(\boldsymbol{r}) - \langle c_{\mathrm{m}} \rangle = V^{-1/2} \sum_{\boldsymbol{k}} C_{\boldsymbol{k}} \exp -\mathrm{i}\boldsymbol{k}\boldsymbol{r} \tag{A.116}$$

when setting

$$C_{\boldsymbol{k}=0} = 0$$

If we choose periodic boundary conditions, the sum includes the wave vectors

$$\boldsymbol{k} = \frac{2\pi}{L} \begin{pmatrix} n_1 \\ n_2 \\ n_3 \end{pmatrix} \tag{A.117}$$

where the n_i's are integer numbers. The density of the wave vectors in $\boldsymbol{k}$-space is given by

$$L^3/(2\pi)^3 = V/(2\pi)^3 \tag{A.118}$$

Since $c_{\mathrm{m}}(\boldsymbol{r})$ is a real function, we have

$$C_{\boldsymbol{k}} = C^*_{-\boldsymbol{k}} \tag{A.119}$$

Scattering intensities are proportional to the mean-squared amplitude

$$\begin{aligned} \langle | C(\boldsymbol{q}) |^2 \rangle \quad = \quad & \frac{1}{V} \sum_{\boldsymbol{k},\boldsymbol{k}'} \langle C_{\boldsymbol{k}} \cdot C^*_{\boldsymbol{k}'} \rangle \cdot \\ & \int\limits_V \int\limits_V \exp \mathrm{i}(\boldsymbol{q}-\boldsymbol{k})\boldsymbol{r} \cdot \exp -\mathrm{i}(\boldsymbol{q}-\boldsymbol{k}')\boldsymbol{r}' \mathrm{d}^3\boldsymbol{r} \mathrm{d}^3\boldsymbol{r}' \end{aligned} \tag{A.120}$$

We have

$$\langle C_{\boldsymbol{k}} C^*_{\boldsymbol{k}'} \rangle = \langle | C_{\boldsymbol{k}} |^2 \rangle \delta_{\boldsymbol{k},\boldsymbol{k}'} + \langle C^2_{\boldsymbol{k}} \rangle \delta_{\boldsymbol{k},-\boldsymbol{k}'} \tag{A.121}$$

Since

$$\langle C^2_{\boldsymbol{k}} \rangle = 0 \tag{A.122}$$

because $C^2_{\boldsymbol{k}}$ includes a phase factor, and

$$\int\limits_V \int\limits_V \exp \mathrm{i}(\boldsymbol{q}-\boldsymbol{k})(\boldsymbol{r}-\boldsymbol{r}') \mathrm{d}^3\boldsymbol{r} \mathrm{d}^3\boldsymbol{r}' = V(2\pi)^3 \delta(\boldsymbol{q}-\boldsymbol{k}) \tag{A.123}$$

we obtain

$$\langle | C(\boldsymbol{q}) |^2 \rangle = \sum_{\boldsymbol{k}} \langle | C_{\boldsymbol{k}} |^2 \rangle (2\pi)^3 \delta(\boldsymbol{q}-\boldsymbol{k}) \tag{A.124}$$

Scattering experiments average over a finite range $\Delta \boldsymbol{q}^3$. Therefore, applying Eq. (A.118), they result in

$$\langle | C(\boldsymbol{q}) |^2 \rangle = V \langle | C_{\boldsymbol{k}=\boldsymbol{q}} |^2 \rangle \tag{A.125}$$

Now we introduce the scattering function of the lattice model, defined by Eq. (3.153):

$$\begin{aligned} S_{\mathrm{c}}(\boldsymbol{q}) : &= \frac{1}{\mathcal{N}_c} \langle | C(\boldsymbol{q}) |^2 \rangle \\ &= v_{\mathrm{c}} \langle | C(\boldsymbol{q}) |^2 \rangle \end{aligned} \tag{A.126}$$

Replacement of the fluctuations in the monomer concentration by fluctuations in the volume fraction occupied by the A-chains, using

$$\delta\phi = v_{\mathrm{c}} \delta c_{\mathrm{m}} \tag{A.127}$$

hence

$$\langle \phi_{\boldsymbol{k}}^2 \rangle = v_{\mathrm{c}}^2 \langle | C_{\boldsymbol{k}} |^2 \rangle \tag{A.128}$$

leads us to

$$\langle \phi_{\boldsymbol{k}}^2 \rangle = v_{\mathrm{c}} S_{\mathrm{c}}(\boldsymbol{q} = \boldsymbol{k}) \tag{A.129}$$

which is the relation to be derived.

A.4.2 Two-Phase Layer Systems

Isotropic samples of a partially crystalline polymer essentially correspond to an ensemble of densely packed, isotropically distributed stacks of parallel lamellar crystallites. If the extensions of the stacks parallel and normal to the lamellar surfaces are large compared to the interlamellar distance, the scattering behavior can be related to the electron density distribution $\rho_{\mathrm{e}}(z)$ measured along a trajectory normal to the surfaces. This trajectory will pass through amorphous regions with density $\rho_{\mathrm{e,a}}$ and crystallites with a core density $\rho_{\mathrm{e,c}}$. The average density $\langle \rho_{\mathrm{e}} \rangle$ lies between these two limits.

We calculate the scattering cross-section per unit volume by application of Eq. (A.59) and consider at first an ensemble of equally oriented stacks. If we choose the orientation of the surface normals parallel to the z-axis, the electron density distribution depends on z only and we can write

$$\Sigma(\boldsymbol{q}) = r_{\mathrm{e}}^2 \int\limits_{x,y,z} \exp \mathrm{i}(q_x x + q_y y + q_z z)(\langle \rho_{\mathrm{e}}(z) \rho_{\mathrm{e}}(0) \rangle - \langle \rho_{\mathrm{e}} \rangle^2) \mathrm{d}x \mathrm{d}y \mathrm{d}z \tag{A.130}$$

Carrying out the integrations for x and y we obtain

$$\Sigma(\boldsymbol{q}) = r_{\mathrm{e}}^2 (2\pi)^2 \delta(q_x) \delta(q_y) \int\limits_{-\infty}^{\infty} \exp \mathrm{i} q_z z \cdot K(z) \mathrm{d}z \tag{A.131}$$

where $K(z)$ designates the one-dimensional electron density correlation function

$$\begin{aligned} K(z) &:= \langle(\rho_{\mathrm{e}}(z)-\langle\rho_{\mathrm{e}}\rangle)(\rho_{\mathrm{e}}(0)-\langle\rho_{\mathrm{e}}\rangle)\rangle \\ &= \langle\rho_{\mathrm{e}}(z)\rho_{\mathrm{e}}(0)\rangle-\langle\rho_{\mathrm{e}}\rangle^{2} \end{aligned} \tag{A.132}$$

The scattering of an isotropic ensemble of stacks of lamellae, as it is found in melt crystallized samples, follows from Eq. (A.131) by calculation of the isotropic average, i.e. by distributing the intensity at $\pm q_z$ equally over the surface of a sphere with the same radius. The resulting isotropic intensity distribution, $\Sigma(q := |\boldsymbol{q}|)$, is given by

$$\Sigma(q)=\frac{2}{4\pi q^{2}}r_{\mathrm{e}}^{2}(2\pi)^{2}\int_{-\infty}^{\infty}\exp\mathrm{i}qz\cdot K(z)\mathrm{d}z \tag{A.133}$$

The reverse Fourier-relation is

$$\begin{aligned} K(z) &= \frac{1}{2r_{\mathrm{e}}^{2}}\frac{1}{(2\pi)^{3}}\int_{-\infty}^{\infty}\exp\mathrm{i}qz\cdot 4\pi q^{2}\Sigma(q)\mathrm{d}q \\ &= \frac{1}{r_{\mathrm{e}}^{2}}\frac{1}{(2\pi)^{3}}\int_{0}^{\infty}\cos qz\cdot 4\pi q^{2}\Sigma(q)\mathrm{d}q \end{aligned} \tag{A.134}$$

Equation (A.134) enables $K(z)$ to be determined if $\Sigma(q)$ is known.

$K(z)$ has a characteristic shape, allowing an evaluation which leads directly to the main parameters of the stack structure. In order to explain the procedure, we first establish the shape of $K(z)$ for a strictly periodic two-phase system and then proceed by a consideration of the modifications introduced by a stepwise perturbation of the system. Figure A.2 provides an illustration and sketches all the steps.

The periodic structure shows an electron density distribution $\rho_{\mathrm{e}}(z)$ as indicated on the left of part (a). It can be described by specifying the 'long period' d_{ac}, the crystallite thickness d_{c} and the electron density difference $\rho_{\mathrm{e,c}}-\rho_{\mathrm{e,a}}$. The crystallinity $\phi_{\mathrm{c}}=d_{\mathrm{c}}/d_{\mathrm{ac}}$ in this example lies below 50% . We first calculate a special correlation function, denoted $K_{\mathrm{a}}(z)$, defined as

$$K_{\mathrm{a}}(z):=\langle(\rho_{\mathrm{e}}(z)-\rho_{\mathrm{e,a}})(\rho_{\mathrm{e}}(0)-\rho_{\mathrm{e,a}})\rangle \tag{A.135}$$

When using K_{a}, all electron densities refer to the electron density of the amorphous regions. Since the ensemble average is identical with an average over all points z' in a stack, $K_{\mathrm{a}}(z)$ may be obtained by an evaluation of the integral

$$K_{\mathrm{a}}(z)=\frac{1}{\Delta}\int_{-\Delta/2}^{\Delta/2}[\rho_{\mathrm{e}}(z')-\rho_{\mathrm{e,a}}][\rho_{\mathrm{e}}(z+z')-\rho_{\mathrm{e,a}}]\mathrm{d}z' \tag{A.136}$$

The integration range Δ has to be sufficiently large. The functions $\rho_e(z)$ and $\rho_e(z+z')$ are square distributions, and contributions to the integral arise only, if two crystalline regions overlap. Consequently, $K_a(z)$ is proportional to the length of the overlap region, and given by

$$K_a(z) = \begin{cases} (\rho_{e,c} - \rho_{e,a})^2 (d_c - z)/d_{ac} & \text{if } |z| < d_c \\ 0 & \text{if } d_c < |z| < d_{ac} - d_c \end{cases} \tag{A.137}$$

and being periodic, by

$$K_a(z + d_{ac}) = K_a(z) \tag{A.138}$$

Having determined K_a, one obtains $K(z)$ by

$$K(z) = K_a(z) - (\langle \rho_e \rangle - \rho_{e,a})^2 \tag{A.139}$$

The result is shown on the right of part (a). There is a regular sequence of triangles, centered at $z = 0, d_{ac}, 2d_{ac}$ etc., which reflect the correlations within one crystallite, between next neighbors, second neighbors, etc. The 'self-correlation triangle' centered at the origin exhibits some characteristic properties. The value at $z = 0$, denoted Q, is

$$K(z = 0) := Q = \phi_c(1 - \phi_c)(\rho_{e,c} - \rho_{e,a})^2 \tag{A.140}$$

The slope $\mathrm{d}K/\mathrm{d}z$ is

$$\frac{\mathrm{d}K}{\mathrm{d}z} = \frac{\mathrm{d}K_a}{\mathrm{d}z} = -\frac{O_{ac}}{2}(\rho_{e,c} - \rho_{e,a})^2 \tag{A.141}$$

Here, O_{ac} denotes the 'specific internal surface' given by the area per unit volume of the interface separating crystalline and amorphous regions. For the periodic stack it is related to the long period by

$$O_{ac} = \frac{2}{d_{ac}} \tag{A.142}$$

The horizontal 'base-line' between the triangles is located at

$$K = -B = -(\langle \rho_e \rangle - \rho_{e,a})^2 \tag{A.143}$$

$$= -\phi_c^2 (\rho_{e,c} - \rho_{e,a})^2 \tag{A.144}$$

$K(z)$ reaches the base-line at

$$z_1 = d_c \tag{A.145}$$

It is now interesting to recognize that application of these relations is not restricted to the ordered periodic system but can be extended, with slight modifications, to real systems which may show variations in the thicknesses of the crystalline and amorphous regions, and may also possess diffuse phase boundaries. The changes in $K(z)$ resulting from a successive perturbation of

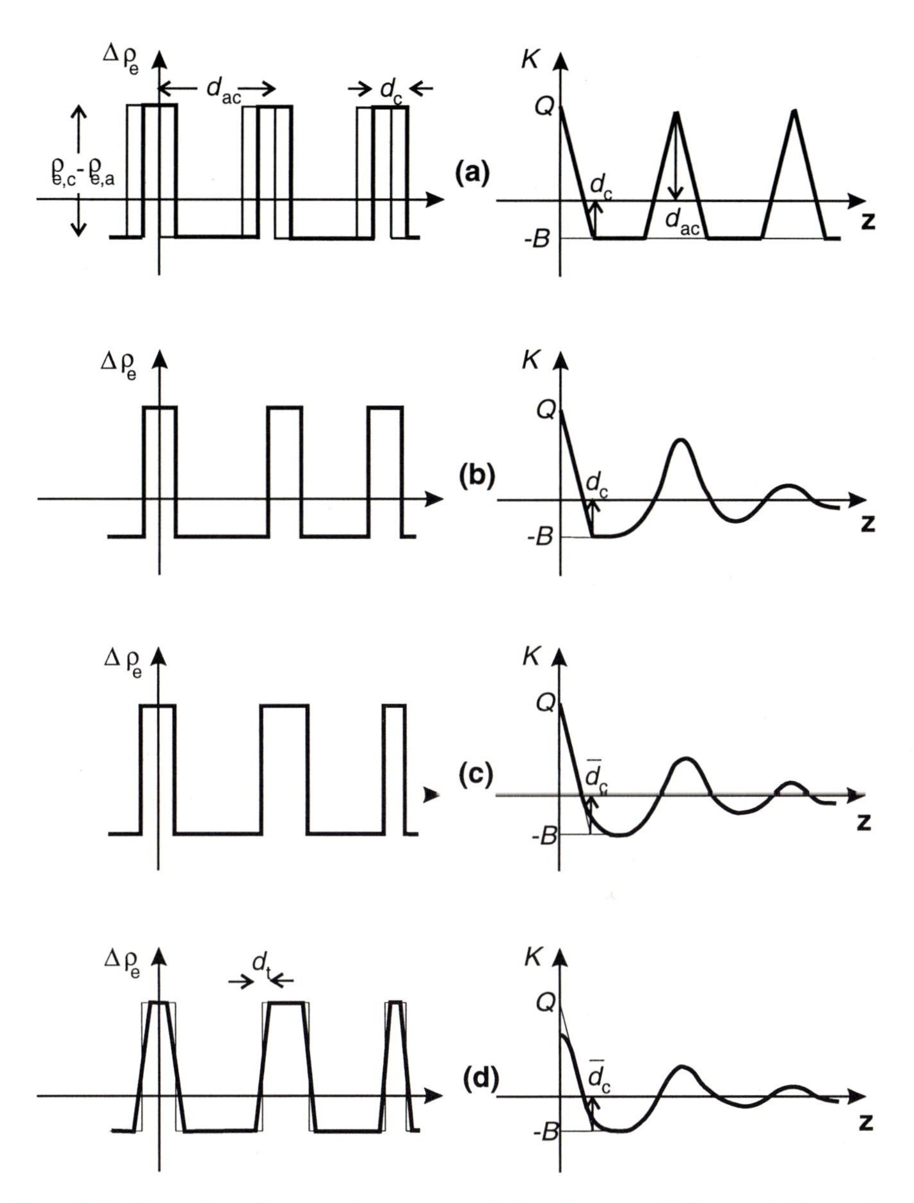

Fig. A.2. Two-phase layer system representative for a partially crystalline polymer. Electron density distribution $\Delta\rho_e(z) := \rho_e(z) - \langle\rho_e\rangle$ and the associated one-dimensional correlation function $K(z)$ for a perfectly ordered system (a). Effects of varying intercrystalline spacings (b), varying crystallite thicknesses (c) and diffuse interfaces (d)

the initial system are schematically indicated in Figs.A.2(b)-(d). All structures in this sequence are understood as having equal crystallinities and equal specific internal surfaces.

First, as indicated in part (b), fluctuations in the intercrystalline spacings are introduced. Since the self-correlation part remains unchanged, the only consequence is a broadening of the peak attributed to next-neighbor correlations. There is a maximum at the position of the most probable distance between neighboring crystallites, and it determines the 'long spacing'.

Secondly, as shown in part (c), we superpose variations in the crystallite thickness. Since $\phi_{\rm c}$ and $O_{\rm ac}$ are assumed to be constant, the value of the correlation function at the origin, Q, the initial slope $\mathrm{d}K/\mathrm{d}z(z=0)$ and the base-line coordinate B are not affected. A modification occurs near to the base of the triangle, where $K(z)$ becomes curved. If we extrapolate the straight part of $K(z)$, it intercepts the base-line at

$$z_1 = \frac{Q+B}{\mathrm{d}K/\mathrm{d}z} = \frac{\phi_{\rm c}}{O_{\rm ac}/2} \tag{A.146}$$

The number average of the crystallite thickness, $\bar{d}_{\rm c}$, is given by

$$\frac{O_{\rm ac}}{2}\bar{d}_{\rm c} = \phi_{\rm c} \tag{A.147}$$

Therefore, we have

$$z_1 = \bar{d}_{\rm c} \tag{A.148}$$

Finally, in the last step, we associate the crystallite surfaces with transition zones (part (d)). We do it under the condition that for each crystallite i the total number of electrons remains unchanged, i.e.

$$\int [\rho_{\rm e}(z') - \rho_{\rm e,a}]\mathrm{d}z' = (\rho_{\rm e,c} - \rho_{\rm e,a})d_{\rm c}^i \tag{A.149}$$

Using this equation in the other direction, one may attribute to each crystallite with a diffuse surface a corresponding lamella with sharp boundaries and thickness $d_{\rm c}^i$. If this replacement is carried out for all crystallites, one returns to a two-phase structure which we address as the 'corresponding two-phase system'. For a transition zone with an extension $d_{\rm t}$ there results a change in shape of the correlation function around the origin, within the range $z < d_{\rm t}$. If $d_{\rm t}$ is small compared to the thicknesses of all crystallites, there still remains a linear portion in the center of the right-hand side of the self-correlation triangle. This allows us to derive directly the parameters of the 'corresponding two-phase system': Extrapolation of the linear section to $z = 0$ gives Q, and a continuation down to the base-line at $K = -B$ yields $\bar{d}_{\rm c}$. The crystallinity $\phi_{\rm c}$, the specific internal surface $O_{\rm ac}$ and the electron density difference then follow by

$$\phi_{\rm c} = \frac{B}{B+Q} \tag{A.150}$$

$$O_{\mathrm{ac}} = \frac{2\phi_{\mathrm{c}}}{\bar{d}_{c}} \tag{A.151}$$

$$(\rho_{\mathrm{e,c}} - \rho_{\mathrm{e,a}})^2 = \frac{Q}{\phi_{\mathrm{c}}(1-\phi_{\mathrm{c}})} . \tag{A.152}$$

So far, we have discussed the case $\phi_{\mathrm{c}} < 0.5$. If we wish to investigate samples with $\phi_{\mathrm{c}} > 0.5$, we have to substitute ϕ_{c} against the volume fraction of the amorphous phase

$$\phi_{\mathrm{a}} = 1 - \phi_{\mathrm{c}} \tag{A.153}$$

and $\bar{d}_{\mathrm{c}}$ against the number average of the thickness of the amorphous layers, $\bar{d}_{\mathrm{a}}$. The substitution rule follows from Babinet's reciprocity theorem which declares that an exchange of the densities in a two-phase structure leaves the scattering function unchanged.

A crucial point of the analysis lies in the knowledge required of the base-line coordinate B. For samples of low or high crystallinity ($\phi_{\mathrm{c}} < 0.3$ or $\phi_{\mathrm{c}} > 0.7$), the base-line usually shows up. The intermediate region is problematic, as here the base-line may not be observed. Then X-ray scattering experiments have to be complemented by other data such as, for example, the density.

In this discussion of the scattering properties of a polymeric layer system, we have dealt with a special two-phase system. Some of the properties are not specific but generally valid for all two-phase systems, independent of their structure. We give here three equations of particular importance.

First, we come back to Eq. (A.140). Use of Eq. (A.134) yields

$$Q = (\rho_{\mathrm{e,c}} - \rho_{\mathrm{e,a}})^2 \phi_{\mathrm{c}}(1-\phi_{\mathrm{c}}) = \frac{1}{r_{\mathrm{e}}^2 (2\pi)^3} \int_{q=0}^{\infty} 4\pi q^2 \Sigma(q) \mathrm{d}q \tag{A.154}$$

Q is often called the 'invariant', for obvious reasons: The total integral, as obtained by an integration over all the reciprocal space, depends only on the volume fractions of the two phases and the electron density difference and is invariant with regard to the detailed structure. Equation (A.154) is not a specific property of layered systems, but generally valid. The proof is simple. One has to formulate the Fourier-transformation reverse to Eq. (A.59), expressing the three dimensional electron density correlation function as a function of $\Sigma(\boldsymbol{q})$

$$\langle \rho_{\mathrm{e}}(\boldsymbol{r}) \rho_{\mathrm{e}}(0) \rangle - \langle \rho_{\mathrm{e}} \rangle^2 = \frac{1}{r_{\mathrm{e}}^2 (2\pi)^3} \int \exp -\mathrm{i}\boldsymbol{q}\boldsymbol{r} \cdot \Sigma(\boldsymbol{q}) \mathrm{d}^3\boldsymbol{q} \tag{A.155}$$

and consider the limit $\boldsymbol{r} \rightarrow 0$

$$\langle \rho_{\mathrm{e}}^2 \rangle - \langle \rho_{\mathrm{e}} \rangle^2 = \frac{1}{r_{\mathrm{e}}^2 (2\pi)^3} \int \Sigma(\boldsymbol{q}) \mathrm{d}^3\boldsymbol{q} \tag{A.156}$$

Direct calculation shows that for a two-phase systems the left-hand sides of Eqs. (A.154) and (A.156) agree.

Secondly, we consider the asymptotic behavior $\Sigma(q \to \infty)$, looking first at the layer system. Due to the reciprocity property of Fourier- transforms, $\Sigma(q \to \infty)$ relates to the limiting behavior $K(z \to 0)$. Therefore, using Eqs. (A.133), (A.140) and (A.141) in a series expansion, we can write

$$\Sigma(q \to \infty) = \frac{1}{2\pi q^2} r_{\mathrm{e}}^2 (2\pi)^2 \lim_{q\to\infty} \int_{-\infty}^{\infty} \cos qz \cdot \left(Q - \frac{O_{\mathrm{ac}}}{2} (\rho_{\mathrm{e,c}} - \rho_{\mathrm{e,a}})^2 z \right) \mathrm{d}z \tag{A.157}$$

For the purpose of a derivation of the asymptotic properties we may employ the following special representation of K valid for small values of z:

$$Q - \frac{O_{\mathrm{ac}}}{2} (\rho_{\mathrm{e,c}} - \rho_{\mathrm{e,a}})^2 z \approx Q \exp - \frac{O_{\mathrm{ac}} (\rho_{\mathrm{e,c}} - \rho_{\mathrm{e,a}})^2}{2Q} z \tag{A.158}$$

With this, the integral can be evaluated

$$\Sigma(q \to \infty) = \frac{2\pi r_{\mathrm{e}}^2}{q^2} \lim_{q\to\infty} \int_{-\infty}^{\infty} \cos qz \cdot Q \exp - \frac{O_{\mathrm{ac}} (\rho_{\mathrm{e,c}} - \rho_{\mathrm{e,a}})^2 z}{2Q} \mathrm{d}z \tag{A.159}$$

which yields

$$\Sigma(q \to \infty) = \frac{2\pi r_{\mathrm{e}}^2}{q^4} O_{\mathrm{ac}} (\rho_{\mathrm{e,c}} - \rho_{\mathrm{e,a}})^2 \tag{A.160}$$

Equation (A.160), known as 'Porod's law', is generally valid for arbitrary two-phase systems. Indeed, an asymptotic law $\Sigma(q) \sim 1/q^4$ is the characteristic signature of two-phase systems with sharp boundaries. According to Eq. (A.160), the asymptotic behavior depends only on the interface area per unit volume, multiplied by the square of the density difference.

A third interesting parameter, l_{c}, is obtained by a combination of Q and O_{ac} in the form of the exponent in Eq. (A.158)

$$l_{\mathrm{c}} := \frac{2Q}{O_{\mathrm{ac}} (\rho_{\mathrm{e,c}} - \rho_{\mathrm{e,a}})^2} = \frac{2\phi_{\mathrm{c}}(1 - \phi_{\mathrm{c}})}{O_{\mathrm{ac}}} \tag{A.161}$$

l_{c} characterizes the length scale of the two-phase structure. Equation (A.161) is a generalization of Eq. (A.147) which concerns $\bar{d}_{\mathrm{c}}$ and thus also a characteristic length of the layer system. There is one technical advantage in the determination of l_{c}: As the electron density difference becomes eliminated, there is no need for intensity measurements in absolute units.

A.5 Further Reading

L.E. Alexander: *X-Ray Diffraction Methods in Polymer Science*, John Wiley & Sons, 1969

B.J. Berne, R. Pecora: *Dynamic Light Scattering*, John Wiley & Sons, 1976

B. Chu: *Laser Light Scattering* Academic Press, 1991

O. Glatter, O. Kratky: *Small Angle X-ray Scattering*, Academic Press, 1982

A. Guinier: *X-Ray Diffraction in Crystals, Imperfect Crystals and Amorphous Bodies*, W.H. Freeman & Co, 1963

J.S. Higgins, H.C. Benoît: *Polymers and Neutron Scattering*, Clarendon Press, 1994

S.W. Lovesey: *Theory of Neutron Scattering from Condensed Matter Vol.1, Nuclear Scattering*, Clarendon Press, 1984

Glossary of Symbols

a_0 — effective length per monomer for an ideal chain (Eq. (2.35))

a_{b} — size of one monomer in the hypothetical limit of freely jointed bonds (Eq. (2.33))

a_{F} — effective length per monomer for an expanded chain (Eq. (2.83))

a_{K} — Kuhn length (Eq. (2.30))

a_{m} — length per monomer in the crystalline state

a_{pr} — segment length of primitive chain (Eq. (6.120))

a_{R} — size of Rouse-segment

$\boldsymbol{a}_l$ — segment of freely jointed chain

a_{s} — average length of segments of freely jointed chain (Eq. (2.16))

$\mathrm{a}(t)$ — general time dependent modulus (Eq. (5.33))

a_T — WLF reduction factor (Eq. (5.122))

A_2 — second virial coefficient (Eq. (2.77))

$\tilde{A}_2$ — modified second virial coefficient (Eq. (3.13))

b_i — scattering length for neutrons of atoms i

b_{m} — lateral monomer diameter (Eq. (4.25))

b_{R} — force constant of springs in Rouse-chain (Eq. (6.25))

$\mathbf{B}, (B_{ij})$ — Finger strain tensor (Eq. (7.81))

$\mathbf{B}(t, t')$ — time dependent Finger strain tensor (Eq. (7.127))

$c_{\mathrm{m}}(\boldsymbol{r})$ — number density of monomers (or other particles)

c_{m}^* — monomer density at the overlap-limit (Eq. (3.2))

c_{p} — number density of polymers

c_{w} — density by weight of a polymer in solution

c_α — heat capacity associated with α-modes (Eq. (5.139))

C_∞ — characteristic ratio (Eq. (2.32))

$C_{\boldsymbol{k}}$	Fourier component of the monomer (particle) density distribution function $c_{\mathrm{m}}(\boldsymbol{r})$
$C(\boldsymbol{q})$	scattering amplitude (Eq. (A.14))
C_1, C_2	WLF-parameters (Eq. (5.129))
$\mathbf{C}, (C_{ij})$	Cauchy strain tensor (Eq. (7.40))
C_{opt}	stress-optical coefficient (Eq. (7.159))
d	(fractal) dimension of a polymer chain
d_i	layer thicknesses in partially crystalline polymers and block-copolymers
d_{ac}	long period of partially crystalline structure
d_{AB}	long period of block copolymers in a layered structure
D	self-diffusion coefficient of chains (Eq. (6.130))
$\hat{D}$	curvilinear diffusion coefficient of reptating chains
D_{coll}	collective diffusion coefficient (Eq. (3.186))
$D(t)$	time dependent tensile compliance (Eq. (5.1))
$D^*(\omega)$	dynamic tensile compliance (Eq. (5.6))
$\boldsymbol{D}$	dielectric displacement vector
e_{p}	internal energy of a polymer chain
$\mathcal{E}$	internal energy of a sample
$\boldsymbol{E}$	electric field strength
$E(t)$	time dependent tensile modulus (Eq. (5.3))
$E^*(\omega)$	dynamic tensile modulus (Eq. (5.7))
$\mathbf{e}, (e_{ij})$	linear strain tensor (Eq. (7.46))
$\mathbf{E}, (E_{ij})$	Eulerian strain tensor (Eq. (7.45))
$\dot{e}_{\mathrm{H}}$	Hencky rate of extension (Eq. (7.115))
$\mathbf{f}$, f	force
$f_{\mathrm{p}}^{\mathrm{e}}$	intramolecular excluded volume interaction energy per polymer chain
f_{p}	Helmholtz free energy of a polymer chain
$\mathcal{F}$	Helmholtz free energy of a sample
F_i	universal functions describing properties of dilute and semi-dilute solutions ($i{=}\Pi$: Eqs. (3.21) and (3.24); $i{=}\xi$: Eq. (3.63); $i{=}\mathrm{R}$: Eq. (3.69))
$g(\boldsymbol{r})$	pair distribution function of monomers (or other particles)
$g_{\mathrm{s}}(\boldsymbol{r})$	pair distribution function of chain segments

$\hat{g}(\boldsymbol{r})$	intramolecular part of the pair distribution function in a semidilute solution
$g(\boldsymbol{r},t)$	time dependent pair distribution function
$\hat{g}(\boldsymbol{r},t)$	self-correlation part of the time dependent pair distribution function
$G(t)$	time dependent shear modulus
$G^*(\omega)$	dynamic shear modulus
G_{pl}	plateau modulus in polymer melts
$\mathrm{G_I}$	strain energy release rate (Eq. (8.49))
$\mathrm{G_{Ic}}$	critical value of $\mathrm{G_I}$ at the onset of uncontrolled crack growth (Eq. (8.50))
g	Gibbs free energy density
g_{p}	Gibbs free energy of a polymer chain
$g_{\mathrm{m}}^{\mathrm{a}}$	chemical potential of a monomer in the melt
$g_{\mathrm{m}}^{\mathrm{c}}$	chemical potential of a monomer in an infinite crystal
$\mathcal{G}$	Gibbs free energy of a sample
$\Delta\mathcal{G}_{\mathrm{b}}$	activation barrier (Eq. (3.125))
$\Delta\mathcal{G}_{\mathrm{loc}}$	local part of $\Delta\mathcal{G}_{\mathrm{mix}}$
$\Delta\mathcal{G}_{\mathrm{mix}}$	Gibbs free energy of mixing for a binary polymer system (Eqs. (3.76) and (3.77))
$h(\mathrm{z})$	osmotic pressure coefficient (Eq. (3.27))
$\Delta h_{\mathrm{m}}^{\mathrm{f}}$	heat of fusion per monomer
h_{p}	enthalpy of a polymer chain
$\Delta\mathcal{H}_{\mathrm{f}}$	heat of fusion
$\Delta\mathcal{H}_{\mathrm{mix}}$	heat of mixing of a binary polymer blend
$\mathrm{H}(\log\hat{\tau})$	relaxation time spectrum (Eq. (5.84))
(H_{ij}^{Os})	Oseen tensor (Eq. (6.146))
I	scattering intensity
I_i, II_i, III_i	invariants of tensor i (Eq. (7.53))
j	rate of transfer
$J(t)$	time dependent shear compliance
$J^*(\omega)$	dynamic shear compliance
J_{e}^0	recoverable shear compliance (Eq. (5.103))
k	Boltzmann constant
$\boldsymbol{k}$	wave vector

$K(z)$	one-dimensional electron density correlation function (Eq. (A.132))
K_i	contrast factors for light scattering experiments (i=l, Eq. (A.51)), X-ray scattering experiments (i=x, Eq. (A.56)) and neutron scattering experiments (i=n, Eq. (A.62))
$\mathrm{K_I}$	stress intensity factor (Eq. (8.52))
$\mathrm{K_{Ic}}$	critical value of $\mathrm{K_I}$ at the onset of uncontrolled crack growth (Eq. (8.56))
l_c	characteristic length in a nonperiodic two-phase system (Eq. (A.161))
l_ext	chain length in the extended conformation
l_ps	persistence length (Fig.2.6)
l_ct	contour length of a chain
l_pr	contour length of primitive path (Eq. (6.120))
$\mathrm{L}(\log\tau)$	retardation time spectrum (Eq. (5.73))
$m(t-t')$	memory function (Eq. (7.194))
M	molecular weight of polymer
$\overline{M}_\mathrm{n}$	number average molecular weight
$\overline{M}_\mathrm{w}$	weight average molecular weight
M_c	critical molecular weight at the entanglement limit (Eq. (5.96))
M_e	average molecular weight of the sequences between entanglement points (Eq. (8.41))
M_m	molecular weight of a monomeric unit
n_i	principal indices of refraction
$\Delta\mathrm{n}$	birefringence coefficient
$\tilde{n}_i$	moles of polymer i in a sample
$\tilde{n}_\mathrm{c}$	moles of structure units (lattice cells) in a binary mixture
$N_{(i)}$	degree of polymerization (of polymer i) = number of monomeric units
N_b	number of backbone bonds of a chain
N_R	number of Rouse-segments in a chain
$N_\mathrm{R,c}$	critical number of Rouse-segments at the entanglement limit
N_s	number of segments in a chain
N_su	number of ideal subunits in an expanded chain
$\mathcal{N}_\mathrm{m}$	number of monomers (or particles) in a sample
$\mathcal{N}_\mathrm{c}$	number of cells in a binary mixture

$\mathcal{N}_\mathrm{p}$	number of polymers (colloids) in a sample
$\mathrm{N_L}$	Avogadro-Loschmidt number
O_{ij}	interface area per unit volume (between phases i, j)
o_p	interface area per polymer in the microphase-separated state of a block copolymer
p	distribution function (for $\boldsymbol{R}, \boldsymbol{r}_{ij}, M$ etc.)
$\boldsymbol{P}$	polarization
$\boldsymbol{p}$	dipole moment
$\boldsymbol{q}$	scattering vector ($q := \lvert \boldsymbol{q} \rvert = 4\pi \sin \vartheta_B / \lambda$)
$\mathcal{Q}$	heat
Q	SAXS invariant for two-phase system (Eq. (A.154))
r_e	classical electron radius (Eq. (A.53))
$\boldsymbol{r}_{ij}$	vector connecting the junction points i, j in a freely jointed chain
$\boldsymbol{R}$	end-to-end distance vector of a polymer chain
R	size ($\langle \lvert \boldsymbol{R} \rvert^2 \rangle^{1/2}$) of a polymer chain in general
R_0	size of an ideal chain
R_F	size of an expanded chain
R_g	radius of gyration of a polymer (or colloid) (Eqs. (A.70) and (A.72))
R_h	hydrodynamic radius (Eq. (6.144))
$\tilde{R}$	perfect gas constant
$S(\boldsymbol{q})$	scattering function ('scattering law'), referred to one monomer (Eq. (A.5))
$S_\mathrm{c}(\boldsymbol{q})$	scattering law, referred to one lattice cell (Eq. (3.153))
$S_\mathrm{D}(R_0 q)$	Debye structure function of an ideal chain of size R_0 (Eq. (2.61))
$S(\boldsymbol{q}, t)$	intermediate scattering law (Eq. (A.28))
$\mathrm{S_{or}^{m}}$	orientational order parameter of monomers (Eq. (8.17))
$s_\mathrm{p}(T)$	entropy of a polymer chain
$\mathcal{S}(T)$	entropy of a sample
$\Delta s_\mathrm{m}^\mathrm{f}$	entropy of fusion per monomer
$\Delta \mathcal{S}_\mathrm{mix}$	entropy of mixing of a binary polymer blend
$\Delta \mathcal{S}_\mathrm{t}$	part of $\Delta \mathcal{G}_{mix}$ describing the increase in the translational entropy
T_c	critical temperature

T_{f}^{∞}	equilibrium melting temperature of an infinite crystal
T_{g}	glass transition temperature
T_{sp}	temperature on spinodal
T_{A}	activation temperature (Eq. (5.126))
T_{V}	Vogel-temperature (Eq. (5.126))
$\mathbf{T}, (t_{ij})$	statistical weight matrix in RIS-model (Eq. (2.127))
u	growth rate of a spherulite
$\boldsymbol{u}$	velocity of a particle
$\tilde{u}(\varphi)$	rotational potential of a backbone bond (per mol)
v	subvolume
$\tilde{v}_i$	molar volume (of a polymer i, a reference unit, a van-der-Waals gas)
v_{c}	volume of a structure unit (lattice cell) in a binary mixture
v_{e}	excluded volume parameter (Eq. (2.78))
v_{m}	volume of a monomer
V	sample volume
$\boldsymbol{v}(\boldsymbol{r})$	flow field in a liquid
w	surface parameter in fracture mechanics (Eq. (8.48))
$\mathcal{W}$	work
$\hat{x}_i, \hat{y}_i, \hat{z}_i$	extension of Rouse-segment i (Eq. (6.66))
X_m, Y_m, Z_m	normal coordinates of Rouse-modes m (Eq. (6.46))
x	overlap-ratio (Eq. (3.16))
X_v	general extensive thermodynamic variable referred to a subvolume v (Eq. (6.10))
z_{eff}	effective coordination number (Eq. (3.82))
z	parameter determining $f_{\mathrm{p}}^{\mathrm{e}}$ (Eq. (2.90)) and N_{su} (Eq. (2.118))
Z	partition function
$\alpha(t)$	time dependent susceptibility (Eq. (5.29))
$\alpha^*(\omega)$	dynamic susceptibility (Eq. (5.42))
β	mean polarizability per monomer
$\Delta\beta$	anisotropy of the polarizability per monomer
γ	shear strain (Eq. (7.99))
γ_{e}	recoverable shear strain

Γ	relaxation rate
δ	phase shift
$\tan\delta(\omega)$	loss tangent (Eq. (5.58))
$\epsilon(t)$	time dependent dielectric function (Eq. (5.16))
$\epsilon^*(\omega)$	complex dielectric function
ζ	friction coefficient of spherical colloid (Eq. (6.142))
ζ_{R}	friction coefficient of a Rouse-segment (Eq. (6.24))
η	steady state shear viscosity
η_0	zero shear rate viscosity
$\hat{\eta}$	extensional viscosity (Eq. (7.125))
$\eta^+(t)$	time dependent shear viscosity (Eq. (7.123))
$\hat{\eta}^+(t)$	time dependent extensional viscosity (Eq. (7.124))
η_{s}	solvent viscosity
$[\eta]$	intrinsic viscosity of a dissolved polymer (Eq. (6.159))
$2\vartheta_{\mathrm{B}}$	scattering angle
θ_i	orientation angle
κ_{osm}	osmotic compressibility (Eq. (A.82)
λ	extension ratio (Eq. (7.1))
$\Lambda(\phi)$	effective interaction parameter (Eq. (3.131))
$\mu(t-t')$	primary response function (Eq. (5.27))
ν_{nuc}	nucleation rate (Eq. (3.125))
ξ_{s}	screening length in semi-dilute solutions (Fig.3.8)
ξ_{t}	thermic correlation length (Eq. (2.117))
ξ_ϕ	correlation length of concentration fluctuations in a binary mixture (Eq. (3.175))
$\boldsymbol{\Omega}$	rotation matrix
Π	osmotic pressure
ρ	density
ρ_{a}	density of amorphous regions in a partially crystalline structure
ρ_{c}	density of crystallites
$\rho_{\mathrm{e}}(\boldsymbol{r})$	electron density
$\rho_{\mathrm{e},i}$	electron density (of monomers, solvents, crystalline and amorphous regions in partially crystalline polymers)
$\rho_{\mathrm{n}}(\boldsymbol{r})$	scattering length density for neutrons

σ	excess free energy per unit area of interface
σ_{m}	excess free energy of monomers at the lateral crystallite surface (Eq. (4.55))
$\sigma_{\mathrm{m,e}}$	excess free energy of monomers at the basal crystallite surfaces (Eq. (4.21))
$\sigma_1, \sigma_2, \sigma_3$	principal stresses
$\boldsymbol{\sigma}, (\sigma_{ij})$	stress tensor
$\hat{\sigma}_{ij}$	nominal stresses
$\Sigma(\boldsymbol{q})$	scattering cross section per unit volume ('Rayleigh ratio')
$\tau_{(i)}$	characteristic time of a relaxation process
τ_{d}	disentangling time
τ_{R}	Rouse-time (Eq. (6.42))
φ_i	rotation angle of bond i
ϕ_i	fraction (volume fraction of components in a blend, of crystalline and amorphous regions, of molecules in different rotational isomeric states)
$\phi_{\boldsymbol{k}}$	Fourier-component of a concentration fluctuation $\delta\phi(\boldsymbol{r})$
χ	Flory-Huggins parameter (Eq. (3.81))
χ_{c}	critical value of χ (Eq. (3.109))
χ_{sp}	value of χ along the spinodal (Eq. (3.130))
ψ	general field or potential
$\psi_{\mathrm{m}}^{\mathrm{e}}$	potential produced by excluded volume forces (per monomer, Eq. (2.78))
Ψ_1	primary normal stress coefficient (Eq. (7.118))
$\Psi_{1,0}$	primary normal stress coefficient at zero shear rate
Ψ_2	secondary normal stress coefficient (Eq. (7.119))

Figure References

[1] P.J. Flory. *Statistical Mechanics of Chain Molecules*, page 40. Wiley-Interscience, 1969.

[2] J. Des Cloiseaux and G. Jannink. *Polymers in Solution*, page 650. Oxford Science Publications, 1990.

[3] Y. Miyaki, Y. Einaga, and H. Fujita. *Macromolecules*, 11:1180, 1978.

[4] R. Kirste, W.A. Kruse, and K. Ibel. *Polymer*, 16:120, 1975.

[5] M. Wintermantel, M. Antonietti, and M. Schmidt. *J. Appl. Polym. Sci.*, 52:91, 1993.

[6] B. Farnoux. *Ann. Fr. Phys.*, 1:73, 1976.

[7] B. Farnoux, F. Bou , J.P. Cotton, M. Daoud, G. Jannink, M. Nierlich, and P.G. de Gennes. *J. Physique*, 39:77, 1978.

[8] W. Gawrisch, M.G. Brereton, and E.W. Fischer. *Polymer Bulletin*, 4:687, 1981.

[9] I. Noda, N. Kato, T. Kitano, and M. Nagesowa. *Macromolecules*, 14:668, 1981.

[10] J.P. Cotton. *J. Physique Lett.*, 41:L–231, 1980.

[11] M.D. Lechner, K. Gehrke, and E.H. Nordmeier. *Makromolekulare Chemie*, page 238. Birkhäuser Verlag, 1993.

[12] T. Koch, G.R. Strobl, and B. Stühn. *Macromolecules*, 25:6258, 1992.

[13] F. Hamada, S. Kinugasa, H. Hayashi, and A. Nakajima. *Macromolecules*, 18:2290, 1985.

[14] M. Daoud, J.P. Cotton, B. Farnoux, G. Jannink, G. Sarma, H. Benoit, R. Duplessix, C. Picot, and P.G. de Gennes. *Macromolecules*, 8:804, 1975.

[15] R.-J. Roe and W.-C. Zin. *Macromolecules*, 13:1221, 1980.

[16] T. Hashimoto, M. Itakura, and H. Hasegawa. *J. Chem. Phys.*, 85:6118, 1986.

[17] G.R. Strobl, J.T. Bendler, R.P. Kambour, and A.R. Shultz. *Macromolecules*, 19:2683, 1986.

[18] D. Schwahn, S. Janssen, and T. Springer. *J. Chem. Phys.*, 97:8775, 1992.

[19] T. Hashimoto, J. Kumaki, and H. Kawai. *Macromolecules*, 16:641, 1983.

[20] T. Koch and G.R. Strobl. *J. Polym. Sci., Polym. Phys. Ed.*, 28:343, 1990.

[21] A. Sariban and K. Binder. *Macromolecules*, 21:711, 1988.

[22] C.C. Han, B.J. Bauer, J.C. Clark, Y. Muroga, Y. Matsushita, M. Okada, T. Qui, T. Chang, and I.C. Sanchez. *Polymer*, 29:2002, 1988.

[23] G.R. Strobl. *Macromolecules*, 18:558, 1985.

[24] M. Takenaka and T. Hashimoto. *J. Chem. Phys.*, 96:6177, 1992.

[25] F.S. Bates and G.H. Frederickson. *Ann. Rev. Phys. Chem.*, 41:525, 1990.

[26] T. Hashimoto, M. Shibayama, and H. Kawai. *Macromolecules*, 13:1237, 1980.

[27] A. Lehmann. Diplomarbeit, Universität Freiburg, 1989.
[28] B. Stühn, R. Mutter, and T. Albrecht. *Europhys. Lett.*, 18:427, 1992.
[29] R. Eppe, E.W. Fischer, and H.A. Stuart. *J. Polym. Sci.*, 34:721, 1959.
[30] B. Kanig. *Progr. Colloid Polym. Sci.*, 57:176, 1975.
[31] A.S. Vaughan and D.C. Bassett. *Comprehensive Polymer Science, Vol.2*, page 415. Pergamon Press, 1989.
[32] T. Albrecht and G.R. Strobl. *Macromolecules*, 28:5827, 1995.
[33] T. Albrecht. PhD thesis, Universität Freiburg, 1994.
[34] R. Mutter, W. Stille, and G. Strobl. *J. Polym. Sci., Polym. Phys. Ed.*, 31:99, 1993.
[35] C.W. Bunn. *Trans. Farad. Soc.*, 35:482, 1939.
[36] M. Kimmig, G. Strobl, and B. Stühn. *Macromolecules*, 27:2481, 1994.
[37] G.H. Michler. *Kunststoff-Mikromechanik*, page 187. Carl Hanser Verlag, 1992.
[38] E. Ergoz, J.G. Fatou, and L. Mandelkern. *Macromolecules*, 5:147, 1972.
[39] G.S. Ross and L.J. Frolen. *Methods of Experimental Physics, Vol.16B*, page 363. Academic Press, 1980.
[40] J.D. Hoffmann, L.J. Frolen, G.S. Ross, and J.I. Lauritzen Jr. *J. Res. NBS*, 79A:671, 1975.
[41] P.J. Barham, R.A. Chivers, A. Keller, J. Martinez-Salazar, and S.J. Organ. *J. Mater. Sci.*, 20:1625, 1985.
[42] A.J. Kovacs, C. Straupe, and A. Gonthier. *J. Polym. Sci., Polym. Symp. Ed.*, 59:31, 1977.
[43] R. Günther. PhD thesis, Universität Mainz, 1994.
[44] M. Cakmak, A. Teitge, H.G. Zachmann, and J.L White. *J. Polym. Sci., Polym. Phys. Ed.*, 31:371, 1993.
[45] G.R. Strobl, T. Engelke, H. Meier, and G. Urban. *Colloid Polym. Sci*, 260:394, 1982.
[46] G.R. Strobl, M.J. Schneider, and I.G. Voigt-Martin. *J. Polym. Sci*, 18:1361, 1980.
[47] K. Schmidt-Rohr and H.W. Spieß. *Macromolecules*, 24:5288, 1991.
[48] H. Seiberle, W. Stille, and G. Strobl. *Macromolecules*, 23:2008, 1990.
[49] J. Heijboer. *Kolloid Z.*, 148:36, 1956.
[50] J. Heijboer. In *Molecular Basis of Transitions and Relaxations (Midland Macromolecular Monographs, Vol.4)*, page 75, Gordon and Breach Science Publishers, 1978.
[51] F.R. Schwarzl. *Polymermechanik*, page . Springer Verlag, 1990.
[52] G.C. Berry and T.G. Fox. *Adv. Polymer Sci.*, 5:261, 1968.
[53] E. Castiff and T.S. Tobolsky. *J. Colloid Sci.*, 10:375, 1955.
[54] S. Onogi, T. Masuda, and K. Kitagawa. *Macromolecules*, 3:109, 1970.
[55] D.J. Plazek, X.D. Zheng, and K.L. Ngai. *Macromolecules*, 25:4920, 1992.
[56] Y. Ishida, M. Matsuo, and K. Yamafuji. *Kolloid Z.*, 180:108, 1962.
[57] N.G. McCrum, B.E. Read, and G. Williams. *Anelastic and Dielectric Effects in Polymeric Solids*, page 305. Wiley & Sons, 1967.
[58] D. Boese and F. Kremer. *Macromolecules*, 23:829, 1990.
[59] A.J. Kovacs. *Fortschr. Hochpolym. Forsch.*, 3:394, 1966.
[60] M. Heckmeier. Diplomarbeit, Universität Freiburg, 1995.
[61] H.A. Flocke. *Kolloid-Z. Z. Polym.*, 180:188, 1962.
[62] K. Schmieder and K. Wolf. *Kolloid Z.*, 134:149, 1953.
[63] B. Holzer and G.R. Strobl. *Acta Polymer.*, 47:40, 1996.

[64] H. Meier. Diplomarbeit, Universität Mainz:, 1981.
[65] H.W. Spiess. *Colloid Polym. Sci.*, 261:193, 1983.
[66] U. Eisele. *Introduction to Polymer Physics*, page . Springer-Verlag, 1990.
[67] D. Richter, B. Farago, L.J. Fetters, J.S. Huang, B. Ewen, and C. Lartigue. *Phys. Rev. Lett.*, 64:1389, 1990.
[68] J. Klein. *Nature*, 271:143, 1978.
[69] T. Koch. PhD thesis, Universität Freiburg, 1991.
[70] P.J. Flory. *Principles of Polymer Chemistry*, page 311. Cornell University Press, 1953.
[71] L.R.G. Treloar. *The Physics of Rubber Elasticity*, page . Clarendon Press, 1975.
[72] J. Meissner. *Kunststoffe*, 61:576, 1971.
[73] J.J. Burke and V. Weiss. *Characterisation of Materials in Research*, page . Syracuse Univ. Press, 1975.
[74] R.L. Anthony, R.H. Caston, and E. Guth. *J.Phys.Chem.*, 46:826, 1942.
[75] R.S. Rivlin and D.W. Saunders. *Phil. Trans. R. Soc.*, A243:251, 1951.
[76] H.M. Laun. *Rheol. Acta*, 21:464, 1982.
[77] J. Meissner. *J. Appl. Polym. Sci.*, 16:2877, 1972.
[78] H.M. Laun and H. Münstedt. *Rheol. Acta*, 15:517, 1976.
[79] H. Münstedt and H.M. Laun. *Rheol. Acta*, 20:211, 1981.
[80] T. Matsumoto and D.C. Bogue. *J. Polym. Sci., Phys. Ed.*, 15:1663, 1977.
[81] C.Wales in H.Janeschitz-Kriegl. *Polymer Melt Rheology and Flow Birefringence*, page 115. Springer-Verlag, 1983.
[82] R.Morbitzer in U.Eisele. *Introduction to Polymer Physics*, page 104. Springer-Verlag, 1990.
[83] W. Retting. *Rheol. Acta*, 8:259, 1969.
[84] C.B. Bucknall. *Advances in Polymer Science, Vol.27*, page 121. Springer-Verlag, 1978.
[85] R. Hiss. Data and micrographs from work on thesis. Freiburg, 1996.
[86] P.J. Lemstra, R. Kirschbaum, T. Ohta, and H. Yasuda. In I.M. Ward, editor, *Developments in Oriented Polymers - 2*, page 39, Elsevier Applied Science, 1987.
[87] M. Iguchi and H. Kyotani. *Sen-i Gakkaishi*, 46:471, 1990.
[88] H.W. Spiess. In I.M. Ward, editor, *Developments in Oriented Polymers - 1*, page 47, Applied Science Publishers, 1982.
[89] I.L. Hay and A. Keller. *Kolloid Z. u. Z. Polym.*, 204:43, 1965.
[90] E.W. Fischer, H. Goddar, and G.F. Schmidt. *Makromol. Chem.*, 118:144, 1968.
[91] R.P. Kambour. *Macromol. Rev.*, 7:1, 1973.
[92] A.S. Argon and J.G. Hannoosh. *Phil. Mag.*, 36:1195, 1977.
[93] S.S. Sternstein. In J.M. Schultz, editor, *Treatise on Materials Science and Technology*, page 541, Academic Press, 1977.
[94] E.J. Kramer. *Adv. Polym. Sci.*, 52:1, 1983.
[95] A.S. Argon and M.M. Salama. *Phil. Mag.*, 36:1217, 1977.
[96] J.G. Williams. *Polym.Engn.Sci.*, 17:144, 1977.
[97] W. Döll. In H.H. Kausch, editor, *Polymer Fracture*, page 267, Springer-Verlag, 1978.
[98] G.H. Michler. *Kunststoff-Mikromechanik*, page 100. Carl Hanser Verlag, 1992.
[99] M.J. Doyle. *J. Polym. Sci., Polym. Phys. Ed.*, 13:2429, 1975.
[100] W. Döll. In A.C. Roulin-Moloney, editor, *Fractography and Failure Mechanisms of Polymers and Composites*, pages 405,416, Elsevier, 1989.
[101] J.F. Fellers and B.F. Kee. *J. Appl. Polym. Sci.*, 18:2355, 1974.

Bibliography

There follows a bibliography for supplementary studies which collects and further extends the suggestions for reading in the text. Included in particular is a selection of treatises on methods and experimental techniques which are of importance for all those actually working in the field.

Textbooks

F.W. Billmeyer: *Textbook on Polymer Science*, John Wiley & Sons, 1984

R.H. Boyd, P.J. Phillips: *The Science of Polymer Molecules*, Cambridge University Press, 1993

J.M.G. Cowie: *Polymers: Chemistry and Physics of Modern Materials*, International Textbook Co, 1973

U. Eisele *Introduction to Polymer Physics*, Springer, 1990

P.J. Flory: *Principles of Polymer Chemistry*, Cornell University Press, 1953

U.W. Gedde: *Polymer Physics*, Chapman & Hall, 1995.

P. Munk: *Introduction to Macromolecular Science*, John Wiley & Sons, 1989

W. Retting, H.M. Laun: *Kunststoff-Physik*, Hanser, 1991

L.H. Sperling: *Introduction to Physical Polymer Science*, John Wiley & Sons, 1992

Comprehensive Treatments and Books with Broader Coverage

G. Allen (Ed.): Comprehensive Polymer Science Vol.1, *Polymer Characterization*, Pergamon Press, 1989

G. Allen (Ed.): Comprehensive Polymer Science Vol.2, *Polymer Properties*, Pergamon Press, 1989

R.W. Cahn, P. Haasen, E.J. Kramer, E.L. Thomas (Eds.): Materials Science and Technology Vol.12 *Structure and Properties of Polymers*, VCH Publishers, 1993

H.F. Mark (Ed.): Encyclopedia of Polymer Science and Engineering Vols.1-17 John Wiley & Sons, 1985-89

J.E. Mark, A. Eisenberg W.W.Graessley, L.Mandelkern, J.L.Koenig: *Physical Properties of Polymers*, Am.Chem.Soc., 1984

J.M. Schultz (Ed.): Treatise on Materials Science and Technology Vol.10, *Properties of Solid Polymeric Materials*, Academic Press, 1977

Theoretically Oriented Treatments

R.B. Bird, R.C. Armstrong, O. Hassager: *Dynamics of Polymeric Liquids* Vol.1 *Fluid Mechanics*, John Wiley & Sons, 1977

R.B. Bird, R.C. Armstrong, O. Hassager: *Dynamics of Polymeric Liquids* Vol.2 *Kinetic Theory*, John Wiley & Sons, 1977

J. des Cloizeaux, G. Jannink: *Polymers in Solution: Their Modelling and Structure*, Oxford Science Publishers, 1990

M. Doi, S.F. Edwards: *The Theory of Polymer Dynamics*, Clarendon Press, 1986

P.J. Flory: *Statistical Mechanics of Chain Molecules*, John Wiley & Sons, 1969.

K.F. Freed: *Renormalization Group Theory of Macromolecules*, John Wiley & Sons, 1987

P.-G. de Gennes: *Scaling Concepts in Polymer Physics*, Cornell University Press, 1979

A.Y. Grosberg, A.R. Khokhlov: *Statistical Physics of Macromolecules*, AIP Press, 1994

R.G. Larson: *Constitutive Equations for Polymer Melts and Solutions*, Butterworths, 1988

W.L. Mattice, U.W. Suter *Conformational Theory of Large Molecules - The Rotational Isomeric State Model in Macromolecular Systems*, John Wiley & Sons, 1994

Specialized Treatments

J.F. Agassant, P. Avenas, J.P. Sergent, P.J. Carreau: *Polymer Processing*, Hanser, 1991

E.H. Andrews, P.E. Reed, J.G. Williams, C.B. Bucknall: Advances in Polymer Science Vol.27 *Failure in Polymers*, Springer, 1978

D.C. Bassett: *Principles of Polymer Morphology*, Cambridge University Press, 1981

A.W. Birley, B. Haworth, J. Batchelor: *Physics of Plastics*, Hanser, 1992

R.M. Christensen: *Mechanics of Composite Materials*, John Wiley & Sons, 1979

E.-J. Donth *Relaxation and Thermodynamics in Polymers: Glass Transition*, Akademie Verlag, 1992

J. Ferguson, Z. Kembowski: *Applied Fluid Rheology*, Elsevier, 1991

J.D. Ferry: *Viscoelastic Properties of Polymers*, John Wiley & Sons, 1970

H. Janeschitz-Kriegl: *Polymer Melt Rheology and Flow Birefringence*, Springer, 1983

H.H. Kausch: *Polymer Fracture*, Springer, 1978

H.H. Kausch (Ed.): Advances in Polymer Science Vols.91/91 *Crazing in Polymers - 2*, Springer, 1990

A.J. Kinloch, R.J. Young: *Fracture Behaviour of Polymers*, Applied Science Publishers, 1983

S. Matsuoka: *Relaxation Phenomena in Polymers*, Hanser, 1992

N.G. McCrum, B.E. Read, G. Williams: *Anelastic and Dielectric Effects in Polymeric Solids*, John Wiley & Sons, 1967

G.H. Michler: *Kunststoff-Mikromechanik*, Hanser, 1992

L.E. Nielsen: *Polymer Rheology*, Marcel Dekker, 1977

L.E. Nielsen: *Mechanical Properties of Polymers and Composites* Vols.1/2, Marcel Dekker, 1974

F.R. Schwarzl: *Polymermechanik*, Springer, 1990

H. Tadokoro: *Structure of Crystalline Polymers*, John Wiley & Sons, 1979

L.R.G. Treloar: *The Physics of Rubber Elasticity*, Clarendon Press, 1975

I.M. Ward: *Structure and Properties of Oriented Polymers*, Applied Science Publishers, 1975

I.M. Ward: *Mechanical Properties of Solid Polymers*, John Wiley & Sons, 1971
A.E. Woodward: *Atlas of Polymer Morphology*, Hanser, 1989
J.G. Williams: *Fracture Mechanics of Polymers*, Ellis Horwood Publishers, 1984
B. Wunderlich: *Macromolecular Physics* Vol.1 *Crystal Structure, Morphology, Defects*, Academic Press, 1973
B. Wunderlich: *Macromolecular Physics* Vol.2 *Crystal Nucleation, Growth, Annealing*, Academic Press, 1976
B. Wunderlich: *Macromolecular Physics* Vol.3 *Crystal Melting*, Academic Press, 1980

Methodically Oriented Treatments

L.E. Alexander: *X-Ray Diffraction Methods in Polymer Science*, John Wiley & Sons, 1969
B.J. Berne, R. Pecora: *Dynamic Light Scattering*, John Wiley & Sons, 1976
D.I. Bower, W.F. Maddams: *The Vibratonal Spectroscopy of Polymers*, Cambridge University Press, 1989
B. Chu: *Laser Light Scattering*, Academic Press, 1991
O. Glatter, O. Kratky: *Small Angle X-ray Scattering*, Academic Press, 1982
J.S. Higgins, H.C. Benoît: *Polymers and Neutron Scattering*, Clarendon Press, 1994
L. Marton, C. Marton (Ed.): Methods of Experimental Physics Vol.16, *Polymers Part A: Molecular Structure and Dynamics*, Academic Press, 1980
L. Marton, C. Marton (Ed.): Methods of Experimental Physics Vol.16, *Polymers Part B: Crystal Structure and Morphology*, Academic Press, 1980
L. Marton, C. Marton (Ed.): Methods of Experimental Physics Vol.16, *Polymers Part C: Physical Properties*, Academic Press, 1980
V.J. McBrierty, K.J. Packer: Cambridge Solid State Science Series, *Nuclear Magnetic Resonance in Solid Polymers*, Cambridge University Press, 1993
P.C. Painter, M.M. Coleman, J.L. Koenig: *The Theory of Vibrational Spectroscopy and its Application to Polymeric Materials*, John Wiley & Sons, 1982
A.C. Roulin-Moloney (Ed.): *Fractography and Failure Mechanisms of Polymers and Composites*, Elsevier, 1989
K. Schmidt-Rohr, H.W. Spiess: *Multidimensional Solid-State NMR and Polymers*, Academic Press, 1994
H.W. Siesler, K. Holland-Moritz: *Infrared and Raman Spectroscopy of Polymers*, Marcel Dekker, 1980
S.J. Spells (Ed.): *Characterization of Solid Polymers*, Chapman & Hall, 1994
B.K. Vainshtein: *Diffraction of X-Rays by Chain Molecules*, Elsevier, 1966

Development Reports

D.C. Bassett (Ed.): Developments in Crystalline Polymers Vol.1, Applied Science Publishers, 1982
D.C. Bassett (Ed.): Developments in Crystalline Polymers Vol.2, Elsevier, 1988
I. Goodman: Developments in Block Copolymers Vol.1, Applied Science Publishers, 1982
I. Goodman: Developments in Block Copolymers Vol.2, Applied Science Publishers, 1985
L.A. Kleintjens, P.J. Leemstra: *Integration of Fundamental Polymer Science and Technology*, Elsevier, 1886

D.J. Meier (Ed.): *Molecular Basis of Transitions and Relaxations*, Gordon and Breach, 1978

R.M. Ottenbrite, L.A. Utracki, S. Inoue (Eds.): Current Topics in Polymer Science Vol.2, Hanser, 1987

D.R. Paul, S. Newman (Eds.): *Polymer Blends* Vols.1/2, Academic Press, 1978

I.M. Ward (Ed.): Developments in Oriented Polymers Vol.1, Applied Science Publishers, 1982

I.M. Ward (Ed.): Development in Oriented Polymers Vol. 2, Elsevier, 1987

Handbooks and Data Sources

J. Brandrup, E.H. Immergut: *Polymer Handbook*, John Wiley & Sons, 1989

D.W. Van Krevelen: *Properties of Polymers*, Elsevier, 1972

Subject Index

Springer-Verlag and the Environment

We at Springer-Verlag firmly believe that an international science publisher has a special obligation to the environment, and our corporate policies consistently reflect this conviction.

We also expect our business partners – paper mills, printers, packaging manufacturers, etc. – to commit themselves to using environmentally friendly materials and production processes.

The paper in this book is made from low- or no-chlorine pulp and is acid free, in conformance with international standards for paper permanency.

Printing: Mercedesdruck, Berlin
Binding: Buchbinderei Lüderitz & Bauer, Berlin